GUIDE TO THE PARASITES OF FISHES OF CANADA

PART IV

Publication Proposals: Proposals for the NRC Monograph Publishing Program should be sent to Gerald J. Neville, Head, Monograph Publishing Program, National Research Council of Canada, NRC Research Press, 1200 Montreal Road, Building M-55, Ottawa, ON K1A 0R6, Canada. Telephone: (613) 993-1513; fax: (613) 952-7656; e-mail: gerry.neville@nrc.ca.

ISBN 0-660-16403-5
ISSN 0706-6481
NRCC No. 40083

Canadian Cataloging In Publication Data

Main entry under title:
Guide to the parasites of fishes of Canada

(Canadian special publication of fisheries and aquatic sciences; no. 124)
Partial contents: V. 4 Trematoda / David I. Gibson.
Includes an abstract in French.
Includes bibliographical references and an index.
Issued by the National Research Council of Canada.
ISBN 0-660-16403-5

1. Fishes — Parasites — Canada — Identification.
I. Kabata, Z. 1924–. II. Margolis, L. 1927–. III. Gibson, David I. Trematoda
IV. Canada. Dept. of Fisheries and Oceans. V. National Research Council Canada
VI. Title. VII. Series.

SH175.G81 1996 597'.0233'0971 C85-075080-6 rev.

This publication is available from: Subscription Office, NRC Research Press, National Research Council of Canada, Ottawa, Ontario K1A 0R6, Canada.

Enquiries: Telephone, 613-993-9084; fax, 613-952-7656

Correct citation for this publication:

Gibson, D.I. 1996. Trematoda. *In* L. Margolis and Z. Kabata (ed). Guide to the parasites of fishes of Canada. Part IV. Can. Spec. Publ. Fish Aquat. Sci. 124. 373 pp.

Canadian Special Publication of Fisheries and Aquatic Sciences 124

Guide to the Parasites of Fishes of Canada

Edited by L. Margolis and Z. Kabata

Department of Fisheries and Oceans
Pacific Biological Station
Nanaimo, British Columbia V9R 5K6

Part IV

Trematoda

David I. Gibson

Department of Zoology
The Natural History Museum
Cromwell Road
London SW7 5BD
United Kingdom

NRC-CNRC

NRC Research Press
Ottawa 1996

CONTENTS

Guide to the Parasites of Fishes of Canada
Part IV

PREFACE

In Part IV of this series, *Guide to the Parasites of Fishes of Canada*, the Trematoda are treated by a renowned helminthologist, Dr. David I. Gibson, from the Natural History Museum, London. He has studied these parasites of fishes from the "four corners" of the globe, including the Pacific and Atlantic coasts of Canada. We are indeed fortunate, as will be subsequent users of the Guide, that Dr. Gibson agreed to undertake the task of authoring this publication.

The three previous parts* in this series dealt with the Monogenea, Turbellaria, Crustacea, Acanthocephala, and Cnidaria. Part IV provides practical keys for identification of the taxa of Trematoda represented in Canadian fishes down to the species level, accompanied by detailed line drawings. Although maintaining the general format of Parts I–III, in keeping with our policy of permitting the authors some degree of latitude in how they approached their task, Part IV has three departures from the general plan. Firstly, it discusses major taxonomic problems, when required, for genera and higher taxa on a global basis. The large number of species included in this work dictated that certain measures be taken to keep the volume within a manageable size. This led to the second and third departures from the basic plan: figures are provided for only one species in a genus, rather than all species, and full descriptions are not provided for individual species. These omissions are compensated for by detailed generic descriptions and inclusion of important distinguishing characters in the keys to species.

The first three parts in this series were published by the Canadian Department of Fisheries and Oceans. Publication responsibilities have now been assumed by the National Research Council of Canada to whom we express our appreciation for continuing this series within its monograph publication program.

Financial assistance for publication was provided by the Pacific Biological Station, Department of Fisheries and Oceans.

The Editors

* Part I. Beverley-Burton, M. 1984. Monogenea and Turbellaria. Part II. Kabata, Z., F. Rafi, and E.L.Bousfield. 1988. Crustacea. Part III. Arai, H. P. 1989. Acanthocephala; and Arai, M. N. 1989. Cnidaria.

ABSTRACT

GIBSON, D.I. 1996. Trematoda. *In:* L. Margolis and Z. Kabata (ed.) Guide to the parasites of fishes of Canada. Part IV. Can. Spec. Publ. Fish. Aquat. Sci. 124: 373 pp.

More than 80 keys leading to the identification of more than 220 species of trematode parasites of Canadian fishes are provided. Diagnoses of more than 80 family-group taxa and 130 genera are given. Each genus is illustrated. For each species of trematode a Canadian host list, locality data, and any relevant remarks are included. In addition, a host-parasite list is given plus parasite and host indices.

RÉSUMÉ

GIBSON, D.I. 1996. Trematoda. *Dans* L. Margolis et Z. Kabata (éd.) Guide to the parasites of fishes of Canada. Part IV. Can. Spec. Publ. Fish Aquat. Sci. 124: 373 pp.

On y retrouve plus de 80 clefs qui permettent d'identifier quelque 220 espèces de trématodes parasites des poissons du Canada. Le livre comprend la description de plus de 80 taxums de famille/groupe et de 130 genres. Chaque genre y est illustré. À chaque espèce de trématode est associée une fiche signalétique, contenant des données sur l'emplacement et des remarques particulières à l'espèce. De plus, on y retrouve une liste des paires hôte-parasite ainsi qu'un index des hôtes et des parasites.

INTRODUCTION

The first record of trematode parasites in Canadian fishes appears to be that of Wright (1879), who recorded the presence of metacercariae of *Distomum gracile* (Leidy), which is probably *Clinostomum complanatum* (Rud.), from perch *Perca flavescens* probably in Ontario. The first specialist who carried out detailed work on the trematode fauna of fishes was, however, Stafford, who between 1900 and 1907, after training in Germany, produced a series of papers on the trematodes of marine and freshwater fishes of eastern Canadian waters. Subsequently, although numerous authors added information, it was not until the second half of the 20th century that a knowledge of this fauna has been built up. Great advances were made, especially during the 1960s, 1970s, and early 1980s, after which a noticeable decline in faunistic studies became apparent. Although parts of the Canadian fauna are now relatively well known, particularly in relation to freshwater and commercial marine fishes, other areas, and particularly in the deep-sea and Arctic regions, are less so.

The arrangement of the text follows largely that of other contributions in the series. The classification (see below) is based at the superfamily level, with definitions of the various taxa down to the generic level. Keys, usually dichotomous, to the superfamilies, families, subfamilies (where included), genera, and species are included. There are no descriptions of individual species because of the large number of taxa involved, but illustrations of members of each genus are included.

Species (especially metacercarial forms), which cannot be identified, are usually excluded from this work. Complete lists of such records can be obtained from Margolis and Arthur (1979) and McDonald and Margolis (1995).

Families, subfamilies, and genera of Digenea are treated in alphabetical order. Data relevant to each species are included as follows:

Scientific name. This is the correct name for the species, which is, in most cases, the name in current use.

Synonym(s). These are other names that have been used for the species in the past: they may be totally different names that have been synonymised with the present species or the same epithet with a different combination of generic name. These synonymy lists are usually not complete, tending to include only those cited in publications that relate to the Canadian fauna.

Site. This is the position or organ in or on the fish where the parasite has been reported in Canadian fishes. In some cases, where this information is not given, its expected site is given in brackets.

Host(s). These are cited using currently accepted nomenclature. The numbers in parentheses following the host name are the numbers assigned to the relevant record, i.e., the author(s) responsible for recording the parasite from the particular host. Question marks associated with these numbers indicate that the record is questionable.

Distribution. The distribution of the records of the various species of parasite in Canadian waters is indicated by abbreviations. Marine records are listed under three geographical regions: Atlantic (Atl), Pacific (Pac), and eastern Arctic (EArc). Freshwater records are indicated in relation to the various Provinces or Territories of Canada: Alberta (Alta); British Columbia (BC); Labrador (Lab); Manitoba (Man); New Brunswick (NB); Newfoundland (Nfld); Nova Scotia (NS); Northwest Territories (NWT); Ontario (Ont); Prince Edward Island (PEI); Quebec (Que); Saskatchewan (Sask); Yukon Territory (YT). In cases where the records are from euryhaline fishes in brackish water, these are indicated by "-b" following the provincial abbreviation.

Records(s). The authors responsible for the published records of a particular parasite are listed and numbered in chronological order, based on the date of the publication. In cases where the parasite is recorded from more than one geographical region, the abbreviated geographical regions of records are listed in parentheses after the references. In cases where there is only a single record, no numbering and no geographical data are included. In addition, in some cases, records of Canadian material in the British Museum (Natural History) (BM(NH)) collections at The Natural History Museum, London, are also listed.

Remarks. These are comments on the systematics, nomenclature, records, or other aspects of the biology of the parasite.

Classification

The classification of the Trematoda is problematical, with several different versions in current use; all different and none satisfactory. The most recent cladistic analysis by Brooks *et al.* (1985, 1989) has been heavily criticised (Gibson 1987; Pearson 1992). Brooks' work has been a great contribution in that it has caused workers to think about the subject and the characters, but it suffers from an inadequate data set and some misinterpretation. A major failing is the selection of the aberrant Heronimidae, parasites of the lungs of chelonians, as the sister group of the remainder of the Digenea based on a misinterpretation of the life-history (see Gibson 1987; Pearson 1992): this view has recently been supported by molecular biological studies (Barker *et al.* 1993a). For a detailed criticism of Brooks' contribution, see Pearson (1992). Brooks' classification is reiterated, little changed, in Brooks and McLennan (1993).

There are two subgroups of the class Trematoda, the subclass Aspidogastrea and the subclass Digenea. The major difference between these two groups is that aspidogastreans lack the alternation of generations that occur in the Digenea, i.e., in all digeneans the sexual generations, which normally occur in a vertebrate, alternate with two asexual generations, normally in a mollusc (in some cases one asexual generation may apparently be lost by secondary reduction but evidence of its former existence is normally apparent). Aspidogastreans, like digeneans, are primarily parasites of molluscs, but many mature in vertebrates. The Aspidogastrea is divided into two orders, the Aspidogastrida and the Stichocotylida (see Gibson and Chinabut 1984). The Digenea can be divided tentatively into three orders, the Strigeida, the Echinostomida, and the Plagiorchiida, based on features of the life-history and larval stages. Essentially, in the Echinostomida the cercariae normally (but not always) encyst in the open, whereas in the other two groups they normally encyst in the next host (see Gibson and Bray 1994). Strigeida and the Plagiorchiida differ in that the cercariae of the former tend to have a furcate tail. These ordinal groups are not included in this work, since it is aimed at keying down digenean parasites of fishes (i.e., in most cases, sexual adults). In this compilation, therefore, the major divisions are at the superfamily level. Within the three orders the superfamilies, families, subfamilies, and genera are dealt with in alphabetical order. The order of the species is dependent upon how they key out.

The groups of trematodes included in this work as occurring in Canadian waters are as follows:

CLASS TREMATODA

Subclass ASPIDOGASTREA

Order ASPIDOGASTRIDA
- Aspidogastridae

Order STICHOCOTYLIDA
- Rugogastridae

Subclass DIGENEA

Order STRIGEIDA

CLINOSTOMOIDEA
- Clinostomidae

DIPLOSTOMOIDEA
- Diplostomidae
- Strigeidae

GYMNOPHALLOIDEA
- (incl. Brachylaimoidea)
- Bucephalidae
- Fellodistomidae

HEMIUROIDEA
- Accacoeliidae
- Azygiidae
- Bunocotylidae

Derogenidae
Didymozoidae
Hemiuridae
Hirudinellidae
Lecithasteridae
Ptychogonimidae
Syncoeliidae

SCHISTOSOMATOIDEA
Sanguinicolidae
Spirorchiidae

Order ECHINOSTOMIDA

ECHINOSTOMATOIDEA
Echinostomatidae
Psilostomidae (= Cathaemasiidae)

PARAMPHSTOMOIDEA
Cladorchiidae

Order PLAGIORCHIIDA

ALLOCREADIOIDEA
Allocreadiidae
Opecoelidae

LEPOCREADIOIDEA
Acanthocolpidae
Deropristidae
Homalometridae
Lepocreadiidae

OPISTHORCHIOIDEA
Cryptogonimidae
Heterophyidae
Opisthorchiidae

PLAGIORCHIOIDEA
Gorgoderidae
Lecithodendriidae
Macroderoididae
Microphallidae
Troglotrematidae

ZOOGONOIDEA
Lissorchiidae
Monorchiidae
Zoogonidae

One group omitted is the Transversotrematidae. Dechtiar and Christie (1988) claimed to have found *Prototransversotrema* sp. under the scales of *Luxilis (Notropis) cornutus* in Lake Ontario. However, this is far from the tropical and subtropical regions of the world where this group normally occurs (Cribb *et al.* 1992: p. 926 and 930). One possible explanation could be that these authors misinterpreted a penetrating cercaria.

Diagnoses and keys

Diagnoses are presented for all the taxa down to generic level, where they are relevant. In the cases where the life history stage that occurs in fish is the metacercaria only, the diagnoses presented relate only to the metacercarial stage. The amount of detail in these diagnoses varies, depending upon the size and complexity of the group. Features of a taxon common to all members of the group may be found in the diagnoses of the higher taxa. A representative of each genus is illustrated.

Keys are included down to the species level. They are so designed that forms known to occur in Canadian fishes can be identified and may well in many cases, therefore, be unsuitable for keying down material from other areas. These keys are normally, but not always, dichotomous and based entirely on features of the parasitic stages that occur in fish. In some instances, they may also require geographical or host data where these help clarify matters.

Anatomical terminology

The terminology used in the keys and diagnoses are standard, although specific groups do have their own specialised features. In the latter cases (e.g., the Diplostomoidea (Strigeoidea)), some explanations are given after the diagnoses. The standard terminology for various organs is illustrated in Fig 1. Further details on the anatomical features of some of the groups with specialised features can be found elsewhere: for the Diplostomoidea, see Dubois (1968, 1970); for the Heterophyidae, see Pearson (1973); and for the Hemiuroidea, see Gibson and Bray (1979). The glossary in the latter work also explains some of the features common to many groups.

Collecting and fixation

Specimens should, where possible, be collected and fixed live. When collected from long-dead or frozen hosts the worms are usually partly macerated, attenuated, and may have shed body spines. Those collected from fixed hosts or fixed guts are normally very contracted. Live worms are best fixed either by heat in either boiling water or almost boiling conventional fixative, e.g., formalin, alcohol, etc.) or cold-fixed in Berland's fluid (95% glacial acetic acid + 5% formalin), and then stored in 70–80% alcohol. Berland's fluid may occasionally not be suitable for metacercariae, or some marine adult digeneans, with large numbers of calcareous corpuscles in the excretory system that react with the acid to form a gas bubble. Flattening during fixation should not generally be used, as it causes considerable distortion; in some cases, however, if the facility of serial sectioning is not available, the flattening of a few specimens may elucidate certain structures. Features not visible in wholemounts are best seen in serial sections: for technique, see Cooper (1988). Specimens intended as wholemounts should be stained in a good carmine stain, such as Mayer's paracarmine, dehydrated, cleared in beechwood creosote (or another suitable clearing agent), and mounted ventral side uppermost on a slide in Canada balsam (plastic mountants are not suitable for material intended to be archival).

CLASS TREMATODA RUDOLPHI, 1808

Platyhelminthes. Permanent parasites generally of internal organs primarily of molluscs, which harbour larval, asexual, or occasionally sexual adult stages, and vertebrates, which harbour adult sexual stage. Molluscan host virtually always involved in life history. Primarily parasites of tissues in molluscs and gut of vertebrates, but may occur in wide variety of body cavities, organs, and tissues in vertebrates. Sexual adult usually with two organs of attachment (normally suckers or sucker-like structures) (occasionally one, rarely none), one generally anterior and one ventral or posterior. Syncytial tegument armed with spines or smooth. Gut always present. Mouth usually within anterior attachment organ. Prepharynx linking mouth and pharynx present or absent. Muscular pharynx normally present. Intestine usually bifurcate, sometimes with single caecum; normally blind, occasionally open via anus(i) or into excretory vesicle. Normally hermaphroditic, rarely dioecious. Testes one to many, commonly two. All or part of male terminal genitalia often enclosed in muscular sac; cirrus sac if enclosing male duct only; sinus sac if enclosing hermaphroditic duct (union of male and female ducts). Male terminal duct normally includes seminal vesicle, pars prostatica, and ejaculatory duct. Genital atrium present or absent. Male and female terminal reproductive ducts usually open through common pore. Ovary normally single. Sperm storage apparatus (seminal receptacle) normally present; seminal disposal apparatus, commonly Laurer's canal, usually present. Mehlis' gland and oötype normally present. Eggs stored in uterus. Ciliated larva (miracidium) develops within egg. Vitellarium variable, exhibiting all forms between follicular and single compact mass. Excretory pore normally single in sexual adult, usually terminal; excretory vesicle normally saccular, I-, V-, or Y-shaped; linked to system of tubules and flame cells.

KEY TO THE SUBCLASSES OF THE TREMATODA

1 Large ventral holdfast (attachment organ) covers most of ventral surface of body; holdfast subdivided into alveoli ....Aspidogastrea

Ventral attachment organ a sucker, occasionally absent, occasionally postero-ventral; not normally covering more than half of ventral surface of body; not subdivided into alveoli ....Digenea

Subclass ASPIDOGASTREA Faust and Tang, 1936

Syn.: Aspidobothria Monticelli, 1888; Aspidocotylea Monticelli, 1892.

Trematoda. Lacking alternation of generations (asexual reproductive stages). Body oval to tubular, with large ventral holdfast (organ of attachment). Holdfast usually discoid to elongate and subdivided into alveoli (Aspidogastrida), or occasionally linear and divided into longitudinal series of alveoli separated by septa or series of independent "suckers" (Stichocotylida). Marginal organs present or absent on holdfast. Mouth present within sucker- or disc-like anterior organ. Pharynx well developed. Intestine blind, usually with single blind caecum, rarely bifurcate (*Rugogaster*). Testes one or two, post-ovarian. Cirrus sac present or absent; sinus sac and hermaphroditic duct rare (*Rohdella*). Common genital pore mid-ventral in forebody (i.e., anterior to holdfast). Ovary single, pre-testicular. Seminal receptacle present only occasionally. Laurer's canal usually present, occasionally blind. Uterine seminal receptacle usually present. Uterus coiled in pre- and post-ovarian or mainly pre-ovarian fields. Eggs operculate; medium-sized to large. Vitellarium follicular, usually in lateral fields, rarely in mid-dorsal field. Excretory system usually V-shaped, sometimes with two vesicles opening by common or independent pores; pore(s) terminal or

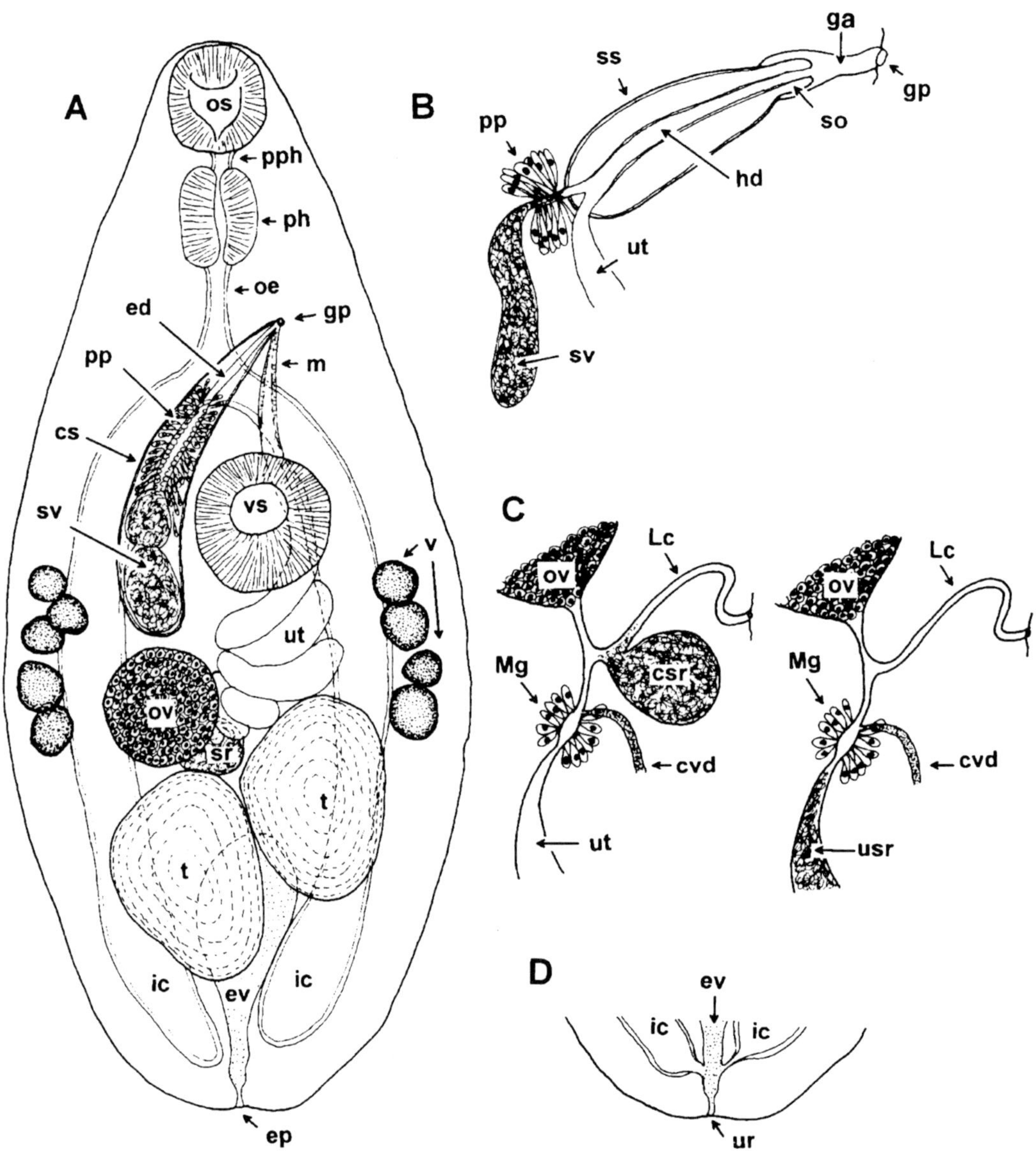

FIG. 1. Basic anatomical features that may be used as taxonomic criteria. A. Entire worm. B. Terminal genitalia where male and female ducts unite to form an hermaphroditic duct. C. Forms of the proximal female genitalia, with two frequently encountered forms of sperm-store, a canalicular seminal receptacle and a uterine seminal receptacle. D. Proximal extremity of body where the intestinal caeca and excretory system unite to form a uroproct. *Abbreviations*: **cs**, cirrus sac (usually encloses ejaculatory duct, pars prostatica and none of, part of, or entire seminal vesicle; contraction causes expulsion of spermatozoa from seminal vesicle); **csr**, canalicular seminal receptacle (formed from base of Laurer's canal); **cvd**, common vitelline duct (links vitelline reservoir and oviduct); **ed**, ejaculatory duct (terminal part of male reproductive duct; may evaginate to form temporary or permanent intromittent organ called a cirrus); **ep**, excretory pore; **ev**, excretory vesicle (may be saccular, I-, V- or Y-shaped); **ga**, genital atrium (region linking male and female ducts with genital pore; sometimes contains intromittent organ); **gp**, genital pore (male and female systems generally open through the same pore); **hd**, hermaphroditic duct (formed in some groups where male and female ducts unite to form common duct; sometimes difficult to distinguish from genital atrium); **ic**, intestinal caecum; **Lc**, Laurer's canal (drain for excess sperm/vitelline products; may occasionally act as vagina; may terminate in resorbtive structure called Juel's organ); **m**, metraterm (muscular terminal region of the uterus); **Mg**, Mehlis' gland (glandular structure

dorso-subterminal. Usually parasitic in molluscs, but also occur in teleosts, elasmobranchs, holocephalans, and chelonians.

Comments

The classification of the Aspidogastrea used here follows Gibson and Chinabut (1984). Only two species of this subclass have been recorded in Canadian waters. Freshwater forms of the Order Aspidogastrida often mature in molluscs. It is very possible that other members of the Order Stichocotylida Gibson and Chinabut, 1984 (e.g., *Stichocotyle*, *Multicalyx*) occur in elasmobranchs off the Canadian coast.

Key to the orders of the Aspidogastrea

1 In freshwater teleosts ....Aspidogastrida

In holocephalans or elasmobranchs ....Stichocotylida

Order ASPIDOGASTRIDA Dollfus, 1958

Aspidogastrea. Holdfast a very large, oval to elongate ventral disc subdivided into marginal ring of alveoli surrounding medial alveoli. Intestine with single caecum. Parasites of molluscs, teleosts, and chelonians.

Family ASPIDOGASTRIDAE Poche, 1907

Aspidogastrida. With features of order.

Subfamily COTYLASPIDINAE Chauhan, 1954

Aspidogastridae. Medial alveoli of holdfast undivided longitudinally.

COTYLOGASTER Monticelli, 1892

Cotylaspidinae. Body oval to elongate. Forebody (region anterior to holdfast) may be telescopically retractile. Marginal ring of alveoli of holdfast surrounds single row of wide medial alveoli. Marginal organs

associated with fertilisation and egg-formation that normally surrounds the oötype); **oe**, oesophagus; **os**, oral sucker; **ov**, ovary; **ph**, pharynx; **pp**, pars prostatica (region of terminal male reproductive duct between seminal vesicle and ejaculatory duct; surrounded by gland cells whose ducts pass through its wall and form small blebs on its lining that burst, releasing a secretion; associated with the lubrication and activation of spermatozoa); **pph**, prepharynx; **so**, sinus organ (temporary or permanent intromittent organ, analogous to a cirrus, formed by eversion of an hermaphroditic duct; **sr**; seminal receptacle (a store of spermatozoa in the female reproductive system; in three common forms: canalicular (q.v.) — distinct structure linked to Laurer's canal, blind — distinct structure when Laurer's canal is absent, and uterine (q.v.) — when sperm fills proximal region of uterus; types are mutually exclusive); **ss**, sinus sac (muscular sac, analogous to a cirrus sac, surrounding terminal genitalia when male and female terminal ducts unite to form an hermaphroditic duct); **sv**, seminal vesicle (a store of spermatozoa in the male reproductive system; may be outside (naked), inside or partly inside a cirrus sac; shape variable; with or without muscular wall); **t**, testis; **ur**, uroproct; **usr**, uterine seminal receptacle (store of spermatozoa in proximal region of uterus — q.v. seminal receptacle); **ut**, uterus; **v**, vitellarium ("yolk" producing organ; variable in structure from follicular, tubular, through to one or two compact masses, often bilaterally symmetrical; **vs**, ventral sucker.

present. Posterior extremity of hindbody may flex dorsally to form dorsal "cone". Mouth within funnel-shaped, often penta-lobed, anterior attachment organ. Prepharynx short to long; pharynx well developed, may be protrusible; caecum simple, reaching close to posterior extremity. Testes two, post-ovarian, in posterior half of body. Seminal vesicle tubular; pars prostatica long, broad, surrounded by dense mass of gland cells; short ejaculatory duct opens into genital atrium. Cirrus sac absent. Common genital pore mid-ventral in posterior forebody. Ovary oval to pyriform, pre-testicular. Laurer's canal dilates to form rudimentary seminal receptacle and opens close to excretory pore. Uterine seminal receptacle present. Uterus long, winding medially through posterior half of body in both pre- and post-ovarian fields. Eggs operculate, large. Vitellarium an inverted U-shaped rectilinear chain of follicles surrounding gonads and much of uterine field. Excretory system with two small lateral vesicles that open together through common pore; pore in middle of dorsal "cone". Parasitic in molluscs and teleosts.

Cotylogaster occidentalis Nickerson, 1902 (Fig. 2)

Site: (Intestine).
Host: *Aplodinotus grunniens.*
Distribution: Ont.
Records: Dechtiar 1972a; Dechtiar and Nepszy 1988; Dechtiar and Christie 1988.
Remarks: Ip and Desser (1984) recorded this parasite in Ontario from the bivalve *Elliptio complanata.*

Order STICHOCOTYLIDA Gibson and Chinabut, 1984

Aspidogastrea. Holdfast a linear row of suckers or alveoli separated by septa. Parasites of elasmobranchs and holocephalans.

Family RUGOGASTRIDAE Schell, 1973

Stichocotylida. Body elongate. Long holdfast with longitudinal series of alveoli separated by raised transverse septa (rugae). Sucker-like anterior attachment organ weakly developed. Short prepharynx, pharynx, and short oesophagus present. Two long, blind intestinal caeca reach close to posterior extremity; intestinal bifurcation close to level of genital pore. Testes numerous, in symmetrical pairs, present inter-caecally along much of length of body. Terminal region of paired sperm ducts act as seminal vesicles; vas deferens short. Small cirrus sac and small genital atrium present. Genital pore mid-ventral just posterior to level of intestinal bifurcation. Ovary entire, dextro-lateral; immediately pre-testicular in middle of anterior third of body. Laurer's canal present, dilated, contains spermatozoa. Uterine seminal receptacle (?). Uterine field extends throughout much of body between levels of ovary and caecal extremities, especially dorsally. Eggs numerous, operculate. Vitellarium follicular, in two lateral fields extending extra-caecally between levels of ovary and caecal extremities. Excretory system (?). Parasitic in rectal glands of holocephalans.

Comments

In the original description of *Rugogaster*, Schell (1973) referred to the anterior-most alveolus as a questionable vestigial ventral sucker. The distinctiveness of this feature is variable. It has been accepted as a ventral sucker by some (e.g., Machida and Araki 1992; Rohde and Watson 1992), but others (e.g., Shvetsova 1990) are sceptical. I consider it more likely to be a slightly modified alveolus. Rohde and Watson suggest this in their comment that the "ventral sucker" and holdfast of *Rugogaster* is equivalent to the holdfast of aspidogastrids and the ventral sucker of digeneans.

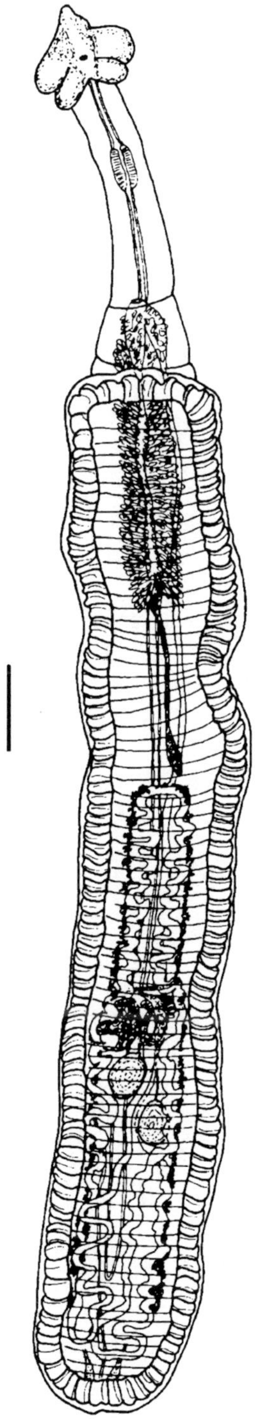

FIG. 2. *Cotylogaster occidentalis* (after Fredericksen (1972)). Scale bar: 1 mm.

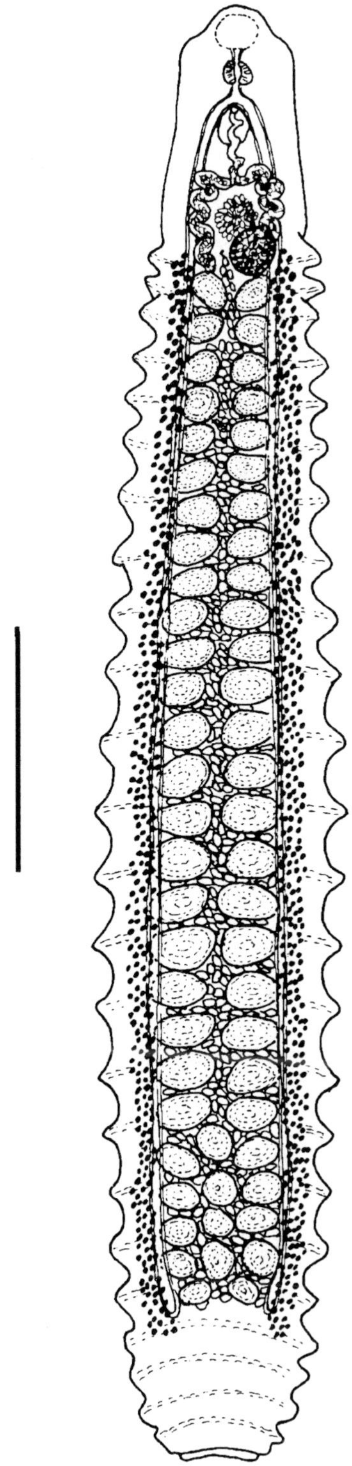

FIG. 3. *Rugogaster hydrolagi* (after Schell (1973)). Scale bar: 2 mm.

RUGOGASTER Schell, 1973

Rugogastridae. With features of family.

Rugogaster hydrolagi Schell, 1973 (Fig. 3)

Site: Rectal gland.
Host: *Hydrolagus colliei.*
Distribution: Pac.
Record: Rohde *et al.* 1992.

Subclass DIGENEA Carus, 1863

Trematoda. With alternation of generations; normally two asexual generations in mollusc and single sexual generation in vertebrate (occasionally invertebrate). Normally hermaphroditic, occasionally partly or entirely dioecious. Generally small, elongate-oval to tubular worms, but various other forms, including round and filamentous occur. Body smooth or armed with spines. Usually with two muscular suckers, sometimes one, occasionally none. Mouth usually, but not always present within oral sucker. Prepharynx present or absent. Muscular pharynx normally present. Oesophagus normally present. Intestine usually bifurcate, sometimes moniliform or saccular; caeca normally blind, occasionally unite, open to exterior through anus(i) or into excretory vesicle. Form of testes variable; one to many, often two. Cirrus sac, occasionally sinus sac, often envelopes all or part of male terminal genitalia. Genital atrium present or absent. Common genital pore normally present; position variable, usually ventral on anterior body, occasionally terminal. Ovary normally single, occasionally multiple; position variable. Seminal receptacle present or absent; usually canalicular or uterine. Seminal disposal apparatus (e.g., Laurer's canal) present or absent. Uterus variable in size, form, and distribution; terminal portion often forms muscular metraterm. Eggs normally oval and operculate; normally tanned; occasionally with spine or filament(s). Vitellarium variable in shape and distribution, exhibiting all forms between follicular and single compact mass. Excretory pore single; usually terminal or subterminal, occasionally dorsal; vesicle saccular, I-, V-, or Y-shaped. Parasitic as sexual adult in all vertebrate groups, occasionally in invertebrates; usually present in gut or other body cavities, occasionally in blood or other tissues, rarely ectoparasitic.

Key to the superfamilies of digeneans known to occur in Canadian waters

1 Adult (or juvenile) worms from gut or occasionally other body cavities, such as gall bladder, and tissues; rarely on gills or in branchial cavity 7

Encysted, encapsulated, or occasionally free metacercarial or juvenile forms from tissues (occasionally lumen of gut: these and forms from eye may be free and active) 2

2 Body armed with spines, usually small, covering part or most of surface (NB. Spines are very easily lost in frozen or poorly fixed material) 4

Body unarmed 3

3 Body composed of two parts; posterior part may be distinct or indistinct (i.e., difficult to distinguish or reduced to small terminal cone); ventral forebody with spoon-shaped concavity or deep, cup-shaped depression; glandular Brandes' organ present posterior to ventral sucker; pseudosuckers may be present lateral to oral sucker; especially common in eye (often free and active) or encysted on pericardium, swimbladder, peritoneum, gill-arches, etc.; more common in freshwater fishes .... Diplostomoidea

Body not obviously composed of two parts; without broad concavity or depression in ventral forebody; normally in body-cavity amongst viscera or free in rectum or branchial cavity; normally in marine fishes ....Hemiuroidea (*Lecithochirium*, *Otodistomum*, *Paraccacladium*, *Lampritrema*)

4 Body often large; pharynx absent ....Clinostomoidea

Body usually small; pharynx present 5

5 Ventral sucker absent; mouth isolated and well apart from anterior extremity; gut with single caecum; primordium of terminal genitalia close to posterior extremity....................................... ....Gymnophalloidea (Bucephalidae)

Ventral sucker present; mouth within oral sucker...6

6 Ventral sucker may be associated with genito-atrial complex that may be armed; oesophagus simple; excretory vesicle I-, V-, or Y-shaped; testicular primordia, if present, often symmetrical or oblique ....Opisthorchioidea

Ventral sucker simple; with or without head collar and circum-oral crown of larger spines, but, when without latter, oesophagus with lateral diverticula; excretory vesicle Y-shaped; testicular primordia, if present, tend to be tandem ....Echinostomatoidea

Ventral sucker simple; without head collar, circum-oral crown of spines, and diverticulate oesophagus; excretory vesicle saccular; testicular primordia, if present, tend to be symmetrical; in anadromous salmonids off Pacific coast (in freshwater salmonids, etc. in Northwest States of United States)...Plagiorchioidea (Troglotrematidae)

7 Parasitic in blood system, especially heart and branchial vessels.............................Schistosomatoidea

Parasitic in gut, body-cavity, gall bladder, urinary system, mouth, branchial cavity, or other tissues 8

8 Mouth isolated and well apart from anterior extremity; gut a simple caecum; ventral sucker absent; terminal genitalia close to posterior extremity.......................Gymnophalloidea (Bucephalidae)

Mouth close to anterior extremity 9

9 Mouth within oral sucker; ventral sucker not at posterior extremity 10

Mouth within pharynx (though this resembles oral sucker); ventral sucker large and at posterior extremity of body (worm appears, therefore, to have "suckers" at both ends); in lampreys.......... ....Paramphistomoidea

10 Body spined over at least part of its surface (NB. Spines easily lost in frozen or poorly fixed material) 11

Body surface without spines 17

11 Uterus normally pre-testicular, occasionally extending more posteriorly (or even (very rarely) into post-testicular region) (Deropristidae) ....Lepocreadioidea

Uterus normally extending into post-testicular region (except *Steganodermatoides*: Zoogonoidea) 12

12 Parasitic in marine fishes..13

Parasitic in freshwater fishes..14

13 Cirrus sac absent; testes two; genital pore immediately pre-acetabular..Opisthorchioidea (Cryptogonimidae, *Metadena*)

Cirrus sac present: testes one and genital pore immediately pre-acetabular or testes two and genital pore lateral to sublateral.. Zoogonoidea (Monorchiidae, Zoogonidae)

14 Genital pore anterior to ventral sucker..15

Genital pore lateral to ventral sucker..16

15 Cirrus sac present..Plagiorchioidea (Macroderoididae)

Cirrus sac absent..Opisthorchioidea (Cryptogonimidae)

16 Excretory vesicle I-shaped: caeca long, reaching at least to level of testes..Zoogonoidea (Lissorchiidae)

Excretory vesicle V- or Y-shaped; caeca short, not reaching to, or at least not posterior to, level of testes..Plagiorchioidea (Microphallidae)

17 Parasitic in urinary system; hindbody often spatulate (lanceolate to discoid)..Plagiorchioidea (Gorgoderidae)

Parasitic in gut and other cavities and tissues..18

18 Male and female terminal ducts united to form hermaphroditic duct; latter often enclosed in sinus sac; cirrus sac absent (except in Hirudinellidae); prepharynx absent; testes normally pre-ovarian (except Azygiinae); vitellarium variable, follicular to compact masses; excretory vesicle Y-shaped..Hemiuroidea

Male and female ducts open into genital atrium or directly through common genital pore; cirrus sac present or absent; prepharynx normally present; ovary normally pre-testicular (except Baccigerinae); vitellarium usually in two lateral fields, usually follicular but follicles may coalesce to form two masses..19

19 Excretory system V- or Y-shaped; distribution of lateral vitelline fields tends to be restricted, sometimes to extent that they coalesce to form two compact masses; cirrus sac normally present; pars prostatica well developed..Gymnophalloidea (Fellodistomidae)

Excretory vesicle I-shaped; lateral vitelline fields normally extend full length of hindbody but occasionally more restricted..Allocreadioidea

SUPERFAMILY CLINOSTOMOIDEA LÜHE, 1901

Strigeida. Present in fish as encysted metacercaria. Body medium-sized to very large (up to 2cm in length); elongate-oval. Collar-like fold may envelope anterior extremity. Tegument armed. Oral sucker small. Ventral sucker in anterior half of body. Typical pharynx normally absent (in some genera included in the family a typical prepharynx and pharynx are present). Oesophagus short, muscular. Caeca long, simple or diverticulate; may unite with excretory system to form uroproct. Reproductive system may be well developed. Testes two, in tandem, in hindbody. Ovary submedian, inter-testicular. Uterus inter-caecal between anterior testis and ventral sucker. Excretory vesicle not prominent, V- to Y-shaped. Parasitic as adults in birds, mammals, and reptiles.

Family CLINOSTOMIDAE Lühe, 1901

Clinostomoidea. With features of superfamily.

CLINOSTOMUM Leidy, 1856

Clinostomidae. Present in fish as encysted metacercaria. Cyst large, subglobular. Body large, stout, elongate-oval; forebody ventrally concave. Collar-like fold may envelope anterior extremity in fixed material. Tegument armed with dense covering of small spines (easily lost in fixed material). Suckers well developed. Oral sucker small, ventrally subterminal. Ventral sucker large, near middle of anterior half of body. Typical pharynx absent. Oesophagus short, muscular, narrow anteriorly, saccular posteriorly. Intestinal bifurcation close to anterior extremity. Caeca long, broad, diverticulate, reaching close to posterior extremity and terminating against excretory vesicle (supposed to unite with excretory vesicle to form uroproct, but I have not seen an actual connection in the metacercaria). Reproductive system well developed. Testes two, in tandem, near middle of hindbody, irregularly oval. Cirrus sac small, containing seminal vesicle, antero-dextral to anterior testis. Large muscular cirrus often formed. Genital pore closely anterior to anterior testis. Ovarian complex inter-testicular. Ovary small, subglobular, dextral. Developing uterus curves sinistrally to anterior testis and unites with long, straight uterine diverticulum that extends anteriorly medially in hindbody. Vitellarium not developed. Excretory vesicle small, V-shaped (or Y-shaped with very short stem); pore subterminal. Parasitic in tissues of freshwater teleosts and less often in amphibians. Adults in oesophagus and buccal cavity of piscivorous birds.

Comments

The so-called "yellow grub" has been recorded in Canadian fishes under the names of *Clinostomum complanatum*, *C. marginatum*, and *Clinostomum* sp. metacercariae. Most authors, following Baer (1933), now consider *C. marginatum* a synonym of *C. complanatum* (see also Dowsett and Lubinsky 1980). Feizullaev and Mirzoeva (1983, 1986), in their revision of the superfamily Clinostomoidea, have gone further and consider *C. complanatum* the only valid species in the genus. Considering the previous difficulty in distinguishing both adult and especially metacercarial forms, most workers are more than happy to accept this state of affairs.

Clinostomum complanatum (Rudolphi, 1814) Braun, 1899 (metacercaria) (Fig. 4)

Syn. *Clinostomum marginatum* (Rudolphi, 1819) Braun, 1899; ?*C. gracile* of Stafford (1904); ?*Distomum gracile* of Wright (1879); *Clinostomum* sp. of authors.

Site: Musculature, gills, gill cavity, fins, mesenteries, viscera.

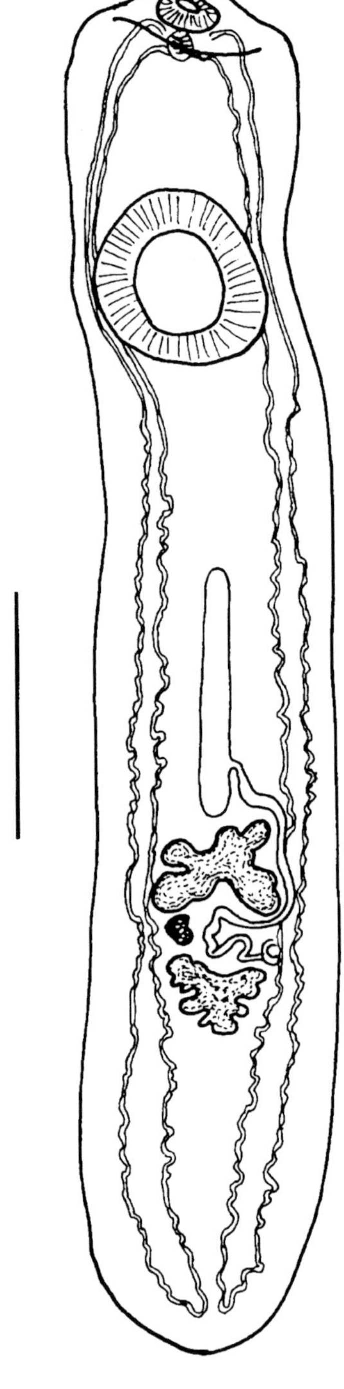

FIG. 4. *Clinostomum complanatum* metacercaria (after Van Cleave and Mueller (1934; as *C. marinatum*)). Scale bar: 1 mm.

Hosts: *Ambloplites rupestris* (4, 5, 9, 12, 17, 18, 23, 27, 29, 31;32;34); *Ameiurus nebulosus* (4, 6, 12, 17, 18, 32); *Aplodinotus grunniens* (34); *Catostomus commersoni* (5, 6, 9, 35); *Esox lucius* (9, 18); *Etheostoma flabellare* (4); *E. nigrum* (4, 17); *Fundulus diaphanus* (19); *Ictalurus melas* (4); *I. punctatus* (33); *Lepomis gibbosus* (3, 4, 6, 8, 12, 17, 25, 29, 32, 34, 37); *L. macrochirus* (4, 29); *Luxilis cornutus* (6, 8, 34); *Micropterus dolomieui* (3, 4, 6, 8, 9, 12, 13, 17, 23, 27, 32, 37); *M. salmoides* (29, 34); *Morone americana* (15); *M. chrysops* (28); *Mylocheilus caurinus* (11); *Notemigonus crysoleucas* (35); *N. heterolepis* (28); *N. hudsonius* (17, 31); *Oncorhynchus clarki* (20); *O. mykiss* (23, 27, 30); *Perca flavescens* (1, 2, 3, 4, 5, 6, 8, 9, 10, 12, 13, 14, 16, 17, 18, 22, 23, 24, 26, 27, 32, 34, 36, 37); *Percina caprodes* (4); *Phoxinus eos* (35); *P. neogaeus* (32); *Pimephales notatus* (17, 35); *P. promelas* (8); *Pomoxis nigromaculatus* (29); *Rhinichthys atratulus* (28); *Richardsonius balteatus* (11); *Salvelinus fontinalis* (7, 9); *Semotilus atromatriculatus* (6, 8); *S. corporalis* (8); *Stizostedion canadense* (17, 26, 27); *S. vitreum vitreum* (13, 17, 21, 22, 23, 24); unspecified trout (20).

Distribution: BC, Man, NB, NS, Ont, Que.

Records: 1. Wright 1879 (Ont?); 2. Stafford 1904 (Que?); 3. Cooper 1915 (Ont); 4. Bangham and Hunter 1939 (Ont); 5. Lyster 1939 (Que); 6. Bangham 1941 (Ont); 7. MacLulich 1943 (Ont); 8. Bangham and Vernard 1946 (Ont); 9. Fantham and Porter 1948 (Que); 10. Worley and Bangham 1952 (Que); 11. Bangham and Adams 1954 (BC); 12. Bangham 1955 (Ont); 13. Dowsett and Lubinsky 1966 (Man); 14. Tedla and Fernando 1969a (Ont); 15. Tedla and Fernando 1969b (Ont); 16. Tedla and Fernando 1972 (Ont); 17. Dechtiar 1972b (Ont); 18. Molnar *et al.* 1974 (Ont); 19. Wiles 1975 (NS); 20. Hoskins *et al.* 1976 (BC); 21. Anthony 1976 (Ont); 22. Evans in Lubinsky 1976 (Man); 23. Anon in Lubinsky 1976 (Man); 24. Dickson in Lubinsky 1976 (Man); 25. Cone and Anderson 1977 (Ont); 26. Anthony 1978b (Ont); 27. Stewart-Hay, Evans, and Dickson, cited in Lubinsky and Loch 1979 (Man); 28. Anthony 1984 (Ont); 29. Anthony 1985 (Ont); 30. Szalai and Dick 1988 (Man); 31. Dechtiar and Lawrie 1988 (Ont); 32. Dechtiar *et al.* 1988 (Ont); 33. Dechtiar and Nepszy 1988 (Ont); 34. Dechtiar and Christie 1988 (Ont); 35. Dechtiar *et al.* 1989 (Ont); 36. BM(NH) collection (Ont); 37. BM(NH) collection (NB).

Remarks: This cosmopolitan species is found in a wide range of freshwater fishes and some amphibians in North America. It is especially common in perciforms, the greatest number of records being in percids, especially *Perca flavescens* and centrarchids. The worms are large, reaching to more than 11 mm in length and 4 mm in breadth, although the majority of Canadian material that I have seen is not larger than 8.5 mm × 2.5 mm.

SUPERFAMILY DIPLOSTOMOIDEA POIRIER, 1886*

Syn. Strigeoidea Railliet, 1919

Strigeida. Found as larva (metacercaria) in fishes. Metacercaria encysted, encapsulated, or free. Cyst wall, when present, normally double-layered; inner layer (usually referred to as "cyst" or "parasite cyst") of parasite origin; outer layer (usually referred to as "capsule" or "host cyst") of host origin. Body composed of two parts; delimitation of two parts varies from distinct and marked by constriction to indistinct. Anterior part of body usually oval or elongate-oval; ventral surface concave; this concavity varies from slight to a deep, cup-shaped invagination. Posterior part of body distinct to indistinct, usually narrower than anterior body; may be reduced to small conical primordium (this region greatly enlarges in adult). Tegument unarmed. Brandes' organ present; may be lobed or unlobed and embedded or protruded. Pseudosuckers present or absent. Oral sucker normally present. Ventral sucker occasionally absent. Prepharynx very short. Pharynx usually present. Oesophagus usually short. Caeca long, usually terminating blindly close to posterior extremity. Primordium of gonads occasionally visible. Excretory system Y-shaped. Paranephridial system present. Encysted in tissues or on serous membranes of fishes, or free in eyes of fishes; occasionally encysted in amphibians and invertebrates or encapsulated in tissues of teleosts. (Adults in birds and occasionally mammals or reptiles.)

Comments

Terminology: Specialized structures that occur in diplostomoids often have several names.

Pseudosuckers, otherwise known as *lappets* or *cotylae*, refer to a pair of muscular structures that occur in the "shoulder" region.

Brandes' organ, otherwise known as the *tribocytic organ* or *adhesive organ*, is a glandular structure found posterior to the ventral sucker but within the ventral concavity of the anterior part of the body (in the case of metacercariae this is often well within the posterior half of the body).

The *paranephridial system*, otherwise known as the *reserve* or *secondary excretory system*, *lymphatic system*, etc., is defined as a series of outgrowths of the excretory vesicle or the primary collecting ducts that may ramify or anastomose but do not possess flame cells (see Pearson 1986). Different diplostomoid larval types have different arrangements of the paranephridial system (see below).

Larval types: Three types of larval diplostomoids are commonly recognized, *diplostomulum* and *neascus* for diplostomids and *tetracotyle* for strigeids. Although not now officially recognized, these names have been used as larval group names and even used at the generic level. The common use of these names, especially "diplostomulum", is diminishing. The following definitions are based on the work of Erasmus (1972), which was, in turn, based on the work of Hoffman (1960):

"Tetracotyle": In fish, amphibians, leeches, and snails. Encysted, with well-defined cyst wall of parasite origin; cyst normally tightly fitting. Anterior body oval and relatively thick; concave ventrally or cup-shaped. Posterior body as round, inconspicuous prominence. Paranephridial system consisting of large lacunae covering lateral and dorsal regions of body; spherical refractile bodies free in system. Pair of lappets present, usually invaginated in encysted worm.

*The appellation Diplostomoidea, rather than the more familiar Strigeoidea, is used because it is an older family-group name.

"Neascus": In fish. Encysted, with cyst wall of parasite origin; cyst usually but not always loosely fitting. Foliaceous anterior body; posterior body well developed. Paranephridial system consisting of narrow channels displayed in complex pattern, with refractile bodies; lacking diverticula. Lappets absent.

"Diplostomulum": In fish, amphibians, and snails. Free or encapsulated, without cyst of parasite origin. Anterior body oval, slightly concave ventrally. Posterior body small and conical. Paranephridial system of well-defined tubules in regular pattern and possessing spherical refractile bodies in small diverticula arising from main trunks. Lappets usually present.

Cyst: The cyst wall of diplostomoid metacercariae, when present, normally consists of two layers: the outer layer, known as the "capsule" or "host-cyst", is of host-origin produced to isolate the host from the parasite; the inner layer, known as the true "cyst" or "parasite cyst", is of parasite origin and produced to isolate the parasite from the protective responses of the host. Some species of *Diplostomum* occurring in the tissues are encapsulated (but not encysted), i.e., they are enclosed in a capsule of host origin only.

KEY TO THE FAMILIES OF THE DIPLOSTOMOIDEA

Comments

Although readily distinguishable as adults, the metacercariae of the diplostomoids have often not developed some of the diagnostic characters. This poses problems in keying down these forms, especially at the family level: keying at this level has not been attempted by Sudarikov (1971) and other authors.

1 Metacercariae encysted or encapsulated in tissues or free in eyes; cyst (inner parasite cyst), when present, usually thin-walled and loose-fitting; ventral depression of forebody usually shallow (but not so in *Crassiphiala*) ....Diplostomidae

Metacercariae usually encysted on serous membranes (peritoneum, pericardium, swimbladder, mesenteries, etc.); cyst usually thick-walled and tight-fitting; ventral depression in forebody usually forming deep posteriorly oriented cavity (but not so in *Ichthyocotylurus*) ....Strigeidae

Family DIPLOSTOMIDAE Poirier, 1886

Diplostomoidea. Present as larva (metacercaria) in fishes. Metacercaria encysted (larval type "neascus"), encapsulated or free (larval type "diplostomulum"). Cyst, when present, usually but not always loosely fitting; parasite cyst wall thin. Body composed of two parts; delimitation of two parts varies from distinct and marked by constriction to indistinct; anterior part with distinct or indistinct ventral concavity; posterior part (although well developed in adult) may be reduced to rudiment at posterior extremity. Brandes' organ present; embedded to protruded. Pseudosuckers present or absent. Oral sucker normally present, ventral sucker may be small or occasionally absent. Primordium of gonads occasionally visible. Paranephridial system present, with three main longitudinal vessels with few to numerous transverse commissures and other small anastomoses present or absent. In skin, eyes, musculature, central nervous system, etc. of freshwater fishes; occasionally in amphibians. (Adults in piscivorous birds; occasionally in piscivorous mammals.)

Comments

Keys: Diplostomid metacercariae, and especially those of the genus *Diplostomum*, are very difficult to distinguish, even using the more recent works of Shigin (1976, 1986), Sudarikov (1971), etc. Thus **all** records should be treated with at least some degree of caution. It has become increasingly apparent that the Nearctic forms cannot be treated in isolation: some sort of rationalisation of the Nearctic and Palaearctic fauna is necessary. This cannot be undertaken, however, until views on the status of species in Europe and Russia have stabilized.

Unidentified species assigned to various genera or larval group names are not included here, as they are listed by Margolis and Arthur (1979) and McDonald and Margolis (1995). Identified species assigned to larval group names, such as "*Neascus*", are listed with the species of the genera to which they appear to key down to most comfortably.

The keys presented below should enable forms to be identified to the generic level; but, until more is known of the host and site specificity, life cycles, etc., the key to species of *Diplostomum* is presented without conviction. The features used in these keys are relevant to known Canadian forms but not necessarily to all forms within the group. Works worth mentioning in relation to the identification of species are those of Hoffman (1960), Sudarikov (1971, translation 1982), and Shigin (1986).

Classification: The diplostomid genera occurring in Canadian fishes split readily into two groups, depending upon whether or not the metacercaria is free or encysted. Dubois (1938, 1970), followed by Yamaguti (1971), listed all of these genera in the subfamily Diplostominae (as Diplostomatinae) Poirier, 1886, although in two different tribes. Sudarikov (1960) listed them in two subfamilies, the Diplostominae (as Diplostomatinae) and Crassiphialinae Sudarikov, 1960 (not Shoop, 1989). Shoop (1989) also listed them in these two subfamilies, but removed the Crassiphialinae from the Diplostomidae into a new family, the Neodiplostomidae. Since Shoop's study is a cladistic analysis and uses only the interpretation of the arrangement of the branches of the paranephridial (secondary excretory) system of the metacercaria as the diagnostic character of the new family, I have retained the earlier conception of the Diplostomidae in this work.

Although the Canadian genera would appear to split readily into two groups, there has been some difference of opinion over the position of *Posthodiplostomum* and *Ornithodiplostomum*. Most workers have included them with *Diplostomum*, but Shoop has placed them in the Crassiphialinae, an arrangement that I have followed here, since, at least as far as the metacercariae are concerned, they fit more comfortably into this group.

Key to the subfamilies of the Diplostomidae

1 Pseudosuckers (lappets) present; body not clearly divided into two parts (though rudiment of hindbody may be present as posterior conical projection); unencysted, especially in eye ....Diplostominae

Pseudosuckers absent; body divided into two parts; encysted; parasite cyst usually loosely fitting, thin-walled ("neascus" type) ....Crassiphialinae

Subfamily CRASSIPHIALINAE Sudarikov, 1960

Diplostomidae. Metacercaria encysted, with thin-walled cyst (of "neascus" type). Body divided into two parts. Pseudosuckers absent. Paranephridial system with numerous transverse commissures and many small anastomoses. In skin, viscera, muscles, gill arches, etc. of teleosts. (Adults in piscivorous birds.)

Key to the genera of the Crassiphialinae

1 Ventral sucker absent; ventral concavity in anterior body extended posteriorly into deep pit containing enlarged Brandes' organ; cysts pigmented, in skin and occasionally on gills and in musculature .... *Crassiphiala*

Ventral sucker present 2

2 Body linguiform, with indistinct constriction between two parts of body and only slight ventral concavity of anterior body; cysts thin-walled, spacious, on peritoneum and viscera (plus a variety of other sites in heavy infections) .... *Ornithodiplostomum*

Body linguiform or fusiform, with distinct constriction between two parts of body and distinct ventral concavity in anterior body 3

3 Body fusiform; with distinct slightly posteriorly directed ventral concavity in anterior body that contains embedded or protuberant and digitiform Brandes' organ; ventral margin of concavity tending to cover Brandes' organ; cysts pigmented, usually in skin or musculature, occasionally in other sites .... *Uvulifer*

Body linguiform, with ventral concavity of anterior body containing embedded Brandes' organ; ventral margin of concavity tending not to cover Brandes' organ; thin-walled cysts on skin or viscera .... *Posthodiplostomum*

CRASSIPHIALA Van Haitsma, 1925

Crassiphialinae. Metacercaria encysted in spacious cyst. Cyst almost spherical, pigmented. Body clearly divided into two parts by constriction. Anterior body with deep ventral concavity extended into posteriorly oriented cavity containing massive, protuberant Brandes' organ with longitudinal aperture. Pseudosuckers absent. Oral sucker large; ventral sucker absent. Paranephridial system reticulate. On skin, occasionally on gill arches and between myotomes of musculature of freshwater teleosts. (Adults in piscivorous birds (kingfishers).)

Crassiphiala bulboglossa (Van Haitsma, 1925) (metacercaria) (Fig. 5)

Syn.: *Neascus bulboglossa* (Van Haitsma, 1925).

Sites: Mainly skin, occasionally between myotomes of musculature and on gill arches.

Hosts: *Couesius plumbeus* (11); *Ericymba buccata* (1); *Esox americanus vermiculatus* (2); *Fundulus diaphanus* (5); *Luxilis cornutus* (7, 8); *Notemigonus crysoleucas* (8, 10); *Notropis hudsonius* (7); *Perca flavescens* (3, 4, 8, 9, 10); *Pimephales notatus* (10, 11); *P. promelas* (10, 11); *Rhinichthys cataractae* (7, 8); *Semotilis atromaculatus* (1, 8, 11); *S. corporalis* (11); *Stizostedion vitreum vitreum* (9, 10); *Umbra limi* (6).

Distribution: Ont, NS, St Lawrence River.

Records: 1. Bangham and Hunter 1939 (Ont); 2. Crossman 1962 (Ont); 3. Dechtiar 1972a (Ont); 4. Dechtiar 1972b (Ont); 5. Wiles 1975 (NS); 6. Foersch *et al.* 1984 (St Lawrence River); 7. Dechtiar and Lawrie 1988 (Ont); 8. Dechtiar *et al.* 1988 (Ont); 9. Dechtiar and Nepszy 1988 (Ont); 10. Dechtiar and Christie 1988 (Ont); 11. Dechtiar *et al.* 1989 (Ont).

Remarks: Details of the life history of this species were discussed by Hughes (1928a) and Hoffman (1956). Like the related *Uvulifer* spp., this is a major causative agent of "blackspot" (Miller 1940b).

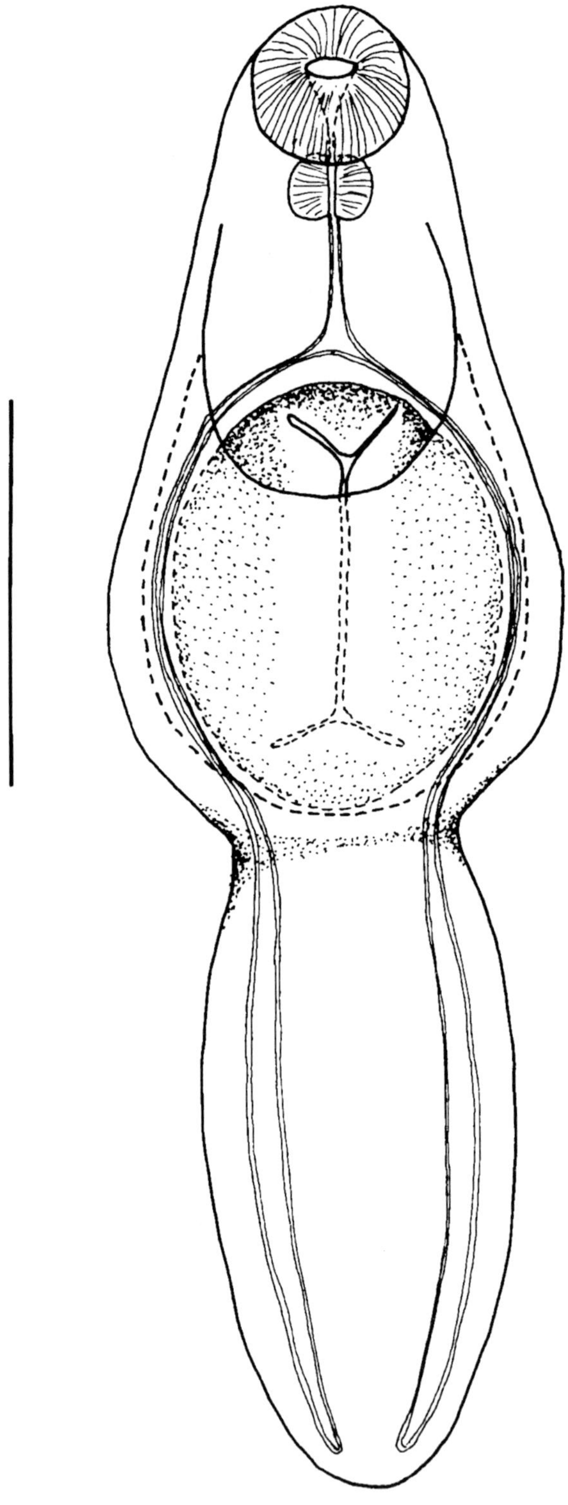

FIG. 5. *Crassiphiala bulboglossa* metacercaria (after Hughes (1928a)). Scale bar (roughly estimated): 100 µm.

ORNITHODIPLOSTOMUM Dubois, 1936

Crassiphialinae. Metacercaria encysted. Cyst very spacious, often much larger than parasite, oval; parasite cyst wall thin. Body linguiform, with indistinct division between spoon-shaped anterior part and relatively short, semi-circular posterior part. Anterior body with indistinct ventral concavity. Brandes' organ oval with central slit or aperture. Pseudosuckers absent. Oral sucker small. Ventral sucker present. Paranephridial system reticulate. Mainly on peritoneum and viscera, but also present in brain and other sites of freshwater teleosts. (Adults in piscivorous birds (anatids, especially mergansers, and ardeids).)

Ornithodiplostomum ptychocheilus (Faust, 1917) Dubois, 1936 (metacercaria) (Fig. 6)

Sites: Brain, viscera.

Hosts: *Fundulus diaphanus* (2); *Hybognathus hankinsoni* (1); *Luxilis cornutus* (1); *Nocomis biguttatus* (1); *Notropis heterolepis* (1); *Perca flavescens* (3); *Pimephales notatus* (1); *P. promelas* (1); *Semotilus atromaculatus* (1); *Stizostedion vitreum vitreum* (3).

Distribution: NS, Ont, Man.

Records: 1. Molnar *et al.* 1974 (Ont); 2. Wiles 1975 (NS); 3. Poole and Dick 1985 (Man).

Remarks: See Hoffman (1958b) for life cycle. The metacercaria was described by Amin (1982a).

POSTHODIPLOSTOMUM Dubois, 1936

Crassiphialinae. Metacercaria encysted. Parasite cyst thin-walled, delicate. Body linguiform, distinctly divided into two regions by annular constriction; anterior part scoop-shaped with distinct ventral concavity; posterior part oval to spherical. Brandes' organ embedded and only slightly protuberant with round or longitudinal aperture. Pseudosuckers absent. Suckers small. Paranephridial system reticulate. In variety of sites, including skin, viscera, eyes, etc. of freshwater teleosts. (Adults in piscivorous birds (ardeids).)

Key to the species of *Posthodiplostomum*

1 Encysted in viscera ....
......................*Posthodiplostomum minimum* (MacCallum, 1921) Dubois, 1936 (metacercaria) (Fig. 7)

Syn.: *Neascus vancleavei* (Agersborg, 1926); *Diplostomum cuticola* (Nordmann, 1832) Diesing, 1850 of Stafford (1904) and Cooper (1915); *Posthodiplostomum cuticola* (Nordmann, 1832) Dubois, 1936 of Margolis and Arthur (1979).

Sites: Mesenteries, liver, kidneys.

Hosts: *Acrocheilus alutaceus* (8); *Alosa pseudoharengus* (15); *Ambloplites rupestris* (1, 3, 4, 6, 12, 13, 15); *Ameiurus nebulosus* (5); *Campostoma anomalum* (4); *Carpiodes cyprinus* (14); *Catostomus catostomus* (6, 8); *C. commersoni* (7, 14); *C. macrocheilus* (8); *Cottus asper* (8, 11); *Couesius plumbeus* (6, 8, 13); *Culaea inconstans* (6, 13); *Esox lucius* (9); *Etheostoma exile* (6); *E. nigrum* (4, 6, 13); *Fundulus diaphanus* (4, 13, 15); *Gasterosteus aculeatus* (8); *Labidesthes sicculus* (15); *Lepomis gibbosus* (4, 5, 6, 7, 8, 9, 10, 13); *L. macrochirus* (4); *L. megalotis* (4); *Lota lota* (8); *Luxilis cornutus* (4, 5, 6, 7, 13); *Micropterus dolomieui* (3, 5, 6, 7, 13, 14); *M. salmoides* (4); *Morone americana* (15); *Mylocheilus caurinus* (8); *Nocomis biguttatus* (6); *N. micropogon* (4); *Notemigonus crysoleucas* (9, 15); *Notropis atherinoides* (4, 6, 13); *N. heterodon* (6); *N. heterolepis* (6, 9, 13); *N. hudsonius* (4, 6, 9, 12, 13); *N. rubellus* (13); *N. volucellus* (4, 6); *Perca flavescens* (5, 6, 7); *Percina caprodes* (9); *Percopsis omiscomaycus* (4, 6, 13); *Phoxinus eos* (6, 7); *P. neogaeus* (9, 13); *Pimephales notatus* (4, 6, 7, 9); *P. promelas* (7, 9, 13); *Pomoxis nigromaculatus* (4, 9, 13); *Ptychocheilus oregonensis* (8); *Rhinichthys atratulus* (4, 9); *R. cataractae* (6, 8, 9, 12, 13); *Richardsonius balteatus* (8); *Salvelinus fontinalis* (8); *Semotilus atromaculatus* (4, 5, 6, 7, 13); *S. corporalis* (7); *Stizostedion vitreum vitreum* (4).

Distribution: BC, Ont, Que.

Records: 1. Stafford 1904 (?Que); 2. Cooper 1915 (Ont); 3. Lyster 1939 (Que); 4. Bangham and Hunter 1939 (Ont); 5. Bangham 1941 (Ont); 6. Bangham 1955 (Ont); 7. Bangham and Venard 1946 (Ont); 8. Bangham and Adams 1954 (BC); 9. Dechtiar 1972b (Ont); 10. Cone

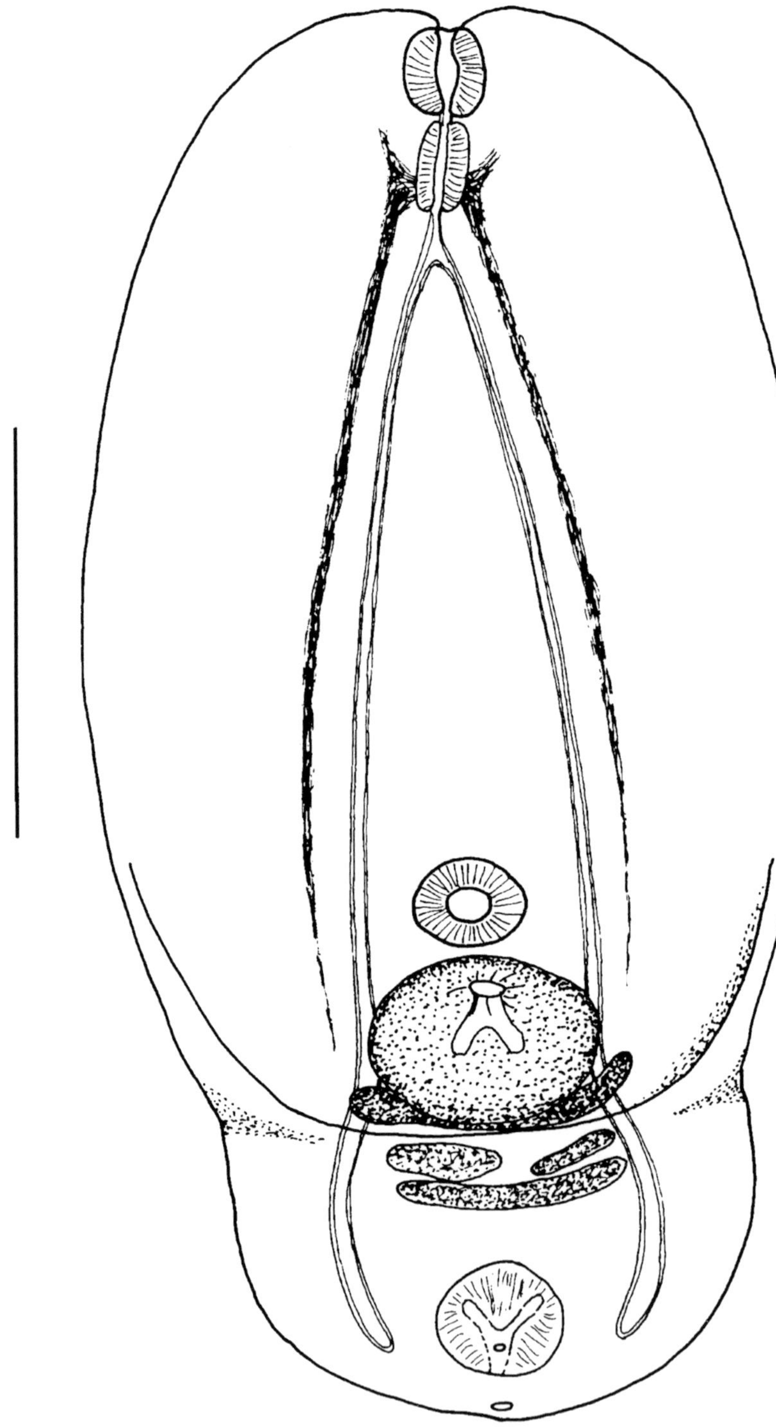

FIG. 6. *Ornithodiplostomum ptychocheilus* metacercaria (after Hughes and Piszczek (1928)). Scale bar (roughly estimated): 100 µm.

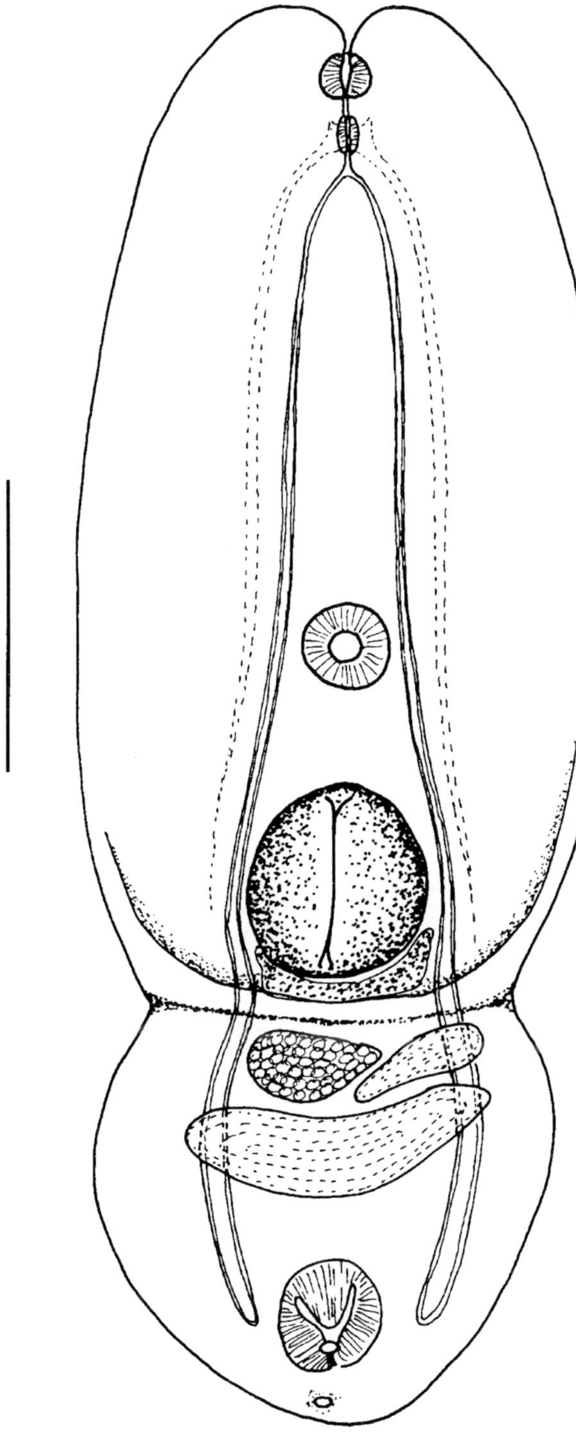

FIG. 7. *Posthodiplostomum minimum* metacercaria (after Hughes (1928b; as *Neascus vancleavei*)). Scale bar (roughly estimated): 200 μm.

and Anderson 1977 (Ont); 11. Anonymous 1981 (BC); 12. Dechtiar and Lawrie 1988 (Ont); 13. Dechtiar *et al.* 1988 (Ont); 14. Dechtiar and Nepszy 1988 (Ont); 15. Dechtiar and Christie 1988 (Ont).

Remarks: See Hoffman (1958a) for details of the life history. Canadian records attributed to *P. cuticola* are *P. minimum* according to Sudarikov (1971), since they are from the viscera and not the skin or muscles, the usual sites for *P. cuticola*. The metacercaria was described by Amin (1982a).

Two subspecies have been recognized:

(a) In centrarchids; on the liver, kidneys, and heart .... *P. minimum centrarchi* Hoffman, 1958 (metacercaria)

Site: Liver.

Hosts: *Ambloplites rupestris* (2); *Lepomis gibbosus* (1, 3); *L. macrochirus* (1, 3); *Micropterus dolomieui* (3); *M. salmoides* (2); *Perca flavescens* (2).

Distribution: Ont.

Records: 1. Dechtiar 1972a; 2. Molnar *et al.* 1974; 3. Dechtiar and Christie 1988.

(b) In cypriniforms; on mesenteries .... *P. minimum minimum* (MacCallum, 1921) Hoffman, 1958 (metacercaria)

Site: Mesenteries.

Hosts: *Carpiodes cyprinus* (1); *Fundulus diaphanus* (3); *Hybognathus hankinsoni* (2); *Luxilis cornutus* (2); *Nocomis biguttatus* (2); *Notropis atherinoides* (1); *N. hudsonius* (1); *Pimephales notatus* (2); *P. promelas* (2); *Semotilus atromaculatus* (2).

Distribution: NS, Ont.

Records: 1. Dechtiar 1972a (Ont); 2. Molnar *et al.* 1974 (Ont); 3. Wiles 1975 (NS).

Encysted in skin and musculature............................*Neascus rhinichthysi* Hunter, 1933 (metacercaria)

Sites: Skin, musculature.

Hosts: *Rhinichthys atratulus*; *R. cataractae.*

Distribution: Ont.

Record: Bangham and Hunter 1939.

Remarks: The adult form of this species is unknown. *P. cuticola* (Nordmann, 1832) would also key down to this position, but there appears to be no valid Canadian record of this species (see "Remarks" on *P. minimum*). See Hughes (1927) for American records of *P. cuticola.*

UVULIFER Yamaguti, 1934

Crassiphialinae. Metacercaria encysted in relatively tightly fitting cyst; cyst oval to pyriform and pigmented. Body clearly divided into two parts by medial constriction. Anterior body with ventral concavity directed posteriorly and containing embedded to digitiform Brandes' organ. Pseudosuckers absent. Oral sucker large; ventral sucker small. Paranephridial system reticulate. In skin, between myotomes of musculature, at base of fins and occasionally in other sites, such as around eyes or in mouth of fishes. (Adults in kingfisher.)

Key to the species of *Uvulifer*

1 Parasite cyst (removed from host capsule) >300 μm in length, oval; Brandes' organ protuberant and digitiform; mainly in musculature and skin .... *Uvulifer ambloplitis* (Hughes, 1927) Dubois, 1938 (metacercaria) (Fig. 8)

Syn.: *Neascus ambloplitis* Hughes, 1927; *Crassiphiala ambloplitis* (Hughes, 1927) Hunter and Hunter, 1931; *Neascus wardi* Hunter, 1928.

Sites: Skin, musculature, fins, gills.

Hosts: *Ambloplities rupestris* (1, 7, 9, 10, 12); *Catostomus commersoni* (7); *Esox lucius* (5, 6, 10, 11, 12); *Etheostoma exile* (7); *Hybognathus hankinsoni* (2, 7); *Lepomis gibbosus* (1, 2, 3, 4,

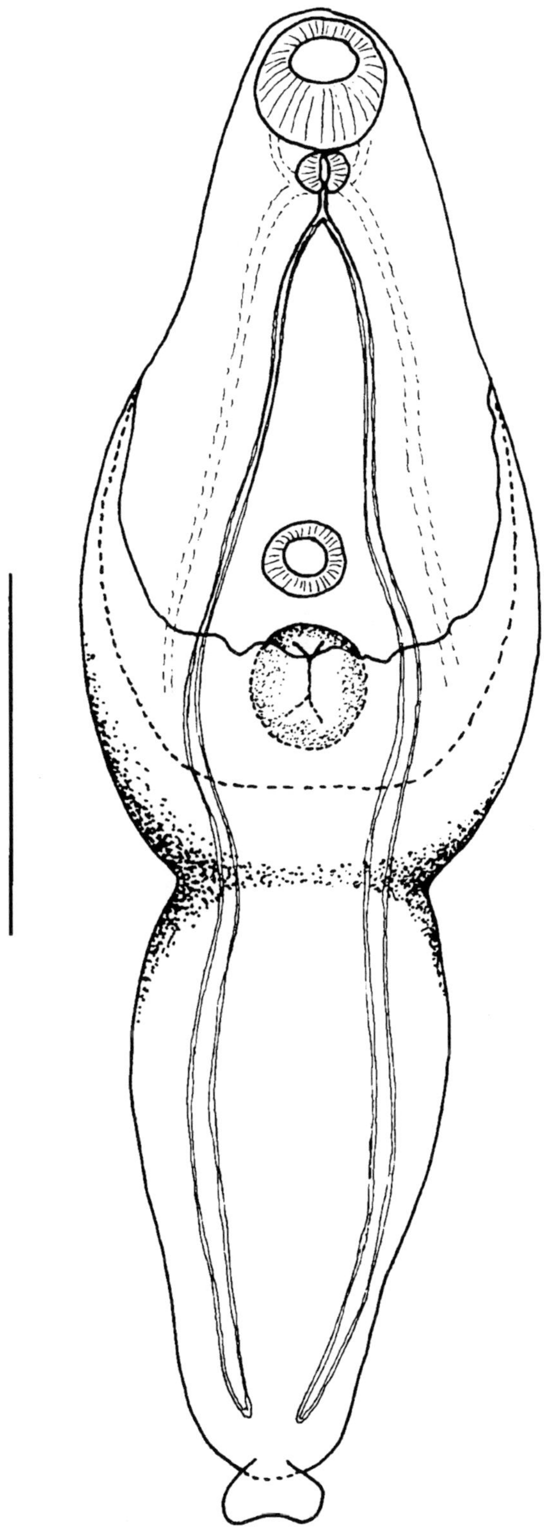

FIG. 8. *Uvulifer amblopliti*s metacercaria (after Hughes (1927)). Scale bar (roughly estimated): 200 μm.

8, 10, 12); *L. macrochirus* (1); *Luxilis cornutus* (2, 7); *Micropterus dolomieui* (2, 3, 4, 6, 10, 11, 12); *Nocomis biguttatus* (7); *Notemigonus crysoleucas* (2); *Notropis heterolepis* (2, 10); *N. hudsonius* (10); *Perca flavescens* (2, 7); *Phoxinus eos* (2); *P. neogaeus* (2); *Pimephales notatus* (2, 7); *P. promelas* (2, 7); *Semotilus atromaculatus* (2, 7); *S. margartia* (2); *Stizostedion vitreum vitreum* (12).

Distribution: Ont, Que.

Records: 1. Bangham and Hunter 1939 (Ont); 2. Bangham 1941 (Ont); 3. Bangham and Venard 1946 (Ont); 4. Lachance 1947 (Que); 5. Dechtiar 1972a (Ont); 6. Dechtiar 1972b (Ont); 7. Molnar *et al.* 1974 (Ont); 8. Cone and Anderson 1977 (Ont); 9. Dechtiar and Lawrie 1988 (Ont); 10. Dechtiar *et al.* 1988 (Ont); 11. Dechtiar and Nepszy 1988 (Ont); 12. Dechtiar and Christie 1988 (Ont).

Remarks: This metacercaria and its cyst were studied *inter alia* by Hughes (1927) and Hunter and Hamilton (1941). These worms may cause "blackspot" in their fish host. The adult occurs in the kingfisher *Megaceryle alcyon*: see Hoffman and Putz (1965) for details of the life-history.

Parasite cyst (removed from host capsule) <300 μm in length, pyriform; Brandes' organ not reported as digitiform; mainly on skin and fins .... *Neascus pyriformis* Chandler, 1951 (metacercaria)

Sites: Fins, skin.

Hosts: *Luxilis cornutus*; *Perca flavescens*; *Semotilus atromaculatus*.

Distribution: Ont.

Record: Molnar *et al.* 1974.

Remarks: This form has been redescribed by Duru *et al.* (1981). Chandler (1951) suspected that it might be the larval stage of *Uvulifer semicircumciscus* Dubois and Rausch, 1950. This metacercaria also causes "blackspot".

Subfamily DIPLOSTOMINAE Poirier, 1886

Diplostomidae. Metacercaria usually free and unencysted, occasionally encapsulated (of "diplostomulum" type). Body not clearly divided into two parts. Pseudosuckers present. Paranephridial system with three or fewer transverse commissures and no small anastomoses. In eyes and central nervous system of fishes and occasionally amphibians. (Adults in piscivorous birds and occasionally mammals.)

Key to the genera of the Diplostominae

1 Body oval (length/breadth ratio <3:1); Brandes' organ round .... *Diplostomum*

Body fusiform (length/breadth ratio >3:1); Brandes' organ elongate-oval .... *Tylodelphys*

DIPLOSTOMUM von Nordmann, 1832

Diplostominae. Metacercaria free and active or encapsulated. Body oval (round to elongate-oval, depending upon fixation), not clearly divided into two parts, although posterior part may be visible as conical terminal process. Ventral concavity of anterior region of body indistinct. Brandes' organ round, with aperture of variable shape. Pseudosuckers well developed. Suckers well developed. Paranephridial system in form of regularly arranged tubules. In eyes and, occasionally, brain of freshwater fishes. (Adults in wide range of piscivorous birds (especially larids and anatids).)

Comments

The identification of metacercariae of the genus *Diplostomum* has been in the past and continues to be a real problem. The reason for this is that, although one or two species, e.g., *D. commutatum* (Diesing,

1850), are readily recognized, the majority are morphologically very similar and distinguishable only if processed in a similar manner and using combinations of metrical data in statistical analyses. The majority of records in the literature are of very little value, most having been referred to as *Diplostomum* sp. or *D. spathaceum*. Detailed studies have been carried out in Europe in recent years by workers such as Shigin and Niewiadomska, and, although these tend to be somewhat contradictory, they are showing signs of solving the problem. Nevertheless, at the present time there is still confusion, for example, regarding the common form from the lens of freshwater fishes, *D. spathaceum*. According to Niewiadomska (1984, 1986) *D. spathaceum sensu* Shigin (1968) is *D. pseudospathaceum* Niewiadomska, 1984, whereas *D. indistinctum sensu* Shigin and *D. helveticum* (Dubois, 1929) are synonyms of *D. spathaceum sensu* Niewiadomska. According to Shigin (1986, 1987) *D. spathaceum sensu* Shigin (1968) and *D. pseudospathaceum* are synonyms of *D. chromatophorum* (Brown, 1931) and *D. helveticum* (syn. *D. indistinctum sensu* Shigin) is a valid species. Recently, Niewiadomska (1989) has countered, indicating that *D. pseudospathaceum* cannot be a synonym of *D. chromatophorum* as the cercariae differ.

It is perfectly clear that, even in the absence of such nomenclatural confusion, the identification of these larvae on morphological grounds will never be easy. Furthermore, in view of the mobility and wide distribution of the avian definitive hosts, it is very likely that many of the species of *Diplostomum* occur throughout the Holarctic region. The situation in North America is likely, therefore, to remain confused until the situation in the Palaearctic is clarified.

If these metacercariae cannot be identified satisfactorily on morphological grounds, we are left with little but host and site to group them. This may not be as inadequate as it might appear, since work in the UK by Brady (1989) has indicated that these metacercariae do tend to be site-specific and, in the case of non-lens forms, exibit host-group specificity. Although *Diplostomum* metacercariae are said to live in an immunologically privileged site, this really only applies to the lens forms, since the retina has a good blood supply and is close to the epichorioid lymph space, and even lens forms have to pass through other parts of the body to reach the lens. In view of the current state of knowledge on the taxonomy of these metacercariae in both Europe and North America, I consider that a key based upon site and host group is the most useful way of dealing with them, although I fully recognize that it is almost certain that more forms than indicated are in fact present in the Canadian fauna and that most records from non-salmonoid, percid, and gasterosteid hosts are still problematical. I have tended to use the North American names in current usage rather than attempt to integrate them into the often older European nomenclature in view of the latter's current instability. Consequent upon the difficulties and confusion within the genus, and especially in the many cases where no detailed site information is presented, the records and hosts associated with the records given below should be treated with a degree of scepticism.

Key to the species of *Diplostomum*

1 In lens of salmonoids and other groups ....
...............................*Diplostomum spathaceum* (Rudolphi, 1819) Olsson, 1876 (metacercaria) (Fig. 9)

Syn.: *Diplostomulum spathaceum* (Rudolphi, 1819); *Diplostomum volvens* Nordmann, 1833 of Cooper (1915).

Site: Lens of eye (records from vitreous humour of eye are probably not *spathaceum*).

Hosts: *Acipenser fulvescens* (27, 30, 35); *Alosa aestivalis* (26, 34); *A. pseudoharengus* (9, 26, 27, 28); *Amboplites rupestris* (7, 28, 30); *Ameiurus nebulosus* (7, 15, 28, 32); *Amia calva* (28, 30); *Anguilla rostrata* (2, 30); *Aplodinotus grunniens* (29, 30); *Carpiodes cyprinus* (28, 29); *Catostomus catostomus* (5, 14, 27, 28, 29, 30); *C. commersoni* (5, 7, 9, 27); *Coregonus artedii* (16, 27, 28, 30, 31, 32); *C. clupeaformis* (6, 9, 14, 16, 20, 28, 30); *C. hoyi* (28); *Cottus asper* (18); *C. bairdi* (27, 28); *C. cognatus* (14); *C. ricei* (28); *Couesius plumbeus* (18, 23, 28, 30); *Culaea inconstans* (28, 32); *Cyprinus carpio* (28, 30); *Dorosoma cepedianum* (28, 30); *Esox lucius* (14, 28, 30); *Etheostoma caeruleum* (30); *E. exile* (7, 30); *E. nigrum* (30); *Fundulus diaphanus* (28, 30); *Gasterosteus aculeatus* (4, 10, 30); *Hybognathus*

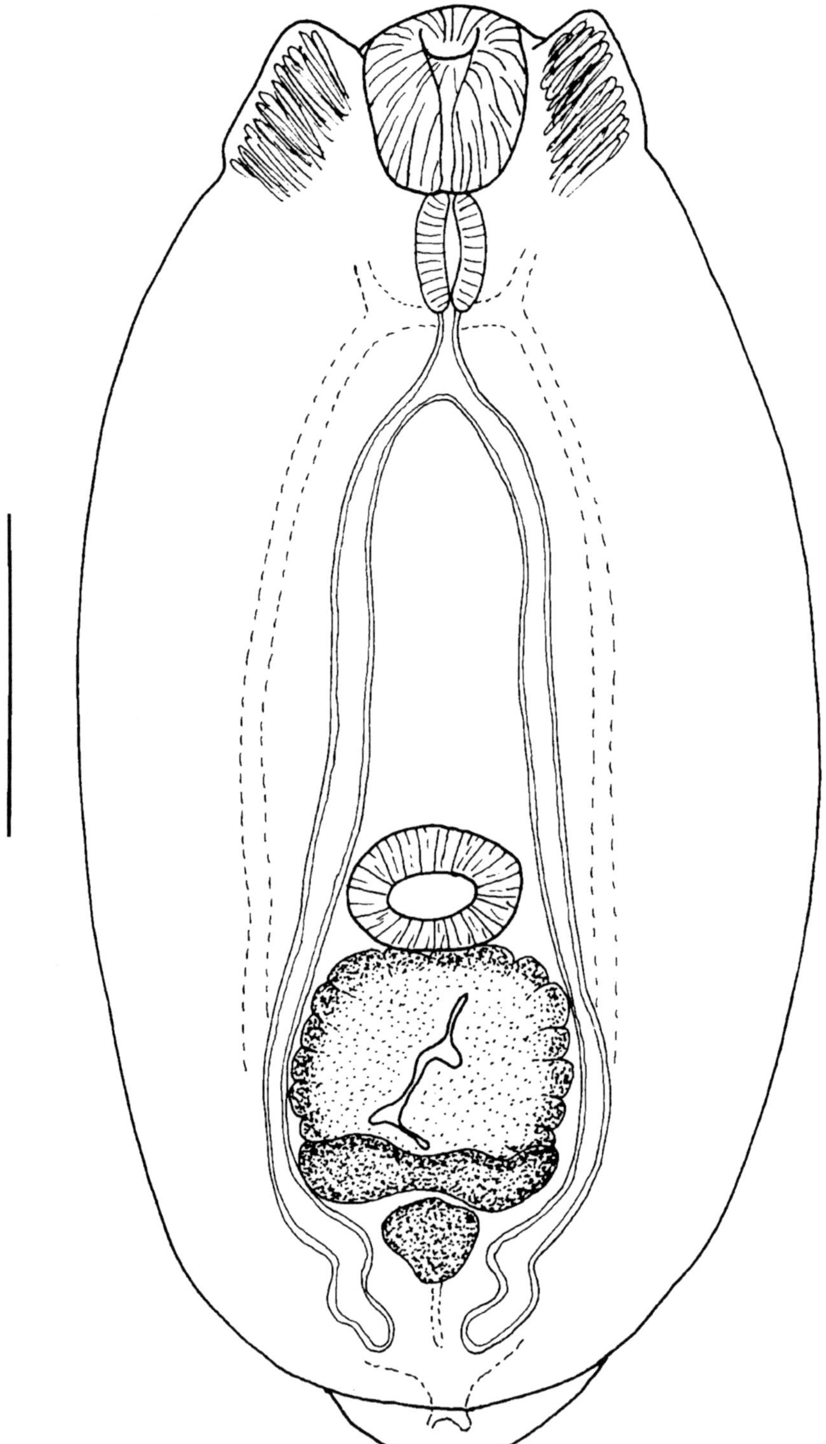

FIG. 9. *Diplostomum spathaceum* metacercaria (after Niewiadomska (1986)). Scale bar: 50 µm.

hankinsoni (7); *Ictalurus punctatus* (28, 29); *Labidesthes sicculus* (30); *Lepisosteus osseus* (30); *Lepomis gibbosus* (28, 30, 32); *Lota lota* (14, 18, 27, 28); *Luxilis cornutus* (7, 27, 32); *Micropterus dolomieui* (1, 28, 29, 30); *M. salmoides* (30); *Morone chrysops* (28, 29, 30); *Nocomis biguttatus* (7); *Notemigonus crysoleucas* (30); *Notropis atherinoides* (28, 30); *N. heterolepis* (7, 28); *N. hudsonius* (27, 28, 30); *N. rubellus* (28); *Noturus gyrinus* (30); *Oncorhynchus kisutch* (27, 28); *O. mykiss* (18, 27, 28, 30); *O. nerka* (18, 28); *O. tschawytscha* (18); *Osmerus mordax* (9, 27, 28, 29, 30); *Perca flavescens* (7, 9, 27, 28, 29); *Percopsis omiscomaycus* (9, 27, 28, 30, 32); *Phoxinus neogaeus* (28); *Pimephales notatus* (7, 28, 32); *P. promelas* (7, 30); *Pomoxis nigromaculatus* (28, 30); *Prosopium cylindraceum* (6, 14, 27, 28); *P. williamsoni* (18); *Ptychocheilus oregonensis* (18); *Pungitius pungitius* (3, 19, 20, 28); *Rhinichthys cataractae* (28); *Salmo salar* (6, 8, 11, 12, 13, 17); *Salvelinus alpinus* (6, 25); *S. fontinalis* (6, 8, 21, 22, 28, 30, 32, 33); *S. fontinalis* × *S. namaycush* (9, 28, 32); *S. malma* (18); *S. namaycush* (6, 14, 20, 28, 30); *Semotilus atromaculatus* (7, 28, 30); *Stizostedion vitreum vitreum* (28, 29, 30); *Thymallus arcticus* (14); *Umbra limi* (28, 30).

Distribution: BC, Lab, NB-b, NB, Nfld-b, Nfld, NS, Ont, Que, Que-b, PEI, YT, Man, Sask, Alta, NWT, Atl.

Records: 1. Cooper 1915 (Ont); 2. Hanek and Threlfall 1970b (Lab, Nfld-b); 3. Hanek and Threlfall 1970c (Nfld); 4. Hanek and Threlfall 1970d (Lab, Nfld-b, Nfld); 5. Threlfall and Hanek 1970a (Lab); 6. Hicks and Threlfall 1973 (Lab); 7. Molnar *et al.* 1974 (Ont); 8. Hare and Frantsi 1974 (NB, NS); 9. Collins and Dechtiar 1974 (Ont); 10. Lester 1975 (BC); 11. Hare 1975 (NB); 12. Hare and Burt 1975a (NB); 13. Hare and Burt 1976 (NB-b, NB); 14. Arthur *et al.* 1976 (YT); 15. Fréchette *et al.* 1978 (Que); 16. Watson and Dick 1979 (Man); 17. Pippy 1980 (Atl, Lab, Nfld, NS, NB, PEI); 18. Anonymous 1981 (BC); 19. Curtis, 1981 (NWT); 20 Leong and Holmes, 1981 (Alta); 21. Black 1981 (Que); 22. Brassard *et al.* 1982 (Que); 23. Anonymous 1984 (BC); 24. Bouvry *et al.* 1984 (Que); 25. Curtis 1984 (Que); 26. Landry *et al.* 1986 (NB); 27. Dechtiar and Lawrie 1988 (Ont); 28. Dechtiar *et al.* 1988 (Ont); 29. Dechtiar and Nepszy 1988 (Ont); 30. Dechtiar and Christie 1988 (Ont); 31. Curtis 1988 (Que/Lab); 32. Dechtiar *et al.* 1989 (Ont); 33. Wright *et al.* 1989 (Que); 34. Landry *et al.* 1992 (NB-b); 35. Choudhoury and Dick 1993 (Man, Sask; as *Diplostomum* sp.).

Remarks: *D. spathaceum* is a name that has been used in the past for any form of *Diplostomum* from any region of the eye, although more usually for forms from the lens. Judging from the work of Shigin (1986), in the Palaearctic several species of *Diplostomum* occur in the lenses of freshwater fishes: other forms include *D. chromatophorum* (Brown, 1931) (?=*D. pseudospathaceum* Niewiadomska, 1984: see comment above); *D. helveticum* (Dubois, 1929); *D. mergi* Dubois, 1932; and *D. parviventosum* Dubois, 1932. It is likely that some of these forms are also present in Canadian fishes.

As indicated above, all records of this species should be treated with caution. As the most common species, it has often been used as a group name for any *Diplostomum* metacercaria found in the eye.

The definitive hosts of *D. spathaceum* (*sensu stricto*) are gulls.

Lens forms have also been recorded in Canada under the names *D. spathaceum indistinctum* and *D. flexicaudum*. These records may well be *D. spathaceum*. These are listed separately below.

D. spathaceum indistinctum (Guberlet, 1923) Hughes, 1929 (metacercaria)

Site: Lens of eye (?external layers of eye and vitreous humour).

Hosts: *Alosa pseudoharengus* (1); *Catostomus catostomus* (1); *C. commersoni* (1); *Coregonus artedii* (1, 2); *C. clupeaformis* (1, 2, 3); *C. hoyi* (1); *Lota lota* (4); *Osmerus mordax* (1); *Perca flavescens* (1); *Pimphales notatus* (5); *Salvelinus fontinalis* × *S. namaycush* (1).

Distribution: Man, Alta, Ont.

Records: 1. Dechtiar and Berst 1978 (Ont); 2. Watson, in Lubinsky and Loch 1979 (Man); 3. Dick and Rosen 1981 (Man); 4. McAllister and Mudry 1983 (Alta); 5. Dechtiar *et al.* 1989 (Ont).

Remarks: Niewiadomska (1986) considered *D. indistinctum sensu* Shigin (1968) a synonym of *D. spathaceum*, whereas Shigin (1986) included it as a synonym of *D. helveticum* Shigin, 1977, another synonym of *D. spathaceum* according to Niewiadomska (1984).

The actual nature of this species is confused; many Canadian records were not from the lens.

Diplostomum flexicaudum (Cort and Brooks, 1928) Van Haitsma, 1931

Syn.: *Diplostomulum flexicaudum* (Cort and Brooks, 1928).

Site: Lens of eye.

Hosts: *Anguilla rostrata* (2); *Catostomus catostomus* (1); *C. commersoni* (1, 2, 3); *Coregonus clupeaformis* (2); *Cyprinus carpio* (2); *Esox masquinongy* (3); *Luxilis cornutus* (2); *Moxostoma anisurum* (3); *M. erythrurum* (2); *M. macrolepidotum* (2); *Notropis atherinoides* (2); *Osmerus mordax* (2); *Percopsis omiscomaycus* (2).

Distribution: Ont.

Records: 1. Bangham 1955 (Ont); 2. Dechtiar 1972a (Ont); 3. Dechtiar 1972b (Ont).

Remarks: A synonym of *D. spathaceum*, according to Shigin (1986).

In other parts of eye or brain 2

2 In brain or retina of gasterosteids (may be encapsulated) .... *D. scudderi* (Olivier, 1941) Dubois, 1966 (metacercaria).

Syn.: *Diplostomulum baeri eucaliae* Hoffman and Hundley, 1957.

Sites: Brain, retina of eye.

Hosts: *Culaea inconstans* (1, 4); *Gasterosteus aculeatus* (2, 3).

Distribution: BC, Ont.

Records: 1. Dechtiar 1972b (Ont); 2. Lester 1975 (BC); 3. Lester and Huizinga 1977 (BC); 4. Dechtiar *et al.* 1988 (Ont).

Remarks: *D. pungiti* Shigin, 1965 occurs under the retina of gasterosteids in the Palaearctic.

In retina or vitreous humour of eye 3

3 In salmonoids ....*D. baeri bucculentum* Dubois and Rausch, 1948 (metacercaria).

Sites: Vitreous humour, retina (records from the retina may belong to another species: see below).

Hosts: *Coregonus clupeaformis* (1, 4); *C. sardinella* (5); *Cottus asper* (2); *Couesius plumbeus* (2); *Lota lota* (2); *Oncorhynchus mykiss* (2, 3, 4); *O. nerka* (2, 3, 4); *O. tshawytscha* (2, 3, 4); *Prosopium williamsoni* (2, 3, 4); *Ptychocheilus oregonensis* (2); *Salvelinus malma* (2, 3, 4); *S. namaycush* (3, 4).

Distribution: BC, Man, NWT

Records: 1. Dick and Rosen 1981 (Man); 2. Anonymous 1984 (BC); 3. Ching 1984 (BC); 4. Ching 1985 (BC); 5. Shostak *et al.* 1987 (NWT).

Remarks: In Britain there is evidence that metacercariae from the retina and vitreous humour of *Oncorhynchus mykiss* are distinct species (Brady 1989), but Canadian workers have assumed, possibly incorrectly, that they are the same. Both Dick and Rosen and Ching obtained adults experimentally in chicks by feeding them with metacercariae from both the humour and the retina. According to Brady (1989) metacercariae from the retina will not develop in chicks, thus the adults obtained by these workers are likely to represent the humoural form only. There appears to be no obvious name to apply to the retinal form from Canadian salmonoids, although it is worth noting that *D. pseudobaeri* Razmashkin and Andreyuk, 1978 occurs under the inner membrane of the back of the eye of coregonids in the Palaearctic.

Shigin (1986) has considered *D. baeri* Dubois, 1937 *sensu* Shigin (1968) a synonym of *D. volvens* Nordmann, 1832, but Niewiadomska (1988) has disagreed with this.

In percids 4

4 In retina..*D. adamsi* Lester and Huizinga, 1977 (metacercaria)

Site: Retina of eye.

Host: *Perca flavescens*.

Distribution: Ont.

Records: Lester and Huizinga 1977; Lester 1977.

Remarks: According to Lester (1977) the specimens reported as *D. huronense* by Tedla and Fernando (1969a) belong to this species; but, since these authors give the site as the humour, this seems unlikely. A similar form occurs under the retina of *Perca fluviatilis* in Britain (Brady, 1989). Possible names for the latter form include *D. volvens* Nordmann, 1832 (see Shigin 1986) or *D. baeri* Dubois, 1937 (see Niewiadomska 1988).

In humour ....*D. huronense* (La Rue, 1927) Hughes, 1929

Syn.: *Diplostomulum huronense* (La Rue, 1927) Hughes and Hall, 1929; *Diplostomum spathaceum huronense* (La Rue, 1927) Dubois, 1966.

Site: Humour of eye; (?lens).

Hosts: *Lepomis gibbosus* (5); *Morone americana* (3, 7); *Osmerus mordax* (6); *Perca flavescens* (2, 4, 7); *Petromyzon marinus* (1).

Distribution: Ont, Nfld.

Records: 1. Wilson and Ronald 1967 (Ont); 2. Tedla and Fernando 1969a (Ont); 3. Tedla and Fernando 1969b (Ont); 4. Tedla and Fernando 1972 (Ont); 5. Cone and Anderson 1977 Ont); 6. Threlfall 1981 (Nfld); 7. Dechtiar and Christie 1988 (Ont).

Remarks: The records from non-percid hosts require confirmation.

According to Shigin (1986) *D. huronense* is a synonym of *D. volvens* von Nordmann, 1832. It is clear that this species should also be compared with *D. numericum* Niewiadomska, 1988 and *D. baeri* Dubois, 1937 *sensu* Niewiadomska (1988) from *Perca fluviatilis* in Europe.

TYLODELPHYS Diesing, 1850

Diplostominae. Metacercaria active and free. Body fusiform, not clearly divided into two regions, although posterior part may be apparent as terminal conical process. Ventral concavity in anterior body very indistinct. Brandes' organ elongate-oval, with long, vertical aperture; close to posterior end of worm. Pseudosuckers present, indistinct. Suckers small. Paranephridial system in form of regularly arranged tubules. In eyes (vitreous humour) and brain of freshwater teleosts and spinal cord of amphibians. (Adults in piscivorous birds (especially Podicipediformes and Ciconiiformes).)

Tylodelphys scheuringi (Hughes, 1929) Dubois, 1938 (metacercaria) (Fig. 10)

Syn.: *Diplostomulum scheuringi* Hughes, 1929; *Diplostomum scheuringi* (Hughes, 1929) Bangham and Hunter, 1939.

Site: Vitreous humour of eye (there are American records from the brain).

Hosts: *Ambloplites rupestris* (5); *Catostomus macrocheilus* (7, 8); *Coregonus clupeaformis* (8); *Cottus asper* (7, 8); *Fundulus diaphanus* (12); *Lepomis gibbosus* (4, 6, 13); *Lota lota* (7, 8, 9); *Luxilis cornutus* (11); *Micropterus salmoides* (5); *Morone americana* (13); *Mylocheilus caurinus* (8); *Oncorhynchus mykiss* (7, 8, 10); *Perca flavescens* (1, 2, 3, 4, 5); *Percina caprodes* (4, 13); *Percopsis omiscomaycus* (14); *Pomoxis nigromaculatus* (4); *Prosopium williamsoni* (7, 8); *Ptychocheilus oregonensis* (8); *Rhinichthys cataractae* (4); *Richardsonius balteatus* (8); *Semotilus corporalis* (14); *Stizostedion vitreum vitreum* (4).

Distribution: Ont, Man, BC, Alta.

Records: 1. Bangham and Hunter 1939 (Ont); 2. Tedla and Fernando 1969a (Ont); 3. Tedla and Fernando 1972 (Ont); 4. Dechtiar 1972b (Ont); 5. Molnar *et al.* 1974 (Ont); 6. Cone and Anderson 1977 (Ont); 7. Anonymous 1981 (BC); 8. Anonymous 1984 (BC); 9. McAllister and Mudry 1983 (Alta); 10. Szalai and Dick 1988 (Man); 11. Dechtiar and Lawrie 1988 (Ont); 12. Dechtiar *et al.* 1988 (Ont); 13. Dechtiar and Christie 1988 (Ont); 14. Dechtiar *et al.* 1989 (Ont).

Comments

The adult form is not known in nature, but aspects of the life history of this worm were discussed by Etges (1961). The same species is claimed to occur in the eyes and brains of newts. In most descriptions the pseudosuckers are not shown, but they are clearly visible in the figure of Van Cleave and Mueller (1934).

Hoffman (1960) indicated that this species is similar to the European form *T. clavata* (Nordmann, 1832).

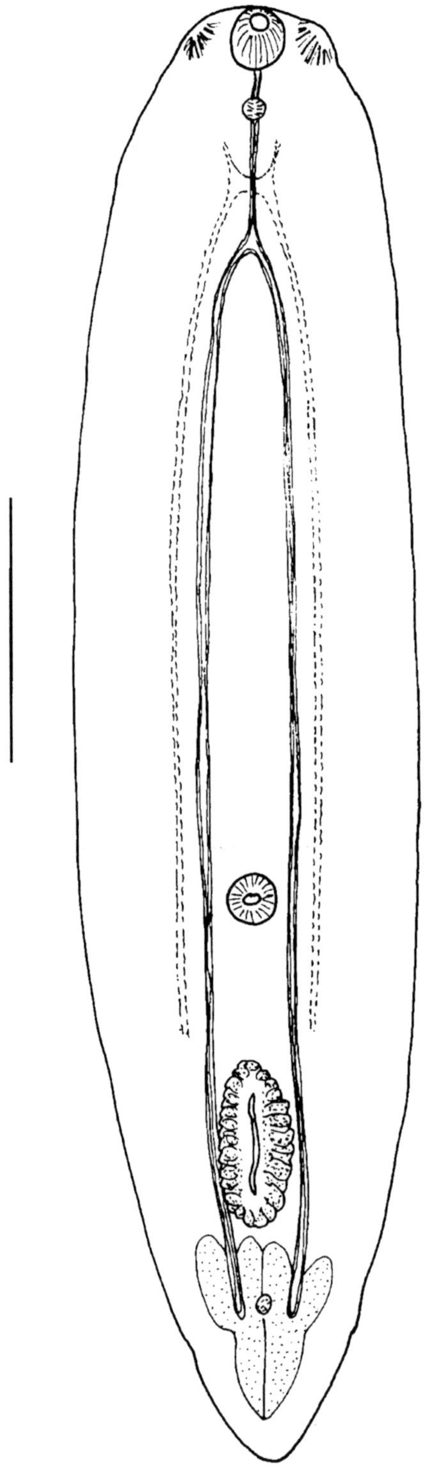

FIG. 10. *Tylodelphys scheuringi* metacercaria (after Van Cleave and Mueller (1934)). Scale bar: 200 µm.

Family STRIGEIDAE Railliet, 1819

Diplostomoidea. Metacercaria encysted (often referred to as "tetracotyle"). Cyst (parasite cyst) wall thin to thick. Body composed of two parts; delimitation of two parts varies from distinct and marked by constriction to indistinct. Anterior part of body with distinct or indistinct ventral concavity, often extended into deep, cup-shaped, posteriorly oriented invagination. Delimitation of posterior part of body may be indistinct; usually narrower than anterior part. Brandes' organ often lobed; sometimes protruded. Pseudosuckers usually large and open on antero-ventral surface, occasionally absent. Oral and ventral suckers present. Paranephridial system composed of lacunae in lateral and dorsal regions of body ("tetracotyle" type). Common in freshwater fishes, especially on serous membranes, occasionally in invertebrates. (Adults in birds, rarely in mammals.)

Comments

Tetracotyle parvula (Stafford, 1904) Mataré, 1909 (syn. *Diplostomum parvulum* Stafford, 1904) was described by Stafford (1904) from the pharynx of *Semotilus corporalis* and as free cysts in the intestine of *Esox lucius* probably in Quebec. In agreement with Margolis and Arthur (1979), it is not possible to recognise this species, although Hughes (1929) suggested that it might be *Ichthyocotylurus platycephalus* (as *Tetracotyle communis*). *T. parvula* is not included in this work.

Key to the genera of the family Strigeidae

1 Ventral depression of forebody deep, extended into saccate posteriorly oriented cavity; Brandes' organ with distinct lobes; cyst wall very thick, difficult to rupture mechanically ....*Apatemon*

Ventral depression of forebody indistinct, present only as slight concavity; Brandes' organ lacking distinct lobes; cyst wall thick (inner parasite cyst may be thin), can be ruptured mechanically ....*Ichthyocotylurus*

APATEMON Szidat, 1928

Strigeidae. Metacercaria encysted, filling tightly fitting, thick-walled cyst; outer layer (capsule) thick, inner layer (parasite cyst) thick and difficult to rupture mechanically. Anterior body with deep, cup-shaped, posteriorly oriented ventral depression. Posterior part of body small. Brandes' organ at base of ventral depression; with well-developed lobes. Encyst in tissues, especially on serous membranes such as pericardium, body cavity, etc., of freshwater teleosts or in leeches or other invertebrates. (Adults in aquatic birds, especially anatids.)

Comments

There has been considerable confusion in the systematics of the common form of *Apatemon* from the Holarctic region. Some authors (e.g., Vojtek 1964; Sudarikov 1971) have believed that the name *A. gracilis* should be reserved for forms that utilise invertebrates, especially leeches, as intermediate hosts, and that the name *A. cobitidis* (Linstow, 1890) should be used for forms that use teleosts. Subspecies of both of these nominal species have been recognized. Experimental work by Lester (1975) and Blair (1976) has, however, indicated that metacercariae from fish develop into adults indistinguishable from *A. gracilis*, thus confirming Dubois' (1968) earlier syonymy of *A. gracilis* and *A. cobitidis*. According to Dubois, species encysting in fishes form the subgenus *Apatemon* and those encysting in leeches and other invertebrates form the subgenus *Australapatemon* Sudarikov, 1959.

Apatemon gracilis (Rudolphi, 1819) Szidat, 1928 (Fig. 11)

Syn.: *A. gracilis pellucidus* (Yamaguti, 1933).

Sites: Eye, brain, body cavity, musculature.

Hosts: *Culaea incostans* (2, 3, 5, 6); *Gasterosteus aculeatus* (1); *Pungitius pungitius* (4).

Distribution: BC, Que, Alta.

Records: 1. Lester 1975 (BC); 2. Rau and Gordon 1977 (Que); 3. Rau and Gordon 1978 (Que); 4. Leong and Holmes 1981 (Alta); 5. Gordon and Rau 1982a (Que); 6. Gordon and Rau 1982b (Que).

ICHTHYOCOTYLURUS Odening, 1969

Strigeidae. Metacercaria encysted, almost filling, but free within, relatively thick-walled cyst; outer layer (capsule) thick; inner layer (parasite cyst) thin to thick, may be ruptured mechanically. Body oval; anterior and posterior parts not normally clearly defined. Anterior body flattened to slightly concave ventrally, but without deep ventral invagination. Brandes' organ without projecting lobes. Encysted mainly on serous membranes, especially pericardium, swimbladder, body cavity, etc., of freshwater fishes. (Adults in wide range of piscivorous birds, e.g., Charadriiformes, Gaviiformes, Podicipediformes and Anseriformes.)

Comments

In addition to the named species below, there are many records listed as *Tetracotyle* sp. from a variety of hosts (see, for example, Margolis and Arthur 1979 and Nepszy 1988).

Key to the species of the genus *Ichthyocotylurus*

1 Parasite cyst wall thin, easy to tear; excysted worm large (510–1300 × 342–900 µm); ventral sucker (150–250 µm) wider than oral sucker (107–155 µm); pseudosuckers as long or slightly longer than oral sucker; especially common in region of heart, peritoneum, swimbladder, etc. of wide range of freshwater teleosts ....*Ichthyocotylurus platycephalus* (Creplin, 1825) Odening, 1969 (Fig. 12)

Syn.: *Tetracotyle communis* Hughes, 1928; *Cotylurus communis* (Hughes, 1928) La Rue, 1932.

Sites: Mesenteries, liver.

Hosts: *Catostomus catostomus* (2, 3, 4); *C. commersoni* (2); *C. macrocheilus* (3, 4); *Etheostoma flabellare* (1); *Percopsis omiscomaycus* (2); *Phoxinus neogaeus* (2); *Prosopium williamsoni* (3, 4); *Ptychocheilus oregonensis* (3, 4); *Stizostedion canadensis* (2).

Distribution: Ont, BC.

Records: 1. Bangham and Hunter 1939; 2. Dechtiar 1972b; 3. Anonymous 1981 (BC); 4. Anonymous 1984 (BC).

Parasite cyst wall thick, difficult to tear; excysted worm small (<600 µm in length) 2

2 Ventral sucker (50–90 µm) a little wider than oral sucker (40–70 µm); pseudosuckers longer than oral sucker; especially common in region of heart and kidneys of salmonoids ....*I. erraticus* (Rudolphi, 1809) Odening, 1969

Syn.: *Tetracotyle intermedia* Hughes, 1928; *Cotylurus erraticus* (Rudolphi, 1809) Szidat, 1928.

Sites: Pericardial cavity, heart, mesenteries.

Hosts: *Alosa pseudoharengus* (2); *Catostomus commersoni* (2, 13); *Coregonus artedii* (1, 4, 5, 6, 9, 10, 12, 14); *C. clupeaformis* (1, 2, 3, 5, 6, 7, 8, 9, 10, 12, 14); *C. hoyi* (4, 10); *Oncorhynchus mykiss* (9); *O. nerka* (2, 10); *Osmerus mordax* (2, 11); *Perca flavescens* (2); *Percopsis omiscomaycus* (2, 10); *Prosopium cylindraceum* (9, 14); *Pungitius pungitius* (4); *Salvelinus fontinalis* (14); *S. fontinalis* × *S. namaycush* (2, 4, 10).

Distribution: Ont, Alta, Que, Lab, Man.

Records: 1. Dechtiar 1972b (Ont); 2. Collins and Dechtiar 1974 (Ont); 3. Leong and Holmes 1974 (Alta); 4. Dechtiar and Berst 1978 (Ont); 5. Watson, in Lubinski and Loch 1979 (Man); 6. Watson and Dick 1979 (Man); 7. Leong and Holmes 1981 (Alta); 8. McAllister and

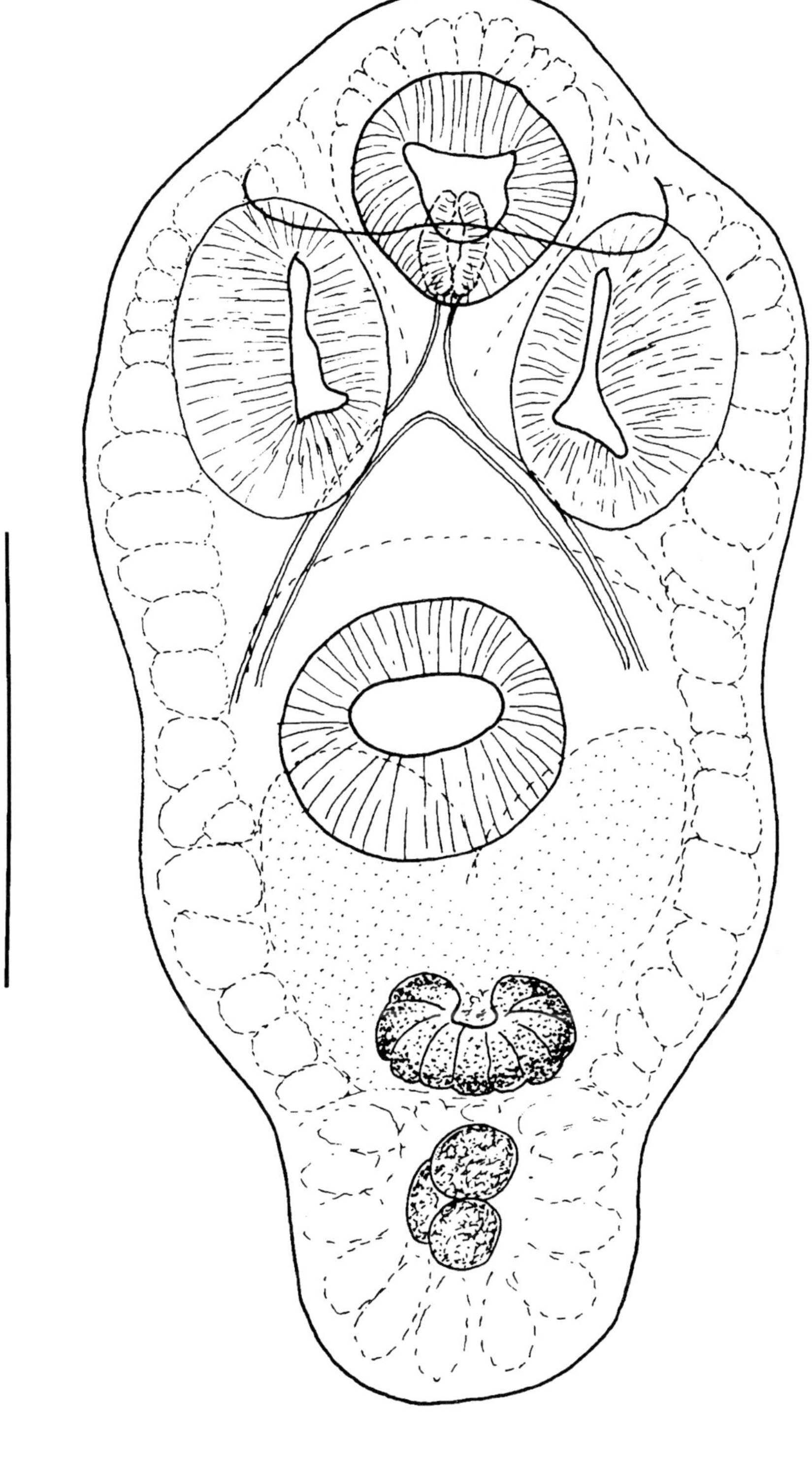

FIG. 11. *Apatemon gracilis* metacercaria (after Vojtek (1964)). Scale bar: 200 µm.

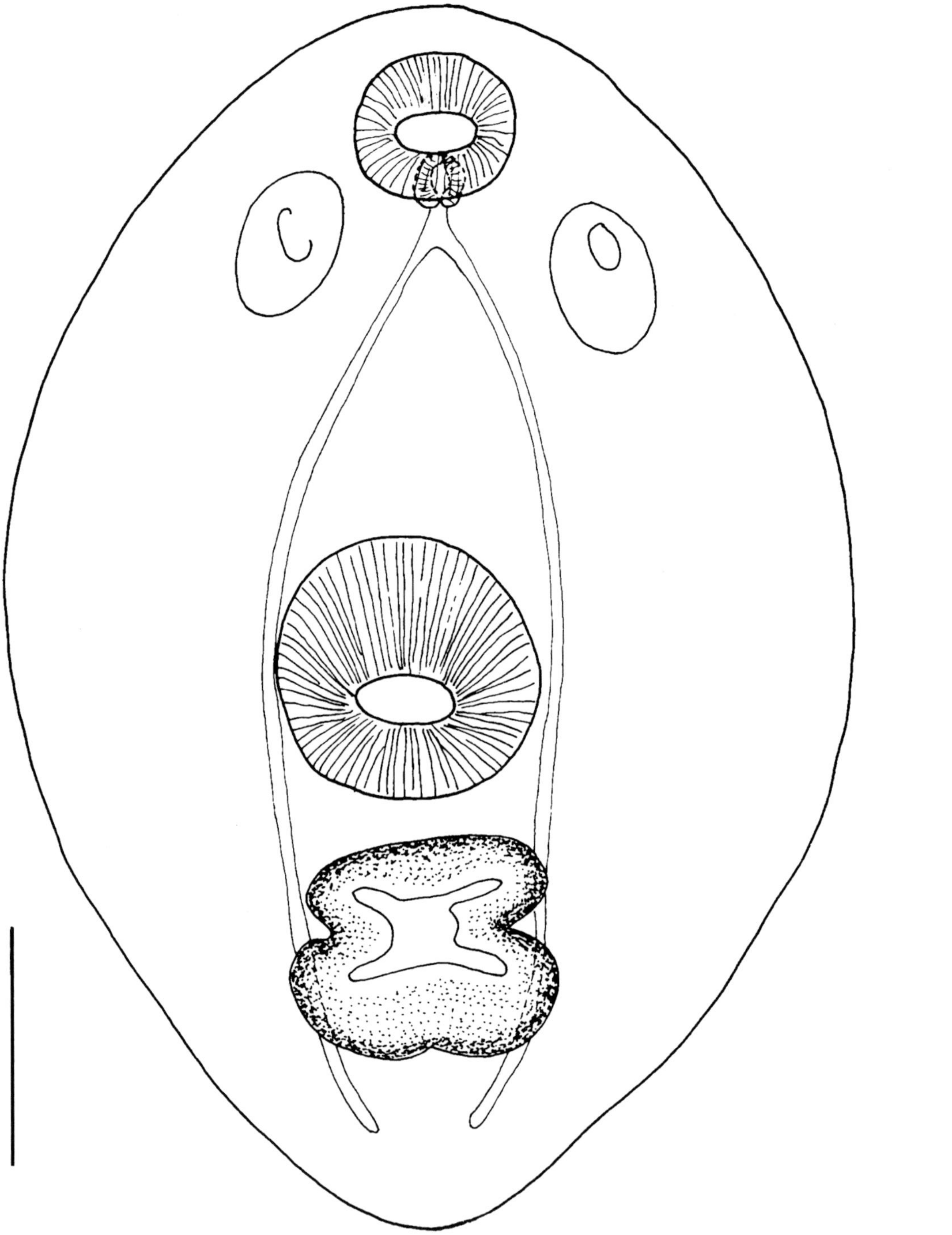

FIG. 12. *Ichthyocotylurus platycephalus* (after Odening (1979)). Scale bar: 200 µm.

Mudry 1983 (Alta); 9. Dechtiar and Lawrie 1988 (Ont); 10. Dechtiar *et al.* 1988 (Ont); 11. Dechtiar and Nepszy 1988 (Ont); 12. Dechtiar and Christie 1988 (Ont); 13. Curtis 1988 (Que/Lab); 14. Dechtiar *et al.* 1989 (Ont).

Remarks: The life cycle of this species in North America was described by Olson (1970). It seems probable that the material recorded by Marcogliese and Cone (1991b) as *Tetracotyle* sp. from *Salvelinus fontinalis* and *Salmo salar* in Newfoundland belonged to this species.

Ventral sucker (35–66 µm) as wide as oral sucker (35–66 µm); pseudosuckers shorter than oral sucker; especially common in region of swimbladder, kidney, peritoneum, and heart of coregonids, cyprinids, percids, and percopsids .... *I. pileatus* (Rudolphi, 1802) Odening, 1969

Syn.: *Tetracotyle diminuta* Hughes, 1928.

Site: Mesenteries.

Hosts: *Perca flavescens* (2, 3, 4, 5); *Percopsis omiscomaycus* (1, 5); *Stizostedion vitreum vitreum* (3).

Distribution: Ont.

Records: 1. Bangham and Hunter 1939; 2. Tedla and Fernando 1972; 3. Dechtiar and Lawrie 1988; 4. Dechtiar *et al.* 1988; 5. Dechtiar and Christie 1988.

SUPERFAMILY GYMNOPHALLOIDEA ODHNER, 1905

Syn.: Brachylaimoidea Joyeux and Foley, 1930

Strigeida. Body small to medium-sized, rarely large; tegument usually but not always spinose. Oral sucker present or absent; when absent replaced by alternative anterior attachment organ (rhynchus). Ventral sucker present or absent. Mouth within oral sucker or isolated and well separated from anterior extremity. Prepharynx absent or short when present. Pharynx well developed. Oesophagus long to almost absent. Caeca blind, two, short to long, or one. Testes two, tandem, oblique, or symmetrical. Cirrus sac present or absent. Seminal vesicle and pars prostatica internal when cirrus sac present; pars prostatica often well developed. Genital atrium usually small or apparently absent, occasionally well developed. Genital pore between forebody and posterior extremity, ventro-medial to ventro-lateral or terminal. Ovary usually pre-testicular, sometimes post- or inter-testicular. Laurer's canal and uterine seminal receptacle usually present; canalicular seminal receptacle rarely present. Uterus usually extensive, filling much of hindbody, occasionally more restricted to pre-gonadal region. Eggs operculate, small, numerous. Vitellarium usually follicular in two lateral fields, but follicles may compact into two (rarely one) masses. Excretory vesicle Y-, V-, or I-shaped; pore normally terminal or almost so. Parasitic as adults in gut, occasionally other body organs, of birds, mammals, fishes, and very occasionally amphibians or molluscs; metacercariae may encyst in fish.

KEY TO THE FAMILIES OF THE GYMNOPHALLOIDEA

1 Ventral sucker absent; mouth isolated and well apart from anterior extremity; intestine with single caecum; genital pore close to posterior extremity .... Bucephalidae

Ventral sucker present; mouth within oral sucker; intestine with one or, more often, two caeca; genital pore in forebody .... Fellodistomidae

Family BUCEPHALIDAE Poche, 1907

Gymnophalloidea. Body oval, fusiform, elongate-oval or elongate, small to medium-sized. Body surface normally armed with small spines. Anterior attachment organ present in form of sucker or rhynchus (lacking true cavity); with or without tentacles, cap-like expansion or other adornments. Ventral sucker absent. Mouth well separated from anterior attachment organ; normally mid-ventral. Prepharynx short. Pharynx well developed. Oesophagus present, sometimes difficult to discern from intestine. Intestine simple (a single caecum), saccular to tubular. Testes two, usually tandem or oblique, occasionally symmetrical; usually in posterior half or middle of body, occasionally in anterior half. Cirrus sac present, in posterior half of body; contains saccular to tubular seminal vesicle, pars prostatica (often large) and ejaculatory duct; papilla with ejaculatory duct at its base may project into genital atrium. Genital atrium small to large. Genital pore terminal or ventro-terminal. Ovary pre-, inter-, or post-testicular; usually close to middle of body. Canalicular seminal receptacle absent. Laurer's canal present. Uterine seminal receptacle present. Uterus long, winding, often filling much of available body space or mainly restricted to anterior or posterior part of body. Metraterm not usually distinctly differentiated. Eggs numerous, small, operculate. Vitellarium follicular, usually in symmetrical lateral fields in anterior or middle of body; fields sometimes confluent anteriorly. Excretory vesicle tubular to saccular, I-shaped. Excretory pore terminal. In gut, especially intestine, and occasionally other body cavities and tissues of teleosts and, very occasionally, amphibians; also present as metacercariae encysted in tissues of fishes.

Comments

The bucephalids in general, and especially those of North American fishes where descriptions are few and far between, are in dire need of revision. Some generic criteria, for example the tentacles of the genus *Bucephalus*, do seem questionable: is their presence affected by fixation or dependent, as has been suggested by Van Cleave and Mueller (1934), upon age? The answer to these questions is probably no, judging by the wide variety of material in the BM(NH) collections and the work of Baturo (1977), who found tentacles in metacercariae. The poor diagnostic features at the generic level (also commented upon by Stunkard 1975) also apply at the subfamily level. Yamaguti (1971) accepted six subfamilies, but I am tempted to follow Kniskern (1952) and accept two, the Bucephalinae, for those genera in which the anterior organ of attachment (the homologue of the oral sucker of most other digeneans) is a sucker, and the Prosorhynchinae Nicoll, 1914, for those in which this organ is a rhynchus. Nevertheless, I have not used them in this work, as it seems possible that the rhynchus may well have evolved from the sucker more than once.

Evidence from the European fauna (e.g., Taskinen *et al.* 1991) indicates that, although these worms can survive, at least for a period, in a number of piscivorous fish hosts, they tend to be rather specific with regard to definitive hosts in which they will become gravid. It is also worth noting that these worms occur both as adults and metacercariae in fishes.

Key to the genera of the Bucephalidae

1 Anterior organ of attachment a sucker 2

Anterior organ of attachment a rhynchus (with no true permanent cavity) 4

2 Anterior sucker simple .... *Prosorhynchoides*

Anterior sucker with projections or a fan-shaped hood 3

3 Anterior sucker with digitate tentacles; small to middle-sized worms; gut saccular; in freshwater and marine teleosts .... *Bucephalus*

Anterior sucker with fan-shaped hood (not always visible); small worms; gut saccular; in freshwater and marine teleosts .... *Rhipidocotyle*

Anterior sucker with small, conical lobes; large worms; gut very long, tubular; in marine teleosts .... *Dolichoenterum*

4 Rhynchus poorly developed; in body cavities, gut, and other tissues of *Hiodon* spp. in fresh water .... *Paurorhynchus*

Rhynchus well developed; in marine teleosts .... *Prosorhynchus*

BUCEPHALUS von Baer, 1827

Bucephalidae. Body elongate-oval to elongate, usually small. Anterior attachment organ a sucker with tentacular appendages. Mouth in middle third of body. Intestine saccular. Testes tandem to oblique; in posterior half of body, occasionally near middle of body. Cirrus sac normally entirely posterior to anterior testis. Ovary pre-testicular. Uterus reaching well anterior to mouth. Vitelline follicles usually in two symmetrical fields in anterior body, occasionally with third asymmetrical field more posterior. Parasitic in gut, especially intestine, of marine and freshwater teleosts.

Bucephalus elegans Woodhead, 1930 (Fig. 13)

Site: (Pyloric caeca).

Host: *Perca flavescens.*

Distribution: Ont.

Records: Bangham 1941; Bangham 1955.

Remarks: *B. elegans* is a parasite of the centrarchid *Ambloplites rupestris.* Its life cycle was described by Woodhead (1930). The species was also described by Van Cleave and Mueller (1934), but their illustration lacks the cephalic tentacles of a *Bucephalus.* According to these authors, their specimens did not mature in *Perca flavescens.* I suspect that Bangham used Van Cleave and Mueller's description rather than that of Woodhead and that the specimens are more likely to have been a *Rhipidocotyle,* possibly *R. papillosa.* It is possible, therefore, that there have been no records of *Bucephalus* in Canada.

There are also two records of *Bucephalus* sp.: Dechtiar *et al.* (1988) in *Morone chrysops* (Ont) and Dechtiar and Nepszy (1988) in *Catostomus commersoni* (Ont).

Bucephalus sp. (metacercaria)

Site: ? (Skin, fins, or gills)

Hosts: *Catostomus catostomus* (3); *Micropterus dolomieui* (5); *Morone chrysops* (1); *Notropis hudsonius* (1, 4); *Percopsis omiscomaycus* (2, 6); *Semotilus atromaculatus* (4); *Stizostedion vitreum vitreum* (5).

Distribution: Ont.

Records: 1. Bangham 1955; 2. Dechtiar 1972a; 3. Dechtiar and Lawrie 1988; 4. Dechtiar *et al.* 1988; 5. Dechtiar and Nepszy 1988; 6.Dechtiar and Christie 1988.

Remarks: More than one species may be involved. Metacercariae of *Rhipidocotyle, Bucephalus,* and *Prosorhynchoides* are very similar, difficult to distinguish, and easily confused.

DOLICHOENTERUM Ozaki, 1924

Bucephalidae. Body very elongate; often large. Anterior attachment organ a sucker adorned with about eight antero-dorsal papilliform projections. Mouth in anterior half of body. Intestine tubular, long, often reflexed, reaching close to posterior extremity. Testes tandem, in posterior part of body, separated by ovary and usually by loops of uterus. Cirrus sac small, usually post-testicular. Ovary inter-testicular. Vitellarium two symmetrical lateral bands of a small number of large follicles approximately between level of posterior testis and just anterior to middle of body. Uterus in posterior half of body. Parasitic in intestine of marine teleosts (normally *Conger* spp.).

Dolichoenterum sp. metacercaria (Fig. 14)

Site: Subcutaneous in head region.

Host: *Coryphaenoides rupestris.*

Distribution: Atl.

Records: Szuks 1975; Szuks 1980.

Remarks: The common species of *Dolichoenterum* in the Atlantic is *D. longissimum* Ozaki, 1924, a parasite of conger eels. There are, however, no records of adults in Canadian waters.

PAURORHYNCHUS Dickerman, 1954

Bucephalidae. Body elongate-oval to elongate, dorso-ventrally flattened, medium-sized. Anterior attachment organ a poorly developed rhynchus. Mouth in anterior third of body. Intestine long, tubular, often reflexed, reaching into posterior half of body. Testes large, oblique, deeply lobed, in posterior half of body. Cirrus sac minute. Ovary at level of anterior testis, irregularly lobed, small in relation to testes. Uterus reaching anterior to mouth. Vitelline follicles small, acinous, pre-testicular, in narrow bands lateral to intestine. Parasitic in various organs and cavities of freshwater teleosts (normally *Hiodon* spp.).

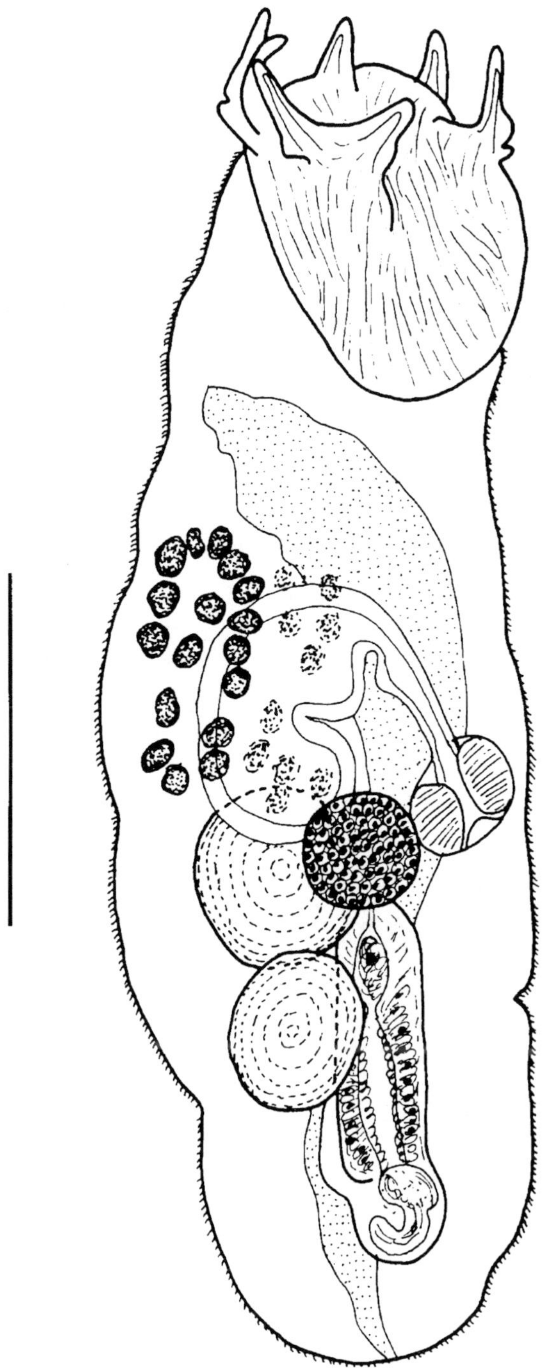

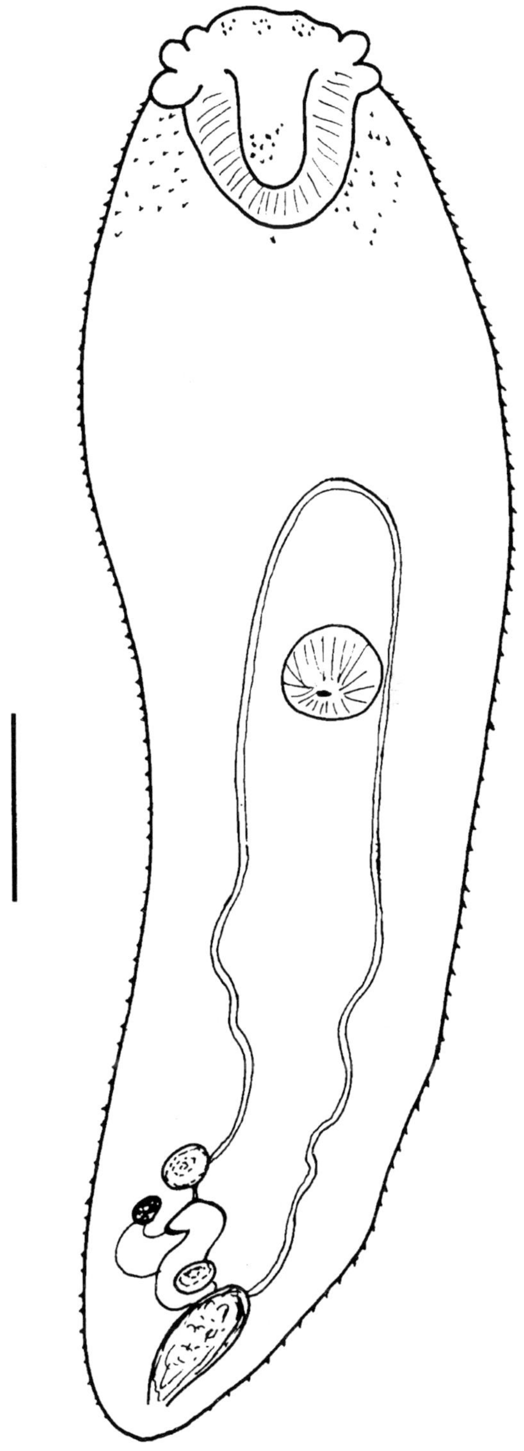

FIG. 13. *Bucephalus elegans* (modified after Woodhead (1930)). Sublateral view with uterus omitted (I doubt whether the "tentacles" are always this obvious). Scale bar: 200 μm.

FIG. 14. *Dolichoenterum* sp. (immature) (after Manter (1934)). Scale bar: 200 μm.

Paurorhynchus hiodontis Dickerman, 1954 (Fig. 15)

Syn.: ?*Bucephalus* sp. of Kennedy and Sprules (1967).

Sites: Body cavity, heart, pericardium, muscle, gills, eyes, stomach, intestine, kidney, swimbladder.

Host: *Hiodon alasoides* (1, 2, 3); *H. tergisus* (4).

Distribution: Alta, Sask, Man.

Records: 1. Margolis 1964 (Sask); 2. Kennedy and Sprules 1967 (?locality); 3. Donald and Kooyman 1974, in Mudry and Anderson 1976, and McAllister and Mudry 1983 (Alta); 4. Glenn 1980 (Man).

Remarks: Margolis and Arthur (1979) considered it likely that Kennedy and Sprules' (1967) record belonged to this species. The species was originally described from the gall bladder, but Glenn (1980) recorded it from numerous sites, noting that gravid worms tended to occur in the intestine, swimbladder, and body cavity.

PROSORHYNCHOIDES Dollfus, 1929

Syn. *Bucephaloides* Hopkins, 1954

Bucephalidae. Body fusiform, oval, elongate-oval or elongate, often widest anteriorly, or subcylindrical, small to medium-sized. Anterior attachment organ a simple sucker, without adornments. Mouth usually in middle third of body, occasionally further anterior or posterior. Intestine short to long, saccular to tubular, may be reflexed. Testes tandem to oblique, near middle of body or in posterior half of body. Cirrus sac well developed, usually large. Ovary pre-testicular or at level of testes. Uterus normally reaches anterior to mouth. Vitelline follicles in symmetrical lateral fields, mainly or entirely pre-ovarian; fields may be confluent anteriorly. In intestine of marine and freshwater teleosts, rarely in amphibians.

Comments

It was once believed, for example by Yamaguti (1971), that *Bucephalopsis* Diesing, 1855 was a senior synonym of *Prosorhynchoides*; but work by Matthews (1973) has indicated that the type species of *Bucephalopsis* is a species of *Bucephalus*.

Key to the species of *Prosorhynchoides*

1 In marine or euryhaline migratory fishes (Pacific coast and British Columbia) ....*Prosorhynchoides basargini* (Layman, 1930) Margolis and Arthur, 1979

Syn.: *Bucephaloides basargini* (Layman, 1930); *Bucephaloides* sp. of Margolis (1956b, 1957); *B. ozakii* Nagaty, 1937 of Bangham and Adams (1954); *Prosorhynchoides ozakii* (Nagaty, 1937) of Margolis and Arthur (1979).

Site: Intestine.

Hosts: *Oncorhynchus gorbuscha* (2, 3, 4, 5); *Salvelinus malma* (1).

Distribution: Pac, BC-b, BC.

Records: 1. Bangham and Adams 1954 (BC); 2. Margolis 1956b (Pac, BC-b); 3. Margolis 1957 (Pac, BC-b); 4. Anonymous 1981 (BC); 5. Anonymous 1984 (BC).

Remarks: This species has been considered a synonym of *Prosorhynchoides gracilescens* (Rudolphi, 1819) by Zhukov (1960) and this name was used by Jennings and Hendrickson (1982) for specimens (adults and metacercariae) from *O. tshawytscha* and *O. kisutch* in Californian rivers. Furthermore, *P. gracilescens* has been regularly recorded from *Oncorhynchus* spp. in Russian Pacific waters. Nevertheless, *P. gracilescens* is considered specific to *Lophius* in European waters, so this synonymy must be in doubt. I have examined specimens provided by Dr. L. Margolis from *O. gorbuscha* in British Columbian waters and *O. nerka* in Alaskan waters. These specimens do resemble the description of *P. basargini*, although whether or not the form from salmonoids is conspecific with that from flatfishes in Russian Pacific waters is open to question. Both forms need to be fully described.

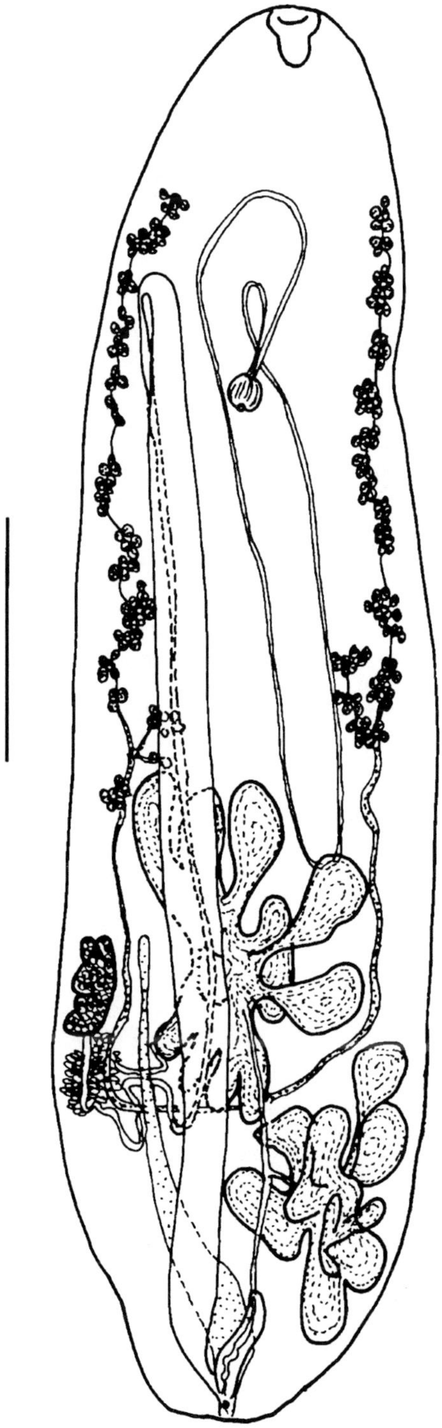

FIG. 15. *Paurorhynchus hiodontis* (after Yamaguti (1971)). Body spines omitted. Scale bar (roughly estimated): 1 mm.

P. ozakii is a parasite of freshwater fishes in Southeast Asia, so its presence in British Columbian waters seems doubtful. In Russian waters there have been several records of *P. gracilescens* and one record of *B. basargini* in *Salvelinus*, so it seems likely that Bangham and Adams' record refers to the same species that occurs in *Oncorhynchus*.

It is also worth noting that specimens of *Prosorhynchoides*, as *Bucephalopsis* sp., were recorded by Arai (1967a) from the pyloric caeca of *Leptocottus armatus* of the Pacific coast of British Columbia.

Reimer (1981) recorded *P. gracilescens* from *Reinhardtius hippoglossoides* in the Eastern Arctic Ocean: in view of the host, perhaps this is more likely to have been *Prosorhynchus squamatus*.

P. basargini (metacercaria)

Sites: Fins (also nasal cavities, mouth, but these may be or include other bucephalid metacercariae).

Hosts: *Clupea pallasi* (1, 2); *Theragra chalcogramma* (3, 4, 5).

Distribution: Pac.

Records: 1. Arthur and Arai 1980a; 2. Arthur and Arai 1980b; 3. Arthur 1983; 4. Arthur 1984; 5. Kabata and Whitaker 1984.

In freshwater fishes. ....*P. pusilla* (Stafford, 1904) (Fig. 16).

Syn.: *Bucephalopsis pusilla* (Stafford, 1904); *Bucephalus pusilla* (Stafford, 1904); *Gasterostomum pusillum* Stafford, 1904.

Sites: Stomach, pyloric caeca, instestine.

Hosts: *Perca flavescens* (8); *Stizostedion canadense* (4, 6); *S. vitreum glaucum* (4); *S. vitreum vitreum* (1, 2, 3, 4, 5, 7, 8, 9, 10).

Distribution: Man, Ont, Que.

Records: 1. Stafford 1904 (Que?); 2. Woodhead 1930 (Man); 3. Lyster 1939 (Que); 4. Bangham and Hunter 1939 (Ont); 5. Bangham 1955 (Ont); 6. Dechtiar 1972b (Ont); 7. Dickson, cited in Lubinsky and Loch 1979 (Man); 8. Poole and Dick 1985 (Man); 9. Dechtiar and Christie 1988 (Ont); 10. Dechtiar *et al.* 1988 (Ont).

Remarks: The morphology and aspects of the life cycle of this species were described by Woodhead (1930, 1931), respectively. This species uses similar definitive and molluscan hosts to European freshwater bucephalids.

PROSORHYNCHUS Odhner, 1905

Bucephalidae. Body stout-fusiform to elongate, usually relatively small. Anterior attachment a rhynchus, usually well developed, without tentacles. Mouth usually near middle of body. Intestine saccular, often small. Testes tandem, oblique, or almost symmetrical, near middle of body or more often in posterior half. Cirrus sac usually large, reaching close to testes. Ovary normally anterior or antero-lateral to anterior, occasionally posterior, testis. Uterus reaches to about level of vitellarium or beyond. Egg shell often thick. Vitelline follicles mainly or entirely anterior to gonads, commonly in single curved field surrounding anterior region of uterine field, sometimes in two symmetrical lateral fields in middle or anterior half of body. Parasitic in intestine of marine and freshwater teleosts.

Comments

The Pacific forms of this genus are in dire need of revision.

Key to the species of *Prosorhynchus*

1 On the Atlantic coast; egg-length 30–38 μm ....*Prosorhynchus squamatus* Odhner, 1905 (Fig. 17)

Syn.: *Gasterostomum armatum* of Stafford (1904, 1907); *Prosorhynchus crucibulum* of Linkletter *et al.* (1977).

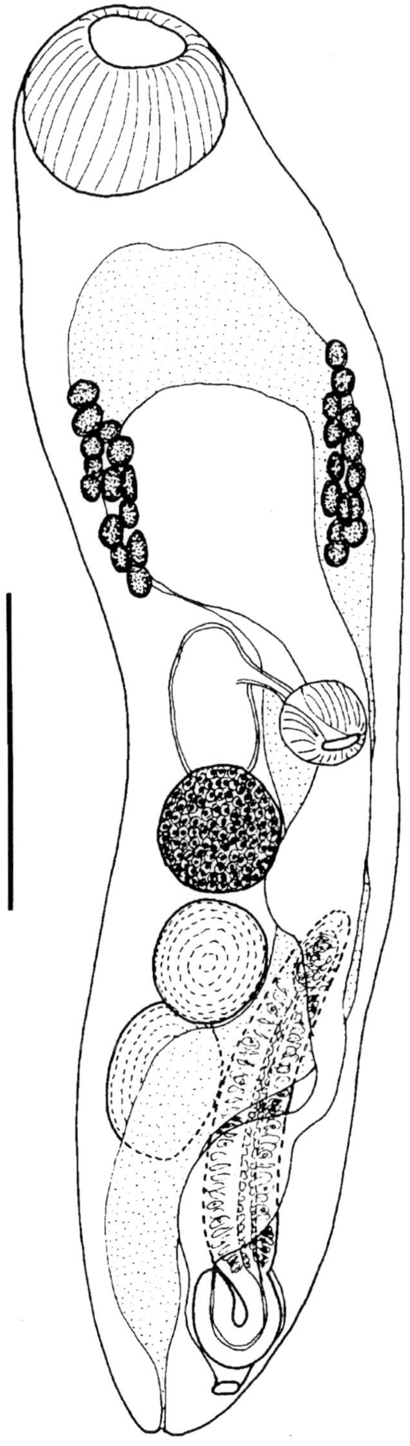

FIG. 16. *Prosorhynchoides pusilla* (modified after Van Cleave and Mueller (1934)). Body spines omitted. Scale bar: 200 μm.

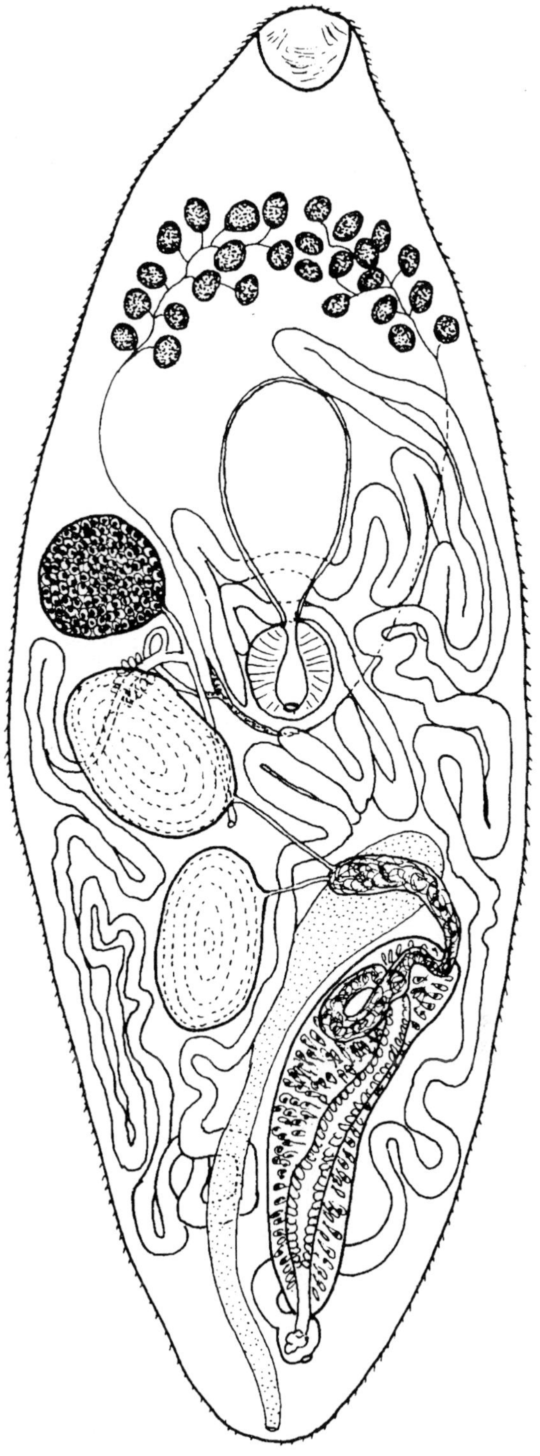

FIG. 17. *Prosorhynchus squamatus* (after Odhner (1905)). Scale bar (roughly estimated): 200 μm.

Sites: Intestine, stomach, pyloric caeca.

Hosts: *Brosme brosme* (1, 2); *Gadus morhua* (6, 11); *Hemitripterus americanus* (1, 2); *Hippoglossoides platessoides* (5, 12); *Hippoglossus hippoglossus* (1, 2, 4); *Melanogrammus aeglefinus* (10); *Myoxocephalus scorpius* (1, 2, 8); *Reinhardtius hippoglossoides* (14); *Salvelinus alpinus* (7, 9, 13); *Squalus acanthias* (3; accidental).

Distribution: Atl, NWT (Baffin Island).

Records: 1. Stafford 1904 (Atl); 2. Stafford 1907 (Atl); 3. Myers 1959 (Atl); 4. Ronald 1960 (Atl); 5. Scott 1975a (Atl); 6. Linkletter *et al.* 1977 (Atl); 7. Dick and Belesovic 1978 (NWT); 8. Bray 1979 (Atl); 9. Dick and Belesovic 1981 (NWT); 10. Scott 1981 (Atl); 11. Appy and Burt 1982 (Atl); 12. Scott 1982 (Atl); 13. Dick 1984 (NWT); 14. Arthur and Albert 1994 (Atl).

Remarks: Unlike some bucephalids, this species appears to mature in a wide range of hosts and has a wide distribution in northern waters of the Northern Hemisphere (Brinkmann 1975). Matthews (1973) discussed its life cycle and distinguished it from its congeners in European waters.

On the Pacific coast 2

2 Eggs <35 μm in length (usually <30 μm); mouth near middle of body 3

Eggs >40 μm (32–40 according to authors: see below); small, slender worms; mouth posterior to middle of body; rhynchus well developed, funnel-shaped ....*P. scalpellus* McFarlane, 1936

Site: Intestine.

Host: *Scorpaenichthys marmoratus.*

Distribution: Pac.

Record: McFarlane 1936.

Remarks: According to Nagaty (1937) this species is a synonym of *P. crucibulum*, but this was not accepted by Nahhas and Krupin (1977), who recorded it from *Liparis callyodon* and *Scorpaenichthys marmoratus* off California. *Liparis* spp. are common host of *P. squamatus*, so it may be that this is also a synonym of the latter species. I have, however, examined the types of *P. scalpellus* and consider the eggs larger and the mouth further posterior than in *P. squamatus*, and the posterior margin of the rhynchus is more pointed. I have also examined one of Nahhas and Krupin's specimens from *L. callyodon*, and, although it is immature, the pharynx is also rather posteriorly situated but not as far back as in the type-specimen. It is worth noting that I measured eggs of up to 50 μm, but McFarlane recorded 32–40 μm and Nahhas and Krupin 33–40 μm.

3 Small, slender worms, normally <2 mm in length; rhynchus poorly developed; vitelline fields extend posterior to pharynx ....*P. apertus* McFarlane, 1936

Site: Intestine.

Host: *Ophiodon elongatus.*

Distribution: Pac.

Record: McFarlane 1936.

Remarks: According to Nagaty (1937) and Pratt and McCauley (1961), this species is a synonym of the Japanese species *P. facilis* (Ozaki, 1924). Although there are some similarities, the eggs of the types, which I have examined, are much smaller (24–27 μm) than those of *P. facilis*. The anterior region of the types appears poorly fixed, attenuated, and the rhynchus rather poorly developed. Ching (1960) also recorded this species in *O. elongatus* off Washington State.

Small to medium-sized, slender to relatively stout worms of up to 3 mm in length; rhynchus well developed; vitelline fields do not extend posterior to pharynx ....*Prosorhynchus* sp.

Syn.: *Prosorhynchus crucibulum* (Rudolphi, 1819) of Sekerak and Arai (1973, 1977), Holmes (1990), and Stanley *et al.* (1992); (?) *Dollfustrema* sp. of Holmes (1990).

Sites: Pyloric caeca, intestine.

Hosts: *Myoxocephaous octodecimspinosus* (3); *Sebastes aleutianus* (1, 2); *S. alutus* (2); *S. babcocki* (2); *S. borealis* (2); *S. brevispinis* (2); *S. crameri* (2); *S. diploproa* (2); *S.*

elongatus (2); *S. entomelas* (2); *S. flavidus* (2, 4); *S. melanops* (5); *S. nebulosus* (3); *S. paucispinis* (2); *S. pinniger* (2); *S. reedi* (2); *S. ruberrimus* (2); *S. zacentrus* (2).

Distribution: Pac.

Records: 1. Sekerak and Arai 1973; 2. Sekerak and Arai 1977; 3. Holmes 1990; 4. Stanley *et al.* 1992; 5. BM(NH) collection: material of J.C. Holmes.

Remarks: The life cycle of *P. crucibulum* in European waters was described by Matthews (1973), who considered that this species only matures in conger eels. It seems very unlikely, therefore, that material from *Sebastes* in the Pacific belongs to this species. It would seem more likely to be *P. squamatus*, considered by some (e.g., Dawes 1947) a synonym of *P. crucibulum*, which occurs in a variety of hosts and, according to Brinkmann (1975), widely in the northern North Atlantic and North Pacific. Nevertheless, I have examined material of both Dr.A. Sekerak, provided by Dr.L. Margolis, and Dr.J.C. Holmes, and consider it morphologically distinct from both *P. squamatus* and *P. crucibulum*. In specimens from *Sebastes* the body is less stout, the eggs are usually, but not always, less than 30 µm in length, the shell is thinner, and in gravid worms the uterus usually breaks through the anterior arch of vitelline cells, whereas in European forms the uterine distribution is restricted within the arch. The rhynchus of the Canadian material resembles that of *P. crucibulum*, being more attenuated posteriorly than in *P. squamatus*. The species most closely resembling the specimens from *Sebastes* appears to be the Japanese species *P. facilis* (Ozaki, 1924), but the latter differs in that the eggs are larger and the cirrus sac does not reach to the level of the testes.

Holmes (1990) recorded *Dollfustrema* sp. from the gut of *Sebastes nebulosus* off Vancouver Island. This single, contracted specimen, now in the BM(NH) collection, is immature and in poor condition. Nevertheless, although under DIC (Nomarski) optics small spines are visible on the rhynchus, there is no sign of the enlarged spines typical of *Dollfustrema* (Dr.Holmes, *in litt.*, suggests that the spines may have been lost as the specimen had been frozen). In view of the condition of the specimen, I prefer to consider it an anomalous specimen of *Prosorhynchus* sp., immature specimens of which are common in *Sebastes* spp. (see below).

Prosorhynchus spp. (metacercaria)

Syn.: *Prosorhynchus crucibulum* (Rudolphi, 1819) metacercariae of Canadian authors.

Sites: Musculature (1, 2, 3, 4); mainly fins (5, 6, 7).

Hosts: *Atheresthes stomias* (6); *Microstomus pacificus* (6); *Pleuronectes vetulus* (6); *Sebastes alutus* (1, 2, 3); *S. flavidus* (8, 9); *S. melanops* (1); *Sebastes* sp. (4; see footnotes 63 and 65 of Margolis and Arthur, 1979); *Theragra chalcogramma* (5, 6, 7).

Distribution: Pac.

Records: 1. Liston *et al.* 1960; 2. Liston and Hitz 1961; 3. Sekerak and Arai 1973; 4. Hoskins *et al.* 1976; 5. Arthur 1983; 6. Kabata and Whitaker 1984; 7. Arthur 1984; 8. Lee *et al.* 1990; 9. Stanley *et al.* 1992.

RHIPIDOCOTYLE Diesing, 1858

Bucephalidae. Body oval or pyriform to elongate, usually relatively small. Anterior attachment organ a sucker, usually adorned with polygonal cap-like expansion. Mouth usually in middle third of body. Intestine saccular, often small. Testes tandem, oblique, or rarely symmetrical, usually in posterior half of body. Cirrus sac well developed, usually reaches to region of testes. Ovary anterior or antero-lateral to anterior testis. Uterus normally reaching into anterior half of body, occasionally only to middle of body. Vitelline follicles usually in two symmetrical lateral fields in about second quarter of body, occasionally fields confluent anteriorly. In intestine of marine and freshwater teleosts.

Key to the species of *Rhipidocotyle*

1 On Pacific coast; vitelline fields extend posterior to pharynx .... *Rhipidocotyle elongata* McFarlane, 1936

Syn.: *Rhipidocotyle* sp. of Arai (1967a).

Site: Intestine.

Hosts: *Leptocottus armatus* (3); *Ophiodon elongatus* (1, 2, 3).
Distribution: Pac.
Records: 1. McFarlane 1936; 2. Arai 1967a; 3. Arai 1969b.
Remarks: Ching (1960) recorded this species in *O. elongatus* off Washington State.
It is worth noting that *Prosorhynchus apertus* is somewhat similar in morphology and occurs in the same host. The two species need to be differentiated using well-fixed material.

Rhipidocotyle sp. (metacercaria)
Site: Fins.
Hosts: *Clupea pallasi* (1, 2); *Theragra chalcogramma* (3, 4, 5).
Distribution: Pac.
Records: 1. Arthur and Arai 1980a; 2. Arthur and Arai 1980b; 3. Arthur 1983; 4. Arthur 1984; 5. Kabata and Whitaker 1984.
Remarks: ?Presumably a larval form of *R. elongata*.

In freshwater teleosts; vitelline fields not extending posterior to pharynx...................................
....*R. papillosa* (Woodhead, 1929) Eckmann, 1932 (adult) (Fig. 18)
Syn.: *Bucephalus papillosus* Woodhead, 1929; *Gasterostomum pusillum* of Cooper (1915).
Sites: Intestine, stomach, pyloric caeca.
Hosts: *Etheostoma nigrum* (1); *Micropterus dolomieui* (1, 2, 3, 4); *Perca flavescens* (5); *Stizostedion vitreum vitreum* (5).
Distribution: Ont, Man.
Records: 1. Cooper 1915 (Ont); 2. Bangham and Hunter 1939 (Ont); 3. Bangham and Venard 1946 (Ont); 4. Bangham 1955 (Ont); 5. Poole and Dick 1985 (Man).
Remarks: This species is morphologically very similar to European forms from percids and esocids; but the cercaria described by Woodhead (1929) is more similar to Finnish material described by Gibson *et al.* (1992) as *R. fennica* than to *R. campanula* (Dujardin, 1845).

R. papillosa (Woodhead, 1929) Eckmann, 1932 (metacercaria)
Syn.: *Gasterostomum pusillum* of Cooper (1915).
Site: Musculature (fins, according to Woodhead 1929).
Hosts: *Ambloplites rupestris*; *Micropterus dolomieui*; *Perca flavescens*; unspecified minnow.
Distribution: Ont.
Record: Cooper 1915.

Rhipidocotyle sp. (metacercaria)
Site: Fins.
Hosts: *Ameiurus nebulosus*; *Hybognathus hankinsoni*.
Distribution: Ont.
Record: Molnar *et al.* 1974.
Remarks: ?Presumably also the metacercaria of *R. papillosa*.

Family FELLODISTOMIDAE Nicoll, 1909

Gymnophalloidea. Body large to small, globular to elongate. Body surface usually smooth, occasionally bearing spines or muscular ornamentation. Oral sucker subterminal, globular. Ventral sucker globular, large to small, in middle or anterior half of body. Prepharynx short to apparently absent. Pharynx well developed, globular to elongate. Oesophagus absent to long. Intestinal bifurcation (when present) in forebody. Gut caeca normally two, narrow to wide, terminating blindly at level of testes or beyond, occasionally with single caecum (? opening into excretory vesicle). Testes two, oval to globular, entire or deeply lobed, symmetrical, oblique, or tandem, in anterior or posterior hindbody, normally post-ovarian. Cirrus sac usually well developed (but occasionally absent), oval to claviform, containing seminal vesicle, pars

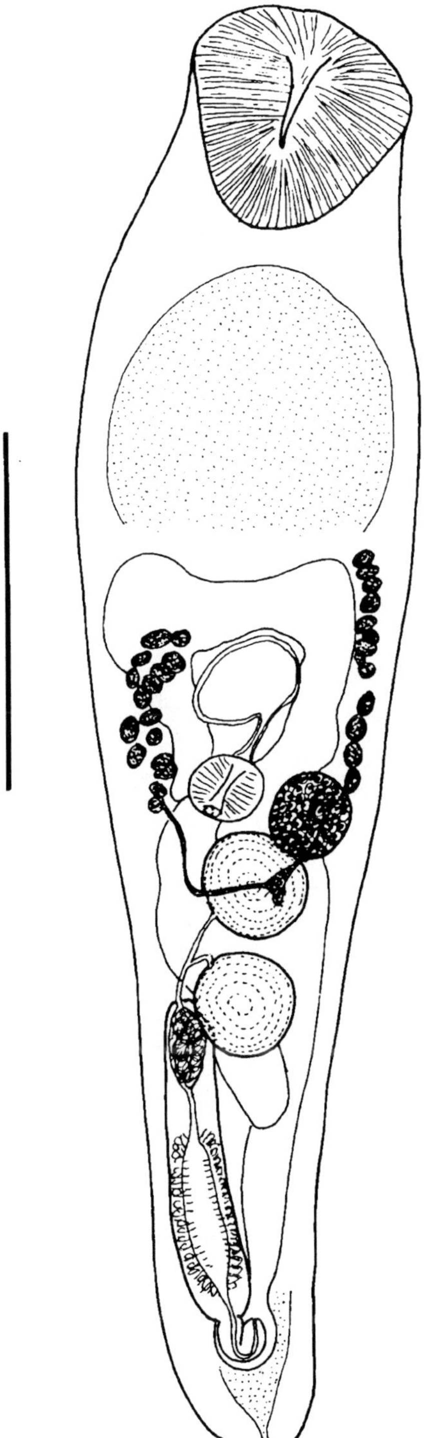

FIG. 18. *Rhipidocotyle papillosa* (after Van Cleave and Mueller (1934)). Body spines omitted. Polygonal cap of oral sucker often more apparent. Scale bar: 200 µm.

prostatica, and ejaculatory duct. Seminal vesicle usually bipartite, occasionally globular or tubular and convoluted. Pars prostatica usually wide with filamentous lining, occasionally greatly reduced. Ejaculatory duct usually wide and convoluted, occasionally small and narrow. Genital atrium short to long. Genital pore ventral, usually sinistral, in middle or posterior half of forebody. Spermatophores may be produced. Ovary entire to multilobate, in hindbody, usually anterior to testes. Laurer's canal present. Uterine seminal receptacle usually present, canalicular seminal receptacle occasionally present. Uterus normally extending into post-testicular zone, bulk of uterus usually in post-testicular zone. Eggs numerous, small, variable, operculate, egg shell occasionally ornamented. Vitellarium follicular, typically in two lateral fields, occasionally amalgamated to form two oval masses, occasionally confluent medially. Excretory pore terminal; vesicle Y- or V-shaped, with anterior arms reaching to about level of pharynx. Parasitic in intestine, pyloric caeca, (?) stomach, bile duct, and gall bladder of marine and occasionally freshwater teleosts; occasionally occurring as adults in molluscs (bivalves and gastropods).

Comments

Detailed descriptions and keys to the Northeast Atlantic forms of this family were published by Bray and Gibson (1980). This work includes all of the Canadian Atlantic species.

Key to the subfamilies of the Fellodistomidae

1 Ovary pre-testicular; uterine seminal receptacle present; Laurer's canal opening dorsally to gonads; vitellarium two (occasionally four) lateral follicular fields; seminal vesicle small relative to size of cirrus sac; genital atrium short, wide ....Fellodistominae

Ovary post- or inter-testicular; canalicular seminal receptacle present; Laurer's canal opening at or near posterior extremity; vitellarium two or several bunches of closely packed follicles; seminal vesicle large, often almost fills cirrus sac; genital atrium long, narrow..............................Baccigerinae

Subfamily FELLODISTOMINAE Nicoll, 1909

Fellodistomidae. Body large and robust to small. Body surface smooth. Ventral sucker usually larger than oral sucker (but occasionally smaller), in middle of body or in anterior half. Prepharynx short or apparently absent. Pharynx large to small, globular to oval. Oesophagus absent, short or long. Intestinal bifurcation in forebody. Caeca two, narrow to wide, reaching to testes or beyond and almost to posterior extremity. Testes two, in anterior hindbody or occasionally filling much of hindbody, post-ovarian. Cirrus sac well developed, oval to claviform, containing many prostatic gland cells. Seminal vesicle normally bipartite. Pars prostatica wide, straight or gently curved. Ejaculatory duct wide, pocketed. Genital atrium small. Genital pore in mid-forebody, sinistral to median line. Spermatophores occasionally seen. Ovary entire to multilobate, usually anterior to right testis. Laurer's canal and uterine seminal receptacle present. Uterus usually extending into post-testicular zone (does not extend into post-testicular zone in type-genus). Metraterm joins genital atrium from left. Eggs numerous, small, often variable, egg shell may be ornamented. Vitellarium follicular, in two (occasionally four) lateral fields, on fore- and/or hindbody. Excretory vesicle V- or Y-shaped, with arms reaching to level of pharynx or oral sucker. Parasitic in intestine, pyloric caeca, (? stomach), bile duct, and gall bladder of marine teleosts.

Key to the genera of the Fellodistominae

1 Excretory vesicle V-shaped; ovary entire or tri-lobed; normally in intestine ....*Steringotrema*

Excretory vesicle Y-shaped; ovary tri-lobed or multilobate; normally in gall bladder, bile duct, or intestine 2

2 Ovary tri-lobed; caeca reach to about anterior margin of testes; vitelline fields entirely in forebody; prepharynx small, but distinct; in intestine .... *Olssonium*

Ovary multilobate; caeca reach to level of testes or, more usually, beyond; vitelline fields not restricted to forebody; prepharynx short, indistinct or apparently absent 3

3 Vitelline fields at about level of ventral sucker; normally in gall bladder or bile duct.... *Fellodistomum*

Vitelline fields mainly in hindbody; normally in intestine .... *Steringophorus*

FELLODISTOMUM Stafford, 1904

Fellodistominae. Body medium-sized and slender to large and robust. Body surface unarmed. Ventral sucker larger, sometimes much larger, than oral sucker, globular, near middle of body or more anterior. Prepharynx short to apparently absent. Pharynx well developed. Oesophagus present or absent. Caeca reaching to level from close to posterior margin of testes to well into post-testicular field. Testes symmetrical or oblique, oval, entire or slightly indented. Cirrus sac claviform. Seminal vesicle bipartite. Pars prostatica wide, surrounded by gland cells. Ejaculatory duct wide, diverticulate. Genital atrium small. Genital pore between ventral sucker and intestinal bifurcation, sinistrally submedian. Spermatophores may be present. Ovary multilobate, just anterior or antero-medial to right testis. Uterine distribution variable, ranging from bulk of coils surrounding ventral sucker to bulk of coils in post-testicular field. Eggs numerous, small, with smooth shells. Vitelline follicles in two lateral fields, lateral to and reaching just anteriorly to ventral sucker. Excretory vesicle Y-shaped, stem reaching forward to level of testes or almost so, arms reaching to deep into forebody. Parasitic in gall bladder or bile duct of marine teleosts.

Comments

This genus was revised by Bray and Gibson (1980) and considered to be monotypic, but subsequent molecular data (Bray *et al.* 1994) has indicated that *F. agnotum* (previously attributed to *Steringophorus*) is more closely related to *F. fellis* than it is to *Steringophorus furciger*. Following Bray *et al.*, I also return *F. sebastodis* to this genus.

Key to the species of *Fellodistomum*

1 Large, robust worms; ventral sucker huge; testes large, filling much of hindbody; uterus coiled mainly around ventral sucker, extends little, if at all, into post-testicular zone; in gall bladder of *Anarhichas* spp .... *Fellodistomum fellis* (Olsson, 1868) Nicoll, 1909 (Fig. 19)

Syn.: *Fellodistomum incisum* (Rudolphi) of Stafford (1904); *Distomum incisum* Rudolphi of Stafford (1907).
Site: Gall bladder.
Hosts: *Anarhichas denticulatus* (3); *A. lupus* (1, 2, 3, 4, 5); *A. minor* (3).
Distribution: Atl.
Records: 1. Stafford 1904; 2. Stafford 1907; 3. Bray 1979; 4. Zubchenko 1980; 5. Bray 1987c.
Remarks: This species was discussed and described in detail in Bray and Gibson (1980).

Large, robust to medium-sized slender worms; ventral sucker not huge; testes in anterior hindbody; uterus coiled mainly in post-testicular field; in bile duct or gall bladder 2

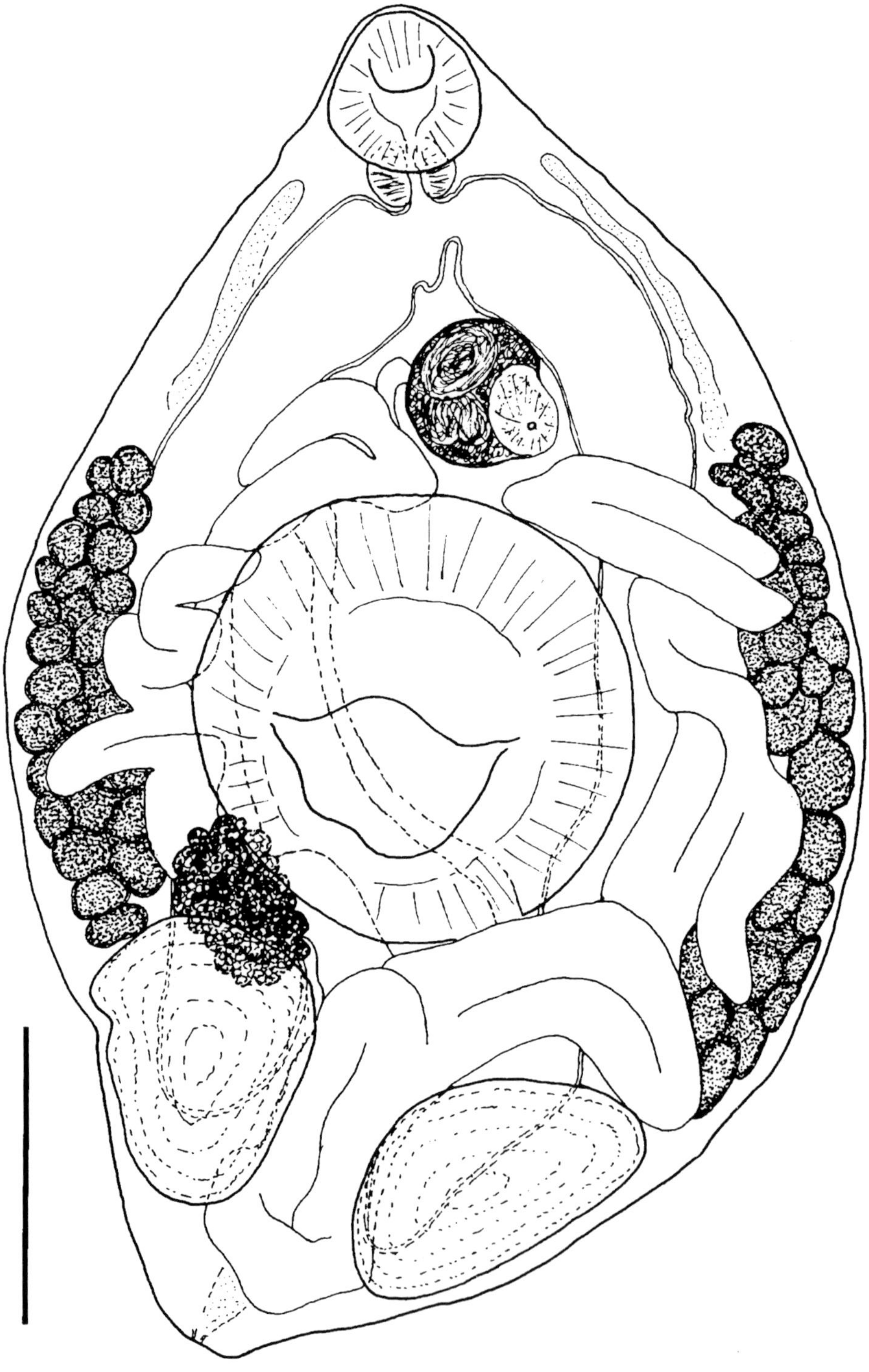

FIG. 19. *Fellodistomum fellis* (after Bray and Gibson (1980)). Slightly flattened specimen. Scale bar: 1 mm.

2 Post-testicular region of uterine field a single convoluted loop (occasionally a few loops); gut caeca terminate close to testes; in bile duct, occasionally gall bladder and (?) anterior intestine of *Anarhichas* spp. on Atlantic coast .... *F. agnotum* Nicoll, 1909

Syn.: ?*Fellodistomum incisum* (Rud.) of Stafford (1904) in part; ?*Distomum incisum* (Rud.) of Stafford (1907) in part; *Steringophorus agnotus* (Nicoll, 1909) Dollfus, 1952.

Sites: Gall bladder, bile duct, (?) anterior intestine.

Hosts: *Anarhichas lupus* (?1,?2, 3, 4, 5); *A. minor* (5).

Distribution: Atl.

Records: 1. Stafford 1904; 2. Stafford 1907; 3. Miller 1941a; 4. Bray 1979; 5. Zubchenko 1980.

Remarks: This species was discussed and described in detail by Bray and Gibson (1980), as *Steringophorus agnotus*.

Post-testicular region of uterine field composed of several convoluted loops; gut caeca extend close to middle of post-testicular zone; in gall bladder of *Sebastes* spp. on Pacific coast .... *F. sebastodis* Yamaguti and Matumura, 1942

Syn.: *Steringophorus sebastodis* (Yamaguti and Matumura, 1942) Dollfus, 1952.

Site: Gall bladder.

Hosts: *Sebastes alutus* (1, 2); *S. elongatus* (2); *S. ruberrimus* (2); *S. zacentrus* (2).

Distribution: Pac.

Records: 1. Sekerak and Arai 1973; 2. Sekerak and Arai 1977.

OLSSONIUM Bray and Gibson, 1980

Fellodistominae. Body small, fusiform. Body surface unarmed. Oral sucker subterminal, globular. Ventral sucker in anterior half of body, similar in size to oral sucker. Prepharynx small but distinct. Pharynx small. Oesophagus short. Caeca narrow, blind, reaching almost to testes. Testes oval, more or less symmetrical, in anterior hindbody. Cirrus sac large, containing bipartite seminal vesicle, wide pars prostatica with filamentous lining and external gland cells, and wide, diverticulate ejaculatory duct. Genital atrium large. Genital pore ventral in posterior forebody, slightly sinistral. Ovary tri-lobed, anterior to dextral testis and postero-dextral to hind margin of ventral sucker. Uterus coils in post-testicular region, in inter-testicular region, and dorsally to ventral sucker. Eggs small, operculate, with smooth shell, but may have boss at one or both poles. Vitellarium follicular, in two lateral fields in forebody. Excretory vesicle Y-shaped, with wide stem bifurcating at level of testes to give rise to dilate arms that extend into forebody. Parasitic in intestine of marine deep-water teleosts.

Olssonium turneri Bray and Gibson, 1980 (Fig. 20)

Site: ? (Normally intestine).

Hosts: *Alepocephalus agassizii*; *A. bairdii*.

Distribution: Atl.

Record: Zubchenko 1984.

Remarks: This species was described by Bray and Gibson (1980) from *Alepocephalus bairdii* in the Northeast Atlantic.

STERINGOPHORUS Odhner, 1905

Fellodistominae. Body large, deep-bodied to dorso-ventrally flattened. Tegument unarmed. Ventral sucker usually larger than oral sucker, in anterior half of body. Prepharynx short or apparently absent. Pharynx well developed. Oesophagus absent to long. Intestinal bifurcation anterior to or at level of genital pore; caeca wide to narrow, reaching to testes, to about middle of post-testicular zone or occasionally beyond.

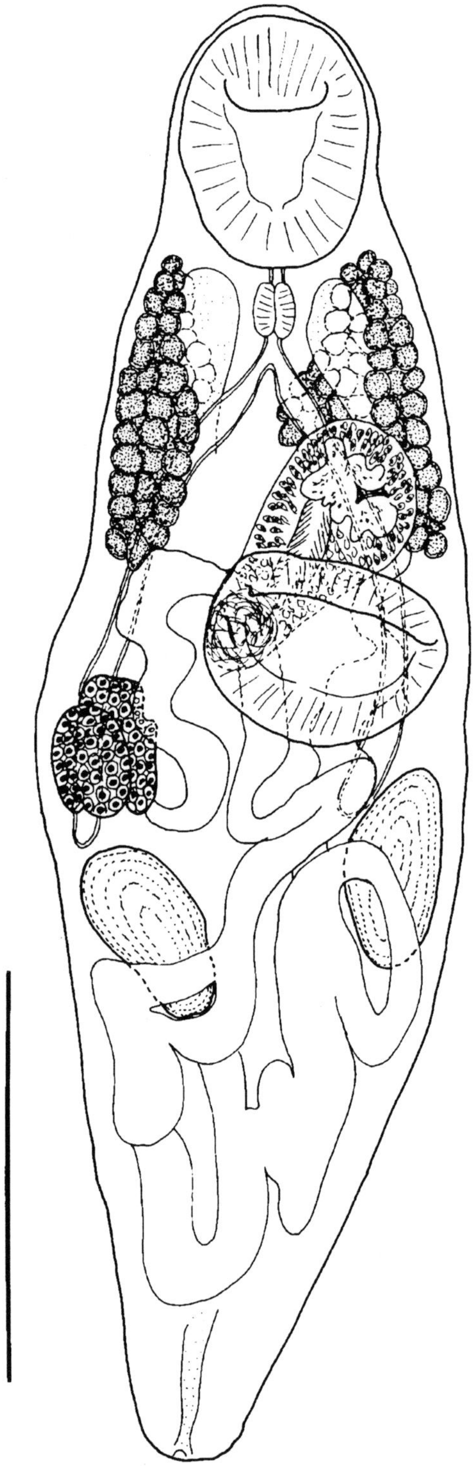

FIG. 20. *Olssonium turneri* (after Bray and Gibson (1980)). Scale bar: 500 μm.

Testes oval, indented or deeply lobed, symmetrical to tandem. Cirrus sac oval. Seminal vesicle bipartite. Pars prostatica wide, surrounded by gland cells. Ejaculatory duct wide, diverticulate. Genital atrium small. Genital pore close to anterior margin of ventral sucker, sinistral to median line. Spermatophores may be present. Ovary multilobed, just pre-testicular. Uterus mainly coiled posterior to testes. Eggs numerous, small, shells occasionally ornamented. Vitelline follicles in two lateral fields between level of ventral sucker and level just posterior to testes. Excretory vesicle Y-shaped, stem reaching to between middle of post-testicular zone and about level of testes, arms extending into forebody. Parasitic in intestine of marine teleosts.

Key to the species of *Steringophorus*

1 Testes tandem; bifurcation of excretory vesicle in middle of post-testicular region ....*Steringophorus pritchardae* (Campbell, 1975) Bray and Gibson, 1980 (Fig. 21)

Syn.: *Abyssotrema pritchardae* Campbell, 1975.

Site: ? (Normally upper intestine or pyloric caeca).

Hosts: *Alepocephalus agassizii*; *A. bairdii*.

Distribution: Atl.

Record: Zubchenko 1984.

Remarks: This species was described in detail by Bray and Gibson (1980). Houston and Haedrich (1986) reported the presence of *Steringophorus* sp. in deep water off Newfoundland in *Coryphaenoides rupestris*, which might belong to this species. *S. blackeri* Bray, 1973 also occurs in deep-sea fishes in other parts of the West Atlantic.

Testes symmetrical to oblique; bifurcation of excretory vesicle further forward than middle of post-testicular region 2

2 Atlantic coast ....*S. furciger* (Olsson, 1868) Odhner, 1905

Syn.: *Distomum furcigerum* Olsson, 1868; *Leioderma furcigerum* (Olsson, 1868) Stafford, 1904; *Fellodistomum furcigerum* (Olsson, 1868) Yamaguti, 1953.

Sites: Intestine, pyloric caeca, (?stomach).

Hosts: *Anarhichas lupus* (13); *A. minor* (11, 13); *Cryptacanthodes maculatus* (1, 2); *Gadus morhua* (17); *Glyptocephalus cynoglossus* (4, 8, 10, 11, 13, 16); *Hippoglossoides platessoides* (1, 2, 5, 6, 10, 11, 12, 16, 17, 19); *Hippoglossus hippoglossus* (4, 13, 23); (?) *Lepidion eques* (14); *Myoxocephalus octodecemspinosus* (11); *Pleuronectes americanus* (1, 2, 3, 4, 9, 13, 16, 18, 20, 22); *P. ferrugineus* (4, 7, 11, 16); *P. putnami* (4); *Reinhardtius hippoglossoides* (1, 2, 13, 15, 23, 24, 25, 26, 27); *Urophycis chuss* (21); *U. musicki* (11, 21).

Distribution: Atl, EArc.

Records: 1. Stafford 1904; 2. Stafford 1907; 3. Heller 1949; 4. Ronald 1960; 5. Scott 1975a; 6. Scott 1975b; 7. Scott 1975c; 8. Scott 1975d; 9. Scott 1976; 10. Redkozubova 1978; 11. Bray 1979; 12. Umnova 1979; 13. Zubchenko 1980; 14. Bray and Gibson 1980; 15. Reimer 1981 (EArc); 16. Scott 1982; 17. Appy and Burt 1982; 18. Khan and Kiceniuk 1983; 19. Zubchenko 1985a; 20. Scott 1985b; 21. Scott 1987; 22. Khan 1987; 23. Scott and Bray 1989; 24. Wierzbicka 1991a; 25. Wierzbicka 1991b; 26. Arthur and Albert 1993; 27. Arthur and Albert 1994.

Remarks: This species was discussed and described in detail by Bray and Gibson (1980). The questionable record from *Lepidion eques* listed above refers to one adult and one immature specimen in the BM(NH) collection from Newfoundland waters which Bray and Gibson (1980) listed as a possible specimen of *S. thulini* Bray and Gibson, 1980. The latter opinion was based upon the lobed nature of the testes in the adult worm, a diagnostic feature of *S. thulini*. I have considered these specimens as possibly being *S. furciger* in view of the sucker ratio and egg size, and the fact that there is no other evidence that *S. thulini* occurs in the Northwest Atlantic region.

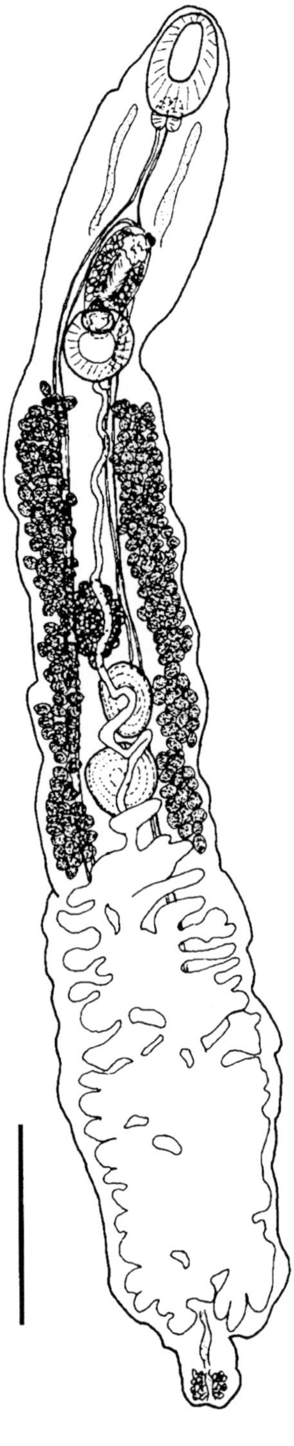

FIG. 21. *Steringophorus pritchardae* (after Bray and Gibson (1980)). Scale bar: 1 mm.

Pacific coast .... *S. brevis* (Ching, 1960) Bray and Gibson, 1980

Syn.: *Fellodistomum brevum* Ching, 1960.

Site: ? (Intestine).

Hosts: *Anoplopoma fimbria* (1, 2); *Microstomus pacificus* (1); *Pleuronectes vetulus* (1).

Distribution: Pac.

Records: 1. Kabata and Whitaker 1984; 2. Kabata *et al.* 1988.

Remarks: Features differentiating this species from *S. furciger* are difficult to assess, as only a single specimen of this species has been described (Ching, 1960). Thus, apparent differences in body shape and the size of the gonads and uterine field may be fixation or developmental artifacts. The most similar Pacific form is *S. melanostigma* (Noble and Orias, 1975) Bray and Gibson, 1980 from a bathypelagic fish off California; the gonads are, however, much smaller in the latter species.

STERINGOTREMA Odhner, 1911

Fellodistominae. Body small to large, globular to oval. Body surface smooth. Ventral sucker large, significantly larger than oral sucker, globular, in middle of body. Prepharynx short. Pharynx well developed. Oesophagus short. Caeca extend back to level of testes or just beyond. Testes two, symmetrical, post-ovarian, in anterior hindbody. Cirrus sac oval. Seminal vesicle bipartite. Pars prostatica wide, with filamentous lining. Ejaculatory duct wide, diverticulate. Genital atrium small. Genital pore median to sinistrally submedian, at about level of intestinal bifurcation. Spermatophores may be present. Ovary entire to trilobate, just anterior or antero-median to right testis. Uterus mainly post-testicular, fills most of hindbody in older specimens. Eggs numerous, often with considerable variation in size, occasionally with ornamented shells. Vitellarium follicular, in two or four lateral fields between mid-forebody and level of testes, occasionally confluent dorsally. Excretory vesicle V-shaped, arms reaching to pharynx. Parasitic in intestine (?occasionally gall bladder and stomach) of marine teleosts.

Key to the species of *Steringotrema*

1 Vitellarium reaches back to level of testes; egg shell smooth or *very* weakly ornamented ....
.... *Steringotrema pagelli* (van Beneden, 1871) Odhner, 1911

Syn.: *Steringotrema cluthense* (Nicoll, 1909) Odhner, 1911.

Sites: Intestine, (?) stomach.

Host: *Hippoglossus hippoglossus.*

Distribution: Atl.

Record: Ronald 1960.

Remarks: This is a Northeast Atlantic species which has only on one occasion been recorded in the Northwest Atlantic. It was discussed and described in detail by Bray and Gibson (1980).

Vitellarium reaches back to middle of ventral sucker; egg shell prominently ornamented ....
.... *S. ovacutum* (Lebour, 1908) Yamaguti, 1953 (Fig. 22)

Syn.: *Rhodotrema ovacutum* (Lebour, 1908) Odhner, 1911; *Stenakron ovacutus* (Lebour, 1908) Strelkov, 1960.

Site: Digestive tract: normally intestine.

Hosts: *Anarhichas lupus* (4, 5, 8); *Hippoglossoides platessoides* (1, 2, 3, 4, 5, 6, 7); *Lycodes vahli* (4).

Distribution: Atl.

Records: 1. Scott 1975a; 2. Scott 1975b; 3. Redkozubova 1978; 4. Bray 1979; 5. Zubchenko 1980; 6. Scott 1982; 7. Zubchenko 1985a; 8. Bray 1987c.

Remarks: According to Bray (1979), ovigerous specimens occur only in *H. platessoides*. This species was discussed and described in detail by Bray and Gibson (1980).

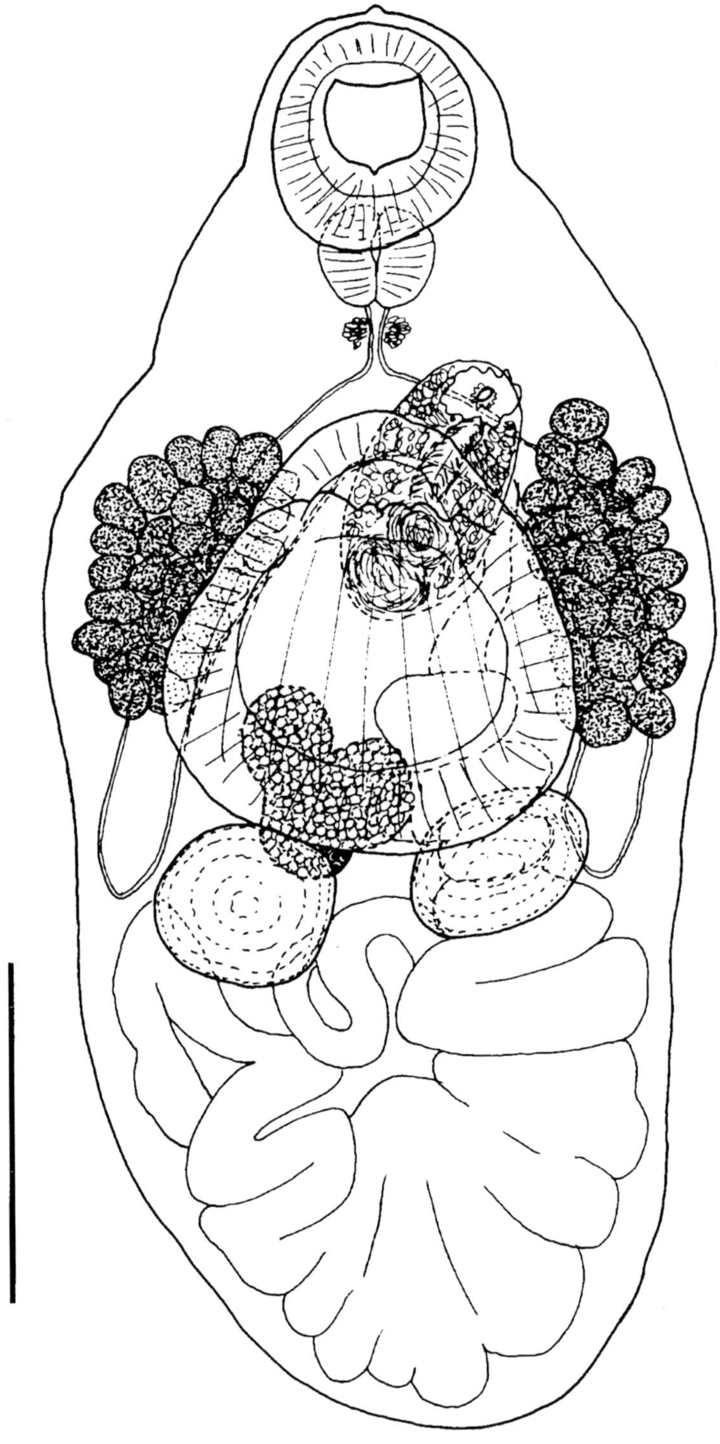

FIG. 22. *Steringotrema ovacutum* (after Bray and Gibson (1980)). Scale bar: 500 µm.

Subfamily BACCIGERINAE Yamaguti, 1958

Fellodistomidae. Body small, oval to fusiform. Body surface smooth or finely spined. Oral sucker subterminal, smaller or larger than ventral sucker. Ventral sucker in middle or anterior half of body. Prepharynx short. Pharynx small, globular to oval. Oesophagus distinct, short to long. Intestinal bifurcation in forebody. Caeca reaching to testicular region or just beyond. Testes two, symmetrical to slightly oblique, in anterior hindbody. Cirrus sac thin-walled or occasionally absent. Seminal vesicle bipartite, large, enclosed in cirrus sac when latter present. Pars prostatica indistinct or apparently absent to well developed. Ejaculatory duct short, narrow, may evert to form small cirrus. Genital atrium deep, narrow. Genital pore median or submedian in posterior forebody. Ovary oval to lobed, inter- or post-testicular. Laurer's canal present, long, opening at or near posterior extremity, proximal region dilated to form canalicular seminal receptacle. Uterine seminal receptacle absent. Uterus almost entirely post-testicular, extensive, opens into genital atrium at base of cirrus sac. Eggs numerous, small, shells smooth, without spines or filaments. Vitellarium in two fields, each with one or several compact clusters of small follicles, fields symmetrical in lateral regions of forebody or anterior hindbody, vitelline ducts wide, conspicuous. Excretory vesicle Y-shaped, with short stem (in fact almost V-shaped), arms reach into forebody. Parasitic in intestine and (?) stomach of marine and brackish water teleosts.

PRONOPRYMNA Poche, 1926

Syn.: *Pentagramma* Chulkova, 1939 *nec* Van Duzee, 1897; *Pseudopentagramma* Yamaguti, 1971.

Baccigerinae. Body small, fusiform. Body surface smooth. Cirrus sac elongate, thin-walled, closely applied to surface of seminal vesicle. Seminal vesicle internal, bipartite. Pars prostatica short, surrounded by few gland cells. Ejaculatory duct short. Permanent cirrus absent. Ovary median, post-testicular, irregularly trilobed. Laurer's canal opening on dorsal surface close to posterior extremity. Canalicular seminal receptacle present. Vitellarium two symmetrical compact masses composed of small follicles, masses may occasionally be dispersed, lateral or just posterior to ventral sucker. Parasitic in intestine, pyloric caeca, and (?) stomach of marine teleosts.

Comments

The status and history of this genus was discussed in detail by Bray and Gibson (1980).

Pronoprymna petrowi (Layman, 1930) Bray and Gibson, 1980 (Fig. 23)

Syn.: *Pseudopentagramma petrowi* (Layman, 1930) Yamaguti, 1971; *Pentagramma symmetricum* Chulkova, 1939; *Faustula sayori* (Yamaguti, 1942) Yamaguti, 1958.

Sites: Intestine, pyloric caeca.

Hosts: *Clupea pallasi* (1, 4, 5, 7, 8); *Coryphopterus nicholsi* (7); *Hypomesus pretiosus* (1); *Mallotus villosus* (1); *Oncorhynchus gorbuscha* (2, 3); *O. kisutch* (9); *Spirinchus thaleichthys* (1); *Thaleichthys pacificus* (6).

Distribution: Pac.

Records: 1. Margolis and Ching 1965; 2. Boyce 1966; 3. Boyce 1969; 4. Arai 1967a; 5. Arai 1969b; 6. Barraclough 1967; 7. Barraclough and Fulton 1967; 8. Arthur and Arai 1980a,b; 9. BM(NH) collection.

Remarks: This species was reviewed by Margolis and Ching (1965) and its nomenclature discussed by Bray and Gibson (1980).

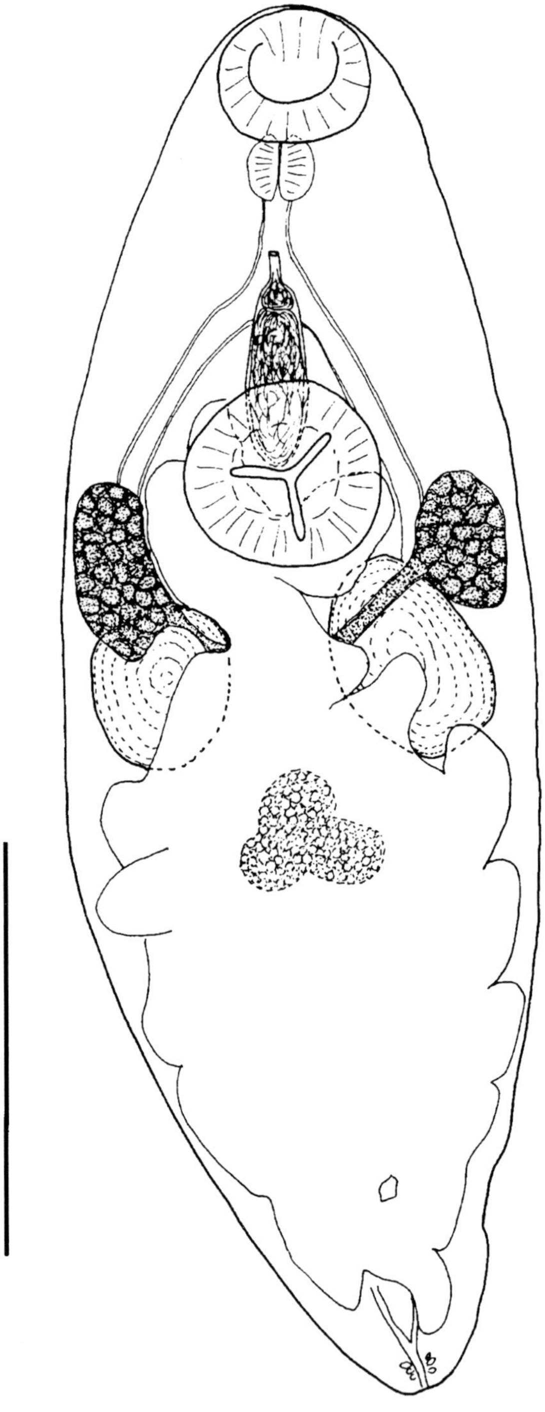

FIG. 23. *Pronoprymna ventricosa* (modified (position of ovary added) after Bray and Gibson (1980)). This northeast Atlantic species is morphologically similar to *P. petrowi*. Scale bar: 500 µm.

SUPERFAMILY HEMIUROIDEA LOOSS, 1899

Strigeida. Body small to large, oval to cylindrical; in histozoic parasites (Didymozoidae) other variations exist and worms may be encysted, sometimes in pairs. Ecsoma present or absent. Body surface smooth, rugate, or plicated (or "scaley"), never spiny, but occasionally papillate. Suckers normally well developed, occasionally small, atrophied or absent (Didymozoidae). Ventral sucker normally in middle or anterior half of body, occasionally just inside posterior half of body, occasionally pedunculate. Prepharynx absent. Pharynx well developed, normally oval, occasionally modified. Oesophagus usually short, occasionally long. "Drüsenmagen" present or absent. Gut caeca usually end blindly near posterior extremity, occasionally form cyclocoel or uroproct. Varying degrees of sexual dimorphism occasionally occur (Didymozoidae). Testes normally two, rarely one or follicular, normally tandem to symmetrical, pre-ovarian and near middle of body, occasionally filamentous, occasionally in forebody or post-ovarian. Seminal vesicle oval to tubular, occasionally constricted into portions, usually thin-walled but occasionally thick-walled, in forebody or hindbody, normally external to sinus sac, rarely partly or entirely internal, occasionally absent as distinct organ. Pars prostatica tubular to vesicular, long or short, normally external to sinus sac, rarely internal, usually in forebody, occasionally entirely inside or extending into forebody, occasionally absent. Ejaculatory duct usually present, usually short, often within sinus sac and/or sinus organ, normally unites with metraterm to form hermaphroditic duct, occasionally within "cirrus sac". Hermaphroditic duct usually present, usually within sinus sac and/or sinus organ. Sinus sac present or absent, oval to cylindrical, normally enclosing ejaculatory duct and part of metraterm and/or hermaphroditic duct, occasionally additionally enclosing ejaculatory (prostatic) vesicle or seminal vesicle and/or pars prostatica. Permanent sinus organ present or absent within genital atrium, conical or tubular, muscular or non-muscular. Temporary sinus organ sometimes formed from hermaphroditic duct. "Cirrus sac", enclosing ejaculatory duct only, and "cirrus" rarely present. Genital atrium large, small, or absent. Common genital pore mid-ventral in forebody. Ovary usually oval, occasionally lobed, rarely tubular or follicular, usually post-testicular, occasionally pre-testicular, rarely inter-testicular, normally in hindbody, rarely in forebody. Mehlis' gland usually post-ovarian, occasionally pre-ovarian. Uterine seminal receptacle plus Laurer's canal and/or Juel's organ *or* blind seminal receptacle alone normally present. Uterine coils usually fill much of hindbody, occasionally extending well into forebody, rarely entirely in forebody. Eggs normally oval, occasionally reniform, usually small, numerous, occasionally with spine, filament(s) or threads. Vitellarium normally follicular, tubular or composed of a small number (often seven) oval to tubular lobes or one to three (usually two) entire or lobed masses, often post-ovarian, occasionally pre-ovarian, sometimes extending throughout hindbody or into forebody, rarely entirely in forebody. Excretory pore terminal, vesicle Y-shaped, arms united in forebody or not. Manter's organ (accessory excretory vesicle) rarely present. Parasitic in gut, especially stomach, primarily of marine teleosts, but commonly in freshwater teleosts and elasmobranchs, occasionally in holosteans, amphibians, reptiles and progenetic in invertebrates; occasionally recorded from gills, skin, body cavity, swimbladder, other organs and body tissues; may also occur as metacercaria or juvenile in fish.

Comments

This superfamily was revised by Gibson and Bray (1979).

KEY TO THE FAMILIES OF THE HEMIUROIDEA

1 As adults in gut and other body organs and cavities; with more or less typical digenean configuration 2

As adults in tissues of body (e.g., pericardium, connective tissue of muscles, operculum, eyes, etc.); usually long and filiform or encysted, often in pairs, and bipartite with narrow anterior region and stout, oval posterior region .... *Didymozoidae*

2 Vitellarium composed of numerous widely distributed follicles 3

Vitellarium otherwise, usually composed of small number of oval to tubular (occasionally branched) lobes or 1–3 distinct oval, lobed, or unlobed masses 4

3 Prostatic sac present; parasitic in elasmobranchs and freshwater fishes .... *Azygiidae*

Prostatic sac absent; parasitic in gut of elasmobranchs .... *Ptychogonimidae*

4 Testes two, occasionally one 5

Testes follicular, 11–18 or many small follicles; usually parasitic in buccal or branchial cavities or on skin of elasmobranchs and marine teleosts .... *Syncoeliidae*

5 Ecsoma* absent (take care with this observation as some hemiurids have a reduced ecsoma and some bunocotylids may retain the vestige of an ecsoma) 6

Ecsoma present (sometimes very reduced; often withdrawn); body surface often plicated; Juel's organ and uterine seminal receptacle present; vitellarium varies between form with seven tubular lobes and form with two distinct oval masses; parasitic mainly in gut of marine teleosts, occasionally present in freshwater teleosts .... *Hemiuridae*

6 Ventral sucker anterior to middle of body; parasitic in marine teleosts; seminal vesicle never enclosed in sinus sac 7

Ventral sucker usually in or near middle of body, occasionally more anterior; significant proportion of uterus usually present in forebody (a small number of marine forms do possess a ventral sucker in the anterior half of the body and uterine coils that extend into the forebody, but these forms also possess a seminal vesicle that is enclosed within the sinus sac); vitellarium one or two masses, entire or lobed (lobes normally shallow, rarely digitate); seminal vesicle in forebody; ovary and vitellarium pre- or post-testicular; parasitic mainly in gut of freshwater and marine teleosts *Derogenidae*

7 "Cirrus" present, enclosed in "cirrus sac"; female duct opens into genital atrium independently; large parasites from gut (?or gills) of marine teleosts (some immature forms occasionally present in salmonids) .... *Hirudinellidae*

'Cirrus" and "cirrus sac" absent; male and female ducts normally unite forming hermaphroditic duct, which is often present within a sinus organ and enclosed in a sinus sac; mainly parasites of marine teleosts 8

8 Vitellarium one, two, or three compact masses; mainly parasitic in stomach .... *Bunocotylidae*

---

* Invagination (usually large) of posterior extremity of body; may be entirely withdrawn or partly to entirely (rare) extruded.)

Vitellarium otherwise 9

9 Vitellarium 6–8 (occasionally twice this number) oval to digitiform lobes, often arranged in rosette, occasionally branched; usually parasitic in intestine ....*Lecithasteridae*

Vitellarium tubular (filamentous); pharynx with narrow anterior extension into base of oral sucker; occasionally present on gills; commonly parasitic in marine sunfish (Molidae) ....*Accacoeliidae*

Family ACCACOELIIDAE Odhner, 1911

Hemiuroidea. Body large or small, commonly elongate. Ecsoma absent. Body surface smooth, but forebody may be papillate. Oral and ventral suckers well developed. Ventral sucker normally in anterior half of body, may be pedunculate. Pharynx well developed, with narrow anterior extension into base of oral sucker, occasionally modified posteriorly. Oesophagus usually long, occasionally short. "Drüsenmagen" present. Gut caeca usually H-shaped, terminate blindly or more commonly form uroproct. Testes two, oblique or in tandem, in hindbody, normally close to middle of body, pre-ovarian. Seminal vesicle thin-walled, tubular, sinuous or convoluted, commonly reaching into hindbody. Pars prostatica tubular, external gland cells may be delimited. Short ejaculatory duct commonly present within sinus sac. Hermaphroditic duct present or absent. Sinus sac and sinus organ present or absent. Genital atrium present. Ovary oval post-testicular. Mehlis' gland pre- or post-ovarian, linked to anterior or posterior region of ovary by oviduct. Laurer's canal and uterine seminal receptacle present. Juel's organ and canalicular or blind seminal receptacle absent. Uterus extensive, coils entirely or almost entirely in hindbody, usually passes close to posterior extremity before looping anteriorly again. Eggs numerous, small, non-filamented. Vitellarium with one or two main collecting ducts, composed of numerous filamentous tubules (?or occasionally chains of follicles) in various parts of the fore- or hindbody. Excretory vesicle Y-shaped, arms initially in dorsal and ventral fields, united in forebody. Metacercariae usually in coelenterates or ctenophores. Parasitic in gut or occasionally on gills of marine teleosts.

Comments

The members of this family occurring in fishes of the Northeast Atlantic were revised and redescribed by Bray and Gibson (1977). This work includes all of the recorded Canadian species plus keys and descriptions to these and forms not recorded from Canadian waters.

Key to the subfamilies of the Accacoeliidae

1 Oesophagus long and narrow; gut caeca H-shaped; uroproct present; Mehlis' gland pre-ovarian; vitellarium with single main collecting duct and system of branching tubules................*Accaceoliinae*

Oesophagus short; gut caeca not distinctly H-shaped and end blindly; Mehlis' gland post-ovarian; vitellarium with symmetrical pair of collecting ducts and system of branching tubules ....*Paraccacladiinae*

Subfamily ACCACOELIINAE Odhner, 1911

Accacoeliidae. Body usually elongate. Lateral flanges occasionally present on ventral sucker. Pharynx occasionally modified to form two muscular bulbs (*Rhynchopharynx*). Oesophagus long and thin. Gut caeca H-shaped. Uroproct present. Mehlis' gland pre-ovarian, linked by oviduct to anterior region of ovary. Vitellarium with single main collecting duct on right side and single system of ramifying branches in forebody and hindbody, left-hand system reduced to small process or small ramifying system. Parasitic on gills or in gut of marine teleosts (especially Molidae).

Key to the genera of the Accacoeliinae

1 Well-defined sinus sac and sinus organ present 2

Well-defined sinus sac and sinus organ absent; sucker-like muscular pads present dorsally in forebody .... *Odhnerium*

2 Ectoparasitic on gills; long proboscis-like sinus organ; strongly developed ventral musculature in hindbody; enormous pars prostatica occupying much of forebody; vitellarium posterior to anterior testis .... *Accacoelium*

Endoparasitic in gut; short cylindrical or dome-shaped sinus organ; vitellarium not usually extending posterior to ovary 3

3 Vitellarium confined to hindbody; ventral sucker on extensible peduncle; pars prostatica relatively short .... *Accacladium*

Vitellarium wholly or partly in forebody; ventral sucker sessile or nearly so; pars prostatica long .... *Accacladocoelium*

ACCACLADIUM Odhner, 1928

Accacoeliinae. body surface smooth. Oesophagus reaches to ventral sucker. Ventral sucker on short peduncle. Pars prostatica well developed. Sinus sac surrounding base of genital atrium. Sinus organ short, cylindrical. Vitellarium between ventral sucker and ovary. Parasitic in intestine of marine teleosts (*Mola*).

Accacladium serpentulum Odhner, 1928 (Fig. 24)

Syn.: *Accacladium nematulum* Noble and Noble, 1937.
Site: Intestine.
Host: *Mola mola*.
Distribution: Atl.
Record: Threlfall 1967.
Remarks: Redescribed in detail by Bray and Gibson (1977).

ACCACLADOCOELIUM Odhner, 1928

Accacoeliinae. Body smooth. Lateral flanges on ventral sucker present or absent (absent in Canadian species). Oesophagus reaches to ventral sucker. Pars prostatica long. Sinus sac present surrounding base of genital atrium. Sinus organ short, cylindrical. Vitellarium reaches anteriorly to oral sucker, may extend posteriorly just past ovary but usually not beyond anterior testis, reduced fraction may branch. Parasitic in intestine of marine teleosts (*Mola*).

Key to the species of *Accacladocoelium*

1 Vitellarium reaching just posterior to ovary .... *Accacladocoelium macrocotyle* (Diesing, 1858) Robinson, 1934

Syn.: *Accacoelium macrocotyle* (Diesing, 1858) Lühe, 1901.
Site: Intestine.
Host: *Mola mola*.

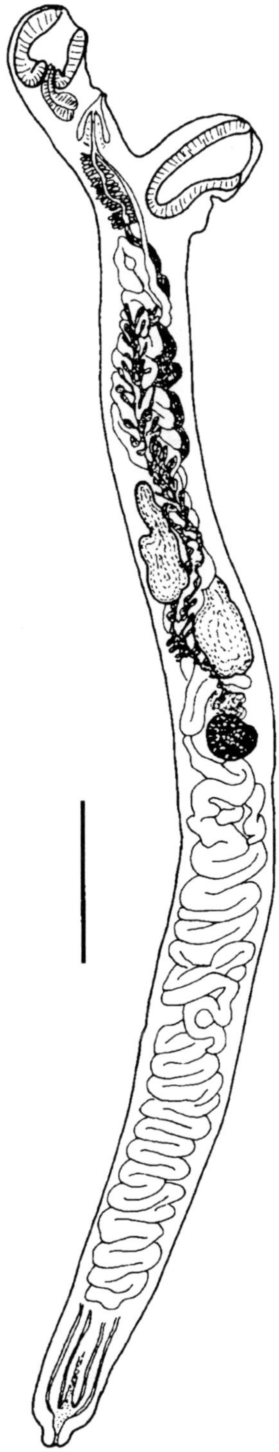

FIG. 24. *Accacladium serpentulum* (after Bray and Gibson (1977)). Scale bar: 2 mm.

Distribution: Atl.
Records: Stafford 1904; Stafford 1907; Threlfall 1967.
Remarks: This species was described in detail by Bray and Gibson (1977).

Vitellarium reaching to anterior margin of anterior testis ....
....*A. nigroflavum* (Rudolphi, 1819) Robinson, 1934 (Fig. 25)

Syn.: *Accacoelium nigroflavum* (Rudolphi, 1819) Lühe, 1901.
Site: Intestine.
Host: *Mola mola*.
Distribution: Atl.
Record: Stafford 1904.
Remarks: This species was redescribed in detail by Bray and Gibson (1977).

ACCACOELIUM Monticelli, 1893

Accacoeliinae. Forebody papillate. Oesophagus reaches to ventral sucker. Ventral sucker on short peduncle. Thick muscular layer in ventral hindbody. Sinus sac present surrounding base of genital atrium. Sinus organ long and strongly muscular, frequently extended through genital pore. Enormous pars prostatica and associated gland cells occupy much of forebody. Vitellarium posterior to anterior testis. Parasitic on gills of marine teleosts (*Mola*).

Accacoelium contortum (Rudolphi, 1819) Looss, 1899 (Fig. 26)

Site: Gills (and pharynx).
Host: *Mola mola*.
Distribution: Atl.
Records: 1. Stafford 1904; 2. Stafford 1907; 3. Threlfall 1967.
Remarks: Redescribed in detail by Bray and Gibson (1977).

ODHNERIUM Yamaguti, 1934

Syn.: *Mneiodhneria* Dollfus, 1935; *Caballeriana* Skryabin and Gushanskaya, 1959.

Accacoeliinae. Body surface smooth, but with muscular sucker-like pads on antero-dorsal surface. Flange-like muscular extensions present on ventral sucker; latter pedunculate. Oesophagus reaches to ventral sucker. Pars prostatica reaches half-way back to ventral sucker. Diffuse muscular region surrounds distal part of metraterm and genital atrium (may be vestige of sinus sac). Male duct enters genital atrium from side through small papilla. Vitellarium tubular, extending from pharynx to ovary. Parasitic in intestine of marine teleosts (*Mola*).

Odhnerium calyptrocotyle (Monticelli, 1893) Yamaguti, 1934 (Fig. 27)

Syn.: *Accacoelium foliatum* (Linton, 1898) Stafford, 1904.
Site: Intestine.
Host: *Mola mola*.
Distribution: Atl, Pac.
Records: 1. Stafford 1904 (Atl); 2. Lloyd 1938 (Pac).
Remarks: This species was described in detail by Bray and Gibson (1977).

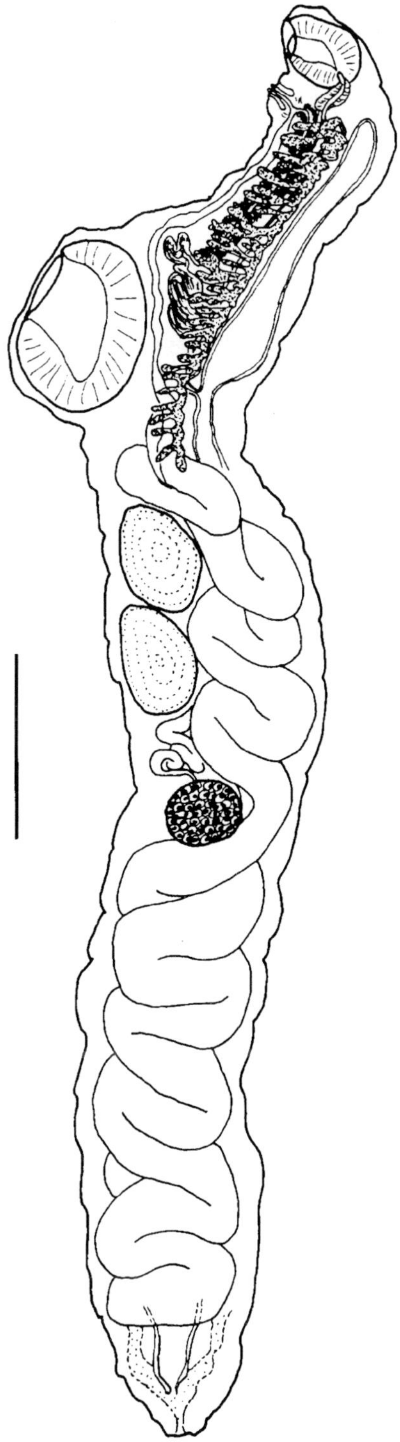

FIG. 25. *Accacladocoelium nigroflavum* (after Bray and Gibson (1977)). Scale bar: 1 mm.

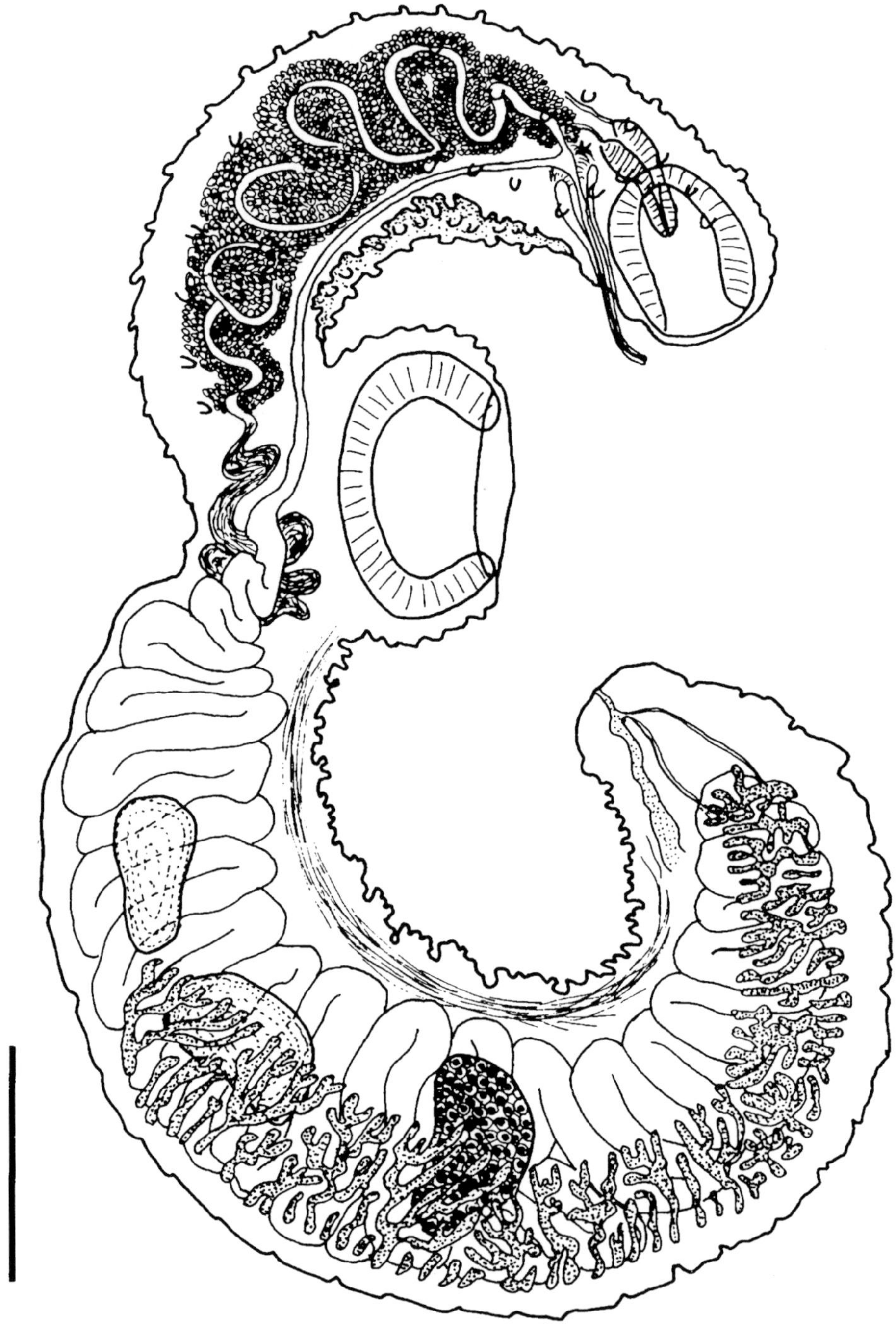

FIG. 26. *Accacoelium contortum* (after Bray and Gibson (1977)). Scale bar: 1 mm.

Subfamily PARACCACLADIINAE Bray and Gibson, 1977

Accacoeliidae. Body elongate, with smooth surface, but with papillae on outer surface of ventral sucker. Ventral sucker on short peduncle. Pharynx extended into base of oral sucker. Oesophagus short, wide. Anterior caecal shoulders small. Gut caeca terminate blindly near posterior extremity. Pars prostatica elongate, convoluted. Sinus sac present surrounding base of seminal genital atrium; musculature diffuse. Sinus organ short, cylindrical. Mehlis' gland post-ovarian. Vitellarium with symmetrical pair of main collecting ducts and ramifying system of tubules, posterior to ovary. Mature forms parasitic in rectum of carnivorous marine teleosts; immature forms parasitic in rectum of medusophagus marine teleosts.

PARACCACLADIUM Bray and Gibson, 1977

Paraccacladiinae. With characters of subfamily.

Paraccacladium jamiesoni Bray and Gibson, 1977 (Fig. 28)

Site: ? (Presumably rectum).
Hosts: *Hippoglossus hippoglossus* (2); *Urophycis chesteri* (1).
Distribution: Atl.
Records: 1. Scott 1987; 2. Scott and Bray 1989.
Remarks: Scott's specimen from *Urophycis chesteri* in the BM(NH) collection is gravid.

Family AZYGIIDAE Lühe, 1909

Hemiuroidea. Body large or small, usually elongate. Ecsoma absent. Body surface smooth, without spines or plications. Oral and ventral suckers well developed; latter in middle or anterior half of body. Pharynx well devloped. Oesophagus usually short. "Drüsenmagen' apparently absent. Gut caeca terminate blindly close to posterior extremity. Testes two, in tandem, oblique, or symmetrical, pre- or post-ovarian in hindbody. Seminal vesicle tubular, usually short, thin-walled, convoluted in forebody. Pars prostatica tubular. Prostatic sac present surrounding pars prostatica and seminal vesicle. Ejaculatory duct usually long and convoluted, but of variable length. Hermaphroditic duct short, at distal extremity of sinus organ. Permanent sinus organ variable in size. Ovary oval, pre- or post-testicular. Mehlis' gland pre-ovarian. Laurer's canal and uterine seminal receptacle present. Juel's organ and canalicular or blind seminal receptacle absent. Uterus entirely or almost entirely pre-ovarian, coiled mainly in hindbody. Eggs numerous, small, nonfilamented. Vitellarium follicular, usually present laterally throughout much of hindbody, occasionally extending into forebody. Excretory vesicle Y-shaped, arms united in forebody or not. Parasitic in stomach or body cavity of elasmobranchs and in stomach of freshwater teleosts and holosteans.

Key to the subfamilies of the Azygiidae

1 Testes post-ovarian .... *Azygiinae*

Testes pre-ovarian .... *Leuceruthrinae*

Subfamily AZYGIINAE Lühe, 1909

Azygiidae. Body normally large, occasionally small. Ventral sucker larger or smaller than oral sucker, in middle or anterior half of body. Testes in tandem, oblique, or symmetrical, post-ovarian. Uterus entirely

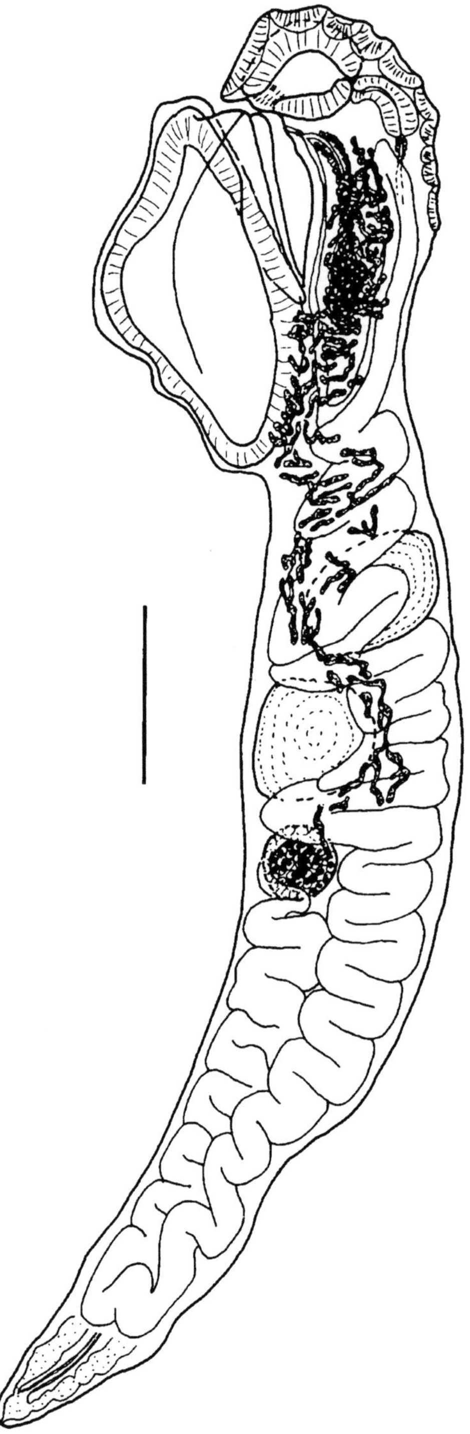

FIG. 27. *Odhnerium calyptrocotyle* (after Bray and Gibson (1977)). Scale bar: 1 mm.

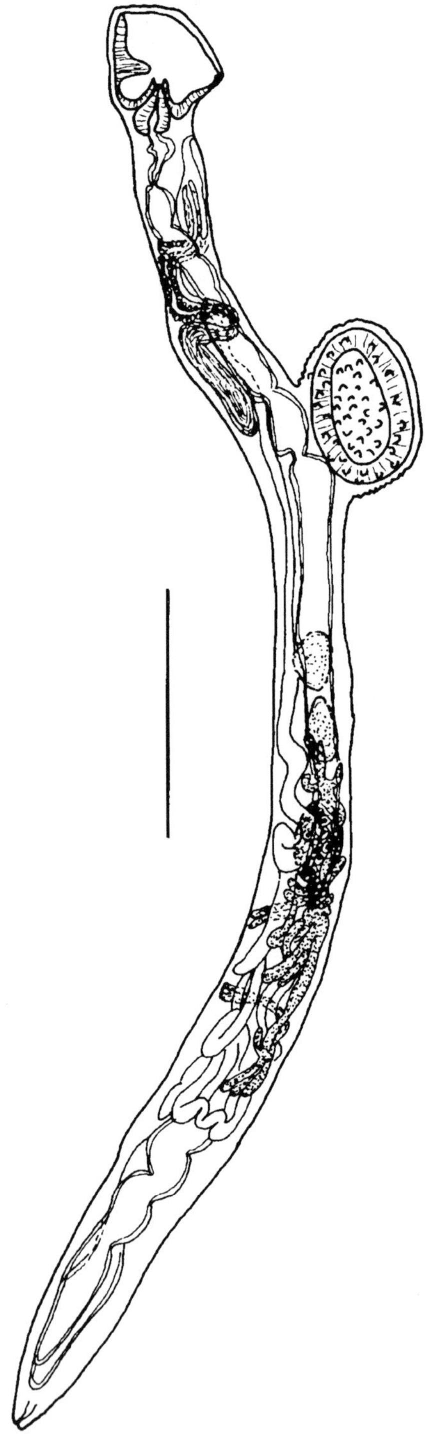

FIG. 28. *Paraccacladium jamiesoni* (after Bray and Gibson (1977)). Scale bar: 1 mm.

pre-testicular. Vitelline field may extend into forebody. Excretory arms may or may not unite in forebody. Parasitic in stomach or body cavity of elasmobranchs and stomach of freshwater teleosts and holosteans.

Key to the genera of the Azygiinae

1 Testes symmetrical, near posterior extremity; vitelline and uterine fields extending into forebody; in freshwater teleosts .... *Proterometra*

Testes tandem, oblique, or occasionally symmetrical, well anterior to posterior extremity; vitelline and uterine fields entirely or almost entirely confined to hindbody 2

2 Lateral vitelline fields confluent posterior to testes; ventral sucker normally larger than oral sucker; parasitic in elasmobranchs .... *Otodistomum*

Lateral vitelline fields not confluent posterior to testes; oral sucker normally larger than ventral sucker; parasitic in freshwater teleosts and holosteans .... *Azygia*

AZYGIA Looss, 1899

Azygiinae. Body medium to large, usually elongate, occasionally oval. Ventral sucker smaller than oral sucker, in anterior half of body. Testes tandem, occasionally symmetrical. Sinus organ a small papilla-like structure. Uterine field between ovary and ventral sucker. Vitelline follicles confined to hindbody, lateral fields not confluent posterior to testes. Excretory arms apparently not united in forebody. Parasitic in stomach of freshwater teleosts and holosteans.

Comments

The two species of this genus occurring in Canada are separated by plastic features, all affected by variation and the degree of contraction, and each of which may not fit in all instances.

Key to the species of *Azygia*

1 Testes normally in posterior quarter of body; ventral sucker usually in second quarter of body; vitelline fields usually reach close to ventral sucker and posterior extremity .... *Azygia angusticauda* (Stafford, 1904) Manter, 1926 (Fig. 29)

Syn.: *Mimodistomum angusticaudum* Stafford, 1904; *Ptychogonimus fontanus* Lyster, 1939.

Site: Stomach (less often intestine).

Hosts: *Ambloplites rupestris* (11, 15); *Ameiurus nebulosus* (11, 15); *Esox americanus vermiculatus* (2, 12); *E. lucius* (14, 15, 18, 21, 22); *E. masquinongy* (10, 15, 18); *Etheostoma exile* (11, 15, 16); *E. nigrum* (15); *Ictalurus punctatus* (11); *Lepomis gibbosus* (6, 8, 11, 15, 22); *Lota lota* (1, 15, 19, 23); *Micropterus dolomieui* (2, 5, 6, 8, 9, 11, 15, 20, 21, 22, 23); *M. salmoides* (11, 16, 22); *Perca flavescens* (3, 4, 6, 8, 9, 11, 13, 16, 17, 20); *Salvelinus fontinalis* (4); *S. namaycush* (7); *Stizostedion vitreum glaucum* (2); *S. vitreum vitreum* (1, 2, 5, 15, 17, 21, 22).

Distribution: Ont, Que.

Records: 1. Stafford 1904 (Que?); 2. Bangham and Hunter 1939 (Ont); 3. Lyster 1939 (Que); 4. Lyster 1940a (Que); 5. Miller 1940a (Que); 6. Bangham 1941 (Ont); 7. MacLulich 1943 (Ont); 8. Bangham and Venard 1946 (Ont); 9. Fantham and Porter 1948 (Que); 10. Choquette 1951 (Que); 11. Bangham 1955 (Ont); 12. Crossman 1962 (Ont); 13. Tedla and Fernando 1972 (Ont); 14. Dechtiar 1972a (Ont); 15. Dechtiar 1972b (Ont); 16. Molnar *et al.* 1974 (Ont); 17. Anthony 1978b (Ont); 18. Anthony 1983 (Ont); 19. Anthony 1987 (Ont); 20. Dechtiar *et al.* 1988 (Ont); 21. Dechtiar and Nepszy 1988 (Ont); 22. Dechtiar and Christie 1988 (Ont); 23. Dechtiar and Lawrie 1988 (Ont).

Remarks: Gibson and Bray (1979) suggested that *Ptychogonimus fontanus* Lyster, 1939, might be a contracted specimen of *Azygia longa*. After re-examining the type specimen of *P. fontanus* deposited in the British Museum (Natural History), I now consider that this specimen is much more likely to be a contracted specimen of *A. angusticauda*.

Testes normally in third quarter of body; ventral sucker usually in first quarter or first third of body; vitelline fields normally separated from ventral sucker and posterior extremity by a significant distance (usually body width or more) .... *A. longa* (Leidy, 1851) Manter, 1926

Syn.: *Azygia lucii* of Cooper (1915); *A. tereticolle* of Stafford (1904).

Site: Stomach (less often intestine).

Hosts: *Acipenser fulvescens* (23); *Ambloplites ruprestris* (7); *Amia calva* (2, 20); *Anguilla rostrata* (3, 4, 6, 7, 11, 20, 24); *Esox lucius* (1, 2, 3, 5, 7, 12, 15, 19); *E. masquinongy* (1, 2, 8, 15, 20); *Ictalurus punctatus* (1); *Lota lota* (1, 16); *Micropterus dolemieui* (2); *Perca flavescens* (2, 13, 22); *Salmo salar* (10, 14, 21); *Salvelinus fontinalis* (9, 17, 21); *S. namaycush* (2); *Stizostedion* sp. (2).

Distribution: Ont, Que, NB, NB-b, Nfld, Man, NS.

Records: 1. Stafford 1904 (Que?); 2. Cooper 1915 (Ont); 3. Lyster 1939 (Que); 4. Bangham 1941 (Ont); 5. Bangham 1955 (Ont); 6. Bangham and Venard 1946 (Ont); 7. Fantham and Porter 1948 (Que); 8. Choquette 1951 (Que); 9. Sandeman and Pippy 1967 (Nfld); 10. Pippy 1969 (NB-b); 11. Dechtiar 1972a (Ont); 12. Dechtiar 1972b (Ont); 13. Anthony 1978b (Ont); 14. Pippy 1980 (NB); 15. Anthony 1983 (Ont); 16. Anthony 1987 (Ont); 17. Frimeth 1987a (NB-b); 18. Frimeth 1987b (NB, NB-b); 19. Dechtiar *et al.* 1988 (Ont); 20. Dechtiar and Christie 1988 (Ont); 21. Marcogliese and Cone 1991b (Nfld); 22. Szalai and Dick 1991 (Man); 23. Choudhury and Dick 1993 (Man); 24. Cone *et al.* 1993 (NS).

Remarks: This species was also reported by Marcogliese and Cone (1991a) in *S. fontinalis*, but this was not an original record.

OTODISTOMUM Stafford, 1904

Azygiinae. Body large, spatulate to elongate. Ventral sucker larger than oral sucker, close to anterior extremity. Testes tandem or slightly oblique. Sinus organ capable of considerable extension or contraction to form small papilla. Uterine field almost entirely between ovary and ventral sucker. Vitelline follicles extend in lateral fields posterior to ventral sucker, reaching back to post-testicular region where fields are confluent. Excretory arms usually unite in forebody, but occasionally do not. Parasitic in stomach or body cavity of elasmobranchs and holocephalans (sharks, rays, and chimaeras).

Comments

All members of this genus occurring in fishes of the Northeast Atlantic were revised and redescribed by Gibson and Bray (1977). This work includes all of the recorded Canadian species.

Key to the species of *Otodistomum*

1 Encysted in visceral region of marine teleosts .... *Otodistomum* spp. metacercariae (Fig. 30)

Syn.: *Xenodistomum melanocystis* Stafford, 1904; *Otodistomum veliporum* of Ronald (1960), Scott (1975a,d).

Sites: Encysted on digestive tract, liver, or gonads.

Hosts: *Eopsetta jordani* (2); *Gadus morhua* (sometimes unencysted) (6); *Glyptocephalus cynoglossus* (3, 5, 6, 7); *Hippoglossoides platessoides* (4, 7); *Hippoglossus hippoglossus* (10); *Lophius americanus* (1); *Microstomus pacificus* (8); *Pleuronectes bilineatus* (2); *P. ferrugineus* (3); *P. vetulus* (8); *Reinhardtius hippoglossoides* (10, 11, 12, 13, 14); *Urophycis chuss* (9); *U. musicki* (9).

Distribution: Atl, Pac.

Records: 1. Stafford 1904 (Atl); 2. Ronald 1959 (Pac); 3. Ronald 1960 (Atl); 4. Scott 1975a (Atl); 5. Scott 1975d (Atl); 6. Appy and Burt 1982 (Atl); 7. Scott 1982 (Atl); 8. Kabata and

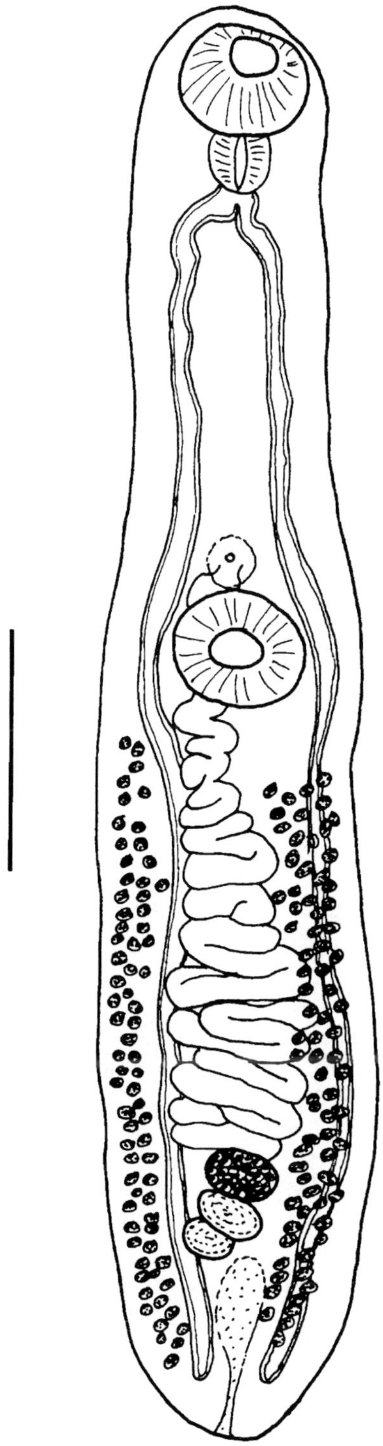

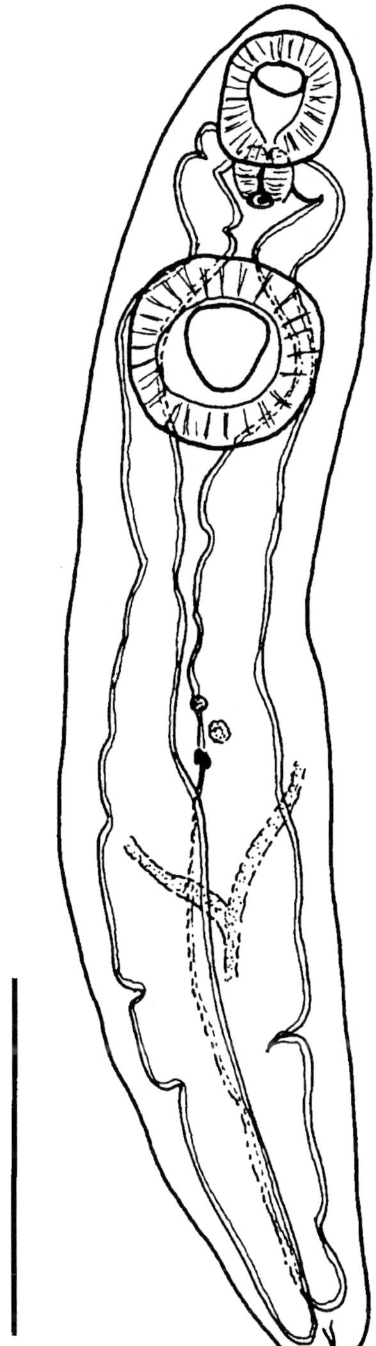

FIG. 29. *Azygia angusticauda* (after Van Cleave and Mueller (1934)). Scale bar: 500 µm. (NB. There appears to be an error in the shape of the excretory vesicle in this illustration).

FIG. 30. *Otodistomum* sp. metacercaria (after Gibson and Bray (1977)). Scale bar: 1 mm.

Whitaker 1984 (Pac); 9. Scott 1987 (Atl); 10. Scott and Bray 1989 (Atl); 11. Arthur and Albert 1992a (Atl); 12. Arthur and Albert 1992b (Atl); 13. Arthur and Albert 1993 (Atl); 14. Arthur and Albert 1994 (Atl).

Remarks: Metacercariae cannot be identified to specific level (Gibson and Bray 1977), although some authors, e.g., Scott and Bray (1989), have referred their specimens to *O. veliporum*.

Free in stomach (or intestine) of elasmobranchs 2

2 Parasitic in hexanchiform, squaliform, and squatiniform sharks; sucker-breadth ratio normally 1:1.5–2.1; vitellarium not usually extending into posterior third of body; uterine field rarely more than one–sixth of total length in mature specimens; eggs 80–130 (usually 90–115) µm, shell-thickness 4–10 (usually 6–7) µm.............................*O. veliporum* (Creplin, 1837) Stafford, 1904

Site: Stomach.

Host: *Squalus acanthias*.

Distribution: Atl.

Records: 1. Myers 1959; 2. Threlfall 1969; 3. Linkletter *et al.* 1977.

Parasitic in rays and some small squaliform sharks; sucker:breadth ratio normally 1:1.2–1.6; vitellarium not usually extending into posterior third of body; uterine field rarely less than one-sixth of total length in mature specimens; eggs 55–102 (usually 65–95) µm, shell-thickness 2.5–7 (usually 3–5) µm ....*O. cestoides* (van Beneden, 1871) Odhner, 1911 (Fig. 31)

Syn: *Distomum veliporum* of Stafford (1907); *Otodistomum veliporum* of Stafford (1904), Myers (1959) in part, Templeman (1965, 1973b), Linkletter *et al.* 1977.

Site: Stomach (occasionally intestine).

Hosts: *Bathyraja richardsoni* (7, 9, 10); *Centroscyllium fabricii* (9, 10); *Raja jenseni* (6, 9, 10); *R. laevis* (1, 2, 3); *R. ocellata* (5); *R. radiata* (4, 5, 8, 11).

Distribution: Atl.

Records: 1. Stafford 1904; 2. Stafford 1907; 3. Cooper 1915; 4. Heller 1949; 5. Myers 1959; 6. Templeman 1965; Templeman 1973b; 8. Threlfall 1969; 9. Gibson 1976; 10. Gibson and Bray 1977; 11. Linkletter *et al.* 1977.

PROTEROMETRA Horsfall, 1933

Azygiinae. Body oval, small. Oral sucker large; ventral sucker small, situated at or just posterior to middle of body. Testes symmetrical at posterior extremity. Sinus organ a small cone. Uterine field extends from ovary into forebody. Vitellarium extends from level of testes or ovary anteriorly well into forebody, in lateral fields. Excretory arms united in forebody. Parasitic in gut of freshwater teleosts (in North America).

Proterometra macrostoma (Faust, 1918) Horsfall, 1933 (Fig. 32)

Sites: (Oesophagus and stomach).

Hosts: *Ambloplites rupestris* (1, 2, 4); *Lepomis gibbosus* (1, 3).

Distribution: Ont.

Records: 1. Bangham 1955; 2. Dechtiar and Lawrie 1988; 3. Dechtiar *et al.* 1988; 4. Dechtiar and Christie 1988.

Subfamily LEUCERUTHRINAE Goldberger, 1911

Azygiidae. Body medium to large, elongate-oval. Ventral sucker smaller than oral sucker, near middle of body. Testes oblique, pre-ovarian, immediately posterior to ventral sucker. Prostatic sac small. Sinus organ small, but well defined. Uterine field between ovary and ventral sucker, passing between testes. Vitelline

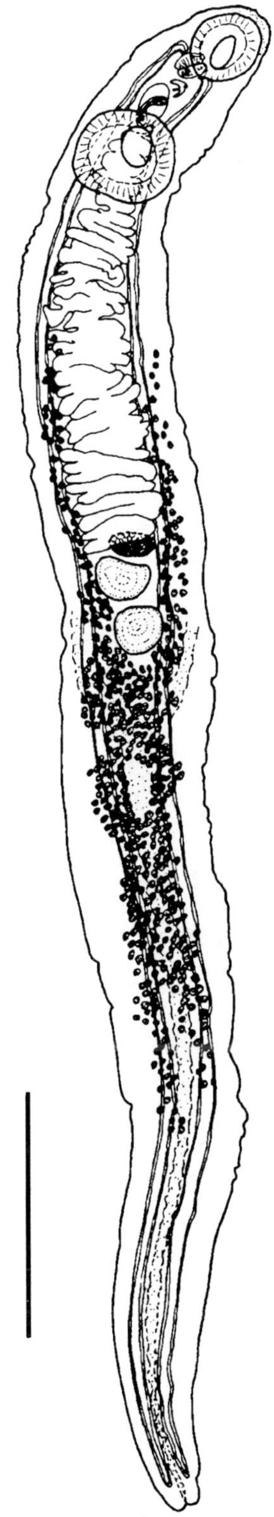

FIG. 31. *Otodistomum cestoides* (after Gibson and Bray (1977)). Scale bar: 5 mm.

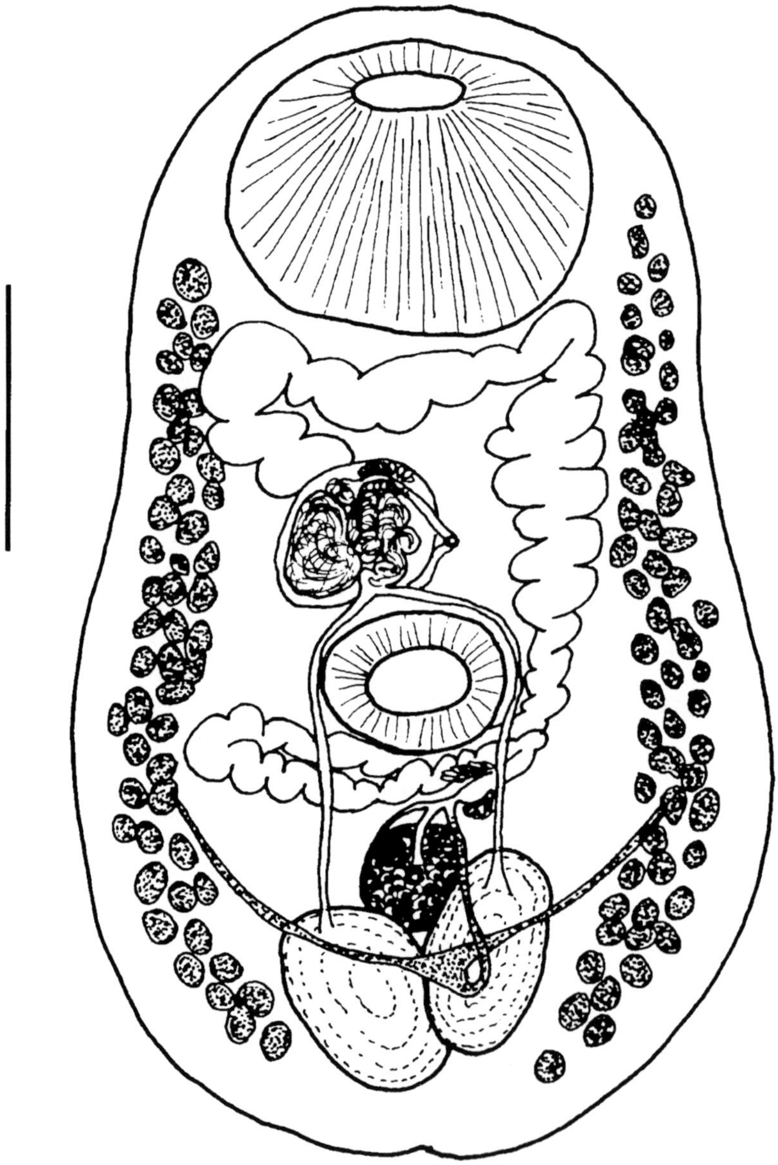

FIG. 32. *Proterometra macrostoma* (after Horsfall (1934)). Scale bar: 500 µm.

follicles in lateral fields, extending almost throughout length of hindbody. Excretory arms unite in forebody. Parasitic in gut of freshwater teleosts and holosteans (in North America).

LEUCERUTHRUS Marshall and Gilbert, 1905

Leuceruthrinae. With features of subfamily.

Leuceruthrus micropteri Marshall and Gilbert, 1905 (Fig. 33)

Site: Stomach (digestive tract).

Hosts: *Ambloplites rupestris* (1); *Amia calva* (1); *Etheostoma nigrum* (1*); *Ictalurus melas* (1); *Micropterus dolimieui* (1, 2); *M. salmoides* (1, 3); *Morone chrysops* (1*,2); *Notropis atherinoides* (1*); *Perca flavescens* (1*); *Percina caprodes* (1*).

Distribution: Ont.

Records: 1. Bangham and Hunter 1939; 2. Dechtiar and Nepszy 1988; 3. Dechtiar and Christie 1988.

Remarks: Some of the above host records (marked with an asterisk) refer to *Leuceruthrus* sp.; but, since *Leuceruthrus* is a monotypic genus, I have included these records under *L. micropteri*.

Family BUNOCOTYLIDAE Dollfus, 1950

Hemiuroidea. Body usually small, fusiform to elongate. Distinct ecsoma absent, but vestige may remain. Body surface smooth or with plications. Ridges around body often present at level of oral sucker and posterior margin of ventral sucker. Ventral sucker normally inside anterior half of worm. Pharynx well developed. Oesophagus normally short. "Drüsenmagen" normally present. Gut caeca normally end blindly near posterior extremity or occasionally form cyclocoel. Testes two, pre-ovarian in hindbody, tandem to symmetrical. Seminal vesicle saccular or tubular, in forebody or hindbody. Pars prostatica tubular or vesicular, short or long, may extend into hindbody. Ejaculatory duct long, short or apparently absent. Sinus sac usually present, occasionally absent. Hermaphroditic duct present, within sinus sac when latter present. Permanent sinus organ normally absent, but temporary sinus organ may form. Genital atrium small or absent. Ovary oval, rarely bilobed, between testes and vitellarium. Mehlis' gland post-ovarian. Laurer's canal and canalicular seminal receptacle absent. Juel's organ and uterine seminal receptacle present or absent. Blind seminal receptacle present or absent. Uterus normally almost entirely in hindbody, mainly pre- to mainly post-ovarian. Eggs numerous, small, without filaments. Vitellarium one or two, occasionally three, entire (rarely slightly lobed) masses, posterior or postero-lateral to ovary. Excretory arms rarely fail to unite in forebody, stem of excretory vesicle often with terminal bulb or with large pore (actually pore may be withdrawn within vestige of ecsoma). Parasitic mainly in stomach of marine teleosts.

Subfamily OPISTHADENINAE Yamaguti, 1970

Bunocotylidae. Body spindle-shaped to elongate with smooth surface. Transverse ridges in body wall present or absent at level of oral sucker and/or posterior margin of ventral sucker (these are often not obvious). Presomatic pit normally absent, reported in one genus. Gut caeca end blindly near posterior extremity. Testes tandem to oblique, usually well posterior to ventral sucker and near ovary, not separated from ovary by large concentration of uterine coils. Seminal vesicle tubular to saccular (? rarely bipartite), in forebody or hindbody. Pars prostatica long or short, tubular or vesicular. Ejaculatory duct long to short or apparently absent. Sinus sac present, oval to elongate-oval, enclosing hermaphroditic duct. Sinus organ (? temporary) occasionally present. Genital atrium usually present; small. Ovary normally oval, occasionally bilobed, normally close to testes. Blind seminal receptacle present, large, usually dorsal or antero-dorsal to ovary. Juel's organ and uterine seminal receptacle absent. Uterus mainly pre- to mainly

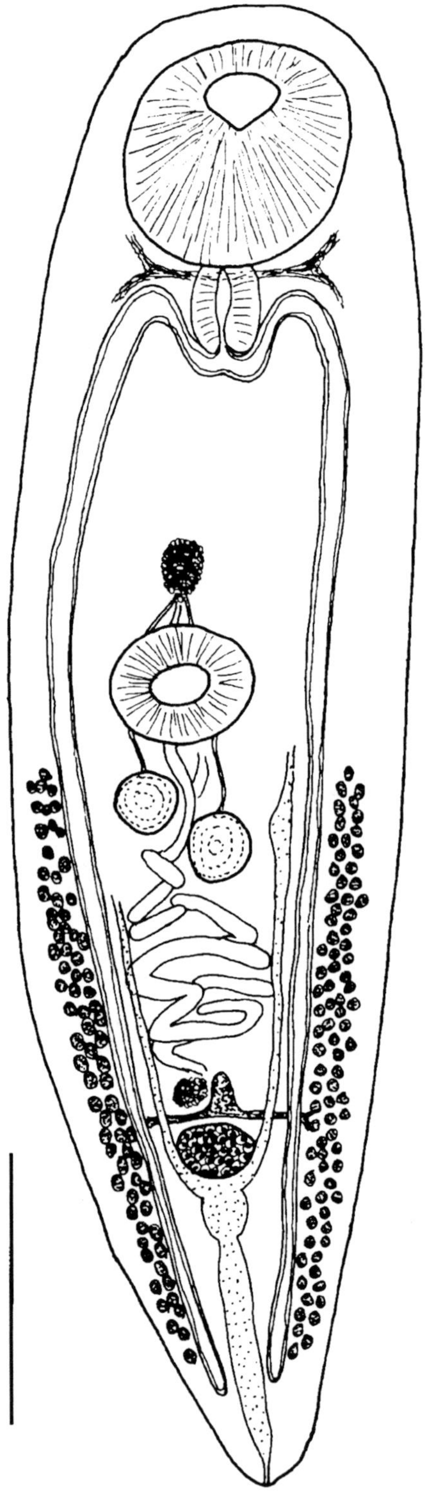

FIG. 33. *Leuceruthrus micropteri* (after Marshall and Gilbert (1905)). Scale bar (roughly estimated): 1 mm.

post-ovarian. Vitellarium two, occasionally three, entire or slightly lobed masses, posterior or postero-lateral to ovary. Excretory arms usually, but not always, united in forebody. Parasitic in stomach, occasionally intestine, of marine teleosts.

Key to the genera of the Opisthadeninae

1 Seminal vesicle in forebody or occasionally dorsal to ventral sucker .... *Genolinea*

Seminal vesicle entirely in hindbody .... *Opisthadena*

GENOLINEA Manter, 1925

Opisthadeninae. Body spindle-shaped to slightly elongate. Transverse ridge usually present around body immediately posterior to ventral sucker (often inconspicuous and frequently not reported), similar ridge may surround oral sucker. Large pre-oral lobe may be present. Ventral sucker normally in anterior half of body, (?) occasionally near middle, sphincter muscles sometimes present around aperture. Testes tandem to oblique, close to ovary. Seminal vesicle small, tubular, convoluted in forebody, occasionally dorsal or postero-dorsal to ventral sucker. Pars prostatica tubular to vesicular, short. Ejaculatory duct short or absent. Sinus sac oval to elongate-oval, small. Sinus organ occasionally present (? temporary). Ovary near middle of hindbody. Uterus usually in both pre- and post-ovarian fields; occasionally post-vitelline distribution is limited. Metraterm reported in some instances to be spinous (?). Vitellarium two compact (occasionally lobed), symmetrical, oblique, or tandem masses, posterior or postero-lateral to ovary. Excretory arms united in forebody. Parasitic mainly in stomach of marine teleosts.

Comments

It is very apparent that the genus *Genolinea* is in dire need of revision, the identity of forms from the Pacific Ocean being especially problematical. After examining specimens of *Genolinea* in the BM(NH) collection from both the Atlantic and Pacific coasts of Canada, I am of the opinion that *G. laticauda* does not occur on the Pacific coast. The species described from the Pacific coast of North America appear to split readily into two forms: (A) *G. montereyensis* Annereaux, 1947, *G. oncorhynchi* Margolis and Adams, 1956, and material in the BM(NH) collection, which are small specimens in which the testes are close to the ventral sucker and the extent of the uterine coils is limited; and (B) large specimens resembling *G. anura* (Layman, 1930), i.e., *G. laticauda* of McFarlane (1936), *G. robusta* Lloyd, 1938, and *G. manteri* Lloyd, 1938, with extensive uterine coils that intersperse between the testes and between the testes and the ventral sucker. I am, however, inclined to consider form (A) as merely young or poorly developed specimens of form (B).

Key to the species of *Genolinea*

1 On Atlantic coast; testes close together and separated from ventral sucker by significant distance even in immature worms; eggs usually 28–34 µm .... *Genolinea laticauda* Manter, 1925

Site: Stomach.

Hosts: *Anarhichas lupus* (5); *Glyptocephalus cynoglossus* (2, 4, 5, 8); *Hemitripterus americanus* (15); *Hippoglossoides platessoides* (5, 11); *Hippoglossus hippoglossus* (1, 13); *Lycodes vahli* (4); *Macrourus berglax* (4, 7); *Melanogrammus aeglefinus* (6); *Myoxocephalus octodecemspinosus* (15); *Nezumia bairdi* (7); *Pollachius virens* (9); *Pleuronectes americanus* (3, 8, 10, 15); *Reinhardtius hippoglossoides* (14); *Urophycis chesteri* (12); *U. chuss* (12, 15); *U. musicki* (12, 15).

Distribution: Atl.

Records: 1. Ronald 1960; 2. Scott 1975d; 3. Scott 1976; 4. Bray 1979; 5. Zubchenko 1980; 6. Scott 1981; 7. Zubchenko 1981a; 8. Scott 1982; 9. Scott 1985a; 10. Scott 1985b; 11. Zubchenko 1985a; 12. Scott 1987; 13. Scott and Bray 1989; 14. Arthur and Albert 1994; 15. BM(NH) collection.

Remarks: Bradstreet *et al.* (1986) recorded a *Genolinea* sp. from *Boreogadus saida* off Cornwallis Island in the eastern Arctic Region.

This parasite appears to occur in both deep-sea and coastal fishes: a detailed study of worms from these different environments is called for. Some, but not all, specimens from *M. berglax* have larger eggs than indicated in the key.

On Pacific coast; testes close together and close to ventral sucker in young worms; separated from each other and from ventral sucker by numerous coils of uterus in fully developed worms; eggs usually 33–38 µm (eggs in proximal uterus or young or poorly developed worms may give misleading results)..*G. anura* (Layman, 1930) Manter, 1934 (Fig. 34)

Syn.: *Genolinea laticauda* of McFarlane (1936), Arai (1969); *G. manteri* Lloyd, 1938; *G. robusta* Lloyd, 1938; *G. montereyensis* Annereaux, 1947; *G. oncorhynchi* Margolis and Adams, 1956.

Site: Stomach.

Hosts: *Artedius fenestralis* (4); *Blepsias cirrhosus* (3, 4); *Hemilepidotus hemilepidotus* (3, 4); *Hexagrammos decagrammos* (8); *H. lagocephalus* (3, 4); *Leptocottus armatus* (4, 8); *Myoxocephalus polyacanthocephalus* (3, 4); *Nautichthys oculofasciatus* (4); *Oligocottus maculosus* (4); *Oncorhynchus gorbuscha* (2); *O. nerka* (4); *O. tshawytscha* (5, 6); *Scorpaenichthys marmoratus* (1, 8); *Sebastes nebulosus* (7).

Distribution: Pac.

Records: 1. McFarlane 1936; 2. Margolis and Adams 1956; 3. Arai 1967a; 4. Arai 1969b; 5. Anonymous 1978; 6. Arai and Mudry 1983; 7. Holmes 1990; 8. BM(NH) collection.

Remarks: See "Comments" above. Some specimens from *S. marmoratus* have eggs in the *G. laticauda* range, but have loops of uterus interposed between the testes in fully developed worms.

OPISTHADENA Linton, 1910

Opisthadeninae. Body elongate. Transverse ridge (fold) of body wall present immediately posterior to ventral sucker. Testes tandem, in posterior half of body. Seminal vesicle in hindbody, tubular and sinuous or apparently saccular, may be enclosed within muscular, ovoid sac. Pars prostatica tubular, not reaching further forward than posterior margin of ventral sucker. Ejaculatory duct long. Sinus sac oval. Hermaphroditic duct may be subdivided. Sinus organ apparently present as small cone, at least temporarily. Ovary close to testes. Uterus mainly pre-ovarian. Vitellarium two symmetrical to oblique masses, post-ovarian. Excretory arms diverticulate, united in forebody. In stomach of marine teleosts (especially *Kyphosus*).

Opisthadena dimidia Linton, 1910 (Fig. 35)

Site: Stomach.

Host: *Xiphias gladius.*

Distribution: Atl.

Record: Hogans *et al.* 1983.

Remarks: This may have been an accidental infection. The record is based on two specimens from one of 300 fishes examined from the Northeast Atlantic. This parasite normally occurs in *Kyphosus* spp. and has been recorded off the southeast coast of the United States on several occasions.

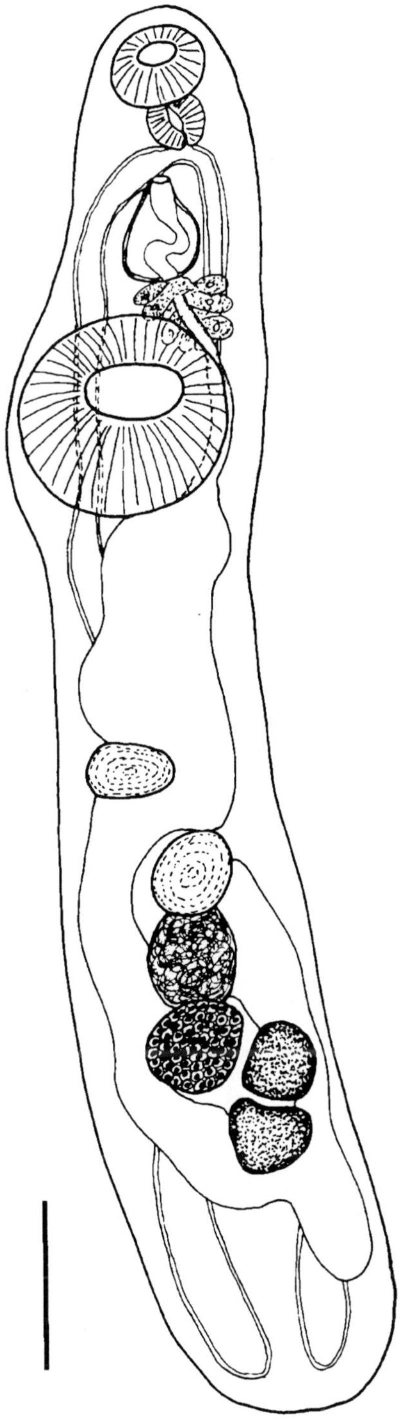

FIG. 34. *Genolinea anura* (after Lloyd (1938)). Scale bar: 250 µm.

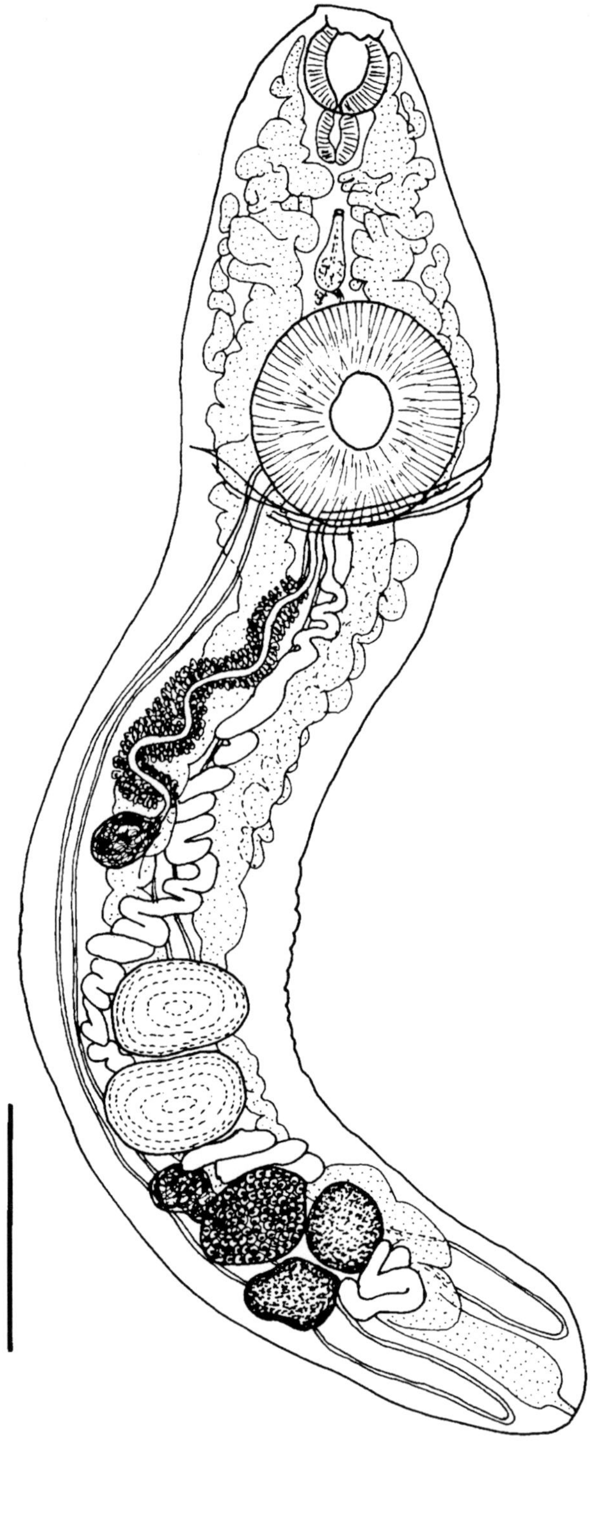

FIG. 35. *Opisthadena dimidia* (after Manter (1947)). Scale bar: 500 µm.

Family DEROGENIDAE Nicoll, 1910

Hemiuroidea. Body normally small, usually spindle-shaped to elongate-oval. Ecsoma absent. Body surface smooth. Oral and ventral suckers well developed, ventral sucker usually near middle of body, occasionally more anterior or posterior. Pharynx well developed. Oesophagus short. "Drüsenmagen' usually present. Gut caeca end blindly or form cyclocoel. Testes two, symmetrical to tandem, pre- or post-ovarian, in hindbody. Seminal vesicle thin-walled, oval, elongate or tubular, not constricted into portions, in forebody, occasionally partly or wholly enclosed within sinus sac. Pars prostatica usually tubular, occasionally vesicular, occasionally enclosed within sinus sac. Latter normally present, occasionally absent, usually small and oval; often weakly developed, may enclose all or part of pars prostatica and seminal vesicle. Permanent sinus organ present as small cone or absent. Hermaphroditic duct normally present, occasionally absent, usually short. Genital atrium present or absent, usually small. Genital pore mid-ventral in forebody. Ovary oval, pre- or post-testicular. Mehlis' gland normally post-ovarian or occasionally at level of ovary. Seminal storage and disposal apparatus variable. Laurer's canal usually present, either opening dorsally to exterior or leading into Juel's organ, often dilated proximally to form small rudimentary seminal receptacle, which is occasionally enlarged to form an apparently functional canalicular seminal receptacle. Juel's organ absent or present in rudimentary or fully developed state. Blind seminal receptacle present rarely. Uterine seminal receptacle rarely absent. Uterus may or may not extend posterior to vitellarium, significant proportion of uterus usually coiled in forebody. Eggs numerous, with or without filaments or threads, rarely with anopercular spine. Vitellarium one or two masses, entire or lobed (lobes normally shallow, rarely digitate), pre- or post-ovarian, symmetrical, oblique, or tandem. Excretory vesicle Y-shaped, arms united in forebody. Parasitic in gut (normally stomach) of freshwater and marine teleosts, but occasionally recorded from amphibians, reptiles, and freshwater shrimps.

Key to the subfamilies of the Derogenidae

1 Testes posterior to ovary and vitellarium .... *Gonocercinae*

Testes anterior to ovary and vitellarium .... *Derogeninae*

Subfamily DEROGENINAE Nicoll, 1910

Derogenidae. Ventral sucker in middle or posterior to middle of body. Gut caeca end blindly or form cyclocoel. Testes pre-ovarian, symmetrical to oblique. Seminal vesicle small, globular to tubular. Pars prostatica usually tubular, occasionally vesicular, short or long. Sinus sac present, globular to cylindrical. Permanent sinus organ present, small, cone-shaped. Hermaphroditic duct normally short. Genital atrium small, often filled by sinus organ. Ovary close behind testes. Laurer's canal present or absent, opening dorsally or into rudimentary Juel's organ, may be dilated proximally forming large rudimentary or functional canalicular seminal receptacle. Blind seminal receptacle present when Laurer's canal and uterine seminal receptacle absent, latter usually present. Uterus coiled throughout hindbody and part of forebody, significant proportion of uterus often present posterior to vitellarium. Eggs without filaments or threads, but may have anopercular spines. Vitellarium two symmetrical to tandem, oval, or slightly indented masses, posterior or occasionally lateral and postero-lateral to ovary. Parasitic in gut (mainly stomach) of marine teleosts.

Key to the genera of the Derogeninae

1 Gut caeca end blindly .... *Derogenes*

Cyclocoel present .... *Progonus*

DEROGENES Lühe, 1900

Derogeninae. Gut caeca end blindly near posterior extremity. Testes symmetrical to oblique. Seminal vesicle globular to tubular and sinuous. Pars prostatica short to long. Sinus sac globular. Male and female ducts unite within sinus organ. Ovary usually close behind testes, may be lateral to posterior testis when latter is oblique. (In *D. varicus* Laurer's canal opens distally into rudimentary Juel's organ and dilates proximally forming large rudimentary seminal receptacle.) Uterine seminal receptacle present. Uterine field usually extends from posterior extremity to region of genital pore, significant proportion of uterus posterior to vitellarium. Eggs without anopercular spine. Vitelline masses symmetrical to oblique, globular or slightly indented, post-ovarian. Parasitic in stomach, oesophagus, or occasionally gall bladder of marine (? and freshwater) teleosts.

Derogenes varicus (O.F. Müller, 1784) Looss, 1901 (Fig. 36)

Syn.: *Derogenes crassus* Manter, 1934 of Zubchenko (1980) and Stewart and Bernier (1984).

Sites: Stomach, oesophagus (? and intestine).

Hosts: *Acipenser oxyrhynchus* (30); *Alosa aestivalis* (71); *A. pseudoharengus* (70); *A. sapidissima* (71); *Ammodytes americanus* (32); *A. dubius* (17); *Anarhichas denticulatus* (32); *A. lupus* (32, 58); *Anguilla rostrata* (1, 2); *Anoplopoma fimbria* (47, 61); *Argentina silas* (16); *Atheresthes stomias* (47); *Boreogadus saida* (32, 48, 55); *Clupea harengus* (1, 2, 3, 50); *Coryphaenoides rupestris* (39); *Cryptacanthodes maculatus* (1, 2); *Gadus morhua* (1, 2, 4, 32, 33, 40, 65); *Gasterosteus aculeatus* (9, 10, 13, 15); *Glyptocephalus cynoglossus* (21, 31, 32, 34, 42); *Hemilepidotus hemilepidotus* (10); *Hemitripterus americanus* (1, 2, 32); *Hippoglossoides platessoides* (1, 2, 7, 18, 19, 31, 32, 34, 42, 53); *Hippoglossus hippoglossus* (1, 2, 7, 31, 32, 34, 63); *Lophius americanus* (1, 2); *Lycodes reticulatus* (32); *L. valhi* (32); *Macrouris berglax* (39); *Mallotus villosus* (49, 54); *Melanogrammus aeglefinus* (1, 2, 38); *Merluccius albidus* (59); *M. bilinearis* (59); *M. productus* (43, 65); *Microstomus pacificus* (47); *Myoxocephalus octodecemspinosus* (32); *M. scorpius* (1, 2); *Nezumia bairdi* (39); *Oncorhynchus gorbuscha* (5, 41); *O. kisutch* (74); *O. nerka* (5, 6); *Osmerus mordax* (1, 2, 4, 24); *Pollachius virens* (1, 2, 51); *Pleuronectes americanus* (7, 22, 42, 52); *P. ferrugineus* (1, 2, 7, 20, 31, 32, 42); *P. putnami* (7, 31); *P. vetulus* (47); *Pungitius pungitius* (14); *Reinhardtius hippoglossoides* (1, 2, 31, 32, 34, 37, 62, 63, 67, 68, 69, 73); *Salmo salar* (35); *Salvelinus alpinus* (25, 40, 44, 48, 64); *S. fontinalis* (25, 56, 57); *Sebastes aleutianus* (28); *S. alutus* (26, 28); *S. babcocki* (28); *S. borealis* (28); *S. brevispinus* (28); *S. caurinus* (28); *S. crameri* (28); *S. diploproa* (28); *S. elongatus* (28); *S. entomelas* (28); *S. fasciatus* (45, 60); *S. flavidus* (28, 66, 70); *S. goodei* (28); *S. helvomaculatus* (28); *S. maliger* (28); *S. marinus* (1, 2, 3, 29, 45); *S. mentella* (45); *S. nebulosus* (65); *S. paucispinis* (28); *S. pinniger* (28); *S. proriger* (28); *S. reedi* (28); *S. ruberrimus* (28); *S. variegatus* (28); *S. zacentrus* (28); *Squalus acanthias* (11); *Tautoglabrus adspersus* (23); *Theragra chalcogramma* (46, 47); *Triglops murrayi* (32); *Urophycis chesteri* (59); *U. chuss* (59); *U. musicki* (32, 59).

Distribution: Atl, Pac, BC-b, Lab-b, NB, NB-b, Nfld-b, Nfld, NS, Que, NWT, EArc.

Records: 1. Stafford 1904 (Alt); 2. Stafford 1907 (Atl); 3. Cooper 1915 (Atl); 4. Heller 1949 (Atl); 5. Margolis 1956b (Pac, BC-b); 6. Margolis 1963 (BC-b); 7. Ronald 1960 (Atl); 8. Sandeman and Pippy 1967 (Nfld); 9. Arai 1967a (Pac); 10. Arai 1969b (Pac); 11. Threlfall 1969 (Atl); 12. Pippy 1969 (Atl, NB, NS, Que); 13. Hanek and Threlfall 1969b (Nfld-b); 14. Hanek and Threlfall 1970c (Lab-b); 15. Hanek and Threlfall 1970d (Nfld-b); 16. Scott 1969c (Atl); 17. Scott 1973 (Atl); 18. Scott 1975a (Atl); 19. Scott 1975b (Atl); 20. Scott 1975c (Atl); 21. Scott 1975d (Atl); 22. Sekhar and Threlfall 1970a (Atl); 23. Threlfall and Hanek 1971 (Lab); 24. Hicks and Threlfall 1973 (Lab); 25. Sekerak and Arai 1973 (Pac); 26. Hanek and Molnar 1974 (Que); 27. Scott 1976 (Atl); 28. Sekerak and Arai 1977 (Pac); 29. Gaevskaya and Umnova 1977 (Atl); 30. Appy and Dadswell 1978 (NB-b); 31. Redkozubova 1978 (Atl); 32. Bray 1979 (Atl); 33. Umnova 1979 (Atl); 34. Zubchenko 1980 (Atl); 35. Pippy 1980 (Atl); 36. Dick and Belosevic 1981 (NWT); 37. Reimer 1981 (EArc); 38. Scott 1981 (Atl); 39. Zubchenko 1981a (Atl); 40. Appy and Burt 1982 (Atl); 41. Margolis 1982 (Nfld); 42. Scott 1982 (Atl); 43. Sankurathri *et al.* 1983 (Pac); 44. Dick 1984 (NWT); 45. Bourgeois and Ni 1984 (Atl); 46. Arthur 1984 (Pac); 47. Kabata and Whitaker 1984 (Pac); 48. Stewart and Bernier 1984 (NWT); 49. Pálsson and Beverley-Burton 1984 (Atl);

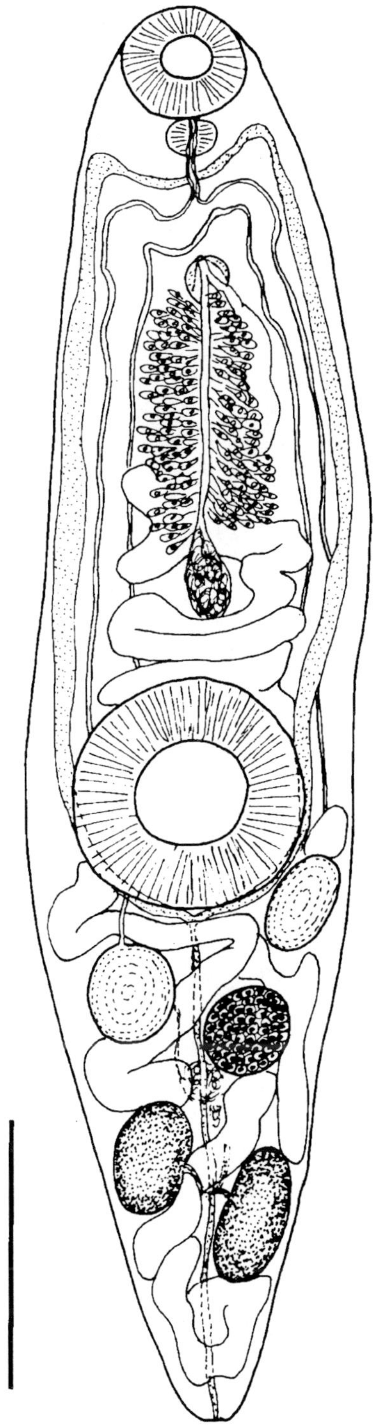

FIG. 36. *Derogenes varicus* (after Odhner (1905)). Scale bar (roughly estimated): 1 mm.

50. McGladdery and Burt 1985 (Atl); 51. Scott 1985a (Atl); 52. Scott 1985b (Atl); 53. Zubchenko 1985a (Atl); 54. Pálsson 1986 (Alt); 55. Bradstreet *et al.* 1986 (EArc); 56. Frimeth 1987a (NB, NB-b); 57. Frimeth 1987b (Atl, NB-b); 58. Bray 1987c (Atl); 59. Scott 1987 (Atl); 60. Scott 1988 (Atl); 61. Kabata *et al.* 1988 (Pac); 62. Wierzbicka 1988 (Atl); 63. Scott and Bray 1989 (Atl); 64. Bouillon and Dempson 1989 (Atl); 65. Holmes 1990 (Pac; some Atl; not all original); 66. Lee *et al.* 1990 (Pac); 67. Wierzbicka 1991a (Atl); 68. Wierzbicka 1991b (Atl); 69. Krzykawski and Wierzbicka 1992 (Atl); 70. Stanley *et al.* 1992 (Pac); 71. Landry *et al.* 1992 (NB-b); 72. Hogans *et al.* 1993 (NS, Atl); 73. Arthur and Albert 1994; 74. BM(NH) collection (Pac).

Remarks: This is a neritic species (Gibson, 1983) prevalent in a wide range of hosts on the continental shelf of both east and west coasts of Canada. The freshwater records are normally in migratory fishes, such as salmonids. Zubchenko's (1981a) records from deep-sea macrourids are, therefore, open to question: according to Dr.R.A. Bray (pers. comm.), *Derogenes* in deep-sea fishes tends to occur in the gall bladder and is likely to be a different species.

Stafford's (1904) species, *Derogenes plenus*, recorded from the intestine of *Anarhichas lupus* from the Atlantic coast may, as suggested by Margolis and Arthur (1979), be a synonym of *D. varicus*. The fact that Stafford would have been familiar with *D. varicus* makes me suspect that it could have been another derogenid, especially as *D. varicus* is only infrequently found in *A. lupus* (cf. Bray, 1987c). A possibility is *Progonus muelleri* (see below), which Bray (1979) also found in *A. lupus* in Canadian waters. It is unlikely to have been *Gonocerca phycidis* (see below), as this is an archybenthal species occurring in deeper waters off the continental shelf (Gibson 1983).

PROGONUS Looss, 1899

Syn. *Genarches* Looss, 1902.

Derogeninae. Cyclocoel present. Testes symmetrical. Seminal vesicle elongate, spindle-shaped, elongate-oval, or globular. Pars prostatica short, slightly vesicular. Sinus sac small, globular. Ovary sinistral, halfway between testes and posterior extremity. Canalicular seminal receptacle present. Laurer's canal ends blindly after passing dorsally through cyclocoel. Rudimentary Juel's organ present as small dilations of Laurer's canal at distal extremity and especially at junction with seminal receptacle. Uterine seminal receptacle absent. Uterus extends posteriorly to vitellarium, fills most of hindbody and some of forebody. Eggs without anopercular spine. Vitelline masses entire, symmetrical, post-ovarian. Parasitic in stomach of marine teleosts.

Progonus muelleri (Levinsen, 1881) Looss, 1899 (Fig. 37)

Syn.: *Genarches muelleri* (Levinsen, 1881) Looss, 1902; *Genarchopsis muelleri* (Levinsen, 1881) Yamaguti, 1953.

Site: Stomach.

Hosts: *Anarhichas lupus* (1, 2); *Artediellus uncinatus* (1); *Glyptocephalus cynoglossus* (1); *Hippoglossoides platessoides* (1); *Lepidion eques* (1); *Lophius americanus* (1); *Lumpenus lampretaeformis* (1); *Lycodes reticulatus* (1); *L. vahli* (1); *Myoxocephalus scorpius* (1); *Pleuronectes ferrugineus* (1); *Reinhardtius hippoglossoides* (1, 3, 4, 5, 6, 7); *Triglops murrayi* (1); *Urophycis musicki* (1).

Distribution: Atl.

Records: 1. Bray 1979; 2. Bray 1987c; 3. Wierzbicka 1988; 4. Wierzbicka 1991a; 5. Wierzbicka 1991b; 6. Krzykawski and Wierzbicka 1992; 7. Arthur and Albert 1994.

Remarks: This is an arctic-boreal species (Gibson 1983) with a circumpolar distribution that extends further south on the western sides of the Pacific and Atlantic Oceans than it does on the eastern sides; hence its apparent absence from British Columbian waters.

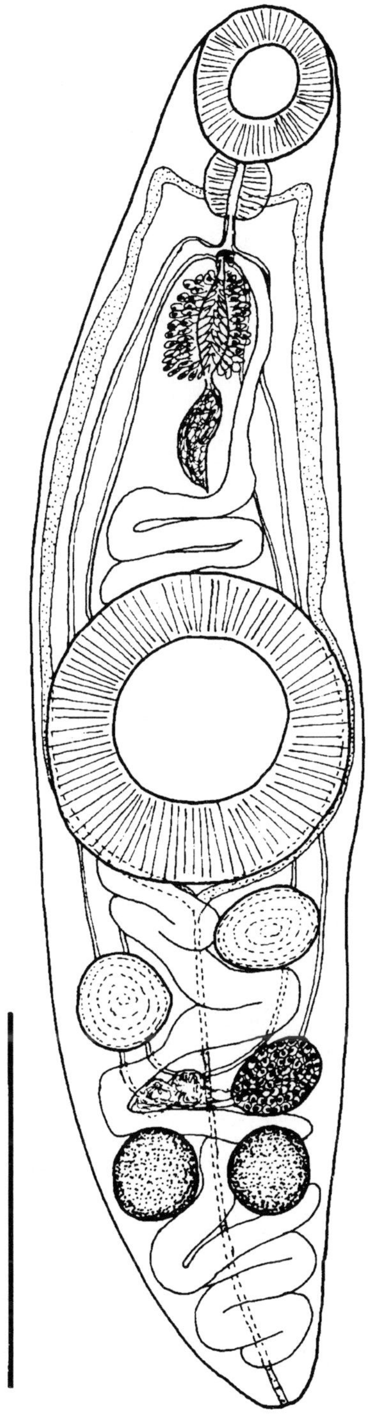

FIG. 37. *Progonus muelleri* (after Odhner (1905)). Scale bar (roughly estimated): 500 μm.

Subfamily GONOCERCINAE Skryabin and Gushanskaya, 1955

Derogenidae. Ventral sucker posterior to middle of body. Caeca end blindly. Testes post-ovarian, tandem to symmetrical, near posterior extremity. Seminal vesicle usually small, oval to tubular, in forebody. Pars prostatica tubular, short, linked to seminal vesicle by short, aglandular duct. Sinus sac absent or poorly developed. Sinus organ absent or present as small, blunt cone. Hermaphroditic duct absent or short. Genital atrium small or apparently absent. Ovary between testes and vitellarium. Laurer's canal present, opening dorsally, dilated proximally forming small rudimentary seminal receptacle. Juel's organ absent. Uterine seminal receptacle present. Uterus entirely pre-ovarian, most of coils usually in forebody. Eggs filamented or not. Vitellarium two entire or indented, oval masses, symmetrical, antero-lateral to postero-lateral to ovary. Parasitic in stomach or branchial cavity of marine teleosts.

GONOCERCA Manter, 1925

Gonocercinae. Testes tandem to almost symmetrical, at posterior extremity of body. Seminal vesicle small, thin-walled, oval, just posterior to genital pore. Pars prostatica short, tubular. Sinus sac absent. Sinus organ normally absent (there are indications that a temporary sinus organ may form). Hermaphroditic duct absent, assuming that small cavity into which male and female ducts open is genital atrium (but short duct may be distinguishable in cases where sinus organ forms). Genital atrium small or apparently absent. Ovary median. Uterus entirely pre-ovarian, largely in forebody. Eggs without filaments. Vitelline masses lateral or antero-lateral to ovary, entire to indented. Parasitic in stomach of marine teleosts (especially in mid-ocean).

Comments

The genus *Gonocerca* is badly in need of critical revision, as there is a plethora of nominal species, many of which are probably synonyms of *G. phycidis* Manter, 1925 (see Gibson 1976). *G. phycidis* is a bipolar species that occurs in a wide range of hosts in archybenthal waters (Gibson 1983). In Canadian waters the reported species are *G. phycidis*, *G. crassa* Manter, 1934, *G. macroformis* Wolfgang and Myers, 1954, and *G. macrouri* Gaevskaya, 1975. The reports of *G. crassa* by Ronald (1960) from *Pleuronectes ferrugineus*, by Szuks (1975) from *Coryphaenoides rupestris*, and by Zubchenko (1981a) from *Macrourus berglax* can probably be attributed to *G. phycidis*, as material in the BM(NH) from these or related hosts from both sides of the Atlantic belong to this species. In fact the distinguishing features of *G. crassa* outlined by Brinkmann (1975) require confirmation, since the position of the genital pore relative to the oral sucker would appear to be a function of contraction. Gibson and Bray (1979) suggested that Brinkmann's material was in fact *Derogenes varicus*: this suggestion has been confirmed by Karlsbakk (1993).

G. macrouri, originally described from *Coryphaenoides rupestris* from the Northeast Atlantic, appears to be a synonym of *G. phycidis*, which occurs commonly in this host on both sides of the Atlantic. *G. macrouri* was reported in Canadian waters by Zubchenko (1981a: also mentioned in 1981b).

With the exception of a single specimen that Zubchenko (1981a: also mentioned in 1981b) claimed to have found in *Coryphaenoides rupestris* off Labrador, there is only the original record of *G. macroformis* from the "ovary" of "cod" (*Gadus morhua*), "witch" (*Glyptocephalus cynoglossus*), and "plaice flounder" (*Hippoglossoides platessoides*) in Newfoundland waters. Considering the numbers of these common species that must have been examined, it is somewhat surprising that it has not been seen again. Nevertheless, I have retained this species as distinct in view of its unusual site and large size, but its validity requires confirmation.

It is also worth noting that Campbell and Munroe (1977) reported three species of *Gonocerca* from macrourids in the Hudson Canyon off northeast United States. These were *G. phycidis* from *Coryphaenoides armatus* and two new species, *G. minuta* from *Nezumia bairdii* and *G. haedrichi* from *C. armatus*. *G. minuta* is extremely small (less than 1mm in length): there are similar specimens from this same host in the northeast Atlantic in the BM(NH) collection. *G. haedrichi* is a large, foliose worm from the ureter and urinary bladder. It remains to be seen how much host- and site-induced variation in size and morphology occurs in *G. phycidis*.

Key to the species of *Gonocerca*

1 Large worms up to 13mm in length; parasitic in ovary (but cf. *G. haedrichi* mentioned above) .... *Gonocerca macroformis* Wolfgang and Myers, 1954

Site: Ovary.
Hosts: *Gadus morhua*; *Glyptocephalus cynoglossus*; *Hippoglossoides platessoides*.
Distribution: Atl.
Record: Wolfgang and Myers 1954.
Remarks: See above.

Worms of variable size up to 10mm in length; parasitic in stomach of wide range of archybenthal fishes .... *G. phycidis* Manter, 1925 (Fig. 38)

Syn.: *Gonocerca crassa* Manter, 1934 auct.; *G. macrouri* Gaevskaya, 1975; *G. macroformis* Wolfgang and Myers, 1954 of Zubchenko (1981a).
Site: Stomach.
Hosts: *Coryphaenoides rupestris* (2, 4, 5, 6, 8); *Gadus morhua* (7); *Hippoglossus hippoglossus* (3, 5, 9); *Macrourus berglax* (3, 6); *Pleuronectes ferrugineus* (1); *Reinhardtius hippoglossoides* (10).
Distribution: Atl.
Records: 1. Ronald 1960; 2. Szuks 1975; 3. Bray 1979; 4. Szuks 1980; 5. Zubchenko 1980; 6. Zubchenko 1981a; 7. Appy and Burt 1982; 8. Zubchenko 1985b; 9. Scott and Bray 1989; 10. Arthur and Albert 1994.
Remarks: Houston and Haedrich (1986) reported *Gonocerca* sp. in the deep-water fishes *Cottunculus microps*, *Macrourus berglax*, *Lycodes esmarkii*, and *Lycodonus mirabilis* off Newfoundland. See also "Comments" on the genus.

Family DIDYMOZOIDAE Monticelli, 1888

Hemiuroidea. Histozoic, encysted or not, often in pairs. Occasionally dioecious to varying degrees. Body shape very variable, common forms range from extremely long, filiform, often tangled to bipartite with small, tubular "forebody" and large, saccular, oval "hindbody", but other often strange variations occur. Tegument unarmed. Oral sucker usually present, sometimes vestigial. Ventral sucker absent or vestigial (may be present in juveniles). Prepharynx absent. Pharynx present, reduced, or absent. Oesophagus present. Caeca long, narrow anteriorly and narrow, dilate, or diverticulate posteriorly, may be atrophied, normally blind. Testes tubular, short to long, one or two. Vas deferens long, tubular, contains spermatozoa but does not form distinct seminal vesicle. Cirrus sac absent. Terminal genitalia normally reduced. Hermaphroditic duct short. Genital pore mid-ventral, close to anterior extremity. Ovary tubular, sometimes branched. Uterine seminal receptacle with Laurer's canal opening into rudimentary Juel's organ, or fully developed Juel's organ with sperm store in Laurer's canal, or blind seminal receptacle present. Uterus tubular or forming large egg-reservoir. Eggs numerous, small, may be reniform or elliptical, often embryonated. Vitellarium tubular, often branched. Excretory vesicle tubular, Y-shaped. In tissues of marine (rarely freshwater) teleosts (especially scombroids); juveniles occasionally found in gut of teleosts.

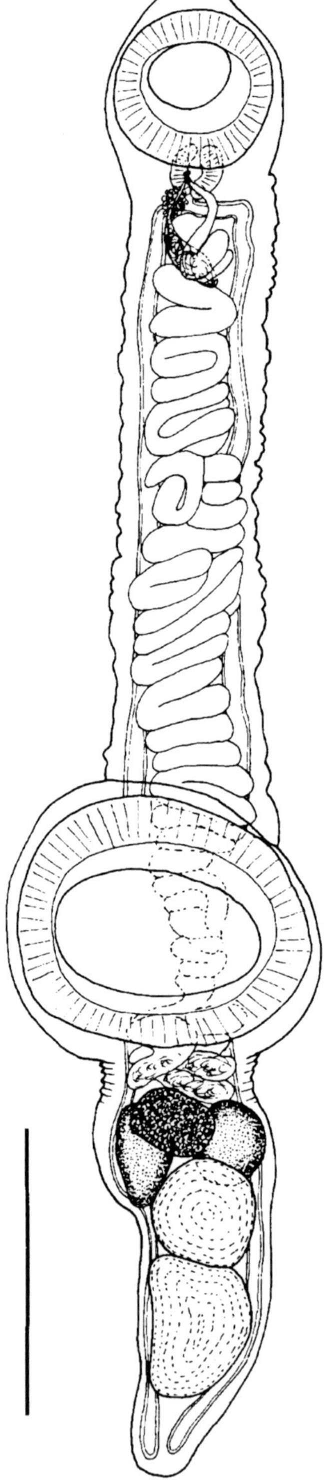

FIG. 38. *Gonocerca phycidis* ex. *Macrourus berglax*, northeast Atlantic. Original. Scale bar: 1 mm.

Comments

This group was considered a superfamily in its own right by Gibson and Bray (1979) in their revision of the Hemiuroidea, but since they are clearly derived from hemiuroids, they are here treated as a family within this group. The relationships of this group have recently been confirmed by molecular studies (Barker *et al.* 1993b).

There are two main subfamilies, the Didymozoinae Monticelli, 1888, which tend to be bipartite, consisting of an oval "hindbody" and a narrow "forebody", and the Nematobothriinae Ishii, 1935, which are very long and filiform. Only one of the three Canadian reports can be attributed to a subfamily. The remaining two reports are as follows.

Didymozoidae: unidentified records

Site: ?(presumably musculature or connective tissue).
Host: "Tuna".
Distribution: ?
Record: Hoskins *et al.* 1976.
Remarks: From a canned or processed product, thus probably not a Canadian record.

Site: (stomach).
Host: *Anoplopoma fimbria.*
Distribution: Pac.
Record: Kabata and Whitaker 1984.
Remarks: The site of this record in the stomach suggests that this was accidental or a juvenile.

Subfamily NEMATOBOTHRIINAE Ishii, 1935

Didymozoidae. Long, filiform, often tangled, dorso-ventrally flattened, not bipartite. Free in tissues or enclosed in pairs. Completely hermaphroditic. Ventral sucker present (vestigial) or absent. Gonads, vitellarium, and uterus, tubular, usually extend along much of length of body.

HALVORSENIUS Gibson, MacKenzie, and Cottle, 1981

Nematobothriinae. Anterior region especially narrow, tapers to point. Oral sucker, pharynx, oesophagus, and caeca greatly atrophied. Ventral sucker absent. Testes two, obliquely tandem. Genital junction (junction between ovary, vitellarium and uterus, i.e., region of oviduct) at two-thirds of body length. Ovary extends between anterior uterine loop and genital junction. Uterus forms three loops: two posterior reaching close to posterior extremity and one anterior reaching close to caecal bifurcation. Vitellarium extending between genital junction and hindmost uterine loop. In connective tissue of scombroid fishes.

Comments

A more detailed definition and description is in Gibson *et al.* (1981).

Halvorsenius exilis Gibson, MacKenzie, and Cottle, 1981 (Fig. 39)

Site: Musculature.
Host: *Scomber scombrus.*
Distribution: Atl.
Record: Morrison *et al.* 1986.
Remarks: Morrison *et al.* (1986) only tentatively suggested that this worm was *Halvorsenius exilis*, as there are several other nematobothriines present in *Scomber* spp. This species is especially

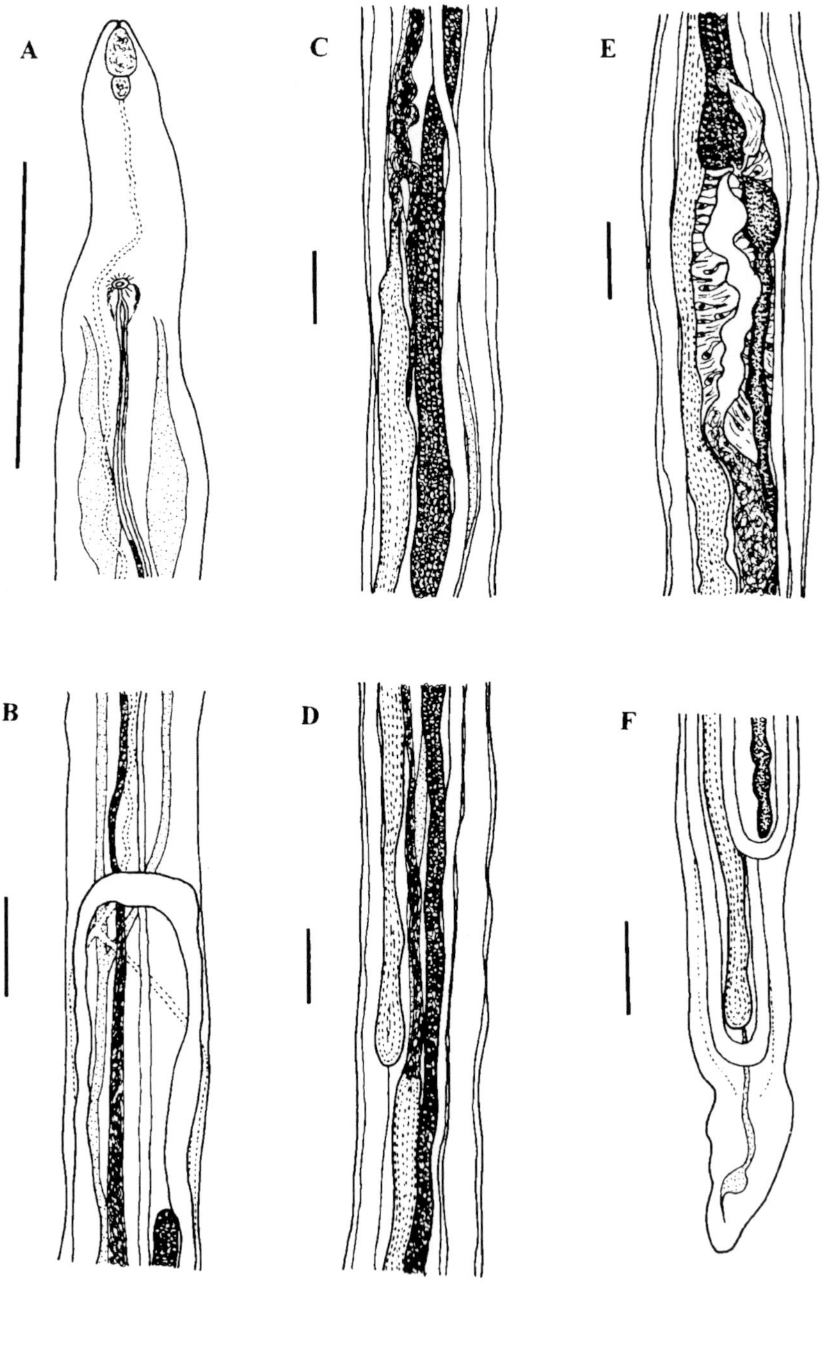

FIG. 39. *Halvorsenius exilis* (after Gibson *et al.* (1981)). A. Anterior extremity. B. Region of anterior uterine loop. C. Region of anterior extremity of anterior testis. D. Region of posterior extremity of anterior testis. E. Region of genital junction. F. Posterior extremity. Scale bars: 100 μm.

common in the pericardial membrane of young mackerel, but may extend into the connective tissue between the muscle bundles and elsewhere. In older fish the worms die and remain only as shadows formed by their eggs.

Family HEMIURIDAE Looss, 1899

Hemiuroidea. Body usually small, but elongate. Ecsoma present, occasionally reduced or vestigial. Body surface unspined, smooth, or with annular plications; latter occasionally crenulate, giving "scaly" appearance. Presomatic pit or ventro-cervical groove occasionally present. Oral and ventral suckers well developed, usually close together. Prepharynx absent. Pharynx well developed. Oesophagus usually short. "Drüsenmagen" normally present. Caeca terminate blindly, usually within ecsoma. Testes two, tandem, oblique, or symmetrical, pre-ovarian, in hindbody. Seminal vesicle tubular, saccular, or constricted into portions, muscular or thin-walled, in forebody or hindbody. Pars prostatica of variable length, usually tubular, but occasionally vesicular, may be linked to seminal vesicle by aglandular duct. Ejaculatory duct, if present, usually short. Sinus sac usually well developed, occasionally reduced or absent. Ejaculatory (prostatic) vesicle occasionally present within sinus sac. Permanent sinus organ and genital atrium well developed, small or absent, temporary sinus organ may form from hermaphroditic duct in some cases. Genital pore mid-ventral at level of oral sucker or pharynx. Ovary oval, usually entire, post-testicular. Mehlis' gland post-ovarian. Laurer's canal and canalicular or blind seminal receptacle absent. Juel's organ and uterine seminal receptacle present. Uterus coiled mainly in pre- and/or post-ovarian region of hindbody, few or no coils present in forebody; initially descending into or towards ecsoma and then ascending towards forebody. Eggs numerous, small, embryonated, rarely with polar filament. Vitellarium varies between forms with seven tubular branches (three on one side of body, four on other) and forms with two distinct, oval masses, mainly post-ovarian. Excretory vesicle Y-shaped, arms united in forebody or not. Parasitic mainly in gut, especially stomach, of marine teleosts; occasionally present in gut of freshwater teleosts and lung of sea snakes.

Comments

The conception of the Hemiuridae used here follows that of Gibson and Bray (1979). Some of the forms that occur on the Atlantic coast of Canada have been described in detail by Gibson and Bray (1986) in their study of hemiurids from the Northeast Atlantic.

Key to the subfamiles of the Hemiuridae

1 Ejaculatory (prostatic) vesicle present within sinus sac, occasionally partly external (this vesicle should not be confused with a pars prostatica which is also present) 2

Ejaculatory (prostatic) vesicle absent 3

2 Long, convoluted hermaphroditic duct and thin-walled permanent sinus organ present; seminal vesicle bipartite, anterior part muscular; vitellarium two irregularly lobed mass .... *Glomericirrinae*

Hermaphroditic duct relatively straight; permanent sinus organ absent; seminal vesicle tubular or saccular and partitioned, usually thin-walled; vitellarium seven digitiform to oval lobes or with tendency to form 2 distinct, often lobed, lateral masses; ecsoma sometimes reduced or apparently absent .... *Lecithochiriinae*

3 Sinus sac absent or poorly developed, when present usually of "open" type; seminal vesicle entirely or mainly thin-walled, usually constricted into portions; ecsoma sometimes poorly developed .... *Plerurinae*

Sinus sac present, usually well developed, occasionally small 4

4 Vitellarium two symmetrical to slightly oblique, entire or lobed masses 5

Vitellarium seven distinct oval to tubular lobes 6

5 Body surface smooth; seminal vesicle in forebody, oval, thick-walled; sinus sac very small .... *Lethadeninae*

Body surface at least partly covered with plications or "scales"; seminal vesicle in hindbody, oval or bipartite, thin-walled or partly to entirely thick-walled .... *Hemiurinae*

6 Seminal vesicle with thick muscular wall, oval; permanent sinus organ normally delicate and amuscular .... *Elytrophallinae*

Seminal vesicle thin-walled; seminal vesicle tubular (in *Tubulovesicula*); permanent sinus organ normally absent (in *Tubulovesicula*) .... *Dinurinae*

Subfamily DINURINAE Looss, 1907

Hemiuridae. Ecsoma well developed, occasionally large. Body surface plicated or smooth (apparently occasionally striated). Presomatic pit absent. Testes symmetrical to tandem, usually oblique. Seminal vesicle thin-walled, oval to tubular, may be constricted into two to four portions, in forebody, dorsal to ventral sucker or in hindbody. Pars prostatica tubular or vesicular, short or long, may be linked to seminal vesicle by aglandular duct. Sinus sac present, small or large, usually oval, not enclosing ejaculatory (prostatic) vesicle. Permanent sinus organ large and muscular, reduced to small papilla or apparently absent. Genital atrium usually well developed, deep or shallow (often depending upon contraction). Ovary usually oval, occasionally reniform or lobed. Terminal portion of uterus may or may not form distinct vesicle just outside sinus sac. Vitellarium normally seven tubular lobes: three on one side, four on other. Excretory arms united or not united in forebody. Normally parasitic in stomach of marine teleosts.

TUBULOVESICULA Yamaguti, 1934

Syn.: *Lecithurus* Pigulewsky, 1938.

Dinurinae. Body surface smooth. Seminal vesicle tubular, sinuous, in hindbody. Pars prostatica with long, wide lumen, sinuous or straight, not connected to seminal vesicle by distinct aglandular duct. Sinus sac present, oval. Permanent sinus organ normally absent, but may occur as small papilla. Ovary oval to round. Vitelline lobes tubular, but often stout. Excretory arms united in forebody. Parasitic in stomach, body cavity and tissues of marine teleosts (also reported from intestine of sea snake).

Tubulovesicula lindbergi (Layman, 1930) Yamaguti, 1934 (Fig. 40)

Syn.: *Dinurus nanaimoensis* McFarlane, 1936.

Site: Stomach (intestine).

Hosts: *Citharichthys stigmaeus* (6, 7); *Cymatogaster aggregata* (7); *Leptocottus armatus* (7); *Myoxocephalus polyacanthocephalus* (6, 7); *Oligocottus maculosus* (7); *Oncorhynchus gorbuscha* (2, 3, 11, 12); *O. keta* (8, 11, 12); *O. kisutch* (12, 13, 14); *O. nerka* (2, 3, 4, 5, 11, 12); *O. tshawytscha* (11, 12); *Ophiodon elongatus* (6, 7); *Pleuronectes vetulus* (1); *Porichthys notatus* (7); *Salvelinus malma* (6, 7); *Scorpaenichthys marmoratus* (1); *Sebastes alutus* (9, 10); *S. borealis* (10); *S. brevispinis* (10); *S. caurinus* (10, 12); *S. crameri* (10); *S. maliger* (10); *S. nebulosus* (13); *S. paucispinis* (10); *S. pinniger* (10); *Syngnathus leptorhynchus* (6, 7).

Distribution: Pac, BC-b, BC.

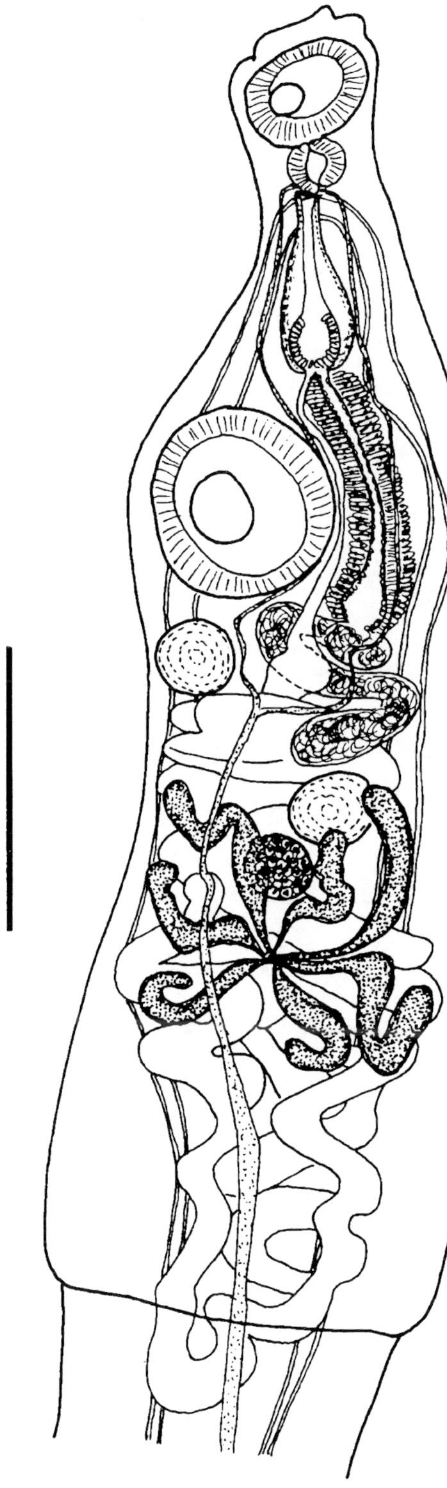

FIG. 40. *Tubulovesicula lindbergi* (after Layman (1930)). Scale bar (roughly estimated): 500 µm.

Records: 1. McFarlane 1936 (Pac); 2. Margolis 1956b (Pac, BC-b); 3. Margolis 1957 (Pac, BC-b,BC); 4. Margolis 1958a (BC); 5. Margolis 1963 (BC-b); 6. Arai 1967a (Pac); 7. Arai 1969b (Pac); 8. Margolis and Boyce 1969 (Pac); 9. Sekerak and Arai 1973 (Pac); 10. Sekerak and Arai 1977 (Pac); 11. Anonymous 1981 (BC); 12. Anonymous 1984 (BC); 13. Holmes 1990 (Pac); 14. BM(NH) collection (Pac).

Subfamily ELYTROPHALLINAE Skryabin and Gushanskaya, 1954

Hemiuridae. Ecsoma well developed. Tegument of soma smooth or plicated. Presomatic pit absent; ventro-cervical groove often present. Testes tandem to symmetrical, usually oblique. Seminal vesicle with exceptionally thick, muscular wall, oval, not constricted into portions, in forebody, dorsal to ventral sucker or in hindbody. Pars prostatica tubular, long or short, usually linked to seminal vesicle by short, aglandular duct. Sinus sac usually tubular, long, not enclosing ejaculatory (prostatic) vesicle. Sinus organ usually well developed, but delicate and amuscular. Genital atrium usually deep (subject to contraction). Ovary oval. Eggs rarely filamented. Vitellarium seven tubular to tear-shaped lobes: three on one side of body, one on other, may form rosette. Excretory arms unite in forebody. Normally in stomach of marine teleosts.

LECITHOCLADIUM Lühe, 1901

Elytrophallinae. Tegument of soma with plications. Oral sucker often funnel-shaped. Pharynx elongate. Seminal vesicle large, in hindbody. Pars prostatica long and sinuous, mainly or entirely in hindbody. Sinus sac tubular, narrow, not reaching level of seminal vesicle and usually entirely or mainly in forebody. Vitelline lobes long and tubular. In stomach of marine teleosts.

Lecithocladium sp. (Fig. 41)

Syn. *Lecithocladium gulosum* (Linton, 1901) of Hogans and Hurlbut (1984).

Site: (?)

Host: *Rhinochimaera atlantica.*

Distribution: Atl.

Record: Hogans and Hurlbut 1984.

Remarks: This record of a single specimen from a holocephalan is almost certainly accidental, and since *R. atlantica* is a deep-sea form, perhaps the worm is more likely to have been a *Glomericirrus* than a *Lecithocladium*. Linton (1901, 1940) recorded *L. gulosum* from a stromateid and scombrids off New England; these are typical hosts of *L. cristatum* (Rud., 1819) and *L. excisum* (Rud., 1819), respectively (see Gibson 1976; Gibson and Bray (1986) for descriptions and Gibson (1976) for diagnostic characters). Unfortunately, the specimen is now lost (W.E. Hogans, pers. comm.). Nahhas and Short (1965) listed *L. gulosum* as a synonym of *L. excisum*, but Gibson and Bray (1986) suggested that it was closer to *L. cristatum*: perhaps Linton's material includes both species. The life cycle of *L. excisum* has recently been worked out by Køie (1991). *L. cristatum* is used to illustrate this genus.

Subfamily GLOMERICIRRINAE Yamaguti, 1958

Hemiuridae. Ecsoma well developed. Tegument of soma plicated. Pre-somatic pit absent. Testes oblique to tandem. Seminal vesicle bipartite, both parts globular to spindle-shaped, anterior part muscular, in hindbody or dorsal to ventral sucker. Pars prostatica linked to seminal vesicle by aglandular duct. Oval sinus sac present in hindbody, dorsal to ventral sucker or occasionally in forebody, enclosing glandular ejaculatory vesicle. Hermaphroditic duct convoluted. Sinus organ present, amuscular, long, convoluted. Genital atrium well developed, long and wide proximally. Vitellarium two irregularly oval to indistinctly lobed, symmetrical masses. Excretory arms united in forebody. Parasitic in stomach of marine teleosts.

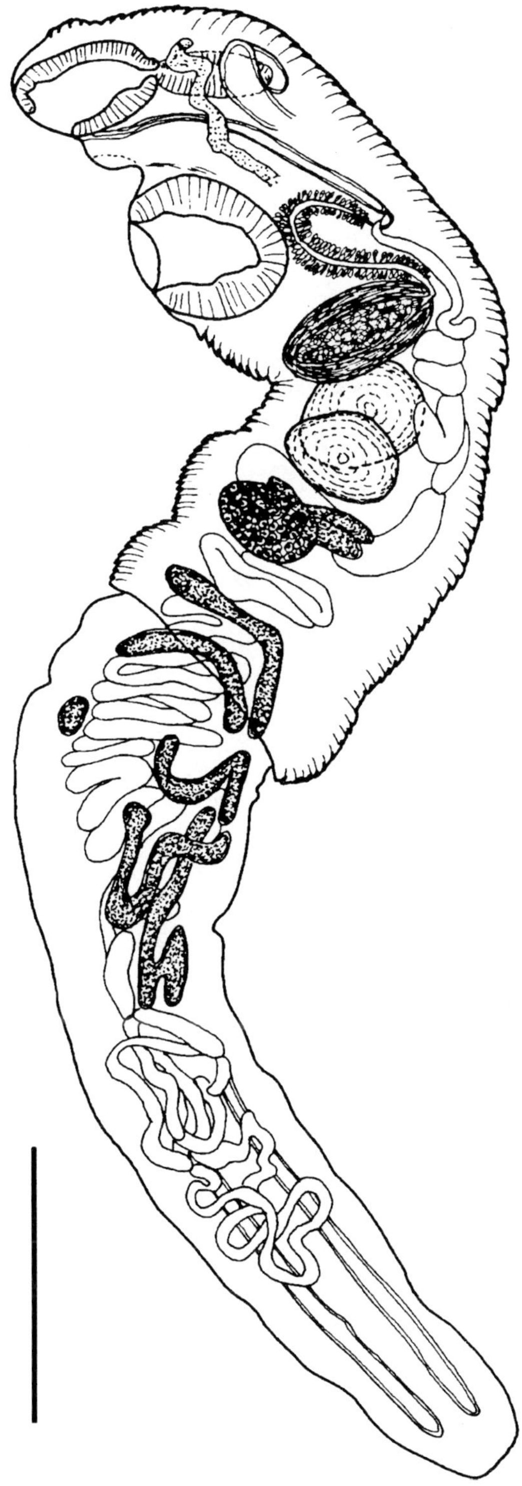

FIG. 41. *Lecithocladium cristatum* (after Gibson (1976)). Scale bar: 500 µm.

GLOMERICIRRUS Yamaguti, 1937

Glomericirrinae. With characters of subfamily.

Glomericirrus macrouri (Gaevskaya, 1975) Gaevskaya, 1979 (Fig. 42)

Syn.: "Genus novum D, species nova" of Armstrong (1974); *Hemiurus macrouri* Gaevskaya, 1975; ?*Dinosoma* sp. of Zubchenko (1975); *Glomericirrus* n. sp. of Haedrich and Polloni (1976); *Glomericirrus ulmeri* Campbell and Munroe, 1977.

Site: Stomach.

Hosts: *Coryphaenoides rupestris* (1, 2, 3); *Nezumia bairdi* (2).

Distribution: Atl.

Records: 1. Zubchenko 1975; 2. Zubchenko 1981b; 3. Zubchenko 1985b.

Remarks: This species was described and discussed in detail by Gibson and Bray (1986). It is worth noting that it appears to have been recorded, as *Dinosoma* sp., in the "North Atlantic" in *Coryphaenoides rupestris* and *Macrourus berglax* by Zubchenko (1975). It was recorded in *C. rupestris* on the Northern North Atlantic Ridge by Gaevskaya (1975) and Zubchenko (1981b). It has also been recorded in several species of *Coryphaenoides* in the Hudson Submarine Canyon to the south of Canadian Atlantic waters by Campbell and Munroe (1977). This species may also have been recorded from *C. rupestris* off Labrador as *Parahemiurus merus* by Szuks (1980).

Subfamily HEMIURINAE Looss, 1899

Hemiuridae. Ecsoma well developed. Body surface plicated or "scaly" (i.e., with crenulate plications). Presomatic pit absent. Testes tandem to oblique. Seminal vesicle thin-walled or partly or slightly muscular, bipartite or oval, in hindbody. Pars prostatica tubular, long, gland cells occasionally delimited by membrane. Sinus sac present, often tubular, not enclosing ejaculatory (prostatic) vesicle. Permanent sinus organ absent, but hermaphroditic duct may be protruded to form temporary sinus organ. Genital atrium usually small, but variable in length. Ovary oval. Vitellarium composed of two distinct oval masses, but these may exhibit slight tendency towards lobation (three lobes on one mass, four on other). Excretory arms united in forebody. Parasitic in stomach of marine teleosts.

Key to the genera of the Hemiurinae

1 Seminal vesicle bipartite .... *Hemiurus*

Seminal vesicle oval .... *Parahemiurus*

HEMIURUS Rudolphi, 1809

Syn.: *Apoblema* Dujardin, 1845; *Pronopyge* Looss, 1899; *Metahemiurus* Skryabin and Gushanskaya, 1954. Hemiurinae. Body surface covered, at least in part, by normal plications. Seminal vesicle constricted into two (?occasionally three) portions, one of which may have thick, muscular wall.

Comments

Although there are reports of *Hemiurus appendiculatus* (Rudolphi, 1802) and *H. communis* Odhner, 1905 from Canadian Atlantic waters, there is no real evidence that these species occur in the western side of the North Atlantic (Gibson and Bray 1986). *H. appendiculatus* is restricted to clupeids of the genus *Alosa* in Mediterranean and European Atlantic waters as far north as southern Norway (i.e., it has a Mediterranean/Lusitanian distribution); in this respect the record claimed by Hogans *et al.* (1993) of a single specimen in *Alosa sapidissima* in the Bay of Fundy requires confirmation. *H. communis* occurs in a wide variety

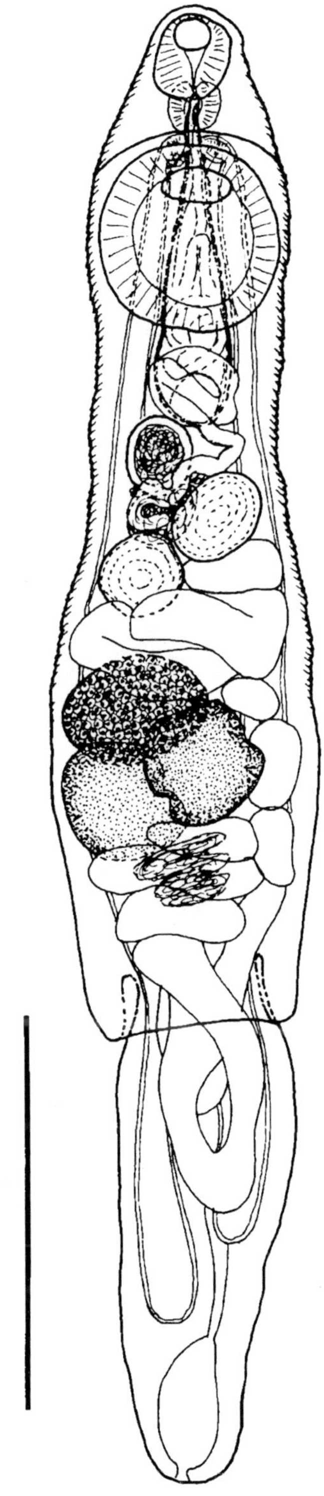

FIG. 42. *Glomericirrus macrouri* (after Gibson and Bray (1986)). Scale bar: 500 μm.

of fishes, but its range is limited to the Atlantic coast of Europe between central Norway and the Bay of Biscay (i.e., it has a boreal distribution). Reports of these species from the Canadian coast are presumably misdeterminations of *H. levinseni* (as noted by Margolis and Arthur 1979) and related hemiurids, such as *Brachyphallus crenatus*. These reports are listed by Margolis and Arthur (1979). There appears, therefore, to be only one species of this genus in Canadian waters. Margolis and Arthur (1979) also report numerous records of *Hemiurus* sp., mainly by Stafford (1902), as occurring in Canadian Atlantic: these too are probably *H. levinseni* or another hemiurid. *Hemiurus* sp. of Bourgeois and Ni (1984) and *H. appendiculatus* of Scott (1982) are treated below as *H. levinseni*.

Hemiurus levinseni Odhner, 1905 (Fig. 43)

Syn.: *Hemiurus appendiculatus* of Stafford (1904, 1907) in part and Scott (1982); *Hemiurus* sp. of Bourgeois and Ni (1984).

Site: Stomach (? and intestine).

Hosts: *Alosa pseudoharengus* (29); *A. sapidissima* (40); *Argentina silus* (11); *Boreogadus saida* (15, 29); *Clupea harengus* (1, 27); *Gadus morhua* (1, 2, 14, 15, 16, 22, 29); *Hemitripterus americanus* (15, 29); *Hippoglossoides platessoides* (13, 16, 21); *Hippoglossus hippoglossus* (9, 15, 18, 29, 35); *Macrourus berglax* (20); *Mallotus villosus* (26, 30); *Melanogrammus aeglefinus* (19); *Merluccius albidus* (31); *M. bilinearis* (31); *Myoxocephalus octodecemspinosus* (15, 29); *Oncorhynchus gorbuscha* (3, 4, 5, 33); *O. keta* (33); *O. kisutch* (33); *O. nerka* (4, 5, 6, 7, 33); *Pollachius virens* (28); *Pleuronectes americanus* (21, as *H. appendiculatus*); *Raja laevis* (8, accidental); *Reinhardtius hippoglossoides* (15, 29, 34, 35, 37, 38, 39, 41); *Salmo salar* (10, 17); *Sebastes fasciatus* (23, 33); *S. marinus* (14, 23); *S. mentella* (23); *S. nebulosus* (37); *Squalus acanthias* (8, accidental); *Tautogolabrus adspersus* (12); *Theragra chalcogramma* (24, 25); *Urophycis chesteri* (29, 31); *U. chuss* (29, 31); *U. musicki* (31).

Distribution: Atl, Pac, BC-b, NS.

Records: 1. Miller 1941a (includes reports of *H. appendiculatus* by Stafford (1904, 1907) (Atl); 2. Heller 1949 Atl); 3. Margolis 1956a (BC-b); 4. Margolis 1956b (Pac, BC-b); 5. Margolis 1957 (Pac, BC-b); 6. Margolis 1958a (Pac, BC-b); 7. Margolis 1963 (Pac, BC-b); 8. Myers 1959 (Atl); 9. Ronald 1960 (Atl); 10. Pippy 1969 (Atl); 11. Scott 1969c (Atl); 12. Sekhar and Threlfall 1970a (Atl); 13. Scott 1975a (Atl); 14. Gaevskaya and Umnova 1977 (Atl); 15. Bray 1979 (Atl); 16. Umnova 1979 (Atl); 17. Pippy 1980 (Atl); 18. Zubchenko 1980 (Atl); 19. Scott 1981 (Atl); 20. Zubchenko 1981a (Atl); 21. Scott 1982 (Atl); 22. Appy and Burt 1982 (Atl); 23. Bourgeois and Ni 1984 (Atl); 24. Arthur 1984 (Pac); 25. Kabata and Whitaker 1984 (Pac); 26. Pálsson and Beverley-Burton 1984 (Atl); 27. McGladdery and Burt 1985 (Atl); 28. Scott 1985a (Atl); 29. Gibson and Bray 1986 (Atl); 30. Pálsson 1986 (Atl); 31. Scott 1987 (Atl); 32. Scott 1988 (Atl); 33. Mortensen and Mothershead 1988 (Pac); 34. Wierzbicka 1988 (Atl); 35. Scott and Bray 1989 (Atl); 36. Holmes 1990 (Pac); 37. Wierzbicka 1991a (Atl); 38. Wierzbicka 1991b (Atl); 39. Krzykawski and Wierzbicka 1992 (Atl); 40. Hogans *et al.* 1993 (includes report of *H. appendiculatus* (see above) (Atl,NS); 41. Arthur and Albert 1994.

Remarks: This is an arctic-boreal species that probably has a circumpolar distribution (Gibson and Bray 1986). It occurs more southerly on the western sides of the Atlantic and Pacific Oceans than on the warmer eastern sides of these oceans, which presumably explains its presence in British Columbian waters only in migratory salmonids.

Cooper (1915) claimed to have found "metacercariae" of this species in the musculature of *Clupea harengus*. This record is very doubtful, as *Hemiurus* spp. are not known to encyst in vertebrates, and hemiurids that do occur outside the gut in fishes do not normally occur in the musculature.

PARAHEMIURUS Vaz and Pereira, 1930

Hemiurinae. Body surface covered, at least in part, with normal plications. Seminal vesicle oval, with muscular wall of variable thickness.

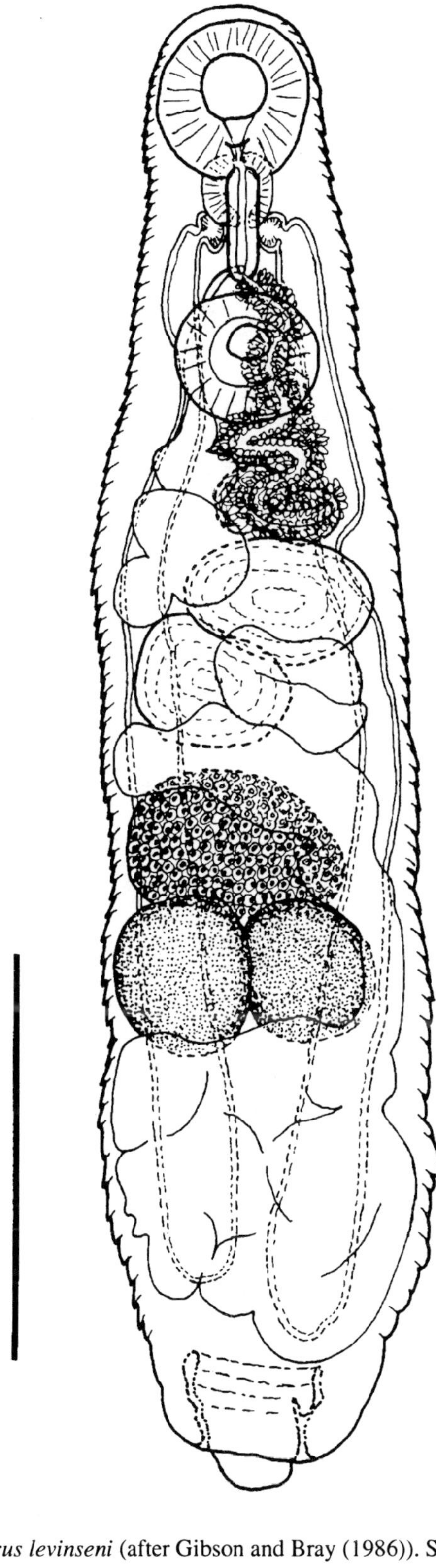

FIG. 43. *Hemiurus levinseni* (after Gibson and Bray (1986)). Scale bar: 1 mm.

Comments

This genus was reviewed by Bray (1990).

Parahemiurus merus (Linton, 1910) Woolcock, 1935 (Fig. 44)

Syn.: *Parahemiurus* sp. of Margolis (1956b, 1957).

Site: Stomach (pyloric caeca, intestine).

Hosts: *Anoplopoma fimbria* (11, 13); *Aprodon cortezianus* (6, 7); *Clupea pallasi* (6, 7, 8, 10); *Eopsetta exilis* (3, 10, 12); *E. jordani* (3); *Gadus macrocephalus* (6, 7); *Hexagrammos stelleri* (6, 7); *Merluccius productus* (9); *Oncorhynchus gorbuscha* (1, 2, 4, 5, 10); *O. kisutch* (6, 7); *O. nerka* (1, 7); *Ophiodon elongatus* (6, 7); *Sebastes caurinus* (15); *S. nebulosus* (14); *Syngnathus leptorhynchus* (7); *Theragra chalcogramma* (11).

Distribution: Pac, BC-b, BC.

Records: 1. Margolis 1956b (Pac, BC-b); 2. Margolis 1957 (Pac); 3. Ronald 1959 (Pac); 4. Boyce 1966 (Pac); 5. Boyce 1969 (Pac); 6. Arai 1967a (Pac); 7. Arai 1969b (Pac); 8. Arthur and Arai 1980a (Pac); 9. Sankurathri, Kabata and Whitaker 1983 (Pac); 10. Love and Moser 1983 (Pac); 11. Kabata and Whitaker 1984 (Pac); 12. Bray 1990 (Pac, BC, BC-b); 13. Kabata *et al.* 1988 (Pac); 14. Holmes 1990 (Pac); 15. BM(NH) collection (Pac).

Remarks: Szuks (1980) claimed to have found this species off the coast of Labrador in the deep-sea fish *Coryphaenoides rupestris*. This requires confirmation: it is much more likely to have been *Glomericirrus macrouri*.

Subfamily LECITHOCHIRIINAE Lühe, 1901

Syn.: Sterrhurinae Looss, 1907; Brachyphallinae Skryabin and Gushanskaya, 1955.

Hemiuridae. Ecsoma usually well developed, occasionally reduced. Body surface usually smooth, but occasionally plicated or rugate. Muscular "shoulder-pads" present or absent. Pre-somatic pit and ventro-cervical groove present or absent. Testes tandem to symmetrical, usually oblique. Seminal vesicle elongate, constricted into two portions, which are occasionally separated by duct, or tubular and convoluted; in bipartite forms anterior half may have thicker wall, normally in forebody, but forms with halves separated by duct may extend into hindbody. Pars prostatica short, vesicular or tubular, may extend slightly into base of sinus sac, linked to seminal vesicle by short aglandular duct. Sinus sac present, rarely of "open" type, enclosing distinct ejaculatory (prostatic) vesicle and metraterm (or part of metraterm). Permanent sinus organ absent. Genital atrium usually small or absent, occasionally well developed. Uterus mainly pre-ovarian or roughly equally distributed in pre- and post-ovarian fields. Eggs without filaments. Vitellarium seven digitiform to oval lobes in lateral groups of three and four, or with tendency to become two distinct lateral masses that often exhibit three and four lobes. Excretory arms united in forebody. Normally parasitic in gut or body cavity of marine teleosts.

Key to the genera of the Lecithochiriinae

1 Vitellarium seven distinct oval to digitiform lobes that may occur as two 3- and 4-lobed masses; body surface smooth .... *Lecithochirium*

Vitellarium two entire masses that may be indistinctly 3- and 4-lobed 2

2 Seminal vesicle in forebody, bipartite but with parts contiguous; body surface with plications .... *Brachyphallus*

Seminal vesicle extends into hindbody, composed of two parts separated by a duct; body surface smooth .... *Dissosaccus*

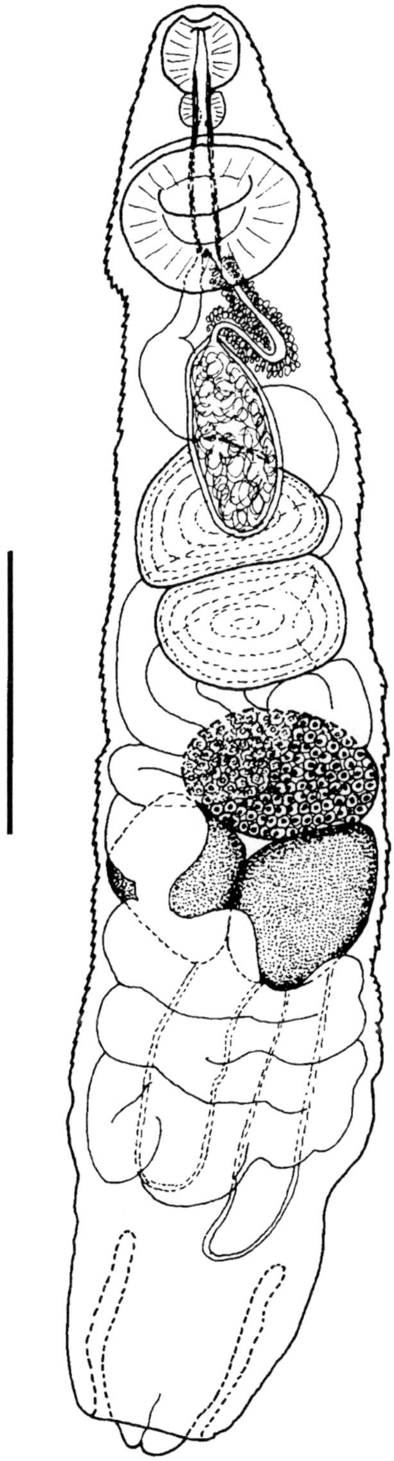

FIG. 44. *Parahemiurus merus* (after Bray (1990)). Scale bar: 200 μm.

Comments

In this particular case, the distinction between the three Canadian species of this subfamily based on the shape of the vitellarium may not be clear-cut, because the vitelline lobes of *Lecithochirium exodicum* are very reduced in comparison with many species of this genus.

BRACHYPHALLUS Odhner, 1905

Lecithochiriinae. Body surface plicated, plications may be crenulate. Presomatic pit present, circular or oval, deep, glandular. Seminal vesicle bipartite, thin-walled, anterior part small, posterior part large; occurring mostly in forebody. Pars prostatica tubular. Temporary sinus organ may be seen. Vitellarium two lateral masses, entire, irregularly lobed, or indistinctly three- and four-lobed. Parasitic in gut (stomach) of marine and migratory teleosts.

Brachyphallus crenatus (Rudolphi, 1802) Odhner, 1905 (Fig. 45)

Syn.: *Hemiurus appendiculatus* of Stafford (1904, 1907) (partim).

Site: Stomach, (intestine).

Hosts: *Alosa aestivalis* (58); *A. pseudoharengus* (50, 58); *A. sapidissima* (59); *Ammodytes dubius* (20); *Anoplopoma fimbria* (53); *Boreogadus saida* (45); *Clupea harengus* (3, 5, 11, 12, 47, 50); *C. pallasi* (32, 33); *Coregonus nasus* (48); *Gadus morhua* (39); *Gasterosteus aculeatus* (15, 17); *Hippoglossoides platessoides* (30, 34); *Hippoglossus hippoglossus* (1, 2, 4, 10, 29, 55); *Oncorhynchus gorbuscha* (7, 8, 37, 40, 42, 54); *O. keta* (37, 42); *O. kisutch* (37, 42); *O. nerka* (7, 8, 9, 54); *O. tshawytscha* (26, 37, 41, 42); *Osmerus mordax* (1, 2, 4, 5, 19, 38); *Pollachius virens* (49); *Pungitius pungitius* (16, 23); *Reinhardtius hippoglossoides* (1, 2, 4, 29, 55, 60); *Salmo salar* (1, 2, 4, 5, 13, 14, 18, 21, 31); *Salvelinus alpinus* (21, 24, 27, 35, 43, 45, 56); *S. fontinalis* (21, 22, 28, 36, 38, 51, 52); *S. malma* (6); *S. namaycush* (21); *Sebastes alutus* (57); *S. marinus* (25, 46); *Theragra chalcogramma* (44); *Trigla pingelii* (45).

Distribution: Atl, Pac, BC, BC-b, Lab-b, Lab, NB, NB-b; Nfld-b, Nfld, NS, PEI, Que, YT, NWT, EArc.

Records: 1. Stafford 1904 (Atl); 2. Stafford 1907 (Atl); 3. Cooper 1915 (Atl); 4. Miller 1941a (Atl); 5. Heller 1949 (Atl, Que); 6. Bangham and Adams 1954 (BC); 7. Margolis 1956b (Pac); 8. Margolis 1957 (Pac, BC-b, BC); 9. Margolis 1963 (Pac, BC-b); 10. Ronald 1960 Atl); 11. Sindermann 1957 (Atl); 12. Sindermann 1961 (Atl); 13. Sandeman and Pippy 1967 (Nfld); 14. Pippy 1969 (Atl, NB, NS, Que, PEI); 15. Hanek and Threlfall 1969a (Nfld-b); 16. Hanek and Threlfall 1970c (Nfld); 17. Hanek and Threlfall 1970d (Atl, Lab-b, Nfld-b, Nfld); 18. Threlfall and Hanek 1970c (Nfld-b); 19. Threlfall and Hanek 1971 (Lab); 20. Scott 1973 (Atl); 21. Hicks and Threlfall 1973 (Lab); 22. Hanek and Molnar 1974 (Que); 23. Dickinson and Threlfall 1976 (Nfld); 24. Mudry and McCart 1976 (YT); 25. Gaevskaya and Umnova 1977 (Atl); 26. Anonymous 1978 (BC); 27. Dick and Belesovic 1978 (NWT); 28. Thompson and Threlfall 1978 (Que); 29. Bray 1979 (Atl); 30. Umnova 1979 (Atl); 31. Pippy 1980 (Atl,NB,NS,Que,PEI); 32. Arthur and Arai 1980a (Pac); 33. Arthur and Arai 1980b (Pac); 34. Zubchenko 1980 (Atl); 35. Dick and Belosevic 1981 (NWT); 36. Black 1981 (Que); 37. Anonymous 1981 (BC); 38. Threlfall 1981 (Nfld); 39. Appy and Burt 1982 (Atl); 40. Margolis 1982 (Nfld); 41. Arai and Mudry 1983 (BC); 42. Anonymous 1984 (BC); 43. Dick 1984 (NWT,EArc); 44. Kabata and Whitaker 1984 (Pac); 45. Stewart and Bernier 1984 (NWT,EArc); 46. Bourgeois and Ni 1984 (Atl); 47. McGladdery and Burt 1985 (Atl); 48. Bond and Ericson 1985 (NWT); 49. Scott 1985a (Atl); 50. Gibson and Bray 1986 (Atl); 51. Frimeth 1987a (NB,NB-b); 52. Frimeth 1987b (NB-b,Atl); 53. Kabata *et al.* 1988 (Pac); 54. Mortensen and Mothershead 1988 (Pac); 55. Scott and Bray 1989 (Atl); 56. Bouillon and Dempson 1989 (Atl); 57. Holmes 1990 (Pac); 58. Landry *et al.* 1992 (NB-b); 59. Hogans *et al.* 1993 (Atl); 60. Arthur and Albert 1994 (Atl.).

Remarks: This species has been redescribed in detail by Gibson and Bray (1986). It is essentially a marine parasite, but its range may extend deep into fresh water as it is common in migratory fishes, such as salmonoids. It is an arctic-boreal species that occurs in both the Atlantic and Pacific oceans, its range extending further south on the colder western sides of these oceans (Gibson 1983), but the lack of records in Canadian and Russian arctic waters suggests that it may not have a circumpolar distribution (Gibson and Bray 1986).

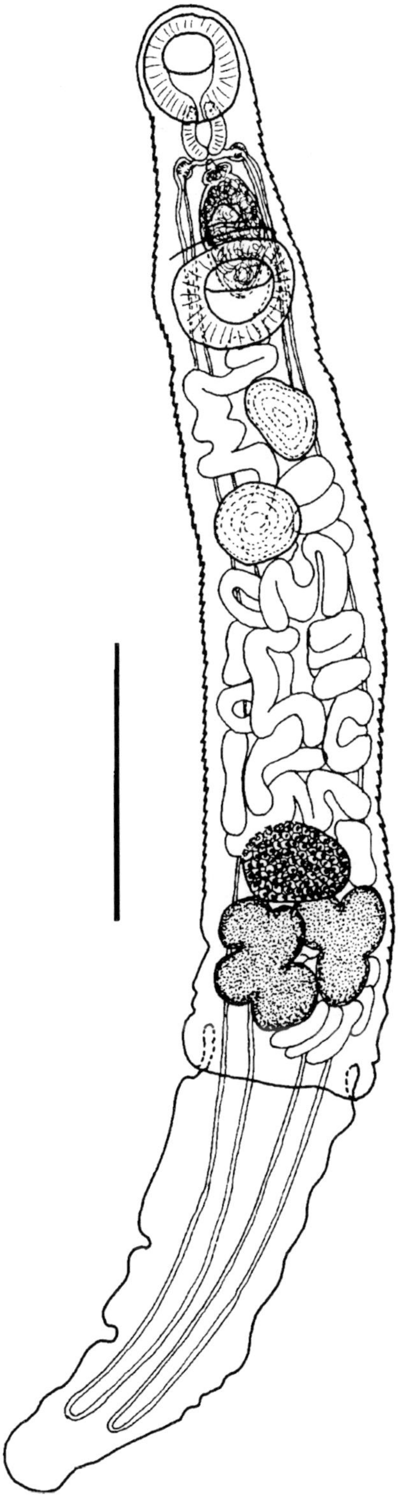

FIG. 45. *Brachyphallus crenatus* (after Gibson and Bray (1986)). Scale bar: 500 µm.

DISSOSACCUS Manter, 1947

Lecithochiriinae. Ecsoma well developed. Body surface smooth. Seminal vesicle in two parts separated by narrow duct, one part normally anterior or dorsal and other mainly posterior to ventral sucker. Pars prostatica (?) tubular. Vitellarium two slightly indented masses. Parasitic in stomach of marine teleosts.

Dissosaccus laevis (Linton, 1898) Manter, 1947 (Fig. 46)

Site: ? (undoubtedly stomach).

Hosts: *Reinhardtius hippoglossoides* (2); *Urophycis chesteri* (1); *U. chuss* (1); *U. musicki* (1).

Distribution: Atl.

Records: 1. Scott 1987; 2. Scott and Bray 1989.

Remarks: This species has been found in deeper waters of the Scotian shelf. Single specimens from *U. chesteri* and *R. hippoglossoides*, collected by Dr.J.S. Scott, are in the BM(NH) collection.

LECITHOCHIRIUM Lühe, 1901

Syn.: *Sterrhurus* Looss, 1907.

Lecithochiriinae. Ecsoma well or poorly developed. Body surface smooth. Pre-oral lobe rarely with two lateral knobs ("horns"). Presomatic pit and/or ventro-cervical groove often present. Seminal vesicle bipartite, (apparently) tripartite, or occasionally coiled, in forebody. Pars prostatica tubular, with wide lumen, to vesicular. Short, narrow extension of pars prostatica and/or ejaculatory duct may be present within sinus sac. Ejaculatory (prostatic) vesicle linked posteriorly to antero-dorsally with pars prostatica or ejaculatory duct. Temporary sinus organ may form. Vitellarium two lateral masses, usually divided into three and four oval to digitiform lobes. Parasitic in gut (mainly stomach) or body cavity of marine teleosts; also recorded from hepatic ducts and gills of marine teleosts and (?) gut of freshwater reptiles; may also occur as encysted juveniles in fishes.

Lecithochirium exodicum McFarlane, 1936 (Fig. 47)

Site: Stomach.

Hosts: *Anoplopoma fimbria* (4); *Eopsetta jordani* (2); *Ophiodon elongatus* (1); *Sebastes caurinus* (3, 8); *S. elongatus* (3); *S. flavidus* (3, 6, 7); *S. nebulosus* (5); *S. paucispinus* (3).

Distribution: Pac.

Records: 1. McFarlane 1936; 2. Ronald 1959; 3. Sekerak and Arai 1977; 4. Kabata *et al.* 1988; 5. Holmes 1990; 6. Lee *et al.* 1990; 7. Stanley *et al.* 1992; 8. BH(NH) collection.

Remarks: There are also several records of this species from these hosts and from *Platichthys stellatus* in Washington State and Californian waters to the south (e.g., Lloyd 1938; Ching 1960). As listed by Margolis and Arthur (1979), this species has during its taxonomic history been transferred to *Sterrhurus* (by Yamaguti 1958) and *Separogermiductus* (by Manter and Pritchard 1960), now synonyms of *Lecithochirium* (see Gibson and Bray 1979), and has also been erroneously associated with the plerurine genus *Adinosoma* (by Skryabin and Gushanskaya 1955) and erroneously synonymized with *Synaptobothrium caudiporum* (Rud.) (by Nasir and Diaz 1971).

Subfamily LETHADENINAE Yamaguti, 1971

Hemiuridae. Ecsoma well developed. Body surface smooth. Pre-somatic pit absent. Testes oblique. Seminal vesicle oval, thick-walled, in forebody. Pars prostatica vesicular, with muscular wall, external gland cells absent or weakly developed, separated from seminal vesicle by aglandular duct and from sinus sac by long ejaculatory duct. Sinus sac small, not containing ejaculatory (prostatic) vesicle. Sinus organ (?) present (?temporary), small. Genital atrium short. Vitellarium two symmetrical, unlobed, oval masses. Excretory arms not united in forebody. Parasitic in stomach of marine teleosts.

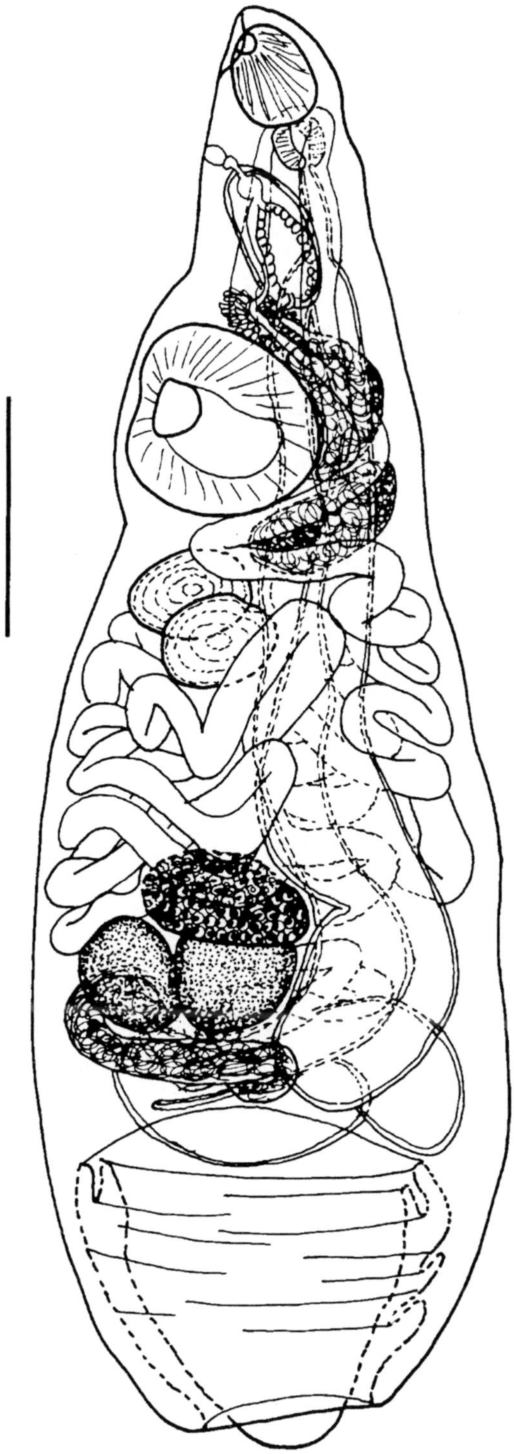

FIG. 46. *Dissosaccus laevis* (after Yamaguti (1971)). Scale bar (roughly estimated): 250 μm.

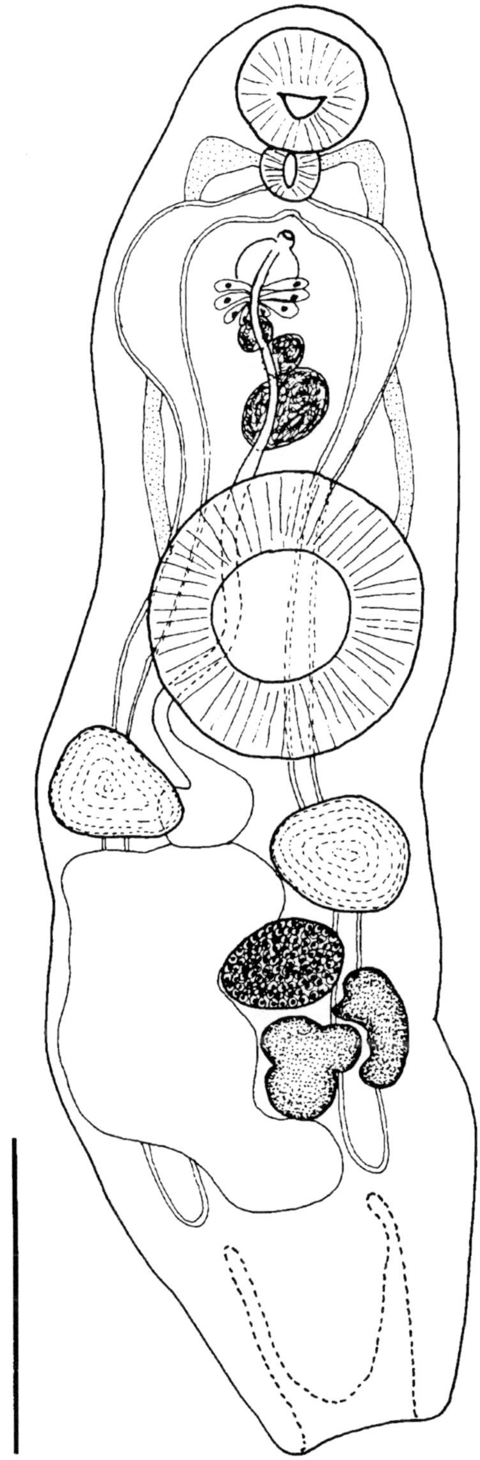

FIG. 47. *Lecithochirium exodicum* (after Lloyd (1938)). Scale bar: 500 μm. More details of the terminal genitalia are figured in Lloyd (1938) and, for related species, in Gibson and Bray (1986)).

LETHADENA Manter, 1947

Lethadeninae. With the characters of the subfamily.

Lethadena profunda (Manter, 1934) Manter, 1947 (Fig. 48)

Site: ? (probably stomach).
Host: *Urophycis chesteri*.
Distribution: Atl.
Record: Scott 1987.
Remarks: This record refers to one immature specimen collected by Dr. J.S. Scott on the Scotian Shelf and now residing in the BM(NH) collection. The specimen is allocated to *L. profunda* because the genus appears to be monotypic and because of its morphological similarity to the original description. This species appears to be relatively rare and occurs in deeper waters off the eastern coast of North America.

Subfamily PLERURINAE Gibson and Bray, 1979

Hemiuridae. Body small, spindle-shaped to cylindrical. Ecsoma reduced or well developed. Body surface smooth, or occasionally with crenulate plications giving a "scaly" appearance. Presomatic pit absent, except in *Synaptobothrium*. Testes symmetrical to tandem, usually oblique. Seminal vesicle elongate, saccular, and constricted into two, three, or four sections, thin-walled, although certain sections may have thicker walls, in forebody to partly in hindbody. Pars prostatica vesicular or tubular, may be partly enclosed by muscles of sinus sac; commonly linked to seminal vesicle by aglandular duct. Sinus sac apparently absent or poorly developed; when present usually of "open" type. Permanent sinus organ absent. Ejaculatory (prostatic) atrium absent. Hermaphroditic duct commonly vesicular proximally and tubular distally. Genital atrium usually deep, but may be shallow or apparently absent. Ovary entire or lobed. Vitellarium in forebody. Parasitic in stomach of marine teleosts.

DINOSOMA Manter, 1934

Plerurinae. Body surface with crenulate plications, giving "scaly" appearance. Testes symmetrical to tandem. Seminal vesicle postero-dorsal to ventral sucker, saccular, bipartite, or wide, sinuous. Pars prostatica vesicular, may be connected to seminal vesicle by long, aglandular duct. Sinus sac apparently absent. Hermaphroditic duct long, narrow, with small vesicle proximally. Ovary oval. Vitellarium two indented or lobed masses.

Dinosoma triangulata Campbell and Munroe, 1977 (Fig. 49)

Site: ? (Probably stomach).
Hosts: *Alepocephalus agassizii* (1); *A. bairdii* (1); *Reinhardtius hippoglossoides* (2).
Distribution: Atl.
Records: 1. Zubchenko 1984; 2. Scott and Bray 1989.
Remarks: This species occurs in deeper waters off the eastern coast of North America (although Gaevskaya and Aleshkina (1983) claim to have found it off the Atlantic coast of Africa). Zubchenko's records are listed as Northwest Atlantic waters, although, judging by his other works, most of his collecting was done off the Canadian coast. The specimen in the BM(NH) collections was collected by Dr.J.S. Scott on the Scotian Shelf. The species was originally described from *Alepocephalus agassizii* and *Antimora rostrata* from the Hudson Submarine Canyon by Campbell and Munroe (1977).

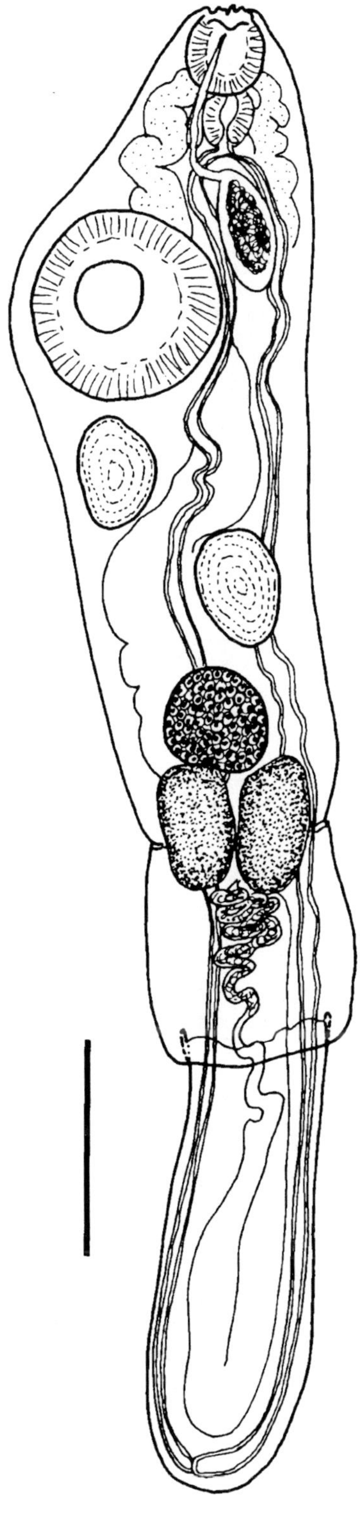

FIG. 48. *Lethadena profunda* (after Manter (1934)). Scale bar: 500 μm.

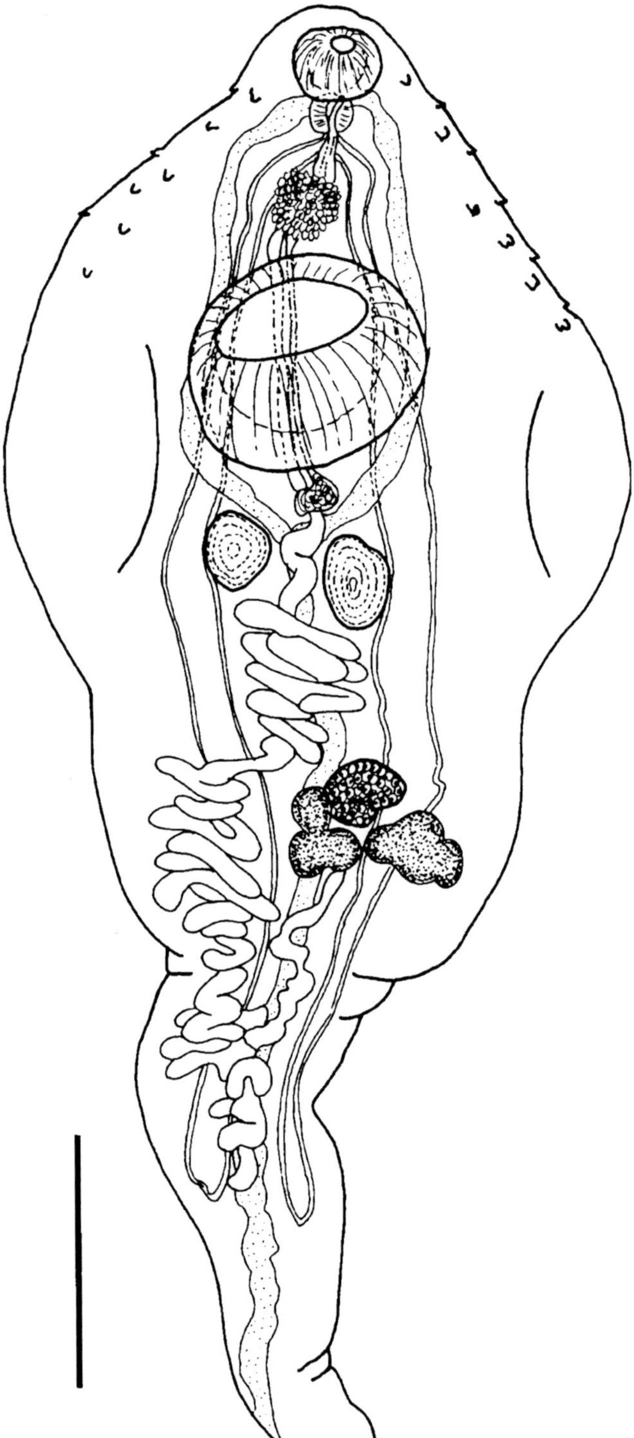

FIG. 49. *Dinosoma triangulata* (after Campbell and Munroe (1977)). Scale bar: 1 mm.

Family LECITHASTERIDAE Odhner, 1905

Hemiuroidea. Body usually small, normally spindle-shaped, occasionally elongate. Ecsoma absent. Body surface smooth. Oral and ventral suckers well developed, ventral sucker normally in anterior half of body. Muscular flange or flanges may be present immediately posterior to ventral sucker. Pharynx well developed. Oesophagus usually short. "Drüsenmagen" normally present. Caeca usually terminate blindly, but occasionally unite forming cyclocoel. Presomatic pit and ventro-cervical groove absent. Testes two, occasionally one, in tandem, oblique, or symmetrical, usually, but not always, pre-ovarian, in hindbody. Seminal vesicle generally thin-walled, occasionally muscular, oval, tubular or constricted into portions, in forebody or hindbody. Pars prostatica usually tubular, occasionally vesicular, may be linked to seminal vesicle by aglandular tube. Ejaculatory duct long, short or absent. Hermaphroditic duct present. Ejaculatory (prostatic) vesicle absent. Sinus sac usually present, well or poorly developed, occasionally absent. Permanent sinus organ normally absent, but hermaphroditic duct often protruded to form temporary sinus organ. Genital atrium large, small, or absent. Ovary usually post-testicular, oval or four-lobed (occasionally three-lobed). Usually only blind seminal receptacle present (normally large, thick-walled, and situated dorsal or antero-dorsal to ovary) and Laurer's canal, Juel's organ and both uterine or canalicular seminal receptacles absent; occasionally only Juel's organ and uterine seminal receptacle present, rarely only Laurer's canal and canalicular seminal receptacle present. Uterus mainly post- to entirely pre-ovarian; main bulk rarely extends into forebody. Eggs numerous, small, rarely filamented. Vitellarium commonly seven-lobed, occasionally six, eight, or double these numbers (sometimes branched) lobes, often in rosette arrangement, usually immediately post-ovarian, occasionally pre-ovarian or at level of ovary. Excretory vesicle Y-shaped, arms united in forebody or not. Parasitic in gut, especially intestine, of marine teleosts.

Subfamily LECITHASTERINAE Odhner, 1905

Lecithasteridae. Caeca terminate blindly. Testes two, occasionally one, pre-ovarian. Seminal vesicle in forebody, dorsal to ventral sucker or, occasionally, in anterior hindbody. Pars prostatica short to medium in length. Ejaculatory duct absent or short. Sinus sac well developed. Permanent sinus organ absent. Ovary entire or four-lobed. Blind seminal receptacle normally large, usually dorsal to ovary. Uterus reaches to post-ovarian region. Vitellarium seven (rarely six or eight) oval to digitiform lobes in rosette or two linked groups of three and four, immediately post-ovarian, occasionally antero-posteriorly oriented. Excretory arms united in forebody or not. Normally parasitic in intestine or stomach of marine teleosts.

Key to the genera of the Lecithasterinae

1 Ovary lobed, usually with four lobes; vitelline lobes tear-shaped .... *Lecithaster*

Ovary oval to globular; vitelline lobes globular .... *Lecithophyllum*

LECITHASTER Lühe, 1905

Lecithasterinae. Testes two, obliquely symmetrical, usually oval, but occasionally lobed. Seminal vesicle saccular to elongate and sinuous, in forebody, dorsal to ventral sucker or, occasionally, in anterior hindbody. Sinus sac oval. Genital atrium short. Ovary normally four-lobed. Seminal receptacle large, globular, dorsal to ovary. Vitellarium a radiating mass of seven tear-shaped lobes. Excretory arms apparently not united in forebody. Parasitic in intestine of marine teleosts.

Key to the species of *Lecithaster*

1 Vitelline lobes short, little longer than wide; eggs usually <19 µm in length .... *Lecithaster confusus* Odhner, 1905

Site: ?Intestine (listed as stomach).

Hosts: *Alosa aestivalis* (3); *A. pseudoharengus* (3); *A. sapidissima* (4); *Clupea harengus* (2); *Osmerus mordax* (1).

Distribution: Atl, NB-b, NS.

Records: 1. Heller 1949 (Atl); 2. McGladdery and Burt 1985 (Atl); 3. Landry *et al.* 1992 (NB-b); 4. Hogans *et al.* 1993 (Atl,NS).

Remarks: This species is rare in Canadian Atlantic waters. It normally occurs in warmer waters to the south. Zubchenko (1980: Atl) and Reimer (1981: EArc) claimed to have found this species in *Hippoglossus hippoglossus* and *Reinhardtius hippoglossoides*, respectively, in northern waters, but I consider it much more likely that they were dealing with *L. gibbosus*.

Vitelline lobes distinctly longer than wide; uncollapsed eggs normally >19 µm in length .... *L. gibbosus* (Rudolphi, 1802) Lühe, 1901 (Fig. 50)

Syn.: *Lecithaster salmonis* Yamaguti, 1934; *L. bothryophorus* of Stafford (1904, 1907).

Sites: Intestine, pyloric caeca (?stomach).

Hosts: *Ammodytes dubius* (24); *A. hexapterus* (9, 10, 11, 12, 14, 15, 16); *Anoplarchus purpurescens* (14); *Anoplopoma fimbria* (49); *Aprodon cortezianus* (9, 11); *Aulorhynchus flavidus* (9, 11); *Blepsias cirrhosus* (9, 11); *Boreogadus saida* (34); *Brachyistius frenatus* (9, 11); *Clupea harengus* (1, 2, 54); *C. pallasi* (9, 10, 11, 12, 13, 14, 15, 37); *Coryphopterus nicholsi* (9, 11); *Cymatogaster aggregata* (9, 10, 11); *Dasycottus setiger* (11); *Gadus macrocephalus* (9, 11); *G. morhua* (32, 34, 42); *Gasterosteus aculeatus* (9, 10, 11, 14, 19, 20, 30); *Glyptocephalus cynoglossus* (28, 41); *Hemilepidotus hemilepidotus* (9, 11); *Hexagrammos decagrammus* (15); *H. lagocephalus* (9, 11); *H. stelleri* (9, 11); *Hippoglossoides platessoides* (25, 26, 34, 35, 36, 41, 53); *Hippoglossus hippoglossus* (36, 63); *Hypomesus pretiosus* (15); *Liparis pulchellus* (14); *Lumpenus sagitta* (14); *Lycodes reticulatus* (34); *L. vahli* (34); *Lyconectes aleutensis* (14); *Mallotus villosus* (47, 56); *Merluccius albidus* (60); *M. bilinearis* (60); *M. productus* (44); *Nautichthys oculofasciatus* (11); *Oligocottus maculosus* (11); *Oncorhynchus gorbuscha* (4, 5, 7, 8, 9, 11, 15, 17, 39, 43, 46); *O. keta* (9, 11, 17, 39, 46, 55); *O. kisutch* (9, 11); *O. nerka* (3, 4, 5, 6, 9, 11); *O. tshawytscha* (9, 11); *Oncorhynchus* sp. (10); *Ophiodon elongatus* (9, 11, 14); *Pholis ornata* (9, 11); *Platichthys stellatus* (9, 11, 14); *Pleuronectes americanus* (29, 41, 52); *P. bilineatus* (9, 11); *P. ferrugineus* (27, 33, 41); *P. vetulus* (49); *Pollachius virens* (51); *Reinhardtius hippoglossoides* (36, 62, 63, 66, 67, 69, 70); *Ronquilus jordani* (9, 11); *Salmo salar* (1, 2, 18, 21, 23, 38); *Salvelinus alpinus* (40, 48, 50, 64); *S. fontinalis* (58, 59); *S. malma* (3, 9, 11); *Sebastes alutus* (31); *S. caurinus* (31); *S. fasciatus* (45, 61); *S. flavidus* (68); *S. marinus* (45); *S. mentella* (45); *S. nebulosus* (65); *S. pinniger* (31); *S. proriger* (31); *S. variegatus* (31); *S. zacentrus* (31); *Syngnathus leptorhynchus* (9, 11); *Tautoglabratus adspersus* (22); *Thaleichthys pacificus* (9, 11, 12, 14); *Theragra chalcogramma* (9, 11, 14); *Triglops murrayi* (34); *Urophycis chuss* (57, 60); *U. musicki* (57, 60).

Distribution: Atl, Pac, BC-b, BC, Lab, NB, NB-b, Nfld, Nfld-b, NS, NWT.

Records: 1. Stafford 1904 (Atl); 2. Stafford 1907 (Atl); 3. Bangham and Adams 1954 (BC); 4. Margolis 1956b (Pac, BC-b); 5. Margolis 1957 (Pac, BC-b, BC); 6. Margolis 1963 (Pac, BC-b); 7. Boyce 1966 (Pac); 8. Boyce 1969 (Pac); 9. Arai 1967a (Pac); 10. Arai 1967b (Pac); 11. Arai 1969b; 12. Barraclough 1967 (Pac); 13. Barraclough and Fulton 1967 (Pac); 14. Robinson *et al.* 1968a (Pac); 15. Robinson *et al.* 1968b (Pac); 16. Robinson 1969 (Pac); 17. Margolis and Boyce 1969 (Pac); 18. Pippy 1969 (Atl,NB,NS); 19. Hanek and Threlfall 1969a (Nfld-b); 20. Hanek and Threlfall 1970d (Nfld-b); 21. Threlfall and Hanek 1970c (Nfld-b); 22. Sekhar and Threlfall 1970a (Atl); 23. Hicks and Threlfall 1973 (Lab); 24. Scott 1973 (Atl); 25. Scott 1975a (Atl); 26. Scott 1975b (Atl); 27. Scott 1975c (Atl); 28. Scott 1975d (Atl); 29. Scott 1976 (Atl); 30. Lester 1975 (Pac); 31. Sekerak and Arai 1977 (Pac); 32. Linkletter *et al.* 1977 (Atl); 33. Redkozubova 1978 (Atl); 34. Bray 1979 (Atl); 35. Umnova 1979 (Atl); 36. Zubchenko 1980 (Atl); 37. Arthur and Arai 1980a,b (BC); 38. Pippy 1980 (Atl,NB,NS); 39. Anonymous 1981 (BC); 40. Dick and Belosevic 1981 (NWT); 41. Scott 1982 (Atl); 42. Appy and Burt 1982 (Atl); 43. Margolis 1982 (Nfld);

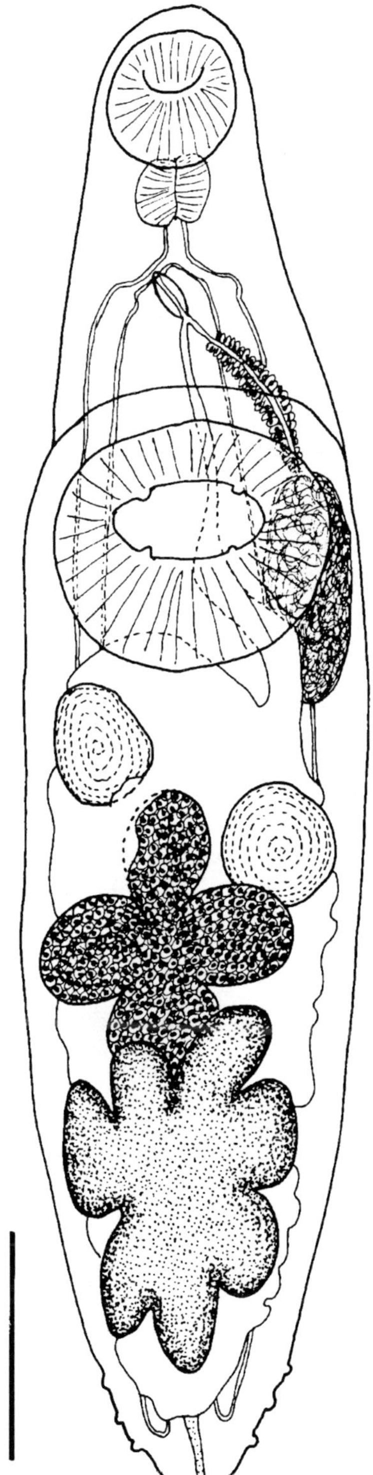

FIG. 50. *Lecithaster gibbosus* (after MacKenzie and Gibson (1970)). Scale bar: 250 µm.

44. Sankurathri *et al.* 1983 (Pac); 45. Bourgeois and Ni 1984 (Atl); 46. Anonymous 1984 (BC); 47. Pálsson and Beverley-Burton 1984 (Atl); 48. Dick 1984 (NWT); 49. Kabata and Whitaker 1984 (Pac); 50. Stewart and Bernier 1984 (NWT); 51. Scott 1985a (Atl); 52. Scott 1985b (Atl); 53. Zubchenko 1985a (Atl); 54. McGladdery and Burt 1985 (Atl); 55. Whitaker 1985 (Pac); 56. Pálsson 1986 (Atl); 57. Scott 1987 (Atl); 58. Frimeth 1987a (NB-b); 59. Frimeth 1987b (NB-b, Atl); 60. Scott 1987 (Atl); 61. Scott 1988 (Atl); 62. Wierzbicka 1988 (Atl); 63. Scott and Bray 1989 (Atl); 64. Bouillon and Dempson 1989 (Atl); 65. Holmes 1990 Pac); 66. Wierzbicka 1991a (Atl); 67. Wierzbicka 1991b (Atl); 68. Stanley *et al.* 1992 (Pac); 69. Krzykawski and Wierzbicka 1992 (Atl); 70. Arthur and Albert 1994 (Atl).

Remarks: This is a common species in arctic-boreal waters in both Atlantic and Pacific waters. Nevertheless, there is some question concerning the conspecificity of specimens from the two oceans, since Køie (1983) recovered the larval stages of *L. gibbosus* in Europe in quite a different host to that recorded by Boyce (1969) in British Columbian waters.

LECITHOPHYLLUM Odhner, 1905

Lecithasterinae. Testes two, obliquely tandem to symmetrical. Seminal vesicle saccular, in forebody or dorsal to ventral sucker. Hermaphroditic duct apparently bipartite. Sinus sac elongate. Genital atrium present, generally deep. Ovary oval or globular. Seminal vesicle large, dorsal to ovary. Vitellarium seven globular lobes, in lateral groups of three and four. Excretory arms united in forebody. Parasitic in stomach of marine teleosts.

Comments

This genus has recently been revised by Gaevskaya (1989).

Lecithophyllum botryophorum (Olsson, 1868) Odhner, 1905 (Fig. 51)

Syn.: *Aponurus* sp. of Margolis (1956b); *Lecithophyllum* sp. of Margolis (1957) and Arai (1967a); *L. anteroporum* Margolis, 1958b; *L. campbellmunroei* Gaevskaya, 1989.

Sites: Intestine, stomach.

Hosts: *Alepocephalus agassizii* (16); *A. bairdii* (16); *Argentina silus* (7, 8, 9,); *Dasycottus setiger* (5, 6); *Hexagrammos lagocephalus* (6); *H. stelleri* (6); *Hippoglossus hippoglossus* (21); *Macrourus berglax* (13); *Merluccius productus* (3, 14); *Oncorhynchus gorbuscha* (2, 3, 12, 15); *O. keta* (12, 15); *O. nerka* (1, 2, 3, 4); *Reinhardtius hippoglossoides* (22); *Sebastes aleutianus* (11); *S. alutus* (10, 11); *S. borealis* (11); *S. crameri* (10, 11); *S. fasciatus* (17, 20); *S. flavidus* (11); *S. marinus* (17); *S. mentella* (17); *S. paucispinis* (11); *S. reedi* (11); *Theragra chalcogramma* (18); *Urophycis chuss* (19); *U. musicki* (19).

Distribution: Atl, Pac, BC-b, BC.

Records: 1. Margolis, 1956b (BC-b); 2. Margolis, 1957 (Pac, BC-b, BC); 3. Margolis 1958b (Pac, BC-b, BC); 4. Margolis 1963 (BC-b); 5. Arai 1967a (Pac); 6. Arai 1969b (Pac); 7. Scott 1969a (Atl); 8. Scott 1969b (Atl); 9. Scott 1969c (Atl); 10. Sekerak and Arai 1973 (Pac); 11. Sekerak and Arai 1977 (Pac); 12. Anonymous 1981 (BC); 13. Zubchenko 1981a (Atl); 14. Sankurathri *et al.* 1983 (Pac); 15. Anonymous 1984 (BC); 16. Zubchenko 1984 (Atl); 17. Bourgeois and Ni 1984 (Atl); 18. Arthur 1984 (Pac); 19. Scott 1987 (Atl); 20. Scott 1988 (Atl); 21. Scott and Bray 1989 (Atl); 22. Arthur and Albert 1994 (Atl).

Remarks: This species is relatively common in deep-water fishes off the edge of the continental shelf in arctic-boreal waters on both sides of the Atlantic. In the Pacific it was described as *Lecithophyllum anteroporum* by Margolis (1958b), but this was synonymised with *L. botryophorum* by Scott (1969b). This synonymy was confirmed by Sekerak and Arai (1973) and accepted by both Margolis and Arthur (1979) and Gaevskaya (1989). Although the Pacific and Atlantic material does not appear to be distinguishable on morphological grounds, the fact that in Pacific waters the parasite occurs only in coastal and migratory fishes, whereas in the Atlantic it is only known from hosts in deeper, offshore waters, does pose a significant ecological question mark against this synonymy. I predict that future work will show that *L. anteroporum* is a valid species.

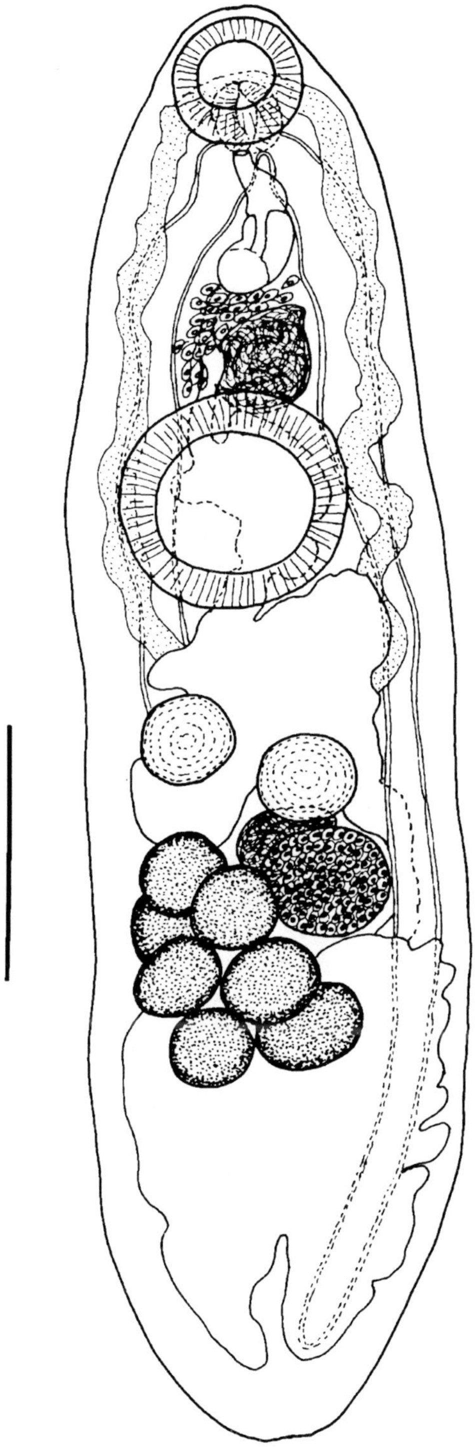

FIG. 51. *Lecithophyllum botryophorum* (after Margolis (1958b: as *L.anteroporum*)). Scale bar: 250 µm.

Gaevskaya (1989) was of the opinion that the worms described by Campbell and Munroe (1977) as *L. anteroporum* from *Alepocephalus agassizii* in deep water off the American Northwest Atlantic coast were a distinct species, which she called *L. campbellmunroei*. Zubchenko's (1984) records of *L. anteroporum* from *Alepocephalus* spp. in Canadian Atlantic waters are presumably conspecific with Campbell and Munroe's material. Gaevskaya distinguished the new form from Margolis' original description of *L. anteroporum* on the basis of the size of its internal organs and the position and shape of the seminal vesicle. After a brief examination of the large collection of material in the BM(NH) collection, including material from *Alepocephalus* spp. from the Northeast Atlantic, I can find no obvious reason for distinguishing this form from *L. botryophorum*.
It is possible that the specimen briefly described as *Leptosoma obscurum* by Stafford (1904), from *Lophius americanus* (as *piscatorius*) off eastern Canada, may belong to this species.

Family HIRUDINELLIDAE Dollfus, 1932

Syn.: Botulidae Guiart, 1938; Lampritrematidae Yamaguti, 1940.

Hemiuroidea. Body large, stout or elongate, contractile. Ecsoma absent. Body surface smooth, may be papillate in forebody or wrinkled. Oral and ventral suckers well developed; latter in anterior half of body. Pharynx well developed. Oesophagus short. "Drüsenmagen" present. Caeca terminate blindly or form uroproct, sometimes fuse subterminally forming cyclocoel, usually diverticulate. Testes two, in tandem, oblique, or symmetrical, pre-ovarian, in hindbody. Seminal vesicle tubular, normally thin-walled, occasionally partly thick-walled, convoluted in forebody. Pars prostatica well developed, tubular, usually long. Ejaculatory duct long, muscular, surrounded by muscular "cirrus sac", opens into genital atrium through well-developed conical to cylindrical "cirrus". Hermaphroditic duct, sinus sac, and sinus organ absent. Genital atrium large, usually capable of being everted. Genital pore mid-ventral in forebody. Ovary oval, post-testicular, in middle or anterior half of hindbody. Laurer's canal and uterine seminal receptacle normally present. Juel's organ and blind or canalicular seminal receptacle absent. Uterus descending ventrally and ascending more dorsally, coiled mainly at level of vitellarium, but often extending more anteriorly in pre-ovarian region, mainly inter-caecal, occasionally reaching extra-caecally; metraterm opens into genital atrium directly or through small papilla-like organ situated immediately posterior to "cirrus". Eggs numerous, small, without filaments. Vitellarium composed of from two to numerous long, straight, or convoluted, branched tubules, mainly post-ovarian, inter- or extra-caecal. Excretory vesicle Y-shaped, arms initially dorso-ventrally oriented, convoluted, united in forebody. Parasitic in stomach (occasionally on gills) of large, carnivorous, marine teleosts.

Comments

Both species of this family that occur in Canadian waters were described in detail by Gibson and Bray (1977) from the Northeast Atlantic.

The "cirrus" and "cirrus sac" are included in quotation marks as they are not considered homologous to these structures in other digenean groups (see Gibson and Bray 1979).

Key to the genera of the family Hirudinellidae

1 Body stout, elongate, or keyhole-shaped; uroproct present; "cirrus sac" small, globular; seminal vesicle entirely thin-walled .... *Hirudinella*

Body slender, elongate; uroproct absent; "cirrus sac" large, elongate; distal part of seminal vesicle with muscular wall .... *Lampritrema*

HIRUDINELLA de Blainville, 1828

Syn.: *Hirudinella* Garcin, 1730; *Uroproctinella* Skryabin and Gushanskaya, 1957.

Hirudinellidae. Body stout, elongate, or keyhole-shaped. Body surface often transversely wrinkled. Uroproct present; caeca may fuse subterminally in older specimens forming cyclocoel. Testes symmetrical to oblique, in anterior hindbody. Seminal vesicle thin-walled throughout its length. "Cirrus sac" relatively small, globular. "Cirrus" cone-shaped to digitiform. Genital atrium capable of being everted through genital pore. Ovary in anterior hindbody. Uterus mainly inter-caecal, coils extending posteriorly from ovary to near posterior limit of vitellarium. Vitellarium in two lateral fields between testes and middle of hindbody. Parasitic in stomach of large, carnivorous, marine teleosts (usually scombroids).

Hirudinella ventricosa (Pallas, 1774) Baird, 1853 (Fig. 52)

Syn.: *Hirudinella marina* Garcin, 1730; *H. clavata* (Menzies, 1791); *H. beebei* Chandler, 1937 (For complete list, see Gibson and Bray 1977).
Site: Stomach.
Hosts: *Thunnus thynnus* (1); *Xiphias gladius* (2, 3).
Distribution: Atl.
Records: 1. Cooper 1915; 2. Iles 1971; 3. Hogans *et al.* 1983.

LAMPRITREMA Yamaguti, 1940

Syn.: *Hirudinelloides* Gaevskaya and Kovaleva, 1977.

Hirudinellidae. Body elongate, slender. Papillae present on forebody. Uroproct absent. Most of seminal vesicle thin-walled, but distal portion forms thick-walled, muscular "pars musculosa". Thick-walled pars prostatica lies ventral to posterior portion of "cirrus sac". Male duct leads into "cirrus sac" some distance from its posterior extremity. "Cirrus sac" large, elongate, club-shaped. "Cirrus" long or short, capable of being extruded some distance through genital pore. Genital atrium deep. Ovary near middle of hindbody. Laurer's canal (?) absent. Seminal receptacle (?) small, enclosed in Mehlis' gland. Uterus extends back to near posterior limit of vitellarium, mainly coiled inter-caecally in post- and pre-ovarian region of hindbody. Vitellarium a pair of lateral tubules with short dorsal branches, mainly extra-caecal, passing posteriorly from ovary to about half-way to posterior extremity. Excretory arms appear to unite in forebody. Parasitic in stomach (? occasionally on gills) of marine teleosts (*Lampris*, *Brama*, and *Thyrsites*). Immature forms recorded from salmonoids (stomach, oesophagus, or gills).

Lampritrema miescheri (Zschokke, 1890) Margolis, 1962 (Fig. 53)

Syn.: *Lampritrema nipponicum* Yamaguti, 1940; *L. atlanticum* Delyamure and Serdyukov, 1970; *L. hawaiiense* Yamaguti, 1970; *Hirudinelloides elongatus* Gaevskaya and Kovaleva, 1977.
Site: Stomach (gills).
Hosts: *Argentina silus* (3, 4); *Brama japonica* (1); *Oncorhynchus tshawytscha* (5, 7); *Salmo salar* (2, 6).
Distribution: Atl, Pac.
Records: 1. Margolis 1962 (Pac); 2. Pippy 1969 (Atl); 3. Scott 1969a (Atl); 4. Scott 1969c (Atl); 5. Anonymous 1978 (BC); 6. Pippy 1980 (Atl); 7. Arai and Mudry 1983 (BC).
Remarks: The normal host of this species is *Lampris guttatus*, although it also occurs in *Brama* spp. in the northern hemisphere. Immature forms are found in salmonoids in which it appears unable to mature (Gibson and Bray 1977).

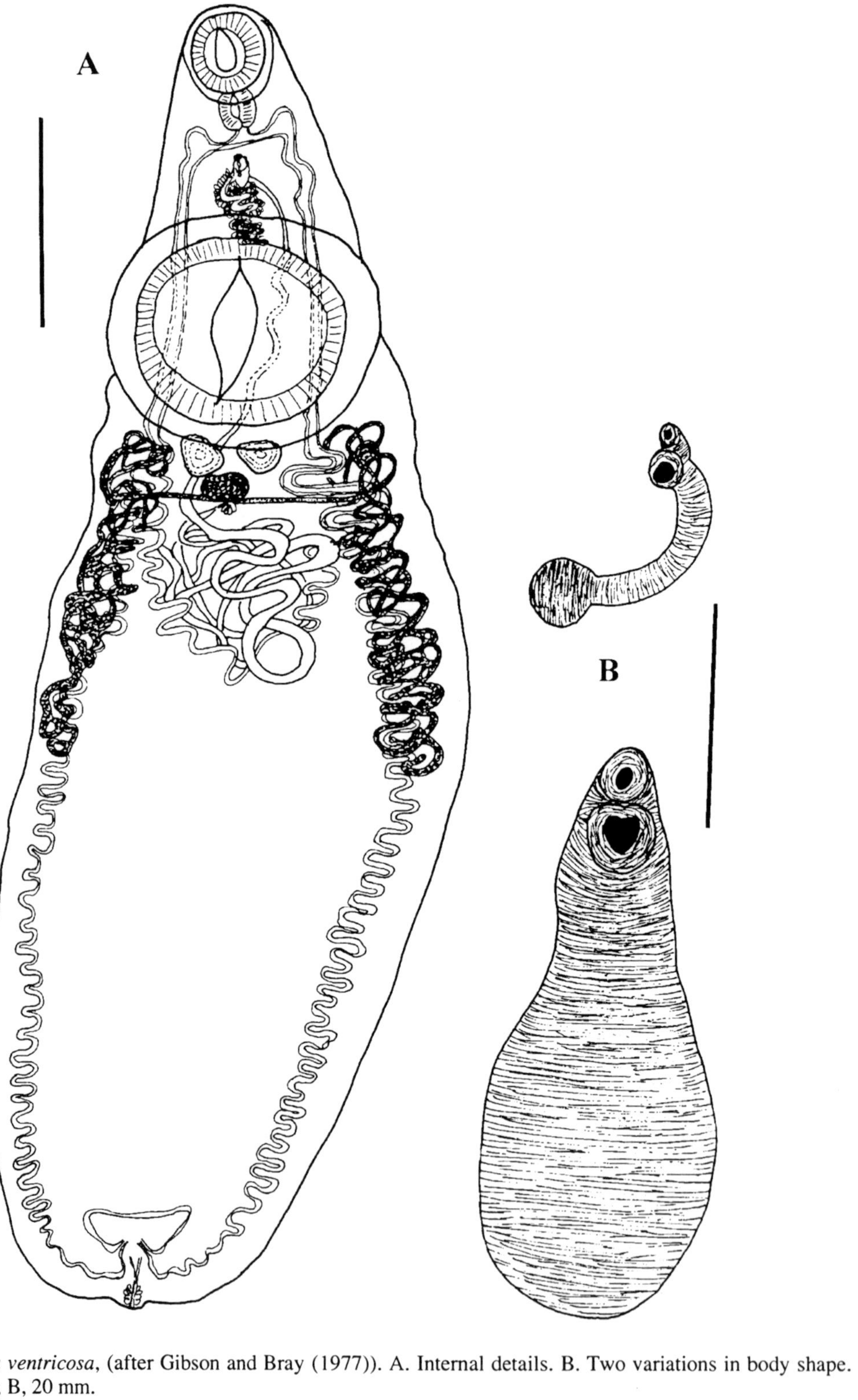

FIG. 52. *Hirudinella ventricosa*, (after Gibson and Bray (1977)). A. Internal details. B. Two variations in body shape. Scale bars: A, 5mm; B, 20 mm.

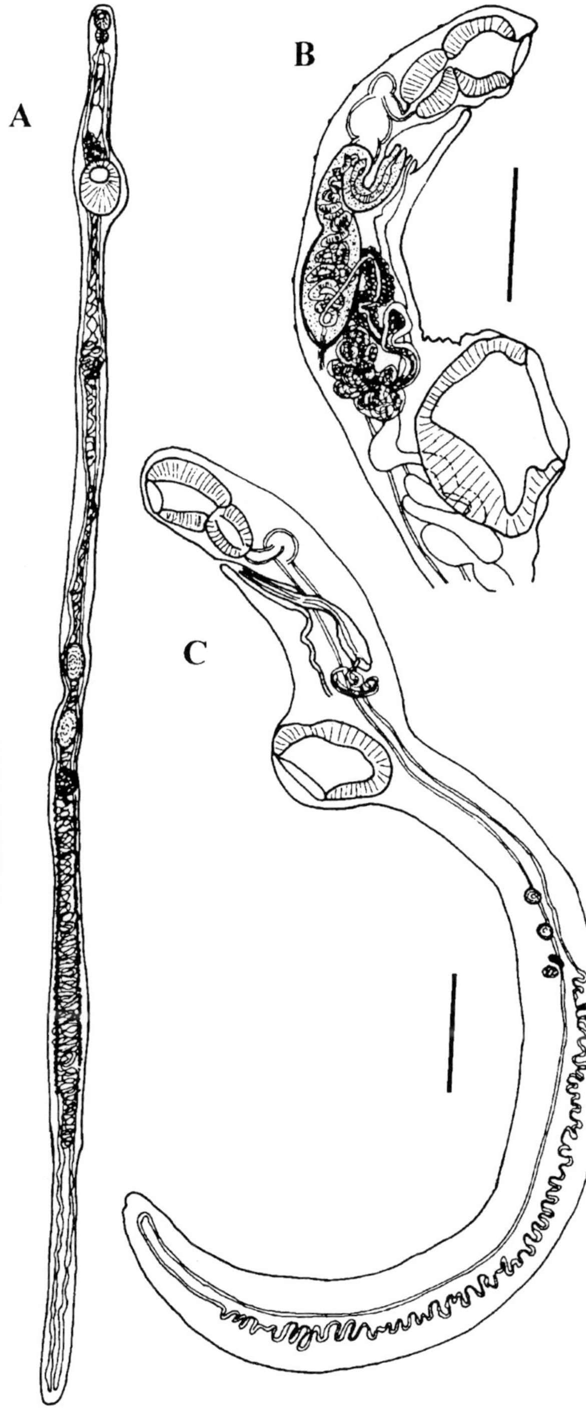

FIG. 53. *Lampitrema meischeri*. A. Adult (after Yamaguti (1940)). B. Sagittal section of adult forebody (after Yamaguti (1940)). C. Immature specimen from salmonid (after Margolis (1962)). Scale bars: A, 5 mm; B, 1 mm; C, 500 μm.

Family PTYCHOGONIMIDAE Dollfus, 1937

Hemiuroidea. Body medium-sized, oval. Ecsoma absent. Body surface smooth, without spines or plications. Suckers well developed, oral larger than ventral, latter in anterior half of body. Pharynx well developed. Oesophagus short. "Drüsenmagen" absent. Gut caeca form uroproct. Testes two, post-ovarian, tandem, in middle of hindbody. Seminal vesicle dilate, tubular, thin-walled, extending posteriorly into anterior hindbody. Par prostatica tubular. Ejaculatory and hermaphroditic duct short. Permanent sinus organ a small cone. Sinus sac absent. Genital atrium contains three distinct concentric folds in its wall that surround sinus organ. Genital pore mid-ventral in forebody. Ovary oval, pre-testicular in hindbody. Mehlis' gland pre-ovarian. Laurer's canal and uterine seminal receptacle present. Juel's organ and canalicular or blind seminal receptacle absent. Uterine field extends between level posterior to testes and ventral sucker. Eggs numerous, without filaments. Vitellarium follicular, occurs in lateral fields extending throughout most of hindbody. Excretory vesicle Y-shaped, arms unite twice in forebody. Parasitic in stomach of elasmobranchs (normally carchariniform sharks).

PTYCHOGONIMUS Lühe, 1900

Ptychogonimidae. With characters of family.

Comments

Ptychogonimus fontanus Lyster, 1939 from *Perca flavescens* and *Salvelinus fontinalis* in freshwater is probably a synonym of *Azygia angusticauda* (q.v.).

Ptychogonimus megastoma (Rud., 1819) Lühe, 1900 (Fig. 54)

Site: (Intestine) (normally stomach).
Host: *Tautoglabrus adspersus.*
Distribution: Atl.
Record: Sekhar and Threlfall 1970a.
Remarks: This species was described in detail by Gibson and Bray (1977) from the Northeast Atlantic. Adults occur in the stomach of carchariniform sharks. Immature specimens are occasionally acquired by teleosts that prey upon crab intermediate hosts: these hosts (which include the Canadian record) are accidental or possibly paratenic (Gibson and Bray 1977).

Family SYNCOELIIDAE Looss, 1899

Hemiuroidea. Body elongate or dorso-ventrally flattened, usually with pedunculate ventral sucker. Ecsoma absent. Body surface smooth, but commonly papillate on forebody and on peduncle, if present. Oral and ventral suckers well developed, may possess accessory suckers around their rim. Glandular cells common in subtegumentary parenchyma and within musculature of suckers. Pharynx well developed. Oesophagus short. "Drüsenmagen" absent. Cyclocoel usually present, but caeca may end blindly or (?) form uroproct. Testes eleven to eighteen distinct, oval masses (usually arranged in pairs), seven to eight transverse rows of small follicles, or just a large number of irregular follicles, pre-ovarian, in hindbody. Seminal vesicle thin-walled, tubular, winding or sinuous, in forebody. Pars prostatica tubular. Ejaculatory duct short. Hermaphroditic duct and genital atrium present, but indistinguishable when sinus organ is absent. Permanent sinus organ and sinus sac present or absent. Genital pore mid-ventral in anterior forebody. Ovary post-testicular, composed of five large, oval, isolated lobes or numerous irregular follicles. Laurer's canal and uterine seminal receptacle present. Juel's organ and canalicular or blind seminal receptacle absent, but rudimentary seminal receptacle may be present. Uterus passes posteriorly but coils mainly in pre-ovarian hindbody. Eggs numerous, small, non-filamented. Vitellarium usually seven (occasionally five or six) isolated, oval

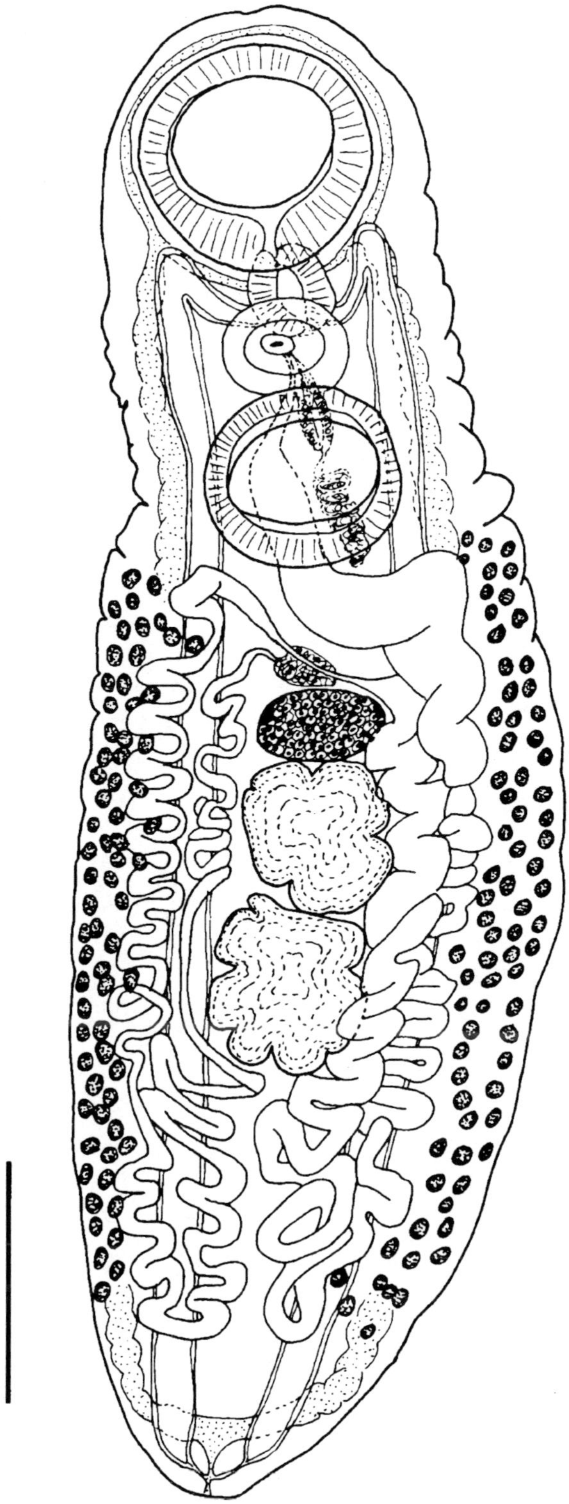

FIG. 54. *Ptychogonimus megastoma* (after Gibson and Bray (1977)). Scale bar: 1 mm.

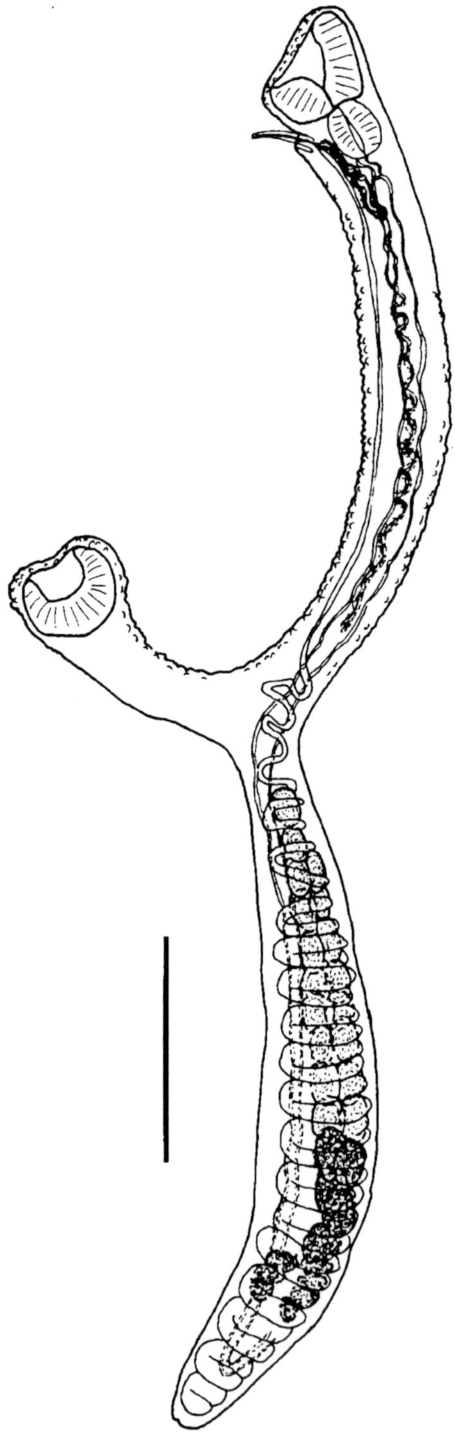

FIG. 55. *Copiatestes filiferus* (after Gibson (1976)). Scale bar: 1 mm.

lobes, or irregular, acinous bunches or rows of follicles, post-ovarian. Excretory vesicle Y-shaped, arms united in forebody, may initially run in dorsal and ventral fields. Free floating metacercarial stage present. Parasitic in branchial and buccal cavities, on skin, in (?) oviduct, and in (?) gut of elasmobranchs and marine teleosts.

Subfamily SYNCOELIINAE Looss, 1899

Syncoeliidae. Small accessory suckers around rim of suckers absent. Peduncle usually present. Cyclocoel present. Testes eleven to eighteen distinct, oval masses, usually arranged in pairs. Permanent sinus organ present or absent. Hermaphroditic duct and genital atrium present, but indistinguishable when sinus organ is absent. Sinus sac absent or rudimentary. Ovary composed of five large, isolated lobes. Rudimentary seminal receptacle may be present as proximal dilation of Laurer's canal. Uterus arranged in large, regular loops dorsally and ventrally to gonads. Vitellarium seven (occasionally five or six) small, isolated, oval lobes. Parasitic in branchial and buccal cavities (? with occasional records from gut) of sharks and marine teleosts.

COPIATESTES Crowcroft, 1948

Syncoeliinae. Hindbody elongate, tubular. Ventral sucker surmounted on well-developed peduncle. Permanent sinus organ present. Hermaphroditic duct and genital atrium easily distinguishable. Parasitic in branchial (especially gill arches and gill rakers) and buccal cavities (? and intestine) of marine teleosts.

Copiatestes filiferus (Leuckart, in Sars, 1885) Gibson and Bray, 1977 (Fig. 55)

Syn.: *Syncoelium filiferum* (Leuckart, in Sars, 1885) Odhner, 1911; *S. katuwo* Yamaguti, 1938; *S. priacanthi* Byrd, 1962.

Sites: Gills, branchial cavity.

Hosts: *Oncorhynchus gorbuscha* (1); *O. nerka* (1, 2); *Sebastes alutus* (3); *S. brevispinus* (3); *S. flavidus* (3, 6); *S. pinniger* (3); *S. proriger* (3); *Theragra chalcogramma* (4, 5).

Distribution: Pac, BC-b.

Records: 1. Margolis 1956b (Pac, BC-b); 2. Margolis 1963 (Pac, BC-b); 3. Sekerak and Arai 1977 (Pac); 4. Arthur 1984 (Pac); 5. Kabata and Whitaker 1984 (Pac); 6. Stanley *et al.* 1992 (Pac).

Remarks: This species was described in detail by Gibson and Bray (1977) from the Northeast Atlantic.

SUPERFAMILY SCHISTOSOMATOIDEA STILES AND HASSALL, 1898

Strigeida. Monoecious or dioecious worms. Body usually elongate, cylindrical or dorso-ventrally flattened; body of male may be widened to form gynaecophoric canal. Tegument armed or not. Oral sucker present or absent. Ventral sucker present or absent. Mouth terminal or subterminal. Pharynx absent. Oesophagus short to long, often bipartite. Intestine blind; X-, H-, or inverted U-shaped, in latter case caeca may be united over posterior part of length. Testes one to numerous, posterior to intestinal bifurcation. Male and female genital pores may open together or separately. Male genital pore variable in position, but posterior to intestinal bifurcation, ventral or dorsal. Ovary variable in shape and position, posterior to intestinal bifurcation. Laurer's canal present or absent. Seminal receptacle uterine, (?) canalicular, oviductal, or apparently absent. Uterus long or short, pre- or post-ovarian. Eggs few to numerous, non-operculate, with thin or thick shell, may be armed with single spine. Female genital pore posterior or anterior to intestinal bifurcation, dorsal or ventral, may be close to male pore. Vitellarium follicular, normally extensive, pre- or post-ovarian. Excretory pore terminal, vesicle often small, normally Y- or V-shaped. In vascular system of fishes, reptiles, birds, and mammals.

Comments

The feature used in the key below is not the only difference, nor is it entirely true for all forms, as a few sanguinicolids may have the vestige of an oral sucker, but it serves well to distinguish Canadian forms.

KEY TO THE FAMILIES OF THE SCHISTOSOMATOIDEA

1 Oral sucker absent .... Sanguinicolidae

Oral sucker present .... Spirorchiidae

Family SANGUINICOLIDAE von Graff, 1907

Syn.: Aporocotylidae Odhner, 1912

Schistosomatoidea. Body lanceolate to elongate and slender, usually symmetrical, occasionally with unilateral notch and lobe near posterior extremity, dorso-ventrally flattened. Tegument usually at least partly spinous, occasionally unarmed. Suckers normally absent, rudimentary oral sucker may be present. Pharynx absent. Oesophagus long, usually bipartite. Intestinal bifurcation in anterior half of body, intestine X-, H-, or inverted U-shaped, in latter case caeca not fused over part of length; anterior caeca long, short, or absent; posterior caeca long or very short. Testes one, two, or numerous, often deeply and irregularly lobed or reticulate, usually between ovary and intestinal bifurcation, occasionally two testes separated by ovary and terminal genitalia. Vasa efferentia and vas deferens may be swollen with spermatozoa. Cirrus sac present or absent. Seminal vesicle usually present, internal or absent, occasionally absent. Male and female genital pores open dorsally close together or separately in posterior half of body. Ovary variable, oval, multilobed, with two symmetrical wings or one irregular, compact, tubular mass. Oviduct long or short. Laurer's canal absent. Canalicular seminal receptacle absent. Uterine seminal receptacle present or absent. Oviductal seminal receptacle may be present. Uterus short to moderately long and coiled; uterine field post-testicular and mainly pre- or post-ovarian. Eggs few to many, non-operculate, unarmed, thin-shelled. Vitellarium follicular, extends between ovary and level anterior to intestinal bifurcation, lateral fields large, fields may be confluent over part or much of length. Excretory pore terminal; vesicle small, V- to Y-shaped, stem and arms usually short. In vascular system of freshwater and marine fishes.

Comments

It is very likely that several other forms than those listed below will be found in marine fishes in Canadian waters, including teleosts, selachians, and holocephalans. Since the presence of these worms, especially those in the blood vessels, is not obvious, they are often missed in parasitological surveys.

In agreement with Van der Land (1967) and Holmes (1971a), no purpose is served in recognizing subfamilies within this group until more is known of their biology and relationships.

Key to the genera of the Sanguinicolidae

1 Body with notch on postero-sinistral margin and lobe immediately posterior to this associated with male pore; testis single, reticulate; uterus post-ovarian; in marine teleosts ....*Psettarium*

Lateral margin of body symmetrical; testes numerous or apparently single and deeply lobed 2

2 Intestinal caeca long; genital pores pre-ovarian; uterus coiled, pre-ovarian; testes numerous; in marine teleosts ....*Aporocotyle*

Intestinal caeca short, reduced to four or five diverticula; genital pores post-ovarian; uterus short, not coiled, post- ovarian; testis appears to be single and deeply lobed or two rows of numerous follicles; in freshwater teleosts..*Sanguinicola*

APOROCOTYLE Odhner, 1900

Sanguinicolidae. Body lanceolate; dorso-ventrally flattened. Tegument with minute spines in irregular groups. Suckers absent. Pharynx absent. Oesophagus long, bipartite. Intestine H- to X-shaped, anterior caeca at least half length of oesophagus, posterior caeca much longer, reaching close to posterior extremity. Testes numerous, fill inter-caecal space between intestinal bifurcation and terminal genitalia. Distal parts of vasa efferentia and vas deferens may be replete with spermatozoa. Cirrus sac present, claviform to retort-shaped, post-testicular, containing simple seminal vesicle, pars prostatica, and ejaculatory duct. Male and female pores open separately into small genital atrium. Genital pore dorsal, medial to sinistral, at about five-sixths of body-length. Ovary oval, near posterior extremity, dextro-medial. Laurer's canal absent. Uterine seminal receptacle present. Uterus moderately coiled, field between ovary and testes. Metraterm between cirrus sac and testes. Eggs numerous, fusiform, non-operculate, thin-shelled, small to large. Vitellarium follicular, lateral fields enveloping lateral regions of caeca from level of ovary to at least level of anterior caeca. Excretory vesicle Y-shaped, stem very short. In vascular system, especially branchial vessels and heart, of marine teleosts.

Comments

Care should be taken in using the number of testes in the key below, since considerable variation has been observed in *Aporocotyle simplex* and it is known that testes begin to degenerate in older worms: nevertheless, the figures given below provide a general guide. Similar care should be taken with egg-size, as the eggs of *Aporocotyle* spp. are soft-shelled and easily deformed.

Key to the species of *Aporocotyle*

1 Fewer than 80 testes 2

More than 80 testes 3

2 About 30–50 testes; anterior intestinal caeca more than half length of oesophagus (both parts) and about one-third length of posterior caeca; in *Merluccius* in the Pacific ....
....*Aporocotyle margolisi* Smith, 1967

Sites: Bulbus arteriosus, heart, branchial vessels.
Host: *Merluccius productus.*
Distribution: Pac.
Records: Smith 1967; Anon (DAFS) 1969; Sankurathri *et al.* 1983.
Remarks: Three other species occur in *Merluccius* spp. in the Atlantic, *A. spinosicanalis* Williams, 1958 in *M. merluccius* from the Northeast Atlantic and *A. argentinensis* Smith, 1969 and *A. australis* Fernandez and Duran, 1985 in *M. hubbsi* and *M. australis*, respectively, from the Southwest Atlantic; but none of these have been reported from *Merluccius* spp. off the Atlantic coast of Canada. The type specimens of the former two species (and *A. simplex*) are in the BM(NH) collection.

About 35–65 testes; anterior intestinal caeca about half length of oesophagus (both parts) and about one-fifth to one-quarter length of posterior caeca; in *Sebastes* in the Pacific ....
....*A.macfarlani* Holmes, 1971

Syn.: *A. simplex* Odhner, 1900 of McFarlane (1936).
Sites: Blood vessels of gills, heart.
Hosts: *Sebastes caurinus* (2, 3, 4); *S. flavidus* (2, 3); *S. maliger* (2, 3, 4); *S. melanops* (2, 3); *S. pinniger* (2, 3); *Sebastes* sp. (1).
Distribution: Pac.
Records: 1. McFarlane 1936; 2. Holmes 1971a; 3. Holmes 1971b; 4. Sekerak and Arai 1977.
Remarks: This species was also reported in Canadian waters as *Aporocotyle* n. sp. in an abstract of the American Society for Parasitology by Holmes (1969) and in *S. maliger* at Friday Harbor, Washington State as *A. simplex* by Ching (1960).

3 About 110–200 testes; vitelline follicles extend anterior to anterior intestinal caeca; eggs in distal uterus ca. 100 µm in length; in flatfishes in Atlantic ....
....*A. simplex* Odhner, 1900

Site: Mesenteric blood vessels (normally branchial vessels and heart).
Hosts: *Glyptocephalus cynoglossus* (2); *Hippoglossoides platessoides* (1, 2, 3, 4).
Distribution: Atl.
Records: 1. Ronald 1960; 2. Zubchenko 1980; 3. Zubchenko 1985a; 4. Morrison *et al.* 1986.
Remarks: This species tends to have a very localized distribution, and, although there are records from other flatfishes, *H. platessoides* appears to be its main host.
Zubchenko (1976, 1981a,b) claimed to have found this species in *Coryphaenoides rupestris* in Canadian waters: this record is listed below.

About 90–125 testes; vitelline follicles do not extend anterior to anterior intestinal caeca; eggs in distal uterus ca. 26–32 µm in length; in *Theragra* in Pacific ....
....*A. theragrae* Ichihara, 1970 (Fig. 56)

Syn.: *Aporocotyle* sp. of Smith (1967).
Site: Branchial vessels.
Host: *Theragra chalcogramma.*
Distribution: Pac.
Records: Smith 1967; Arthur 1983; Arthur 1984; Kabata and Whitaker 1984.
Remarks: There are numerous records of this species in *T. chalcogramma* in Northwestern Pacific waters.

Other species:

Aporocotyle sp.

Syn: *A. simplex* Odhner of Zubchenko (1976, 1981a,b).
Site: ?
Host: *Coryphaenoides rupestris.*
Distribution: Atl.

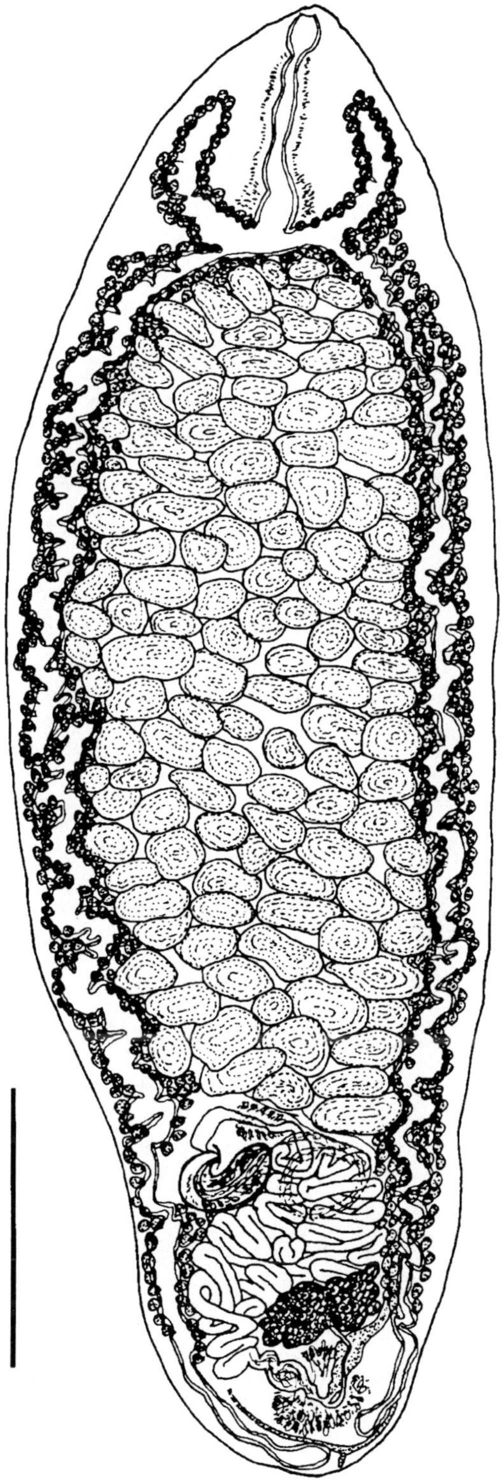

FIG. 56. *Aporocotyle theragrae* (after Ichihara (1970)). Scale bar: 1 mm.

Records: 1. Zubchenko 1976; 2. Zubchenko 1981a; 3. Zubchenko 1981b.

Remarks: Zubchenko's (1976, 1981a,b) record refers to three single specimens recorded from three hosts off Labrador. In view of the fact that *Aporocotyle* spp. appear to be rather or very host specific, the appellation *A. simplex* is undoubtedly a misidentification. As the worms were not described and there are no known blood flukes recorded from this host, it is not possible to key it down.

PSETTARIUM Goto and Ozaki, 1930

Sanguinicolidae. Body narrow, elongate to lanceolate, dorso-ventrally flattened, with notch on sinistral margin close to posterior extremity, with dorsal, conical lobe immediately posterior to notch. Tegument spinose laterally, sometimes ventrally. Suckers absent. Pharynx absent. Oesophagus long, bipartite. Intestine X- to H-shaped; anterior caeca short, posterior caeca extend into posterior half of body but terminate well anterior to ovary. Testis single, indistinct, reticulate, elongate, extending from ovary to inter-caecal field. Vas deferens sinuous, median, replete with spermatozoa. Cirrus sac present or absent, small and indistinct when present. Seminal vesicle present or absent, naked or enclosed in cirrus sac. Pars prostatica indistinct. Ejaculatory duct long or short. Male and female genital pores separate. Male genital pore on dorsal lobe posterior to notch. Ovary an irregular, tubular mass or two symmetrical, reticulate lobes, between uterus and testes. Laurer's canal absent. Uterine seminal receptacle (?). Uterus moderately long, coiled between posterior extremity and ovary, with terminal loop immediately posterior or lateral to ovary. Female genital pore dorsal, median or submedian, just anterior to level of lateral notch. Eggs numerous, oval, non-operculate, thin-shelled. Vitellarium follicular (or composed of tubular acini), lateral fields extend from level of ovary to level between intestinal bifurcation and anterior extremity, fields may be partly or entirely confluent. Excretory vesicle (?). In vascular system of marine teleosts.

Psettarium sebastodorum Holmes, 1971 (Fig. 57)

Site: Heart.

Hosts: *Sebastes aleutianus* (4); *S. alutus* (3, 4); *S. aurora* (1, 2); *S. babcocki* (4); *S. brevispinis* (2); *S. caurinus* (1, 2, 4); *S. crameri* (1, 2, 4); *S. diploproa* (1, 2); *S. flavidus* (1, 2, 4); *S. helvomaculatus* (1, 2, 4); *S. maliger* (1, 2, 4); *S. melanops* (1, 2); *S. nigrocinctus* (1, 2, 4); *S. paucispinis* (2); *S. pinniger* (1, 2, 4); *S. proriger* (2, 4); *S. reedi* (4); *S. variegatus* (4); *S. zacentrus* (1, 2, 4).

Distribution: Pac.

Records: 1. Holmes 1971a; 2. Holmes 1971b; 3. Sekerak and Arai 1973; 4. Sekerak and Arai 1977.

SANGUINICOLA Plehn, 1905

Sanguinicolidae. Body lanceolate, dorso-ventrally flattened, anterior extremity may be protruded to form proboscis. Tegument armed or striated along margins. Suckers absent. Pharynx absent. Oesophagus long. Intestine X-shaped, with four, occasionally five, very short caeca (diverticula). Testis single, elongate, deeply lobed (may appear to be two rows of follicles), median, between ovary and intestine. Vas deferens runs medially full length of testis, swollen with spermatozoa. Cirrus sac present, elongate, fusiform, containing long, simple seminal vesicle, (?) pars prostatica (indistinct if present), and ejaculatory duct. Male and female genital pores close together. Male genital pore dorsal, medial, or submedial; near posterior extremity. Ovary divided into symmetrical wings with median commissure, just posterior to testis. Laurer's canal (?) absent. Oviduct long, oviductal seminal receptacle present. Uterus very short, post-ovarian, containing only one or few eggs. Eggs oval, non-operculate, thin-shelled. Female genital pore close to male genital pore. Vitellarium follicular, lateral fields extend from level of testis, ovary, or further posteriorly to level of intestine or oesophagus, fields may be confluent; main vitelline duct long, runs from about level of intestine to about level of genital pores. Excretory vesicle small, V- to Y-shaped, stem and arms short. In vascular system of freshwater fishes.

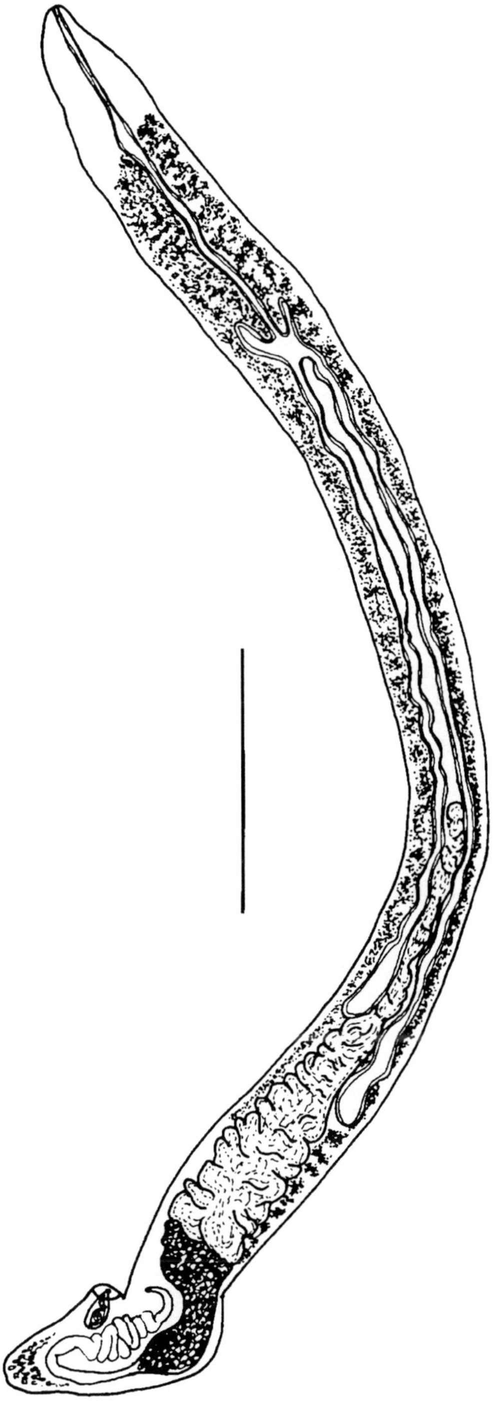

FIG. 57. *Psettarium sebastodorum* (after Holmes (1971a)). Scale bar: 1 mm.

Comments

It seems very likely that further species of this genus will be found in Canadian freshwater fishes. A key to the North American species was given by Hoffman *et al.* (1985).

Sanguinicola occidentalis Van Cleave and Mueller, 1932 (Fig. 58)

Site: Blood.

Hosts: *Perca flavescens* (1, 6, 8); *Stizostedion vitreum* (1, 2, 3, 4, 5, 6, 7).

Distribution: Ont.

Records: 1. Dechtiar 1972a; 2. Anthony 1976; 3. Anthony 1978b; 4. Dechtiar and Lawrie 1988; 5. Dechtiar *et al.* 1988; 6. Dechtiar and Nepszy 1988 (Ont); 7. Dechtiar and Christie 1988; 8. Dechtiar *et al.* 1989.

Sanguinicola spp.

Site: Blood.

Hosts: *Aplodinotus grunniens* (1, 5); *Carpiodes cyprinus* (1, 5); *Catostomus catostomus* (3, 4); *C. commersoni* (4, 5, 6); *Moxostoma anisurum* (2); *M. erythrurum* (1); *M. macrolepidotum* (1); *Notropis hudsonius* (1, 3); *Pimephales notatus* (2); *P. promelas* (2).

Distribution: Ont.

Records: 1. Dechtiar 1972a; 2. Dechtiar 1972b; 3. Dechtiar and Lawrie 1988; 4. Dechtiar *et al.* 1988; 5. Dechtiar and Nepszy 1988; 6. Dechtiar *et al.* 1989.

Family SPIRORCHIIDAE Stunkard, 1921

Schistosomatoidea. Body lanceolate to slender, symmetrical, dorso-ventrally flattened. Tegument smooth or armed. Oral sucker present, usually small, terminal. Ventral sucker usually absent, occasionally present closely posterior to intestinal bifurcation. Pharynx normally absent, occasionally present. Oesophagus normally long, often surrounded by gland cells posteriorly. Intestine blind, usually an inverted U-shape, bifurcation in anterior half of body. Caeca long, occasionally united over entire or part of length to form single caecum. Testes one, two, several, or many, usually situated in inter-caecal field, usually pre-ovarian, but may be post-ovarian or separated by ovary. Cirrus sac present or absent, when present contains seminal vesicle, pars prostatica, and eversible ejaculatory duct. External seminal vesicle present or absent. Genital pore variable in position: anterior, posterior, lateral, ventral, dorsal. Ovary median or submedian: anterior, posterior, or between testes. Seminal receptacle and Laurer's canal present or absent. Uterus short, restricted to ovarian region. Eggs one to few, operculate, usually with polar filaments. Vitellarium follicular, lateral fields extra-caecal or circum-caecal. Excretory pore terminal, vesicle V- to Y-shaped. In vascular system of chelonians.

Spirorchiidae gen. sp. (Fig. 59)

Syn.: *Spirorchis* sp. of Appy and Dadswell (1978).

Site: Mesenteric blood vessels.

Host: *Acipenser brevirostrum*

Distribution: NB-b.

Record: Appy and Dadswell 1978.

Remarks: The finding of a spirorchiid in a chondrostean is remarkable, because, except for a single record of *Spirorchis* sp. from the vascular system of *Rana pretiosa* in Washington State by Lehmann (1965), all records of spirorchiids are from chelonians. In view of the known host specificity of blood flukes, I was very sceptical of this record, especially since one would not expect an accidental parasite to reach or survive in such a site. Having examined the voucher specimen in the US National Museum collection, I can verify that this worm is a spirorchiid, but aspects of its morphology, which are clearer than the figure of Appy and Dadswell (1978) indicates, and taxonomic problems within the group preclude, in my opinion, a generic allocation at this time. Appy and Dadswell placed this form in the genus *Spirorchis* MacCallum, 1919, but the position of the testes

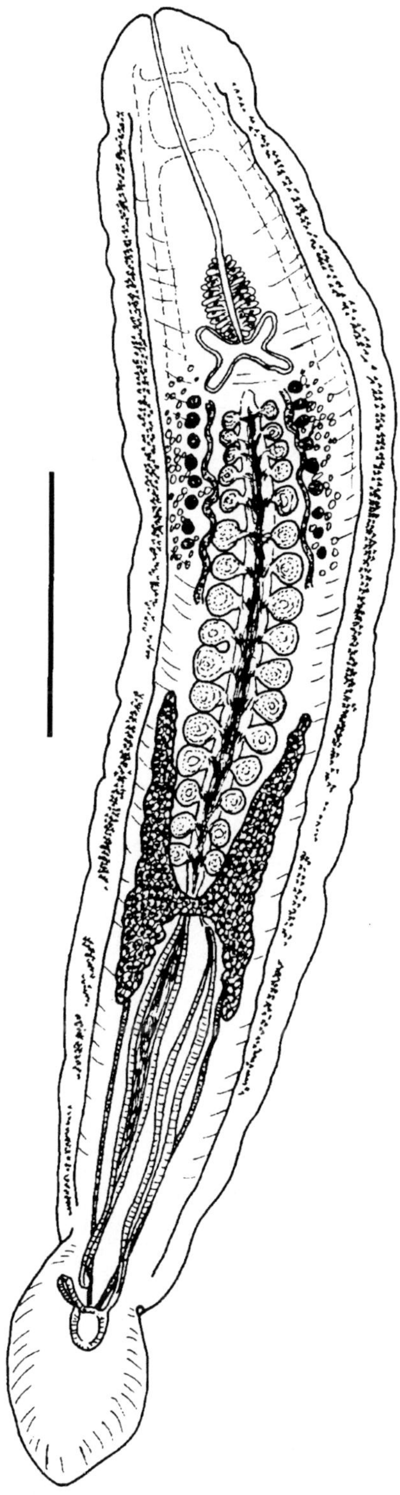

FIG. 58. *Sanguinicola occidentalis* (after Van Cleave and Mueller (1932)). Scale bar: 200 µm.

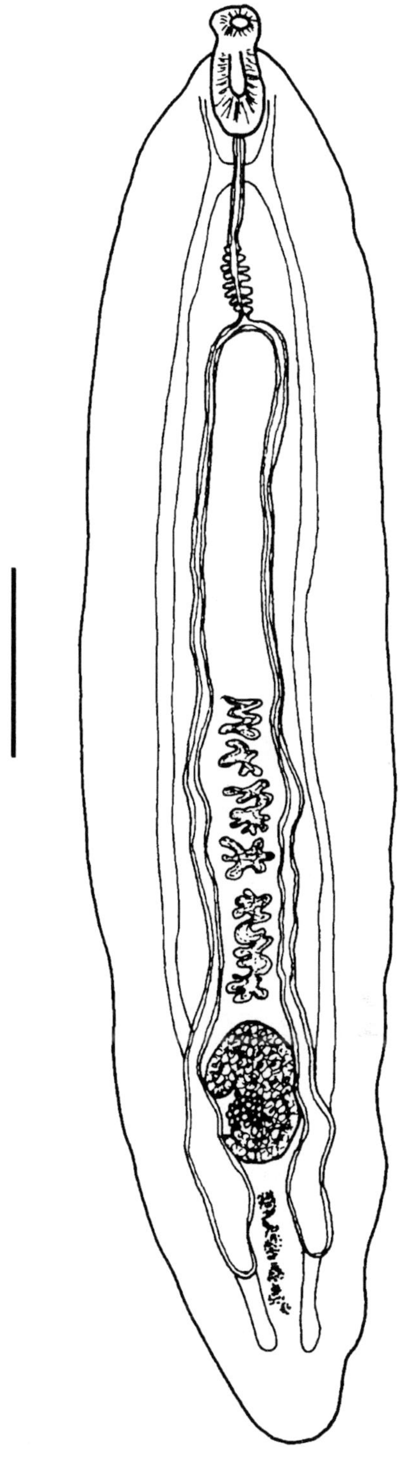

FIG. 59. Spirorchiidae gen. sp. (as *Spirorchis* sp.) (after Appy and Dadswell (1978)). Scale bar: 200 µm.

in the posterior part of the inter-caecal field suggests the genus *Henotosoma* Stunkard, 1922. Furthermore, in the specimen that I examined, there is material in the post-ovarian region that resembles the testes. If this is testicular material, then this suggests a relationship with the genus *Diarmostorchis* Ejsmont, 1927. Although both *Henotosoma* and *Diarmostorchis* were recognized by Yamaguti (1971), they were considered synonyms of *Spirorchis* by Byrd (1939) and Platt (1992) in their revision of the spirorchiids and spirorchiins, respectively.

SUPERFAMILY ECHINOSTOMATOIDEA LOOSS, 1899

Echinostomida. Occasionally occur as larva (metacercaria) in/on fishes. Metacercaria encysted, cyst wall two/three-layered and usually surrounded by capsule of host origin. Body elongate to elongate-oval, with, or occasionally without, head collar that is often armed with one or two crowns of spines. Tegument armed. Oral and ventral suckers well developed. Prepharynx present. Pharynx well developed. Intestinal bifurcation normally in forebody. Gut caeca normally blind, extend close to posterior extremity. Gonadal primordia often present. Excretory vesicle Y-shaped. Occasionally encysted superficially in/on teleosts; more usually in/on molluscs, vegetation, substratum, other invertebrates, and amphibians. (Parasitic as adults in intestine of, especially, aquatic birds and mammals.)

KEY TO THE FAMILIES OF THE ECHINOSTOMATOIDEA

1 Head collar and crown of circum-oral spines present .... Echinostomatidae

Head collar and crown of circum-oral spines absent .... Psilostomidae

Family ECHINOSTOMATIDAE Looss, 1899

Echinostomatoidea. Occasionally present in fishes as encysted metacercaria. Body oval to elongate. Tegument armed with small spines. Head collar present, usually armed with single or double crown of spines. Suckers well developed. Prepharynx present, often long. Pharynx well developed, often well separated from oral sucker. Oesophagus short to long. Caeca normally blind, rarely forming uroproct. Excretory system Y-shaped. Encysted as metacercariae in/on molluscs, annelids, fishes, tadpoles, etc. (Adults mainly in intestine, occasionally other organs, of birds, mammals, and occasionally reptiles.)

Comments

Metacercariae of several echinostomatid genera are occasionally found in/on fishes in the United States.

ECHINOCHASMUS Dietz, 1909

Echinostomatidae. Usually present in/on fishes as encysted metacercariae. Body oval. Tegument armed. Head collar well developed, with single crown of circum-oral spines interrupted dorsally and ventrally. Ventral sucker usually just post-equatorial. Prepharynx long. Pharynx well developed. Oesophagus short to long. Caeca short to long. Stem of excretory vesicle short. Encysts under skin, in/on gills, etc. of freshwater teleosts, occasionally in/on snails and tadpoles. (Adults in intestine of birds and mammals.)

Echinochasmus sp. (metacercaria) (Fig. 60)

Site: Encysted in gills.
Host: *Notemigonus crysoleucas*.
Distribution: Ont.
Record: Dechtiar *et al.* 1989.
Remarks: The metacercaria of *E. milvi* is used to illustrate this genus.

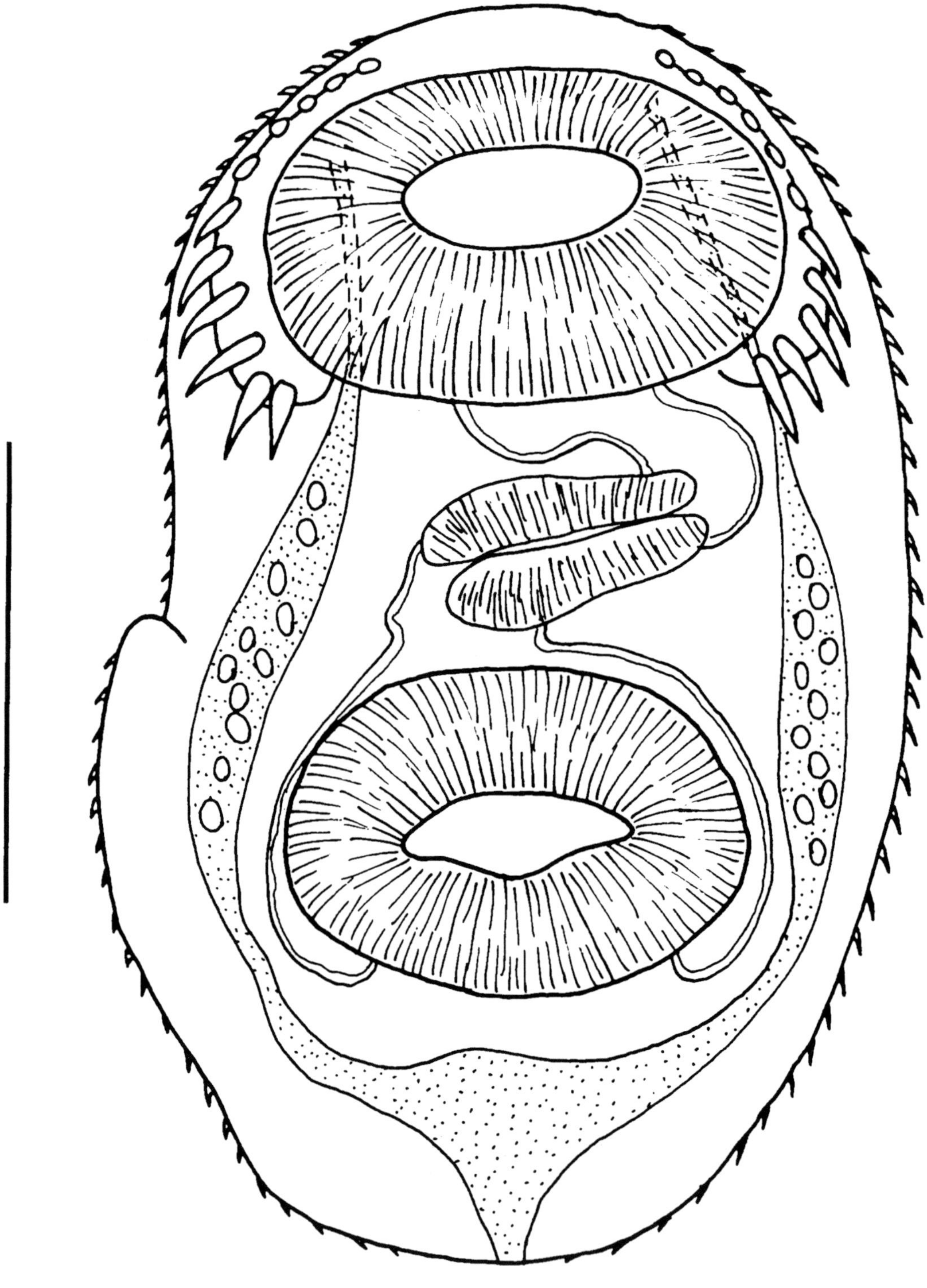

FIG. 60. *Echinochasmus milvi* metacercaria (after Besprozvannykh (1989)). Structures within the excretory vesicle are calcareous corpuscles, features typical in the excretory system of live metacercariae. Scale bar: 50 µm.

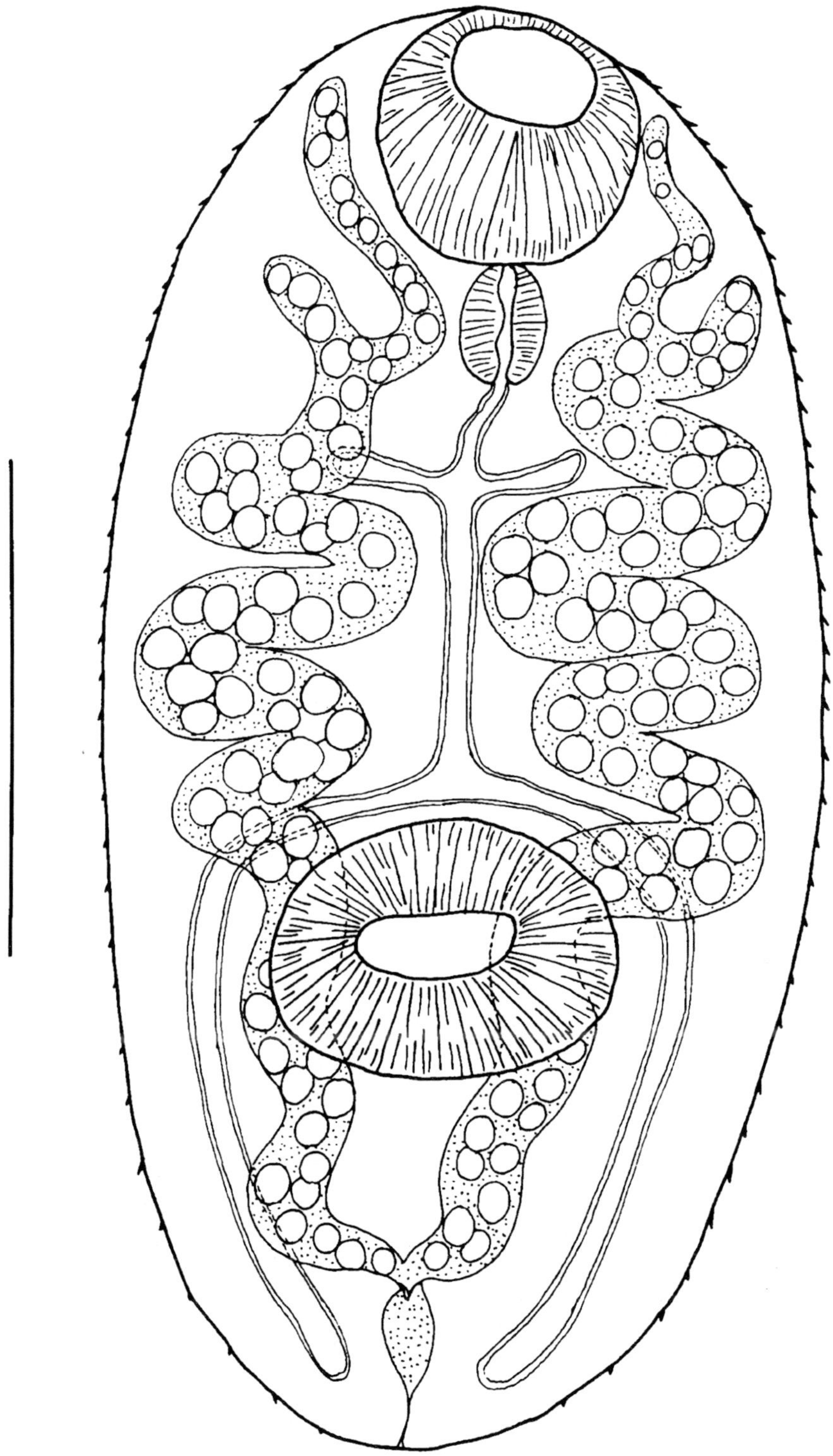

FIG. 61. *Ribeiroia ondatrae* metacercaria (after Beaver (1939)). Scale bar: 100 µm. Spheres in the excretory system are calcareous corpuscles that are readily visible in most metacercariae when examined alive.

Family PSILOSTOMIDAE Looss, 1904

Syn. Cathaemasiidae Fuhrmann, 1928.

Echinostomatoidea. Occasionally present in fishes as encysted larvae (metacercariae). Body oval to elongate-oval. Tegument spined. Head collar and circum-oral crown(s) of spines normally absent (apparently present in some species of *Cathaemasia*). Suckers well developed. Oral sucker occasionally forms pentagonal hood-like expansion. Ventral sucker equatorial to post-equatorial. Prepharynx present, small. Pharynx well developed. Oesophagus with or without lateral diverticula. Caeca blind or forming uroproct, with or without lateral and sometimes branched diverticula. Excretory system Y-shaped. Encysted superficially (under scales, on skin, in lateral line, on gills, in nares, etc.) of teleosts and amphibians, in/on molluscs and other invertebrates, or on vegetation or substratum. (Adults in birds and occasionally mammals.)

Comments

The Cathaemasiidae is tentatively treated as a synonym of the Psilostomidae until the relationships of some of the minor echinostomatoid groups are resolved.

Ribeiroia ondatrae (Price, 1931) may not be the only psilostomid in Canadian fishes. There is material of *Pulchrasoma reticulata* (Wright, 1879) from a Canadian kingfisher *Megaceryle alcyon* in the BM(NH) collection, but perhaps its intermediate host is more likely to be an amphibian.

RIBEIROIA Travassos, 1939

Psilostomidae. Metacercaria encysted; cyst small, elongate-oval, tightly fitting, usually encapsulated by host connective tissue, cyst wall thin. Body oval. Tegument spined. Head collar and circum-oral ring(s) of spines absent. Ventral sucker just post-equatorial. Oesophagus long, with one digitiform lateral diverticulum on each side. Caeca without diverticula, terminate blindly near posterior extremity. Stem of Y-shaped excretory vesicle short. Encysts under scales, especially in lateral line, of freshwater teleosts, occasionally in nares of tadpoles and teleosts. (Adults in proventriculus of wide range of piscivorous birds, birds of prey, and mammals (especially muskrat).)

Ribeiroia ondatrae (Price, 1931) Price, 1942 (metacercaria) (Fig. 61)

Site: Skin (normally under scales).
Host: *Ambloplites rupestris*.
Distribution: Ont.
Record: Molnar *et al.* 1974.
Remarks: The adult has been recorded from muskrats in Ontario (Price, 1931), and Kuntz (1951) experimentally infected muskrats with metacercariae from *A. rupestris* and other fishes.

SUPERFAMILY PARAMPHISTOMOIDEA FISCHOEDER, 1901

Echinostomida. Body stout, often large, usually conical to fusiform, rarely flattened. Tegument unarmed, tegumentary papillae present or absent. Oral sucker absent. Ventral sucker normally occurs ventro-terminally or terminally at posterior end of body, well developed, often very large. Prepharynx absent. Ventral pouch present or absent. Pharynx (referred to by some authors as oral sucker) terminal or ventro-terminal at anterior extremity opening via mouth, sometimes with one or two postero-lateral diverticula. Oesophagus present, with or without muscular bulb or sphincter. Caeca normally two long, straight or sinuous, end blindly. Testes normally two, usually inter-caecal, tandem, oblique, or symmetrical. Tubular seminal vesicle, pars musculosa (seminalis), pars prostatica, and ejaculatory duct normally present. Cirrus sac present or absent, sinus sac rare. Genital pore usually mid-ventral in anterior region of body, occasionally dorsal or opening into ventral pouch. Ovary post-testicular. Laurer's canal and uterine seminal receptacle present, other forms of seminal receptacle absent. Uterus mainly inter-caecal, contains numerous large eggs. Vitellarium normally follicular, usually in lateral fields. Excretory pore dorsal; vesicle short, saccular, or tubular. Lymphatic system usually developed. Parasitic in gut (rarely liver) of wide range of vertebrate groups.

Family CLADORCHIIDAE Southwell and Kirshner, 1932

Paramphistomoidea. Body elongate-oval to pyriform. Ventral sucker ventro-terminal or terminal. Pharynx with pair of postero-lateral diverticula. Oesophagus with muscular thickening. Testes not always inter-caecal. Cirrus sac present. Genital pore mid-ventral at level of intestinal bifurcation. Uterine coils mainly post-testicular. Vitellarium follicular (rarely compact), usually in lateral fields. Eggs large, sometimes embryonated. Normally parasitic in gut of fishes, amphibians, and reptiles.

OPHIOXENOS Sumwalt, 1926

Cladorchiidae. Body fusiform, more tapered anteriorly. Oesophagus with posterior muscular bulb. Caeca long, broad, straight, extending posteriorly to ovary. Testes entire, oblique, inter-caecal, in anterior half of body. External seminal vesicle sinuous. Cirrus sac small; thin walled. Ovary median or submedian between posterior testis and ventral sucker. Uterus inter-caecal, winding around ovary and between and/or dorsal to testes. Vitelline follicles not numerous, mainly inter-caecal in post-testicular region. Excretory pore dorsal at level of anterior margin of ventral sucker. In intestine of lampreys, frogs, and turtles.

Comments

The systematic position recognized here for this genus follows Sey (1991).

Ophioxenos microphagus (Ingles, 1936) Beverley-Burton, 1987 (Fig. 62)

Syn.: *Ophioxenos lampetrae* Beverley-Burton and Margolis, 1982.
Site: Intestine.
Host: *Lampetra richardsoni*.
Distribution: BC.
Record: Beverley-Burton and Margolis 1982.
Remarks: The synonymy was proposed by Beverley-Burton (1987).

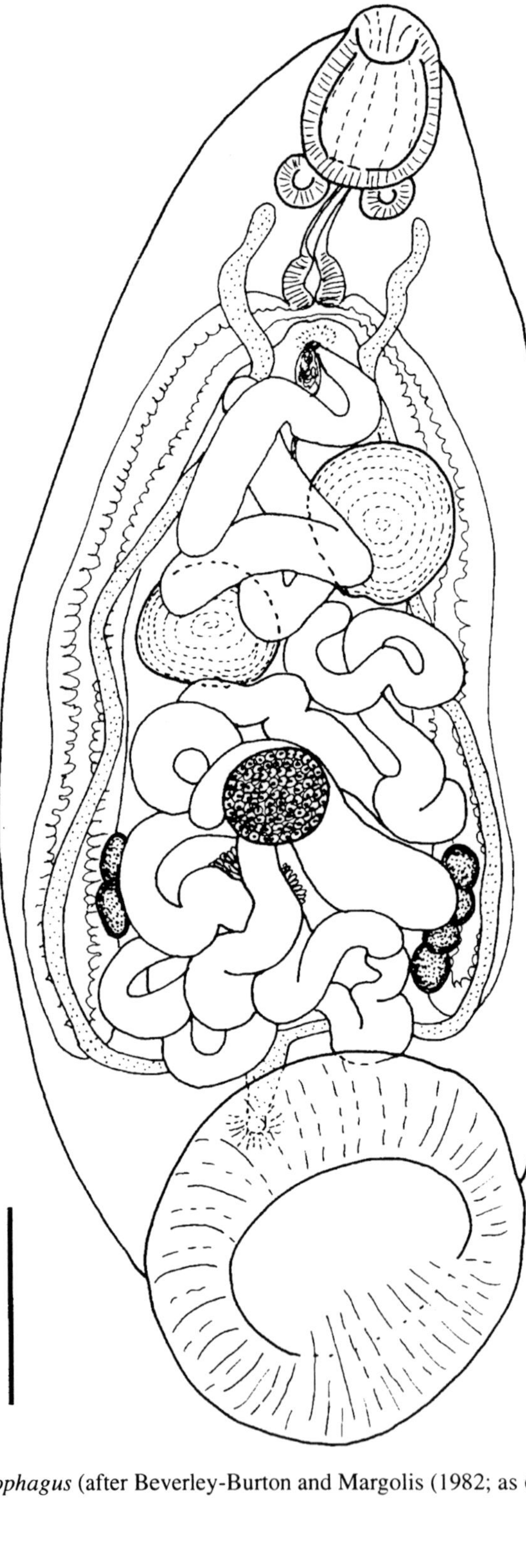

FIG. 62. *Ophioxenos microphagus* (after Beverley-Burton and Margolis (1982; as *O. lampetrae*)). Scale bar: 500 µm.

SUPERFAMILY ALLOCREADIOIDEA LOOSS, 1902

Plagiorchiida. Body small to medium-sized, usually cylindrical to fusiform, occasionally pyriform to oval. Tegument unarmed. Eye-spot pigment occasionally present. Oral and ventral suckers well developed, ventral sucker usually but not always larger and in anterior half of body. Prepharynx, pharynx, and oesophagus present. Intestinal bifurcation in forebody, caeca usually end blindly near posterior extremity, occasionally unite or open via anus, twin ani, or uroproct. Testes normally two, usually in tandem, occasionally oblique to symmetrical, in hindbody. Cirrus sac present or absent. Seminal vesicle, pars prostatica, and ejaculatory duct internal or external. Genital atrium small. Genital pore median to sublateral in forebody. Ovary entire or lobed, normally pre-testicular, occasionally lateral to anterior testis. Laurer's canal present, canalicular or uterine seminal receptacle present (mutually exclusive). Uterine field usually between gonads and ventral sucker, sometimes extending into post-testicular region. Eggs usually large, numerous, operculate. Vitellarium follicular, usually in lateral and post-testicular fields, may extend into and sometimes confluent in forebody. Excretory pore terminal. Excretory vesicle I-shaped, tubular. In intestine, rarely in gall bladder, of marine and freshwater teleosts, occasionally in amphibians and reptiles.

KEY TO THE FAMILIES OF THE ALLOCREADIOIDEA

1 In freshwater; uterus often, but not always extending posterior to ovary; cirrus sac present; canalicular seminal receptacle present; uterine seminal receptacle never present; oral sucker sometimes lobed .... Allocreadiidae

2 In marine, especially, and freshwater *(Plagioporus, Allopodocotyle)* fishes; uterus usually, but not always, anterior to ovary or anterior testis; cirrus sac present or absent; canalicular or uterine seminal receptacle present; oral sucker not lobed .... Opecoelidae

Family ALLOCREADIIDAE Looss, 1902

Allocreadioidea. Body small to medium-sized, elongate, fusiform, or pyriform. Tegument unarmed. Eyespots present or absent (may be present as dispersed pigment). Oral sucker ventrally subterminal, with or without two, four, or six muscular, papilliform lobes. Ventral sucker in anterior half of body. Prepharynx present, short. Pharynx well developed. Oesophagus present, variable in length. Intestinal bifurcation normally in posterior half of forebody, occasionally in anterior half. Caeca blind, usually long, extending close to posterior extremity, but occasionally short, ending at level of testes. Testes normally two, occasionally four, in tandem, oblique, or occasionally almost symmetrical, usually near middle of hindbody. Cirrus sac present, claviform, variable in length, ranging from entirely in forebody to extending into anterior hindbody, containing internal seminal vesicle, pars prostatica, and ejaculatory duct. Temporary unarmed cirrus may form. Genital atrium small. Genital pore median or submedian in forebody, pre-bifurcal, bifurcal, or post-bifurcal. Ovary normally entire, pre-testicular in anterior hindbody. Laurer's canal and canalicular seminal receptacle present. Uterine seminal receptacle absent. Uterine coils in hindbody, vary in distribution, usually pre-testicular but often reaching to posterior extremity and sometimes almost filling hindbody. Eggs numerous, large, without spines or filaments. Vitellarium follicular, lateral fields vary in distribution from extending full length of worm to being restricted to region close to ventral sucker, may be confluent in post-testicular zone and/or occasionally in forebody, main lateral collecting ducts may unite in post-testicular zone. Excretory vesicle I-shaped, short to long, restricted to hindbody. Parasitic in intestine (occasionally gall bladder) of fishes, amphibians, and reptiles, especially in fresh water.

Comments

Adult allocreadiids are often difficult to distinguish from opecoelids, but the parthenitae normally occur in bivalves rather than gastropods and the cercaria is normally an ophthalmo-xiphidiocercariae rather than being non-oculate, cotylomicrocercous, and sometimes lacking a stylet.

The characters in general use for distinguishing the subfamilies of the Allocreadiidae, which are essentially those of the first two parts of the key presented below, appear to me to be criteria of generic significance only. I have, therefore, not included the subfamily level in the classification of the Canadian forms, as I can find no good reason to recognise the Bunoderinae Looss, 1902 or the Crepidostominae Dollfus, 1951. It may well be that future work will show that all allocreadiids are restricted to the single subfamily Allocreadiinae.

Since the completion of the first draft of this section a detailed revision of the North American "papillose" allocreadiids has been published by Caira (1989): the latter supersedes the earlier revision of Hopkins (1934).

Key to the genera of the Allocreadiidae

1 Uterus filling much of hindbody, reaching to posterior extremity in fully developed worms 2

Uterus normally restricted to field between posterior testis (usually anterior testis) and ventral sucker, but may occasionally extend further posteriorly but only in median field 4

2 Oral sucker with four to six muscular, papilliform lobes; lateral vitelline fields extend over greater area than region between ovary and ventral sucker .... *Bunodera*

Oral sucker lacking any kind of lobation 3

3 Lateral vitelline fields small, restricted to region between ovary and ventral sucker .... *Plagiocirrus*

Lateral vitelline fields extensive, extending between levels of pharynx and testes .... *Culaeatrema*

4 Oral sucker normally surmounted by two or six muscular, papilliform lobes (careful examination of well-fixed material is required, especially to determine the presence or absence of the dorsal lobes) 5

Oral sucker without lobes 7

5 Oral sucker with six muscular, papilliform lobes 6

Oral sucker with two lateral, muscular, papilliform lobes .... *Creptotrema*

6 Tandem testes split longitudinally during development to form four masses; in fully developed worms uterus may pass back medially almost to posterior extremity; mainly in ictalurids .... *Megalogonia*

Tandem testes not split longitudinally; uterine field normally between level of anterior testis and ventral sucker even in fully developed worms .... *Crepidostomum*

7 Lateral vitelline fields restricted to hindbody .... *Allocreadium*

Lateral vitelline fields reach forward close to level of pharynx .... *Polylecithum*

ALLOCREADIUM Looss, 1900

Medium-sized worms, fusiform to elongate. Pharynx small. Oesophagus variable in length. Intestinal bifurcation in posterior half of forebody. Caeca end blindly at level between posterior extremity and middle of posterior testis. Testes two, usually tandem, near middle of hindbody. Cirrus sac claviform, often broad, posterior extent normally varying between anterior margin of ventral sucker and level of anterior margin of ovary, containing coiled, saccular seminal vesicle, pars prostatica, and ejaculatory duct. Genital pore ventro-median or submedian, usually at level of oesophagus or intestinal bifurcation. Ovary usually entire or occasionally indistinctly lobed, situated between anterior testis and ventral sucker. Laurer's canal and canalicular seminal receptacle present. Uterine field between anterior testis and ventral sucker, sometimes extending extra-caecally, coils often encroaching between anterior testis and ovary. Eggs few to numerous, large. Vitelline follicles in lateral fields extending from posterior extremity to level between posterior margin of ventral sucker and pharynx, confluent in post-testicular zone. Excretory vesicle short, reaching to near level of posterior testis. In intestine of freshwater teleosts.

Allocreadium lobatum Wallin, 1909 (Fig. 63)

Syn.: *Allocreadium isoporum* (Looss, 1894) of Canadian authors.

Site: Intestine.

Hosts: *Catostomus catostomus* (6, 9); *C. commersoni* (8, 16); *Couesius plumbeus* (6, 7); *Esox lucius* (11); *Luxilis cornutus* (3, 4, 5, 10, 18); *Mylocheilus caurinus* (6, 12, 13, 14); *Nocomis biguttatus* (4); *Notemigonus crysoleucas* (19); *Notropis hudsonius* (4, 17); *Oncorhynchus mykiss* (6); *Pimephales promelas* (8); *Prosopium williamsoni* (6); *Ptychocheilus oregonensis* (6, 12, 13, 14); *Rhinichthys cataractae* (8, 17, 18); *Richardsonius balteatus* (6); *Semotilus atromaculatus* (3, 4, 5, 18); *S. corporalis* (1, 2, 15); *Thymallus arcticus* (11).

Distribution: BC, Lab, Ont, Que, YT, NB.

Records: 1. Stafford 1902 (locality unspecified); 2. Stafford 1904 (Que?); 3. Bangham 1941 (Ont); 4. Bangham 1955 (Ont); 5. Bangham and Venard 1946 (Ont); 6. Bangham and Adams 1954 (BC); 7. Threlfall and Hanek 1971 (Lab); 8. Dechtiar 1972b (Ont); 9. Hanek and Molnar 1974 (Que); 10. Molnar *et al.* 1974 (Ont); 11. Arthur *et al.* 1976 (YT); 12. Anonymous 1978 (BC); 13. Arai and Mudry 1983 (BC); 14. Anonymous 1984 (BC); 15. Rand and Burt 1985 (NB); 16. Courtney 1987 (Ont); 17. Dechtiar and Lawrie 1988 (Ont); 18. Dechtiar *et al.* 1988 (Ont); 19. Dechtiar *et al.* 1989 (Ont).

Remarks: This species is morphologically very similar to *A. isoporum* (Looss, 1894), the type species of the genus from the Palaearctic, differing only in that the testes are consistently lobed and the eggs are a little smaller.

I have attributed the early records by Stafford (1902, 1904) of *A. isoporum* to *A. lobatum*, as suggested by Margolis and Arthur (1979). I have also treated subsequent Canadian records of *A. isoporum* similarly until there is evidence that the latter species does occur in North America.

This species has recently been recorded in *Catostomus commersoni* from American waters of Lake Superior by Hogue *et al.* (1993).

BUNODERA Railliet, 1896

Syn.: *Bunoderina* Miller, 1936; *Allobunodera* Yamaguti, 1971.

Allocreadiidae. Small to medium-sized worms, body elongate to pyriform. Eye-spots present or absent. Oral sucker surmounted by six (occasionally four) muscular, papilliform lobes. Ventral sucker in anterior half of body. Oesophagus usually long. Intestinal bifurcation in posterior half of forebody. Caeca short or long, end blindly in hindbody. Testes two, in tandem, oblique, or almost symmetrical, position in hindbody variable. Cirrus sac small, claviform, broad, contains saccular, often bi-partite, seminal vesicle, pars prostatica, and ejaculatory duct. Genital pore ventrally median or submedian at about level of intestinal bifurcation or post-bifurcal in forebody. Ovary entire, median or submedian, immediately posterior to ventral sucker. Uterine field fills much of hindbody, often extending extra-caecally in post-testicular and testicular

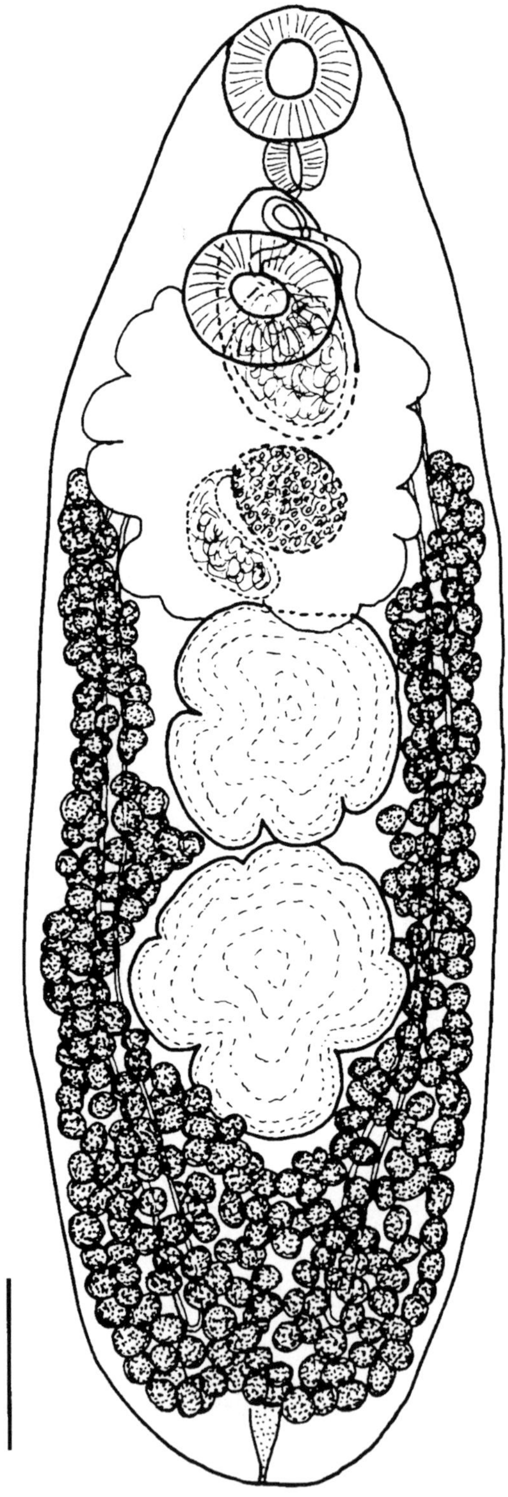

FIG. 63. *Allocreadium lobatum* (after Mueller (1934)). Scale bar: 500 µm.

zone in fully developed worms, coils encroach between gonads; uterus tubular or saccular. Eggs numerous (sometimes exceedingly so), large. Vitelline follicles in lateral fields, anterior limit varies between pharynx and posterior margin of ventral sucker, posterior limit varies between just anterior to anterior testis and posterior extremity. Excretory vesicle reaches to about level of anterior testis or just beyond. In intestine of freshwater fishes.

Comments

The genus *Bunoderina* was erected by Miller (1936) for *B. eucaliae* because the uterus was tubular in mature specimens rather than being saccular as it is in *Bunodera*. He later (1940) considered this genus to be a synonym of *Bunodera* because a saccate uterus was described for *Bunodera sacculata*, which otherwise fitted into *Bunoderina* (and was later transferred to this genus by Yamaguti 1958), and because of its great morphological similarity to *Bunodera*. Nevertheless, *Bunoderina* has still been recognized by authors such as Yamaguti (1958, 1971) and Skryabin and Koval (1966) because of the short caeca and restricted vitelline distribution. Miller's (1940a) view is accepted here because in some specimens of *Bunodera luciopercae* the caeca can end laterally to the testes (although still close to the posterior extremity) and in *B. mediovitellata*, for which Yamaguti (1971) erected the genus *Allobunodera*, the distribution of the lateral vitelline fields is also restricted.

All North American species of this genus have been redescribed by Caira (1989).

Key to the species of *Bunodera*

1 Lateral vitelline fields occupy almost full length of worm (but often difficult to see in fully gravid worms mounted as whole-mounts); caeca long, reaching close to posterior extremity...... ....*Bunodera luciopercae* (Müller, 1776) Stiles and Hassal, 1898

Syn.: *Distomum nodulosum* (Frölich, 1791).

Site: Intestine.

Hosts: *Acipenser fulvescens* (1); *Ambloplites rostrata* (1); *Anguilla rostrata* (1); *Aplodinotus grunniens* (1); *Esox lucius* (10); *Gasterosteus aculeatus* (9); *Lepomis macrochirus* (1); *Lota lota* (12); *Micropterus salmoides* (1); *Oncorhynchus mykiss* (15, 16, 18); *Perca flavescens* (3, 4, 6, 7, 11, 12, 13, 19); *Prosopium cylindracium* (14); *Pungitius pungitius* (8, 16); *Salmo salar* (14); *Salvelinus alpinus* (14, 20); *S. fontinalis* (2, 5, 14, 15); *S. malma* (15); *S. namaycush* (14, 15).

Distribution: Alta, BC, Lab, Lab-b, Nfld-b, Nfld, Ont, Que.

Records: 1. MacCallum 1895 (Ont); 2. Stafford 1904 (unspecified locality); 3. Bangham and Hunter 1939 (Ont); 4. Fantham and Porter 1948 (Que); 5. Sandeman and Pippy 1967 (Nfld); 6. Tedla and Fernando 1969a (Ont); 7. Tedla and Fernando 1972 (Ont); 8. Hanek and Threlfall 1970c (Nfld); 9. Hanek and Threlfall 1970d (Lab-b, Nfld-b, Nfld); 10. Threlfall and Hanek 1970b (Lab); 11. Cannon 1971 (Ont); 12. Cannon 1972 (Ont); 13. Cannon 1973 (Ont); 14. Hicks and Threlfall 1973 (Lab); 15. Mudry and Anderson 1977 (Alta, BC); 16. Anonymous 1978 (BC); 17. Leong and Holmes 1981 (Alta); 18. Arai and Mudry 1983 (BC); 19. Caira 1989 (Ont); 20. Bouillon and Dempson 1989 (Lab).

Remarks: Marcogliese and Cone (1991a) mentioned the occurrence of this species in *Salvelinus fontinalis*, but this is not an original record.

Lateral vitelline fields restricted to within region between pharynx and posterior testis......................2

2 Lateral vitelline fields not extending into forebody; caeca long, reaching into post-testicular field...*B. mediovitellata* Tsimbaluk and Roitman, 1966

Syn.: *Allobunodera mediovitellata* (Tsimbaluk and Roitman, 1966) Yamaguti, 1971.

Site: Intestine.

Host: *Gasterosteus aculeatus*.

Distribution: BC.

Records: Lester 1975; Kennedy 1979; Caira 1989.

Lateral vitelline fields extending into forebody; caeca short, reaching to level of testes only 3

3 Lateral vitelline fields reaching back as far as testes; sucker-ratio 1:1 or <1; mature forms with. saccate uterus..*B. sacculata* Van Cleave and Mueller, 1932 (Fig, 64)

Syn.: *Bunoderina sacculata* (Van Cleave and Mueller, 1932) Yamaguti, 1958.

Site: Intestine.

Hosts: *Lepomis gibbosus* (2); *L. macrochirus* (13); *Lota lota* (8); *Luxilis cornutus* (4); *Perca flavescens* (1, 2, 3, 4, 5, 6, 7, 8, 9, 10, 11, 12, 13, 14, 15); *Stizostedion vitreum vitreum* (10).

Distribution: Ont, Que.

Records: 1. Lyster 1939 (Que); 2. Bangham 1941 (Ont); 3. Bangham 1955 (Ont); 4. Bangham and Venard 1946 (Ont); 5. Tedla and Fernando 1969a (Ont); 6. Tedla and Fernando 1972 (Ont); 7. Cannon 1971 (Ont); 8. Cannon 1972 (Ont); 9. Cannon 1973 (Ont); 10. Dechtiar 1972b (Ont); 11. Poole and Dick 1985 (Man); 12. Dechtiar and Nepszy 1988 1988 (Ont); 13. Dechtiar and Christie 1988 (Ont); 14. Dechtiar *et al.* 1988 (Ont); 15. Caira 1989 (Ont).

Lateral vitelline fields not reaching back as far as testes; sucker-ratio 1:>1; uterus tubular, ... even in fully developed worms..*B. eucaliae* (Miller, 1936) Miller, 1940

Syn.: *Bunoderina eucaliae* Miller, 1936.

Site: Intestine.

Hosts: *Culaea inconstans* (1, 2, 3, 4, 5, 6, 7, 8, 9); *Gasterosteus aculeatus* (4, 8); *Pungitius pungitius* (6); *Umbra limi* (5, 7).

Distribution: BC, Ont, Que.

Records: 1. Miller 1936 (Que); 2. Miller 1940a (Que); 3. Bangham and Hunter 1939 (Ont); 4. Bangham and Adams 1954 (BC); 5. Bangham 1955 (Ont); 6. Dechtiar 1972b (Ont); 7. Dechtiar *et al.* 1988 (Ont); 8. Dechtiar and Christie 1988 (Ont); 9. Brooks 1992 (Ont).

Remarks: This species is morphologically very similar to *B. sacculata*, and initially I considered it possible that specimens recorded under this name might be poorly developed specimens of the latter species. However, after a detailed study, the species has been recognized by Caira (1989). The holotype of this species is in the BM(NH) collection.

CREPIDOSTOMUM Braun, 1900

Allocreadiidae. Body small to medium-sized, oval to elongate. Oral sucker surrounded dorsally and laterally by six muscular, papilliform lobes. Ventral sucker larger, smaller or similar in size to oral sucker, in anterior half of worm. Pharynx small to large. Intestinal bifurcation in posterior half of forebody. Caeca end blindly near posterior extremity. Testes two, tandem, near middle of hindbody. Cirrus sac claviform, usually reaching into anterior hindbody, occasionally ending dorsally to ventral sucker or even extending into hindbody, containing tubular to saccular and often coiled seminal vesicle, pars prostatica, and ejaculatory duct. Genital pore mid-ventral in forebody, at level of pharynx to post-bifurcal. Ovary submedian, between anterior testis and ventral sucker. Uterine field usually between anterior testis and ventral sucker, occasionally overlies anterior testis, often intrudes between anterior testis and ovary. Eggs large, numerous, without filaments. Vitelline fields extend laterally from posterior extremity to level ranging between just posterior to ventral sucker and pharynx, usually confluent in post-testicular zone and sometimes in forebody. Excretory vesicle reaches forward to level close to anterior testis. In intestine of freshwater fishes.

Comments

The following key is modified after Amin (1982b). All North American species of this genus have been redescribed by Caira (1989).

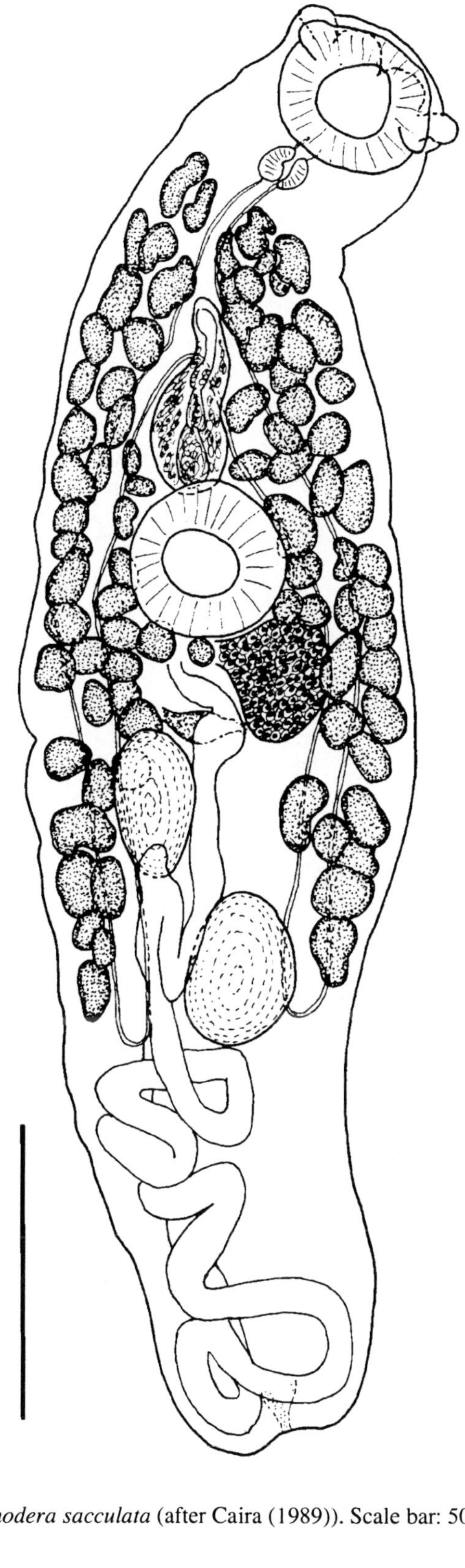

FIG. 64. *Bunodera sacculata* (after Caira (1989)). Scale bar: 500 μm.

Key to the species of *Crepidostomum*

1 Genital pore pre-bifurcal 2

Genital pore ventral or posterior to intestinal bifurcation 3

2 Sucker ratio 1:>1 .... *Crepidostomum farionis* (Müller, 1780) (Fig. 65)

Site: Intestine, pyloric caeca, gall bladder.

Hosts: *Catostomus catostomus* (38, 48); *Coregonus artedii* (30, 34, 35, 39, 59, 61); *C. clupeaformis* (4, 17, 19, 25, 26, 27, 34, 35, 40, 46, 59); *C. hoyi* (59); *C. sardinella* (45, 47); *Cottus cognatus* (29, 46); *Esox lucius* (34, 37); *Gasterosteus aculeatus* (14, 49); *Lota lota* (8, 26, 38, 46, 48); *Oncorhynchus clarki* (8, 26, 43); *O. gorbuscha* (42, 54, 59); *O. keta* (52); *O. kisutch* (8, 11, 12, 20, 38, 40, 43); *O. mykiss* (5, 7, 8, 13, 26, 29, 38, 46, 48, 58); *O. nerka* (8, 9, 10, 19, 29, 38, 46, 48, 55, 60); *O. tshawytscha* (29, 38, 46, 48); *Perca flavescens* (19, 53); *Prosopium cylindraceum* (1, 2, 4, 25, 59, 63); *P. williamsoni* (8, 26, 29, 38, 46, 48); *Salmo salar* (13, 15, 17, 18, 21, 22, 23, 24, 36, 67); *S. trutta* (13, 16); *Salvelinus alpinus* (17, 33, 39, 41, 45, 50, 51, 64); *S. fontinalis* (3, 4, 5, 6, 8, 13, 16, 17, 18, 26, 27, 31, 32, 56, 57, 59, 60, 62, 65, 66, 67); *S. fontinalis* × *S. namaycush* (19, 26, 28, 63); *S. malma* (8, 11, 12, 26, 29, 38, 46, 48); *S. namaycush* (3, 4, 17, 25, 26, 27, 44, 47, 48, 59); *Thymallus arcticus* (8, 25, 29, 46, 63); unspecified salmonid (20).

Distribution: Pac, Alta, BC, Lab, NB-b, NB, Nfld-b?, Nfld, NS, Ont, PEI, Que, YT, NWT, Man.

Records: 1. Bangham 1941 (Ont); 2. Bangham 1955 (Ont); 3. MacLulich 1943 (Ont); 4. Bangham and Venard 1946 (Ont); 5. Fantham and Porter 1948 (Que); 6. Choquette 1948 (Que); 7. Rawson 1953 (Alta); 8. Bangham and Adams 1954 (BC); 9. Margolis 1957 (BC); 10. Margolis 1963 (BC); 11. Arai 1967a (Pac); 12. Arai 1969b (Pac); 13. Sandeman and Pippy 1967 (Nfld); 14. Threlfall 1968 (Nfld); 15. Pippy 1969 (Lab, NB-b, NB, Nfld, NS, PEI); 16. Threlfall and Hanek 1970c (Nfld-b and/or Nfld); 17. Hicks and Threlfall 1973 (Lab); 18. Hare and Frantsi 1974 (NS); 19. Collins and Dechtiar 1974 (Ont); 20. Leong and Holmes 1974 (Alta); 21. Hare and Burt 1975a (NB); 22. Hare and Burt 1975b (NB); 23. Hare 1975 (NB); 24. Hare and Burt 1976 (NB- b, NB); 25. Arthur *et al.* 1976 (YT); 26. Mudry and Anderson 1977 (Alta, BC); 27. Chinniah and Threlfall 1978 (Lab); 28. Dechtiar and Berst 1978 (Ont); 29. Anonymous 1978 (BC); 30. Anthony 1978a (Ont); 31. Gordon *et al.* 1978 (Que); 32. Thompson and Threlfall 1978 (Que); 33. Curtis 1979 (NWT); 34. Watson, in Lubinsky and Loch 1979 (Man); 35. Watson and Dick 1979 (Man); 36. Pippy 1980 (PEI, NB, NS, Nfld, Lab); 37. Watson and Dick 1980 (Man); 38. Anonymous 1981 (BC); 39. Dick and Belosevic 1981 (NWT); 40. Leong and Holmes 1981 (Alta); 41. Curtis 1982 (NWT); 42. Margolis 1982 (Nfld); 43. Margolis and Moravec 1982 (BC); 44. Stewart and Bernier 1982 (NWT); 45. Stewart and Bernier 1983 (NWT); 46. Arai and Mudry 1983 (BC); 47. Stewart and Bernier 1984 (NWT); 48. Anonymous 1984 (BC); 49. Cone and Ryan 1984 (Nfld); 50. Curtis 1984 (Que); 51. Dick 1984 (NWT); 52. Whitaker 1985 (BC); 53. Poole and Dick 1985 (Man); 54. Anthony 1986 (Ont); 55. Bailey and Margolis 1987 (BC); 56. Frimeth 1987a (NB, NB-b); 57. Frimeth 1987b (NB, NB-b); 58. Szalai and Dick 1988 (Man); 59. Dechtiar and Lawrie 1988 (Ont); 60. Dechtiar *et al.* 1988 (Ont); 61. Curtis 1988 (Que/Lab); 62. Baggs and Cowan 1989 (Nfld); 63. Caira 1989 (YT); 64. Bouillon and Dempson 1989 (Lab); 65. Wright *et al.* 1989 (Que); 66. Albert and Curtis 1991 (Que); 67. Marcogliese and Cone 1991b (Nfld).

Remarks: A non-original record of this species in *Salvelinus fontinalis* was also published by Marcogliese and Cone (1991a).

Sucker-ratio 1:1 or <1 .... *C. isostomum* Hopkins, 1931

Syn.: *C. laureatum* of Cooper (1915) in part; *C. canadense* Hopkins, 1931.

Site: Intestine.

Hosts: *Cottus asper* (4); ?*Etheostoma exile* (1); *E. nigrum* (1, 5, 8); *Perca flavescens* (3); *Percina caprodes* (6, 9); *Percopsis omiscomaycus* (2, 5, 6, 7, 8, 9).

Distribution: BC, Ont, Que.

Records: 1. Cooper 1915 (Ont); 2. Bangham and Hunter 1939 (Ont); 3. Lyster 1940a (Que); 4. Bangham and Adams 1954 (BC); 5. Bangham 1955 (Ont); 6. Dechtiar 1972b (Ont); 7. Dechtiar and Lawrie 1988 (Ont); 8. Dechtiar *et al.* 1988 (Ont); 9. Dechtiar and Christie 1988 (Ont).

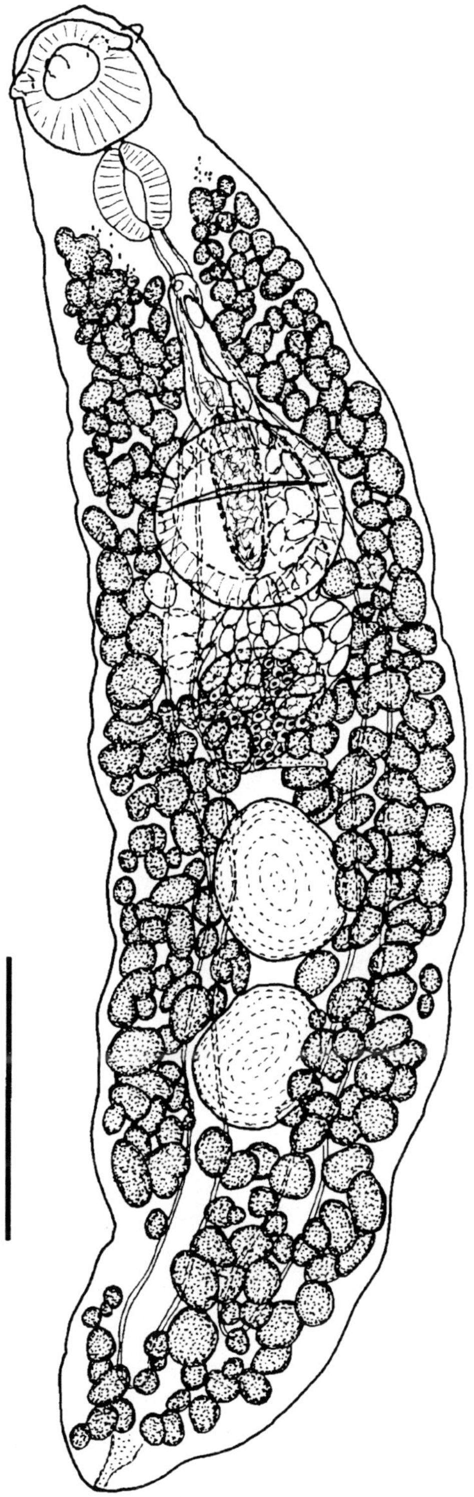

FIG. 65. *Crepidostomum farionis* (after Caira (1989)). Scale bar: 500 µm.

3 Vitelline follicles restricted to hindbody .... *C. brevivitellum* Hopkins, 1934

Site: Intestine.

Host: *Anguilla rostrata.*

Distribution: Lab, Que, NS.

Records: Lyster 1939 (Que); Fantham and Porter 1948 (Que); Hanek and Threlfall 1970b (Lab); Hanek and Molnar 1974 (Que); Cone *et al.* 1993 (NS).

Vitelline follicles extend into forebody 4

4 Cirrus sac stout, rarely extending into hindbody .... *C. auriculatum* (Wedl, 1858) Pratt, 1902

Syn.: *Acrodactyla petalosa* (Lander, in Looss, 1902); *C. lintoni* (Pratt, in Linton, 1901).

Site: Intestine.

Hosts: *Acipenser fulvescens* (1, 2, 3, 4, 5, 6, 7, 9, 10, 11, 13, 14); *A. transmontanus* (8, 12).

Distribution: Ont, Que, BC, Man Sask.

Records: 1. Stafford 1904 (Que?); 2. Cooper 1915 (Ont); 3. Lyster 1939 (Que); 4. Bangham and Hunter 1939 (Ont); 5. Bangham 1955 (Ont); 6. Dechtiar 1972b (Ont); 7. Anthony 1974 (Ont); 8. Margolis and McDonald 1986 (BC); 9. Dechtiar and Lawrie 1988 (Ont); 10. Dechtiar *et al.* 1988 (Ont); 11. Dechtiar and Christie 1988 (Ont); 12. Caira 1989 (BC); 13. Choudhoury *et al.* 1990 (Man, Sask); 14. Choudhoury and Dick 1993 (Man, Sask).

Remarks: Prior to Margolis and McDonald (1986), most Canadian records were under the name *C. lintoni*. This species is morphologically very similar to the Palaearctic species *C. auriculatum*, which also occurs in sturgeons. It is clear that material from Europe and North America still requires detailed comparison, but I have followed Skryabin and Koval (1966), Skryabina (1974), and Caira (1989) in considering these two species synonymous.

Cirrus sac slender, extending into hindbody 5

5 Mid-dorsal pair of oral papillae notched at tip .... *C. illinoiense* Faust, 1918

Syn.: *C. hiodontos* Hunter and Bangham, 1932.

Sites: Intestine, stomach, liver, swimbladder.

Host: *Hiodon tergisus.*

Distribution: Ont, Man.

Records: Hunter and Bangham 1932 (Ont); Bangham and Hunter 1939 (Ont); Glenn 1980 (Man).

Papillae not notched 6

6 Sucker-ratio usually 1:<0.9; lateral vitelline fields not normally confluent in post-testicular zone; pars prostatica in anterior half of cirrus sac; base of dorsal oral papillae contiguous .... *C. cornutum* (Osborn, 1903) Stafford, 1904

Site: Intestine, pyloric caeca, gall bladder.

Hosts: *Ambloplites rupestris* (1, 2, 3, 4, 6, 8, 10, 11, 14); "*Ameiurus nigricans*"* (15); *A. nebulosus* (5, 6); *Amia calva* (4, 12); *Anguilla rostrata* (8, 9); *Ictalurus melas* (4); *I. punctatus* (2); *Lepomis gibbosus* (5, 6, 7, 12, 16); *L. macrochirus* (14); *Micropterus dolomieui* (2, 3, 4, 5, 6, 7, 8, 13, 14, 15); *M. salmoides* (4, 6, 14); *Noturus gyrinus* (10); *Perca flavescens* (11); *Salvelinus fontinalis* (8).

Distribution: NS, Ont, Que, NB.

Records: 1. Stafford 1904 (Que?); 2. Cooper 1915 (Ont); 3. Lyster 1939 (Que); 4. Bangham and Hunter 1939 (Que); 5. Bangham 1941 (Ont); 6. Bangham 1955 (Ont); 7. Bangham and Venard 1946 (Ont); 8. Fantham and Porter 1948 (NS, Que); 9. Dechtiar 1972a (Ont); 10. Dechtiar 1972b (Ont); 11. Dechtiar and Lawrie 1988 (Ont); 12. Dechtiar *et al.* 1988 (Ont); 13. Dechtiar and Nepszy 1988 (Ont); 14. Dechtiar and Christie 1988 (Ont); 15. Caira 1989 (Ont); 16. BM(NH) collection (NB; imm.?).

---

*This is the name used on the USNM index card. It may be a lapsus for *Hypentelium nigricans*, but Caira (in litt.) has suggested that it may refer to *Ictalurus (Ameiurus) melas*.

Sucker-ratio close to 1:1; lateral vitelline fields confluent in post-testicular zone; pars prostatica in posterior half of cirrus sac; base of dorsal oral papillae contiguous...........*C. cooperi* Hopkins, 1931

Syn.: *C. ambloplitis* Hopkins, 1931; *C. solidum* Van Cleave and Mueller, 1932; *C. fausti* Hunninen and Hunter, 1933; *C. laureatum* of Stafford (1904) and Cooper (1915) in part; *Bunodera nodulosa* of Stafford (1904) in part; ?*C. cornutum* of Cooper (1915) in part.

Sites: Intestine, pyloric caeca, gall bladder.

Hosts: *Ambloplites rupestris* (3, 11, 33); *Catostomus commersoni* (12, 21); *Coregonus artedii* (28); *C. clupeaformis* (10, 12, 27); *Cottus cognatus* (27); *Cyprinus carpio* (13); *Esox lucius* (27); *E. niger* (13); *Fundulus heteroclitus* (13); *Gasterosteus aculeatus* (16, 17); *Lepomis gibbosus* (10, 11, 12, 21, 33); *Lota lota* (27); *Micropterus dolomieui* (8, 11, 21, 33); *Perca flavescens* (1, 2, 7, 8, 10, 11, 12, 13, 18, 19, 22, 23, 25, 26, 29, 30, 32, 33, 34, 35, 36); *Prosopium cylindraceum* (10); *Salmo salar* (27); *Salvelinus alpinus* (24); *S. fontinalis* (1, 4, 5, 6, 9, 13, 14, 15, 20, 24, 31); *S. namaycush* (27); *Stizostedion vitreum vitreum* (1); unspecified fishes (6).

Distribution: Lab, Man, NB, Nfld-b?, Nfld, NS, Ont, PEI, Que-b, Que.

Records: 1. Stafford 1904 (Que?); 2. Cooper 1915 (Ont); 3. Hopkins 1931 (Ont); 4. Richardson 1936 (Que); 5. Richardson 1937 (Que); 6. Richardson 1942 (Lab, NB, NS, PEI, Que); 7. Bangham and Hunter 1939 (Ont); 8. Lyster 1939 (Que); 9. Lyster 1940a (Que); 10. Bangham 1941 (Ont); 11. Bangham 1955 (Ont); 12. Bangham and Venard 1946 (Ont); 13. Fantham and Porter 1948 (NS, Que); 14. Choquette 1948 (Que); 15. Choquette 1954 (Que); 16. Hanek and Threlfall 1969b (Nfld); 17. Hanek and Threlfall 1970d (Nfld); 18. Tedla and Fernando 1969a (Ont); 19. Tedla and Fernando 1972 (Ont); 20. Threlfall and Hanek 1970c (Nfld-b and/or Nfld); 21. Dechtiar 1972b (Ont); 22. Cannon 1972 (Ont); 23. Cannon 1973 (Ont); 24. Hanek and Molnar 1974 (Que-b, Que); 25. Stewart-Hay, in Lubinsky 1976 (Man); 26. Dickson, in Lubinsky 1976 (Man); 27. Chinniah and Threlfall 1978 (Lab); 28. Anthony 1978a (Ont); 29. Anthony 1978b; 30. Stewart-Hay and Dickson, in Lubinsky and Loch 1979 (Man); 31. Black 1981 (Que); 32. Poole and Dick 1985 (Man); 33. Dechtiar *et al.* 1988 (Ont); 34. Dechtiar and Nepszy 1988 (Ont); 35. Dechtiar and Christie 1988 (Ont); 36. Szalai and Dick 1991 (Man).

Remarks: This species is morphologically very similar to the Palaearctic species *C. metoecus* (Braun, 1900). According to Skryabin and Koval (1966), these two species differ in the anterior limit of the vitellarium (not reaching the level of the pharynx in *C. cooperi*) and in the presence and absence of vitelline follicles dorsal to the caeca in the anterior hindbody (absent in *C. cooperi*, according to Hopkins 1934). Neither of these differences appears to be clear cut, however, since the anterior limit of the vitelline field may not reach the pharynx in extended specimens of *C. metoecus*, and Van Cleave and Mueller's (1932) figure of *C. cooperi* (as *C. solidum*) clearly shows vitelline follicles dorsal to the caeca. Nevertheless, it is clear that *C. cooperi* occurs much more frequently in non-salmonoid hosts than *C. metoecus*. Initially I considered it likely that North American records of *C. metoecus* were *C. cooperi*, but in the light of Caira's (1989) revision and material that I have seen from *Salvelinus fontinalis*, I accept the presence of *C. metoecus*.

Marcogliese and Cone (1991a) mentioned the presence of this species in Salvelinus fontinalis, but this is not an original record.

Sucker-ratio 1:>1; vitelline fields confluent in post-testicular zone; pars prostatica in anterior half of cirrus sac; base of dorsal oral papillae not contiguous...*C. metoecus* (Braun, 1900) (Braun, 1900)

Site: Intestine.

Hosts: *Coregonus clupeaformis* (1); *Oncorhynchus clarki* (2); *O. kisutch* (2; see "Remarks"); *Salvelinus fontinalis* (3); *S. namaycush* (1); *Thymallus arcticus* (1).

Distribution: YT, NB, BC (see "Remarks").

Records: 1. Arthur *et al.* 1976 (YT); 2. Margolis and Moravec 1982 (BC); 3. BM(NH) collection (NB).

Remarks: See "Remarks" on *C. cooperi*. Caira (1989) referred to specimens from British Columbia: these (Caira, in litt.) were from *Oncorhynchus kisutch*.

CREPTOTREMA Travassos, Artigas, and Pereira, 1928

Allocreadiidae. Body small, elongate-oval to fusiform. Oral sucker with one pair of lateral, muscular, papilliform lobes. Ventral sucker slightly larger than oral sucker, in anterior half of worm. Pharynx and oesophagus well developed. Intestinal bifurcation in posterior half of forebody. Caeca terminate near posterior end of body or lateral to posterior testis (when post-testicular zone is short). Testes two, symmetrical, oblique, or tandem, near middle of hindbody or more posterior. Cirrus sac elongate-claviform, usually reaching close to posterior margin of ventral sucker, containing elongate, saccular, straight, or sinuous seminal vesicle, pars prostatica, and eversible ejaculatory duct. Genital pore ventro-median at level of intestinal bifurcation or just anterior. Ovary entire, dextro-median, in anterior hindbody, separated from testes by coils of uterus in fully developed worms. Uterine field between mid-testicular level and ventral sucker. Vitelline follicles often large, restricted to lateral fields between level of pharynx and level close to posterior extremity, fields not normally confluent. Excretory vesicle reaching to about level of anterior testis. In intestine of freshwater teleosts and amphibians.

Creptotrema funduli Mueller, 1934 (Fig. 66)

Syn.: ?*Allocreadium commune* of Cooper (1915) in part.

Site: Intestine.

Hosts: *Fundulus diaphanus* (1, 3); *Perca flavescens* (6); *Umbra limi* (2, 4, 5).

Distribution: NS, Ont, Man.

Records: 1. ?Cooper 1915 (Ont); 2. Bangham 1955 (Ont); 3. Wiles 1975 (NS); 4. Dechtiar *et al.* 1988 (Ont); 5. Dechtiar and Christie 1988 (Ont); 6. Szalai and Dick 1991 (Man).

Remarks: *Creptotrema* differs from *Crepidostomum* in that there are claimed to be only two lateral rather than two lateral and four dorsal muscular, papilliform lobes associated with the oral sucker. As far as I am aware, this species has only been described by Mueller (1934), so the absence of dorsal lobes requires confirmation, since these lobes were not observed by Surber (1928) in his original description of *Megalogonia ictaluri*. Furthermore, the presence of the lateral lobes is also in doubt, as Manter (1962), who studied syntypes, stated that they are very inconspicuous or lacking since the anterior region of the body is more or less folded to resemble a pair of papillae. Nevertheless, several workers have subsequently been able to identify *C. funduli* without comment. Manter also drew attention to a conspicuous, finger-like lobe that projects inwards from the dorsal wall of the oral sucker. He also suggested that the species might be an opecoelid close to *Plagioporus*; but the typical allocreadiid arrangement of the uterus passing between the ovary and the anterior testis tends to preclude this, although it does have a superficial resemblance to *P. sinitsini*, which was described by Mueller (1934) in the same paper as *C. funduli*.

CULAEATREMA Lasee, Font, and Sutherland, 1988

Allocreadiidae. Small worms, body fusiform. Eye-spots apparently absent. Oral sucker without lobes. Ventral sucker anterior to middle of body, larger than oral sucker. Pharynx well developed. Oesophagus relatively long. Intestinal bifurcation just posterior to middle of forebody. Caeca relatively long, reaching to testes, blind. Testes two, oblique, in hindbody, occasionally absent. Cirrus sac relatively small, claviform, overlaps ventral sucker, contains elongate seminal vesicle. Genital pore ventro-median, close to level of intestinal bifurcation. Ovary entire, pre-testicular, median or submedian. Uterine field fills much of hindbody. Eggs relatively large, not excessively numerous, embryonated. Vitelline follicles irregular, numerous, in lateral fields, extending between levels of pharynx and testes. Excretory vesicle reaches to level of posterior margin of ovary. In intestine of freshwater fishes (*Culaea*).

Comments

Following a cladistic study, Brooks (1992) considered this genus a synonym of *Bunodera*. This has not been followed here simply because to do so would complicate the keying of other allocreadiids. The

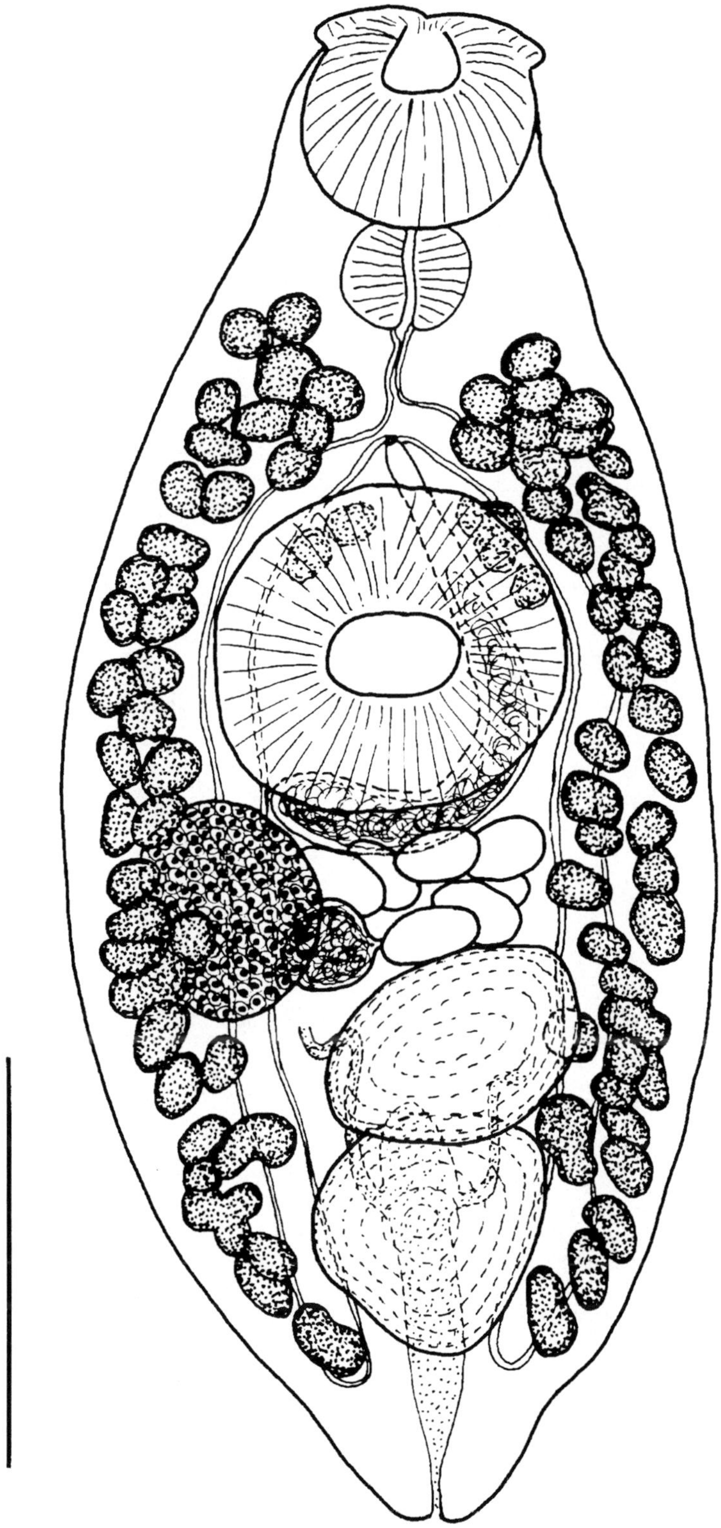

FIG. 66. *Creptotrema funduli* (modified after Mueller (1934)). Scale bar: 250 µm.

absence (when interpreted as an apparent loss) of a character is often very useful in keying down taxa, yet often given little weight in phylogenetic studies.

Culaeatrema inconstans Lasee, Font, and Sutherland, 1988 (Fig. 67)

Syn.: *Bunodera inconstans* (Lasee, Font, and Sutherland, 1988) Brooks, 1992.
Site: (Intestine).
Host: *Culaea inconstans*.
Distribution: Ont.
Record: Brooks 1992.
Remarks: Brooks (1992) considered this species to be closely related to *Bunodera eucaliae*, which also occurs in *Culaea inconstans* in Canadian waters.

MEGALOGONIA Surber, 1928

Allocreadiidae. Body small, elongate-oval to elongate-pyriform. Oral sucker with six muscular, papilliform lobes. Ventral sucker similar in size to oral sucker, in anterior half of body. Oesophagus short to medium-sized. Intestinal bifurcation near middle of forebody. Caeca terminate close to posterior extremity. Testes apparently four (i.e., two tandem testes completely or partly, depending upon state of development, split longitudinally into four masses), near middle of hindbody. Cirrus sac elongate-claviform, sinuous or curved, reaching into anterior hindbody, containing elongate, saccular seminal vesicle, pars prostatica, and ejaculatory duct. Genital pore ventro-medial in forebody, posterior to intestinal bifurcation. Ovary entire, dextro-medial, midway between testes and ventral sucker, often separated from testes by loops of uterus. Uterine field variable in distribution, in young worms, between testes and ventral sucker, in fully developed worms extends medially between lateral testicular masses and may reach close to posterior extremity. Vitelline follicles in lateral fields between level of pharynx and posterior extremity, fields normally confluent in forebody and post-testicular zone. Excretory vesicle reaches posterior margin of anterior testicular masses. In intestine of freshwater teleosts.

Comments

This monotypic genus is clearly very closely related to, and has arisen from, *Crepidostomum*. In fact very young specimens are morphologically very similar to *C. cooperi*. There is a case for regarding *Megalogonia* as a synonym of *Crepidostomum*, as proposed by Van Cleave and Mueller (1934) and Caira (1989); but fully developed worms are so different from *Crepidostomum* spp. that major changes in the generic diagnosis would have to be made to accommodate them.

Megalogonia ictaluri Surber, 1928 (Fig. 68)

Syn.: *Crepidostomum ictaluri* (Surber, 1928) Van Cleave and Mueller, 1934.
Site: Intestine.
Hosts: *Ameiurus nebulosus* (3, 4, 5, 6, 7, 9); *Ictalurus melas* (2); *I. punctatus* (1, 2, 4, 7, 8); *Noturus flavus* (2, 9); *N. gyrinus* (9); *N. miurus* (2).
Distribution: Ont, Que.
Records: 1. Lyster 1939 (Que); 2. Bangham and Hunter 1939 (Ont); 3. Bangham 1941 (Ont); 4. Bangham 1955 (Ont); 5. Dechtiar 1972b (Ont); 6. Fréchette *et al.* 1978 (Que); 7. Dechtiar *et al.* 1988 (Ont); 8. Dechtiar and Nepszy 1988 (Ont); 9. Dechtiar and Christie 1988 (Ont).
Remarks: This species was redescribed by Caira (1989) as *Crepidostomum ictaluri*.

PLAGIOCIRRUS Van Cleave and Mueller, 1932

Allocreadiidae. Body small, oval to elongate-oval. Body surface unarmed. Oral sucker without lobes. Ventral sucker larger than oral sucker, in anterior half of worm. Pharynx and oesophagus well developed.

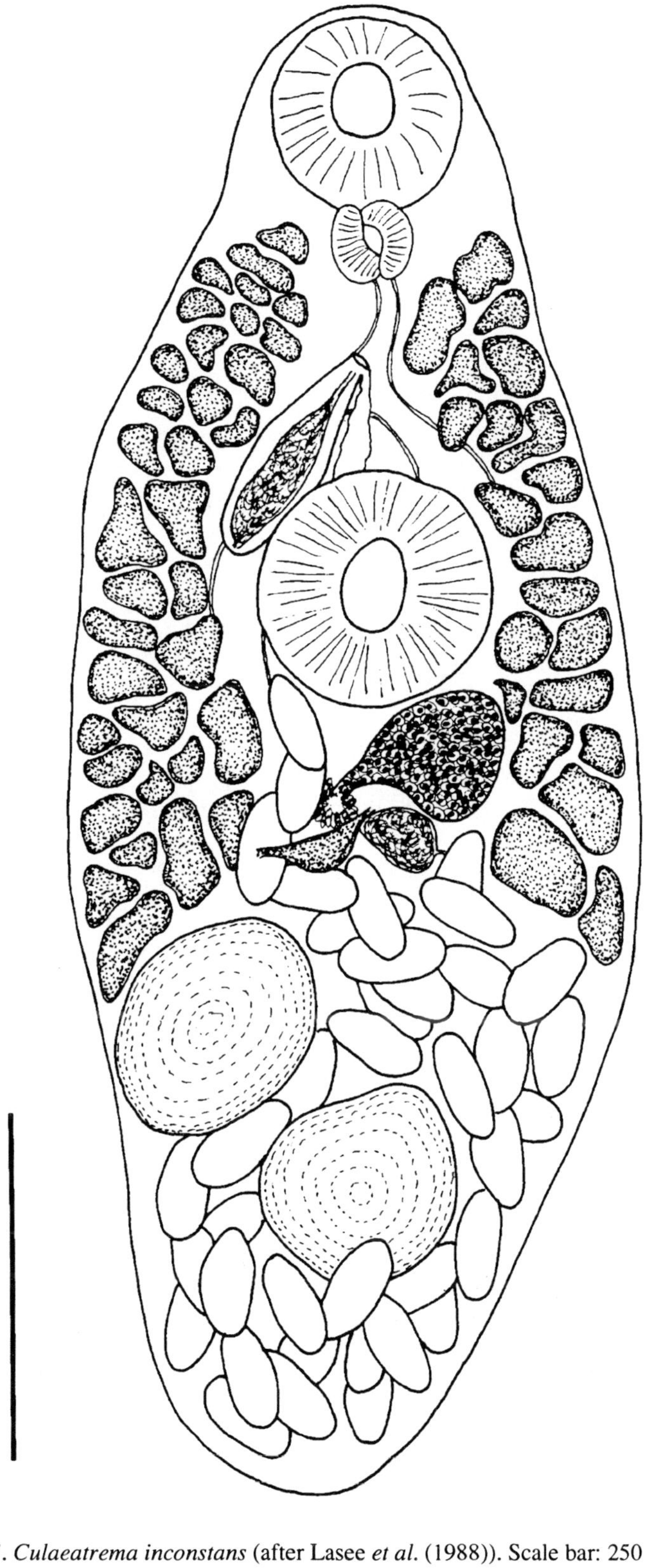

FIG. 67. *Culaeatrema inconstans* (after Lasee *et al.* (1988)). Scale bar: 250 μm.

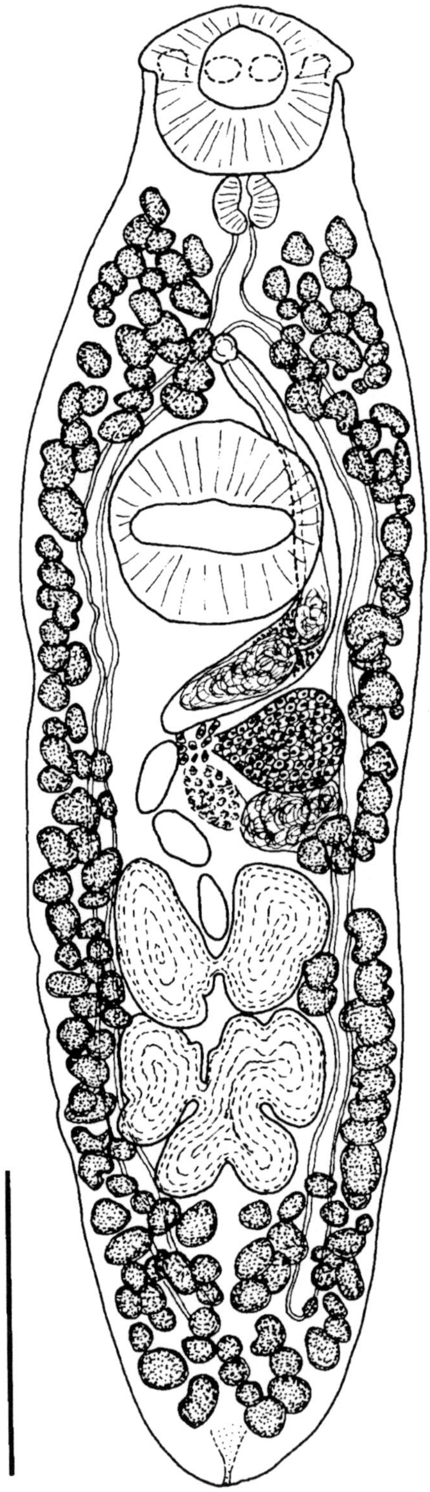

FIG. 68. *Megalogonia ictaluri* (after Caira (1989)). Scale bar: 250 μm.

Intestinal bifurcation in middle or posterior half of forebody. Caeca terminate close to posterior extremity. Testes two, tandem to slightly oblique, in posterior third of body. Cirrus sac elongate, claviform, containing saccular, bipartite seminal vesicle, pars prostatica, and ejaculatory duct, reaching close to posterior margin of ventral sucker. Genital pore sinistrally sublateral at level of oesophagus. Ovary entire, median to dextrally submedian, in middle or just inside posterior half of body, almost contiguous with or separated by loops of uterus from anterior testis. Uterine field extends between ventral sucker and posterior extremity, enveloping gonads. Vitelline follicles restricted to two small lateral fields extending between level of ovary and posterior margin of ventral sucker, fields not confluent. Excretory vesicle reaches almost to ovary. In intestine of freshwater teleosts.

Comments

The position of the genus will be problematical until details of the life history are known, although most authors have included it within the Allocreadiidae. Margolis and Arthur (1979) listed Bangham and Adams' (1954) record discussed below under the Lissorchiidae. In fact, the genus does closely resemble *Lissorchis* spp., which also occur in *Catostomus* spp., apart from the more anterior position of the genital pore and the absence of spines on the body. There may well, therefore, have been some confusion between these forms.

Plagiocirrus primus Van Cleave and Mueller, 1932

Site: (?intestine).
Host: *Notomigonus crysoleucas*.
Distribution: Ont.
Record: Dechtiar and Christie 1988.

Plagiocirrus sp. of Bangham and Adams (1954) (Fig. 69)

Site: (?intestine).
Host: *Catostomus catostomus*.
Distribution: BC.
Record: Bangham and Adams 1954.
Remarks: Bangham and Adams mentioned the presence of long specimens of an undescribed species of *Plagiocirrus*. Two species are known, *P. primus* Van Cleave and Mueller, 1932 and *P. testeus* Fritts, 1959, the latter being recorded from *Catostomus macrocheilus* in Idaho. When compared, these two species are morphologically very similar and may prove to be synonymous.

A *Plagiocirrus* sp. was also reported by Hogue *et al.* (1993) in *Catostomus catostomus* from American waters of Lake Superior.

POLYLECITHUM Arnold, 1934

Allocreadiidae. Body medium-sized, elongate to fusiform. Oral sucker without lobes. Ventral sucker just inside anterior half of body. Intestinal bifurcation in anterior half of forebody. Caeca end close to posterior extremity. Testes two, tandem, near middle of hindbody. Cirrus sac small, entirely or almost entirely in forebody, enclosing coiled seminal vesicle, pars prostatica, and ejaculatory duct. Genital pore median, well posterior to intestinal bifurcation. Ovary submedian, close to posterior margin of ventral sucker, separated from testes by major part of uterine field. Uterine field between posterior testis and ventral sucker, may overlap caeca laterally. Lateral vitelline fields extend from level of pharynx to posterior extremity (? sometimes with gap at level of ventral sucker), fields confluent in forebody and in post-testicular zone. Excretory vesicle (?). In intestine of freshwater fishes.

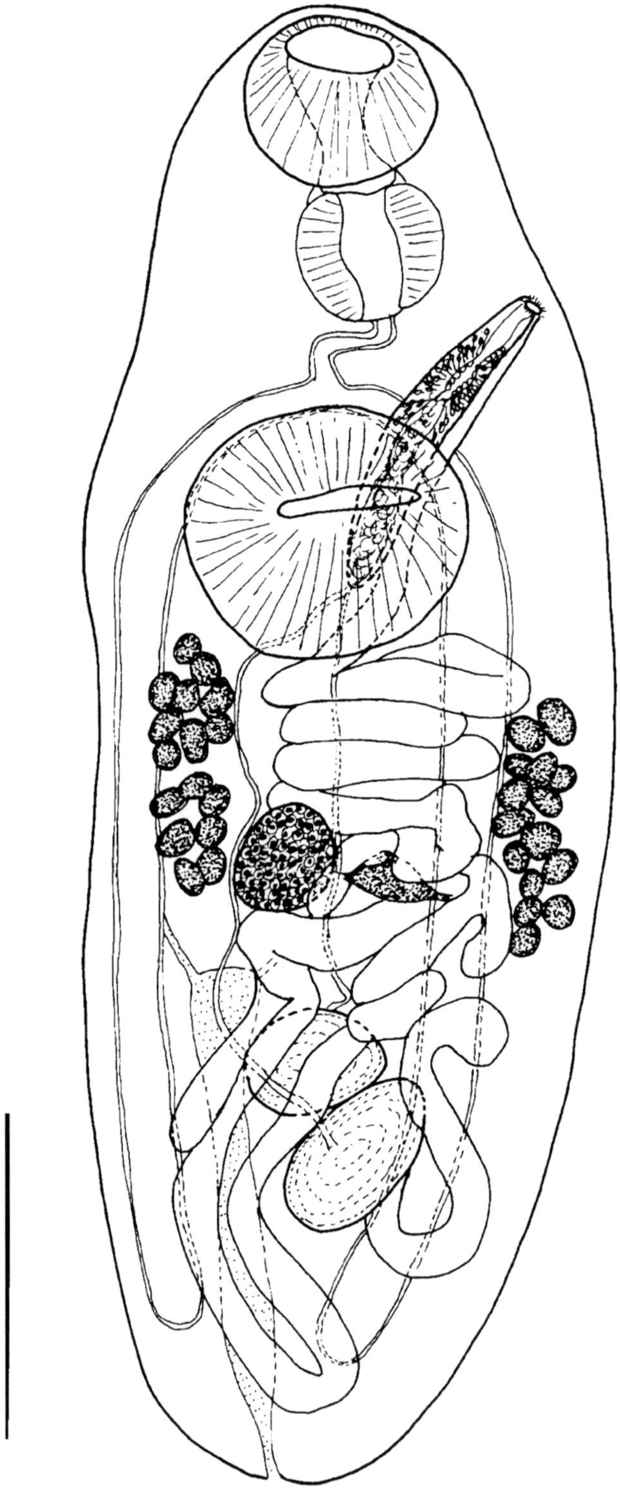

FIG. 69. *Plagiocirrus primus* (after Yamaguti (1971)). Scale bar (roughly estimated): 250 μm.

Comments

This genus differs from *Allocreadium* only in that the anterior limit of the lateral vitelline fields extends into the forebody. The use of this feature as a generic criterion is very inconsistent, being generally accepted in the Opecoelidae but not, for example, in the allocreadiid genus *Crepidostomum*.

Polylecithum ictaluri (Pearse, 1924) Arnold, 1934 (Fig. 70)

Syn.: *Allocreadium ictaluri* Pearse, 1924; *A. halli* Mueller and Van Cleave, 1932.

Site: Intestine.

Hosts: *Ameiurus nebulosus* (1, 2, 4); *Ictalurus melas* (3).

Distribution: Ont, Que.

Records: 1. Bangham 1941 (Ont); 2. Bangham and Venard 1946 (Ont); 3. Worley and Bangham 1952 (Que); 4. Dechtiar 1972b (Ont).

Family OPECOELIDAE Ozaki, 1925

Allocreadioidea. Body small to medium-sized, oval to cylindrical, usually dorso-ventrally flattened. Tegument smooth, unarmed. Non-oculate. Oral and ventral suckers present. Ventral sucker usually larger, embedded to pedunculate, may be lobed or bear papillate or tentacular projections. Prepharynx, pharynx, and oesophagus present, prepharynx short. Intestinal bifurcation in forebody. Caeca usually end blindly in hindbody, but may form cyclocoel or open as uroproct, twin ani or single anus. Testes normally two, in tandem, oblique, or occasionally symmetrical, in hindbody, rarely numerous testes present. Cirrus sac present or absent, restricted to forebody to extending deep into hindbody. Seminal vesicle, pars prostatica, and ejaculatory duct internal or external. Genital atrium normally small. Genital pore ventral in forebody, pre- or post-bifurcal, median to sublateral. Accessory suckers occasionally present. Ovary pre-testicular or occasionally lateral to anterior testis, entire to lobed. Laurer's canal present; opens dorsally. Canalicular seminal receptacle or uterine seminal receptacle present, each mutually exclusive. Uterine field normally between gonads and ventral sucker. Eggs relatively large, usually numerous, operculate, occasionally with polar filaments. Vitellarium follicular, distribution variable, but usually present in lateral fields and post-testicular field (except Stenakrinae), restricted to hindbody or extending into forebody. Excretory vesicle I-shaped, tubular. In intestine of marine and freshwater teleosts.

Comments

The classification of this family is based upon Gibson and Bray (1982, 1984). The group is one that is very difficult to work with, as in gross morphology the numerous taxa are very similar. The divisions used, especially at the generic level, are essentially phenetic.

Three subfamilies are recognized in the Canadian fauna (see the following key). A fourth subfamily, the Opecoelininae Gibson and Bray, 1984, has been recorded as occurring in Washington State waters just to the south of Vancouver Island in the form of the genus *Opecoelina* Manter, 1934, as *O. theragrae* Lloyd, 1938 and *O. radifistuli* (Acena, 1941); but both of these species are in fact species of the plagioporine genus *Podocotyle* Dujardin, 1845 (cf. Gibson and Bray 1984; Gibson 1986). The same is probably also true of *Opecoelina* sp. of Kabata and Whitaker (1984), a single specimen of which these authors recorded from the intestine of *Anoplopoma fimbria* off British Columbia. I have examined this worm; although in very poor condition, it does resemble a *Podocotyle*.

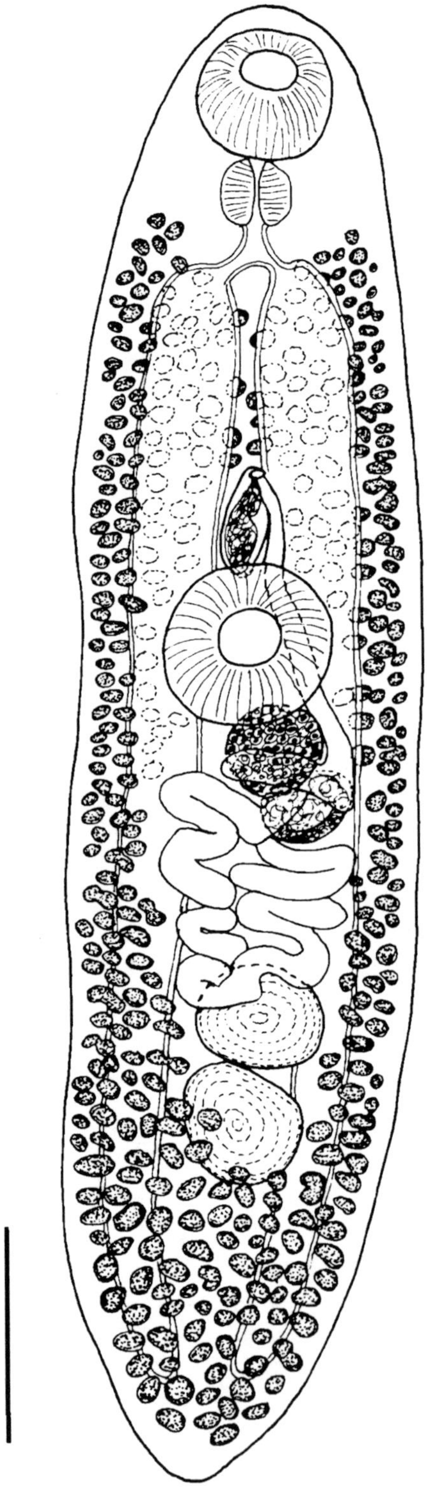

FIG. 70. *Polylecithum ictaluri* (after Mueller and Van Cleave (1932; as *Allocreadium halle*)). Scale bar: 500 µm.

Key to the subfamilies of the Opecoelidae

1 Distinct cirrus sac present surrounding internal seminal vesicle; naked seminal vesicle absent 2

Distinct functional cirrus sac surrounding internal seminal vesicle absent; naked seminal vesicle normally present (cirrus sac occasionally present as membranous sac surrounding seminal vesicle or surrounding distal extremity of male duct); uterine seminal receptacle present; canalicular seminal receptacle absent .... Opecoelinae

2 Uterine seminal receptacle present; canalicular seminal receptacle absent; post-testicular field usually very short .... Stenakrinae

Canalicular seminal receptacle present; uterine seminal receptacle absent; post-testicular field usually (but not always) large and containing vitelline follicles .... Plagioporinae

Subfamily OPECOELINAE Ozaki, 1925

Opecoelidae. Medium-sized oval to cylindrical worms. Ventral sucker often pedunculate, often lobed or bearing papillate or tentacular projections. Accessory sucker occasionally present. Caeca usually open through anus, pair of ani or uroproct, but may end blindly. Testes two, usually tandem, occasionally oblique. Functional muscular cirrus sac absent, vestige may remain surrounding region of ejaculatory duct or as membrane surrounding seminal vesicle and pars prostatica. Seminal vesicle normally external. Ovary pre-testicular, entire or lobed. Canalicular seminal receptacle absent. Uterine seminal receptacle present. Eggs usually numerous, large, without filaments. Vitelline fields usually restricted to hindbody, occasionally extend into forebody, normally confluent in post-testicular region. In intestine of marine, rarely freshwater, teleosts.

Key to the genera of the Opecoelinae

1 Accesssory sucker present (may be indistinct) between genital pore and ventral sucker 2

Accessory sucker absent 3

2 Caeca blind; ventral sucker normally embedded or semi-embedded .... *Genitocotyle*

Caeca open into base of excretory vesicle to form uroproct; ventral sucker normally pedunculate .... *Opecoeloides*

3 Seminal vesicle naked; ventral sucker embedded, without lobed margins; gut caeca end blindly .... *Pseudopecoelus*

Seminal vesicle surrounded by thin, membranous sac; ventral sucker somewhat pedunculate, with muscular lobes on margins; gut caeca open through separate ani ventrally to excretory pore .... *Anomalotrema*

ANOMALOTREMA Zhukov, 1957

Opecoelinae. Elongate, medium-sized worms. Oral sucker ventro-terminal. Ventral sucker larger than oral, semi-embedded to pedunculate, in anterior half of worm, distinctively lobed, with two lobes on anterior and one median interlocking lobe on posterior margin. Accessory sucker absent. Pharynx and oesophagus well developed. Intestinal bifurcation in posterior forebody. Caeca open side by side through separate ani ventrally to excretory pore (?or unite to form common anus). Testes in tandem, large, irregularly oval or indented. Cirrus sac non-functional, reduced to amuscular membrane surrounding seminal vesicle, pars prostatica, and ejaculatory duct. Seminal vesicle tubular, coiled, widest proximally. Genital pore pre-bifurcal, ventrally sinistro-medial at level of oesophagus. Ovary pre-testicular, indistinctly or distinctly tri-lobed to campaniform. Uterine field between ovary and ventral sucker. Vitelline follicles restricted to hindbody, in lateral fields, often interrupted at level of testes and ovary, confluent in post-testicular zone, often between testes and occasionally between ovary and anterior testis. Excretory vesicle reaching to level close to ovary. In intestine of marine teleosts.

Key to the species of *Anomalotrema*

1 Usually less than 3 mm in length (but may reach 4.5 mm); ovary clearly 3–4 lobed; sucker-ratio 1:1.8–2.4; in Atlantic waters............................*Anomalotrema koiae* Gibson and Bray, 1984 (Fig. 71)

Syn.: "Opecoelidae" of Bourgeois and Ni (1984).

Site: Intestine.

Hosts: *Hippoglossus hippoglossus* (5); *Reinhardtius hippoglossoides* (5); *Sebastes fasciatus* (4); *S. marinus* (1, 2); *S. mentella* (1); *Urophycis chesteri* (3).

Distribution: Atl.

Records: 1. Gibson and Bray 1984; 2. Bourgeois and Ni 1984; 3. Scott 1987; 4. Scott 1988; 5. Scott and Bray 1989.

Remarks: The Canadian material described by Gibson and Bray (1984) was collected by Dr. J.S. Scott in the Bay of Fundy and Passamaquoddy Bay. This species occurs in both scorpaenid and gadid fishes in Northeast Atlantic waters. In agreement with Scott (1988), the record of "Opecoelidae" of Bourgeois and Ni (1984) is likely to have been this species.

Up to 7.4 mm in length (but may be much smaller); ovary indistinctly lobed or bell-shaped; sucker-ratio 1:1.47–1.6; in Pacific waters ....*A. putjatini* Zhukov, 1957

Site: Intestine.

Host: *Hemilepidotus hemilepidotus*.

Distribution: Pac.

Record: Gibson and Bray 1984.

Remarks: The material discussed by Gibson and Bray (1984) was collected by Dr. H.P. Arai off Vancouver Island. This species has been recorded from hexagrammid, cottid, gadid, and agonid fishes in the Sea of Japan, off the Chukotsk Peninsula, and in Peter the Great Bay in the northern Northwest Pacific.

GENITOCOTYLE Park, 1937

Opecoelinae. Body small to medium-sized. Ventral sucker larger than oral sucker, in anterior half of body, embedded to semi-embedded, may be surrounded by folds of body wall, occasionally with lobes or papilli-form projections. Accessory sucker present, large to small, between genital pore and ventral sucker. Pharynx oval. Oesophagus long to medium-sized. Intestinal bifurcation in posterior half of hindbody. Caeca end blindly close to posterior extremity. Testes tandem, close together, close to middle of hindbody. Cirrus sac absent. Seminal vesicle external, tubular, long, reaching into hindbody, convoluted or straight. Pars prostatica distinctive, external gland cells may be few or indistinct. Ejaculatory duct short. Genital atrium small, may be glandular. Genital pore ventrally sinistro-submedian at level of oesophagus. Ovary bi-, tri-, or irregularly lobed, closely anterior to testes. Uterine field between ovary and ventral sucker. Uterine seminal receptacle present (canalicular seminal receptacle reported in literature may be small dilation of Laurer's canal). Vitelline follicles in lateral field, confluent in post-testicular region and reach anteriorly as far as ventral sucker, only as far as testes, or into forebody. Excretory vesicle reaches to level of ovary. In intestine of marine and euryhaline migratory teleosts.

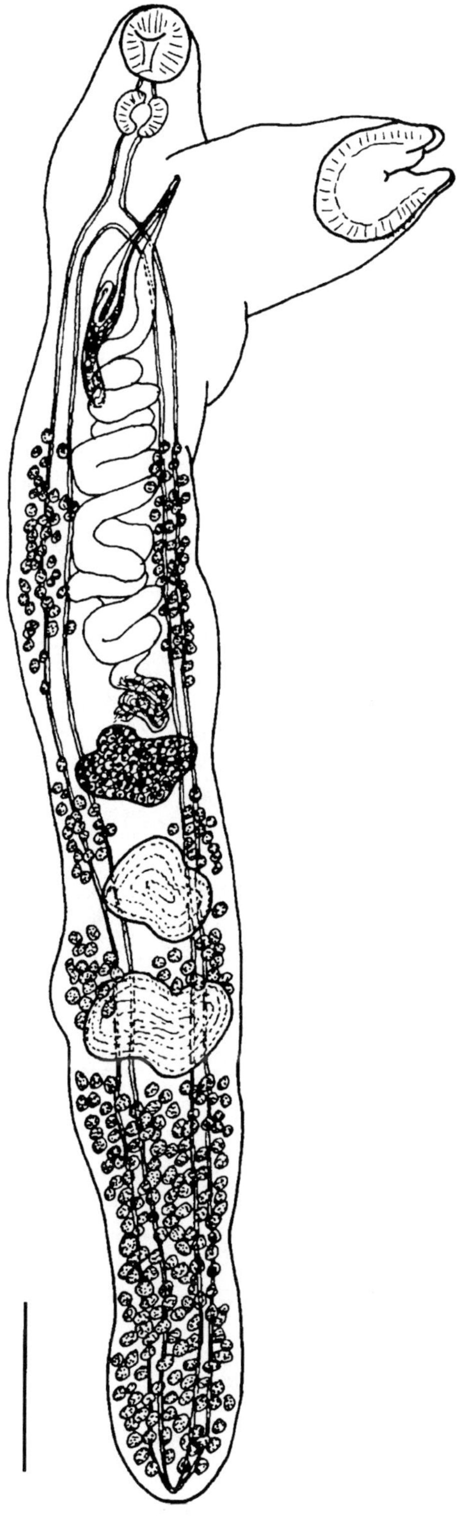

FIG. 71. *Anomalotrema koiae* (after Gibson and Bray (1984)). Scale bar: 500 µm.

Key to the species of *Genitocotyle*

1 Atlantic coast; vitelline follicles extend into forebody .... *Genitocotyle atlantica* Manter, 1947 (Fig. 72)

Site: Intestine.
Host: *Alosa sapidissima.*
Distribution: Atl, NS.
Record: Hogans *et al.* 1993.
Remarks: This is a parasite of marine fishes further south in American waters, but apparently extends its range north into Canadian waters in this migratory clupeid.

Pacific coast; vitelline follicles do not extend into forebody .... *G. acirrus* Park, 1937

Site: Intestine.
Host: *Pleuronectes vetulus.*
Distribution: Pac.
Record: Kabata and Whitaker 1984.
Remarks: Although this species (also called *G. acirra* and *G. acirrata*) occurs in a wide range of hosts off the Pacific coast of the United States, it is especially prevalent in embiotocids.

OPECOELOIDES Odhner, 1928

Syn. *Cymbephallus* Linton, 1934.

Opecoelinae. Body small to medium-sized, elongate. Ventral sucker well developed, usually larger than oral sucker, in anterior half of body, usually pedunculate, occasionally apparently semi-embedded, with digitate or papillate marginal projections (normally six, occasionally five). Accessory sucker present, between genital pore and ventral sucker. Pharynx often large. Oesophagus short to long. Intestinal bifurcation in posterior forebody, sometimes dorsal to peduncle of ventral sucker. Caeca long, narrow, open into excretory vesicle close to posterior end of body to form uroproct. Testes tandem, contiguous or apart, in mid- to posterior half of hindbody, entire to lobed. Cirrus sac absent. Seminal vesicle external, long, tubular, usually extending well into hindbody. Pars prostatica present, external gland cells may be indistinct. Ejaculatory duct short. Genital pore sinistro-submedian, pre-bifurcal. Ovary just anterior to testes, entire to lobed, median or submedian. Uterine field between ovary or anterior testis and ventral sucker. Uterine seminal receptacle present. Lateral vitelline fields normally restricted to hindbody, confluent in post-testicular zone and sometimes between testes. Excretory vesicle (?) usually reaches to about level of ovary. In intestine of marine teleosts.

Comments

Since the connection between the caeca and the excretory vesicle is often obscured by vitelline follicles and is thus difficult to see in wholemounts, there has obviously been some confusion between members of this genus and *Genitocotyle*.

Opecoeloides vitellosus (Linton, 1900) von Wicklen, 1946 (Fig. 73)

Syn.: *Cymbephallus vitellosus* Linton, 1900.
Site: Intestine.
Hosts: *Eopsetta exilis* (1); *Rhinochimaera atlantica* (2).
Distribution: Pac, Atl.
Records: 1. Ronald 1959 (Pac); 2. Hogans and Hurlbut 1984 (Atl).
Remarks: I am extremely sceptical about both of these records. *O. vitellosus* normally occurs off the Atlantic coast of the United States, so the Pacific records may refer to *Genitocotyle acirrus*, which has been recorded from a range of hosts off the Pacific coast of the United States and Canada or to some other opecoeline. The record from the holocephalan is presumably accidental and is perhaps more likely to be a deep-water form, such as *Anomalotrema*.

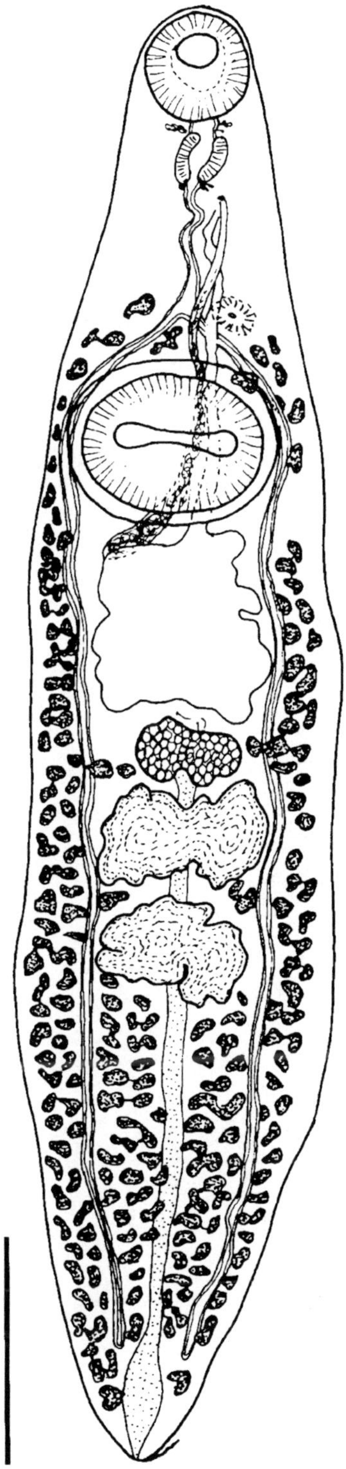

FIG. 72. *Genitocotyle atlantica* (after Manter (1947)). Scale bar: 250 µm.

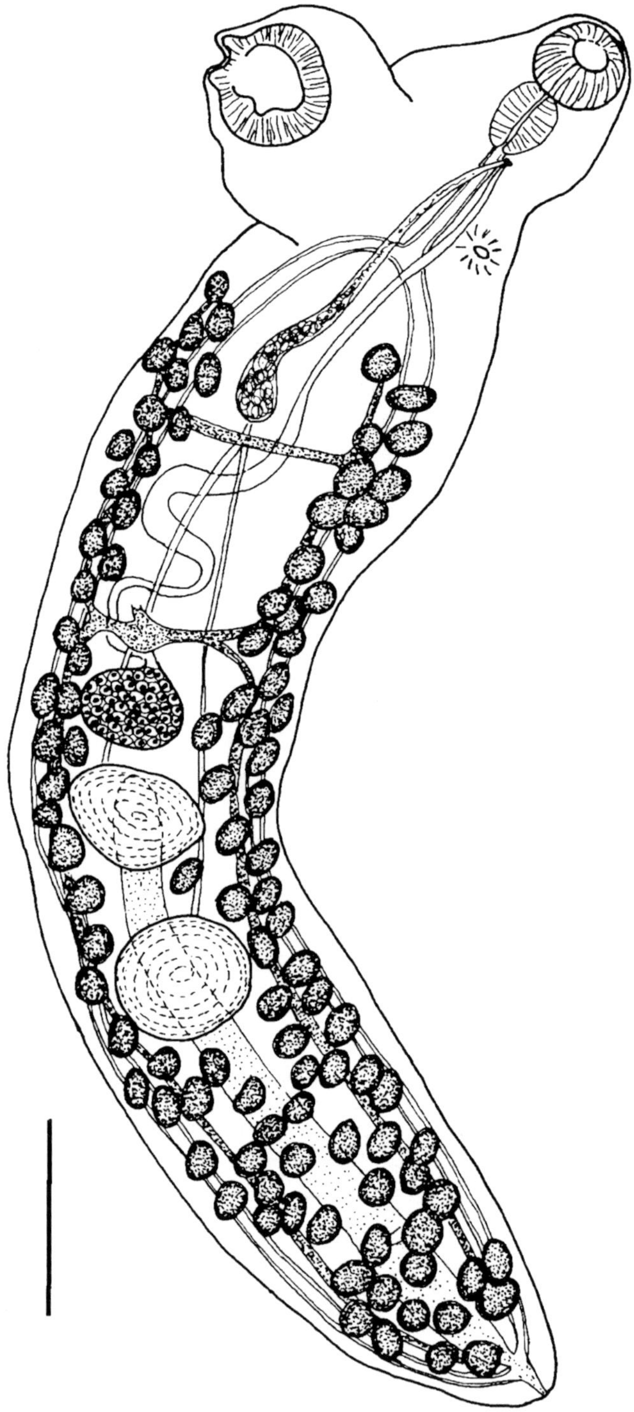

FIG. 73. *Opecoeloides vitellosus* (modified after Hunninen and Cable (1941: as *Anisoporus manteri*)). Presence of transverse vitelline commissures requires confirmation. Scale bar: 250 µm.

PSEUDOPECOELUS von Wicklen, 1946

Opecoelinae. Body small to medium-sized, elongate. Ventral sucker normally larger than oral sucker, embedded or protuberant, in anterior half of worm, without lobes or papilliform projections. Accessory sucker absent. Pharynx oval, often small. Oesophagus short to medium-sized. Intestinal bifurcation in middle or posterior half of hindbody. Caeca end blindly near posterior extremity. Testes normally tandem, occasionally oblique, near middle of hindbody. Cirrus sac normally absent, occasionally present as membranous rudiment surrounding pars prostatica and/or ejaculatory duct. Seminal vesicle external, tubular, long, reaching into anterior hindbody. Pars prostatica and ejaculatory duct short. Genital pore ventrally sinistro-medial, usually at level of pharynx, occasionally lateral to oesophagus. Ovary entire, irregularly lobed, or distinctly tri-lobed, pre-testicular. Uterine field between ovary and ventral sucker. Vitelline follicles normally fill lateral fields in hindbody, occasionally with gaps laterally to testes, follicles occasionally extend short distance into forebody, confluent in post-testicular zone and occasionally between testes. Excretory vesicle reaching to about level of ovary. In intestine of marine teleosts.

Key to the species of *Pseudopecoelus*

1 Eggs <90 µm in length; Pacific coast ....*Pseudopecoelus japonicus* (Yamaguti, 1938) (Fig. 74)

Site: Intestine.
Host: *Aprodon cortezianus.*
Distribition: Pac.
Record: BM(NH) collection.
Remarks: Due to its isolation, this determination is somewhat tentative. *P. japonicus* was originally described from a variety of fishes in Japanese waters, although it is also said to occur in Australasian waters (Manter 1954; Korotaeva 1982) and in the Indian Ocean (Parukhin 1976). The specimens in the BM(NH) collection were collected by Dr. H.P. Arai. The Canadian material is morphologically somewhat similar to *P. nossamani* Kruse, 1977, a species that was originally recorded from *Aptocyclus ventricosus* in the Bering Sea. It differs in that the genital pore is at the level of the pharynx, the typical position in *Pseudopecoelus*, rather than just anterior to the intestinal bifurcation (I have seen some of Kruse's type material and can confirm this).

Eggs >90 µm (but may be smaller in young worms); off Atlantic coast (but see "Remarks")...... ....*P. vulgaris* (Manter, 1934) von Wicklen, 1946

Syn.: *Cymbephallus vulgaris* Manter, 1934.
Site: Intestine.
Hosts: *Anoplopoma fimbria* (2); *Merluccius bilinearis* (1, 3); *Urophycis musicki* (1).
Distribution: Atl, Pac.
Records: 1. Scott 1987 (Atl); 2. Kabata *et al.* 1988 (Pac); 3. BM(NH) collection (Atl).
Remarks: Lloyd (1938) claimed to have found this species in *Lycodopsis pacificus* in Puget Sound; but his two specimens were macerated and his determination provisional. Kabata *et al.* (1988) have also recorded eight specimens from a single sablefish off the Pacific coast.

Subfamily PLAGIOPORINAE Manter, 1947

Opecoelidae. Oval to cylindrical worms. Body surface normally smooth, in one case apparently covered with spiniform papillae. Ventral sucker usually without lobes or projections, occasionally pedunculate, occasionally with circum-acetabular folds or radiating musculature. Caeca usually end blindly, occasionally form cyclocoel or open via anus or ani. Testes normally two, tandem, oblique, or rarely symmetrical, rarely with numerous testes. Functional muscular cirrus sac present, enclosing seminal vesicle, pars prostatica, and ejaculatory duct. External seminal vesicle absent. Ovary entire to lobed, usually pre-testicular, occasionally lateral or antero-lateral to anterior testis. Canalicular seminal receptacle present. Uterine

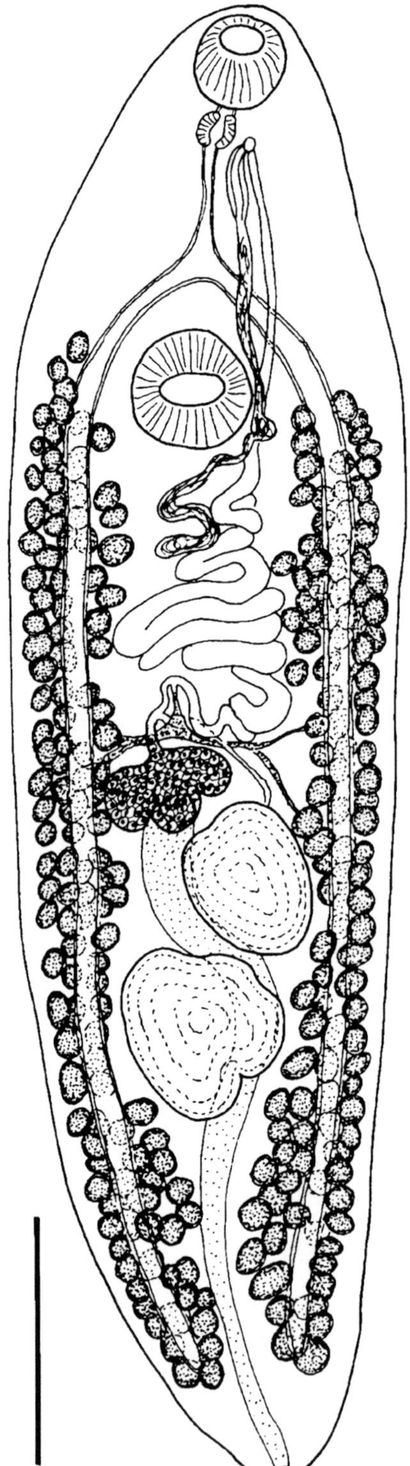

FIG. 74. *Pseudopecoelus japonicus* (after Yamaguti (1971)). Scale bar (roughly estimated): 500 μm.

seminal receptacle absent. Eggs usually numerous, large, occasionally with filaments. Uterine field pre-testicular. Vitelline fields in hindbody or extending into forebody, usually confluent in post-testicular field and sometimes in forebody. Normally in intestine of marine and freshwater teleosts.

Key to the genera of the Plagioporinae

1 Eggs filamented 9

Eggs without filaments 2

2 Vitelline fields normally restricted to hindbody 3

Vitelline fields extend into forebody on both sides of body (often confluent dorsally in forebody) 5

3 Ovary distinctly or indistinctly tri-lobed (occasionally campaniform) 4

Ovary entire .... *Allopodocotyle*

4 Caeca end blindly .... *Podocotyle*

Caeca open through separate ani; ventral sucker with internal muscular ring; Pacific coast .... *Pellamyzon*

5 Genital pore median (or almost so); Atlantic coast .... *Peracreadium*

Genital pore sinistrally submedian to sublateral 6

6 In marine or anadromous teleosts; excretory vesicle long, reaching to level of ovary 7

In freshwater teleosts; excretory vesicle short, reaching to level of posterior testis .... *Plagioporus*

7 Ovary entire .... *Macvicaria*

Ovary indistinctly or distinctly lobed; Pacific waters 8

8 Ovary indistinctly tri-lobed; cirrus sac short, restricted to forebody .... *Neolebouria*

Ovary multi-lobed (with 3–8, usually 3–5, lobes); cirrus sac long, extending into hindbody; in *Oncorhynchus* .... *Plagioporus shawi*

9 Testes 2 .... *Helicometra*

Testes 9 .... *Helicometrina*

ALLOPODOCOTYLE Pritchard, 1966

Plagioporinae. Body small to medium-sized, elongate-oval. Ventral sucker larger than oral sucker, embedded to pedunculate. Oesophagus short to medium-sized. Intestinal bifurcation in middle or posterior half of forebody. Caeca end blindly near posterior extremity. Testes two, tandem to oblique, near middle of hindbody. Cirrus sac short to extending deeply into anterior hindbody, contains seminal vesicle, pars prostatica, and ejaculatory duct. Seminal vesicle tubular, straight or coiled. Genital pore ventrally sinistro-medial at level between pharynx and intestinal bifurcation. Ovary entire, immediately pre-testicular. Uterine field between ovary and ventral sucker. Eggs without filaments. Vitelline follicles laterally in hindbody, confluent in post-testicular zone and sometimes between testes. Excretory vesicle reaching to level of posterior testis or ovary. In intestine of marine and freshwater teleosts.

Comments

There may be a case for splitting this genus in the same way in which *Macvicaria* has been split from *Plagioporus*, because it appears that freshwater forms may have a much shorter excretory vesicle than marine forms.

The three nominal North American species of this genus are morphologically similar and all occur in freshwater. One of the main features used to differentiate them has been egg length; but in fact considerable variation has been found between the egg lengths given in the original descriptions and in subsequent measurements of the type material, i.e., *A. boleosomi* (see Pearse 1924, 40 (?or 160) µm; Pritchard 1966, 64–85 µm), *A. lepomis* (see Dobrovolny 1939a, 70–80 (mean 79) µm; Pritchard 1966, 80–114 µm) and *A. virens* (see Sinitsin 1931, 58 µm; Pritchard 1966, 72–86 µm). Although the morphological similarity between *A. virens* and *A. boleosomi* and between *A. virens* and *A. lepomis* has been pointed out by Gibson (1986) and Pritchard (1966), respectively, *A. virens* does appear to be a much larger worm and it also differs in that the excretory vesicle reaches forward to the level of the ovary rather than to the level of the posterior testis: in any case, this species does not have to be considered further here, as it has not been recorded in Canadian waters.

With regard to *A. boleosomi* and *A. lepomis*, although these two species were synonymised by Peters (1957), they were recognized as distinct (but in different genera) by Pritchard (1966) on the basis of egg size. Kuntz and Font (1984) have dismissed Pritchard's generic distinction but upheld the specific distinction on the basis of egg size, which they list for *A. boleosomi* as being 46–72 (mean 64) µm. Until a more detailed study of these forms is carried out, it would appear that they can only be separated using the rather crude key given below.

Key to the species of *Allopodocotyle*

1 Eggs normally <80 µm ....*Allopodocotyle boleosomi* (Pearse, 1924) Kuntz and Font, 1984 (Fig. 75)

Syn.: *Allocreadium boleosomi* Pearse, 1924; *Plagioporus boleosomi* (Pearse, 1924) Peters, 1957; *Podocotyle boleosomi* (Pearse, 1924) Yamaguti, 1971.
Site: Gut.
Host: *Percina caprodes*.
Distribution: Ont.
Record: Bangham and Hunter 1939.

Eggs normally >80 µm ....*A. lepomis* (Dobrovolny, 1939) Pritchard, 1966

Syn.: *Plagioporus lepomis* Dobrovolny, 1939; *Podocotyle lepomis* (Dobrovolny, 1939) Yamaguti, 1953.
Site: Intestine.
Host: *Petromyzon marinus*.
Distribution: Ont.
Record: Wilson and Ronald 1967.

HELICOMETRA Odhner, 1902

Plagioporinae. Body small to medium-sized, elongate-oval. Ventral sucker larger than oral sucker, in anterior half of worm, embedded or semi-embedded. Oesophagus short to medium-sized. Intestinal bifurcation in about middle of forebody. Caeca terminating blindly near posterior extremity. Testes two, tandem to oblique, near middle of hindbody. Cirrus sac claviform, short to extending into anterior hindbody, contains winding seminal vesicle, pars prostatica, and ejaculatory duct. Genital pore ventro-median at level of intestinal bifurcation. Ovary normally irregularly or distinctly lobed, often irregularly tri-lobed,

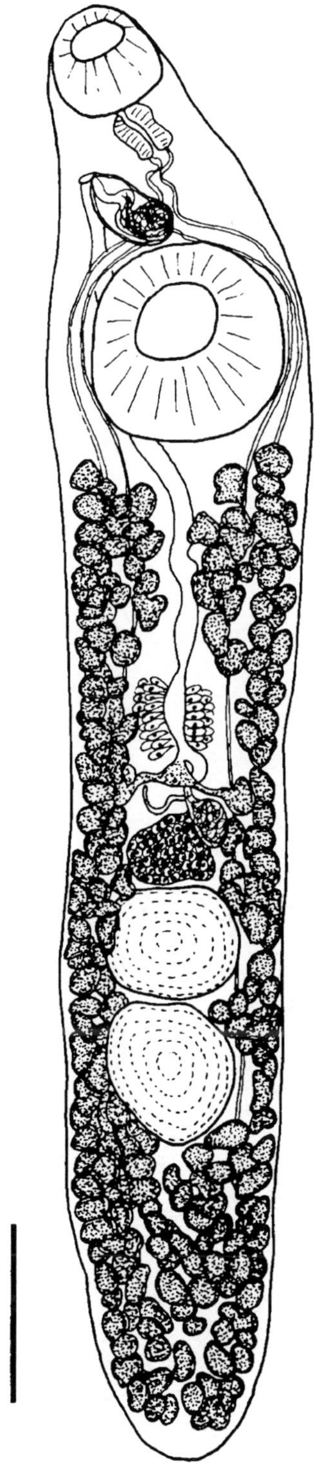

FIG. 75. *Allopodocotyle boleosomi* (after Kuntz and Font (1984)). Scale bar: 250 µm.

occasionally entire, immediately pre-testicular. Uterine field between ovary and ventral sucker. Eggs filamented, regularly arranged in uterus. Vitelline fields usually lateral, between posterior extremity and anterior limit that varies from anterior hindbody to pharynx, fields confluent in post-testicular zone and sometimes dorsally at level of uterine field or in forebody. Excretory vesicle reaching to level of ovary. In intestine of marine teleosts.

Comments

Sekerak and Arai (1974) extended the conception of the genus *Neohelicometra* Siddiqi and Cable, 1960 to include forms, such as *N. sebastis* and *N. insolita*. *Neohelicometra* was originally a monotypic genus erected to include *Helicometra*-like worms with a terminal funnel-shaped oral sucker and twin ani. Sekerak and Arai cast doubt upon the presence of ani as being a feature of generic importance. Bray (1979) did not accept this, indicating that it was the terminal funnel-shaped oral sucker that was not a feature of generic significance. He, therefore, retained only the type-species in *Neohelicometra* because of the presence of the paired ani and transferred *N. sebastis* to *Helicometra*.

Gibson and Bray (1982) have suggested that there may be some value in raising *Metahelicometra* Yamaguti, 1971 to full generic status for species in which the vitelline follicles are restricted to the hindbody.

Key to the species of *Helicometra*

1 Vitellarium not extending into forebody; cirrus sac not extending into hindbody; Atlantic coast 2

Vitellarium extends into forebody; cirrus sac extends into hindbody; Pacific coast......................
....*Helicometra sebastis* (Sekerak and Arai, 1974) Bray, 1979

Syn.: *Helicometra* sp. of Arai (1969b) and Sekerak and Arai (1973); *Neohelicometra sebastis* Sekerak and Arai, 1974.

Sites: Pyloric caeca, intestine.

Hosts: *Sebastes aleutianus* (3, 4); *S. alutus* (2, 3, 4); *S. babcocki* (3, 4); *S. caurinus* (3, 4); *S. diploproa* (3, 4); *S. flavidus* (3, 4, 6); *S. helvomaculatus* (3, 4); *S. maliger* (1, 3, 4); *S. nebulosus* (4, 5); *S. nigrocinctus* (3, 4); *S. reedi* (3, 4); *S. ruberrimus* (3, 4); *S. variegatus* (3, 4); *S. zacentrus* (3, 4).

Distribution: Pac.

Records: 1. Arai 1969b; 2. Sekerak and Arai 1973; 3. Sekerak and Arai 1974; 4. Sekerak and Arai 1977; 5. Holmes 1990; 6. Stanley *et al.* 1992.

2 Oral sucker distinctly funnel-shaped; gonads indistinctly irregularly lobed ....
....*H. insolita* Polyanski, 1955

Syn.: *Neohelicometra insolita* (Polyanski, 1955) Sekerak and Arai, 1974.

Site: Intestine.

Host: *Lumpenus lampretaeformis*.

Distribution: Atl.

Record: Bray 1979.

Oral sucker not distinctly funnel-shaped; gonads deeply and irregularly lobed ....
....*H. plovmornini* Issaichikov, 1928 (Fig. 76)

Site: Intestine.

Hosts: *Anarhichas lupus* (1, 2); *Lycodes reticulatus* (1); *L. vahli* (1); *Triglops murrayi* (1).

Distribution: Atl.

Records: 1. Bray 1979; 2. Bray 1987c.

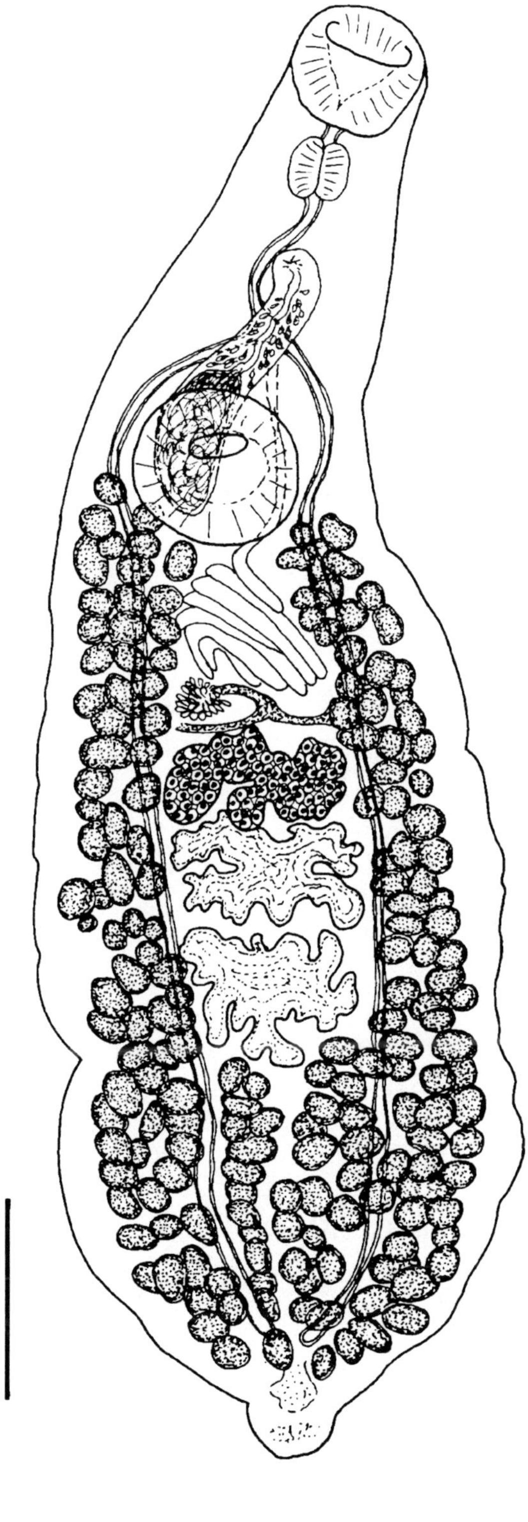

FIG. 76. *Helicometra plovmornini* (after Bray (1979)). Scale bar: 500 µm.

HELICOMETRINA Linton, 1910

Plagioporinae. Body small to medium-sized, fusiform to pyriform. Ventral sucker larger than oral sucker, in anterior half of body, embedded or semi-embedded. Pharynx small to medium-sized. Oesophagus short to long. Intestinal bifurcation in mid- to posterior forebody. Caeca terminate blindly close to posterior extremity. Testes nine, usually in two longitudinal, inter-caecal rows in hindbody. Cirrus sac claviform, usually mainly in forebody, occasionally extending into anterior hindbody, containing winding seminal vesicle, pars prostatica, and ejaculatory duct. Genital pore ventro-medial close to level of intestinal bifurcation. Ovary irregularly lobed, median, immediately pre-testicular. Uterine field between ovary and ventral sucker. Eggs filamented. Vitelline follicles in lateral fields between level of oesophagus and posterior extremity, gap at level of ventral sucker present or absent; fields confluent in post-testicular region and may approach one another dorsally in forebody. Excretory vesicle reaches to level of ovary. In intestine of marine teleosts.

Helicometrina nimia Linton, 1910 (Fig. 77)
Site: (?intestine).
Host: *Sebastes nebulosus.*
Locality: Pac.
Record: Holmes 1990.

MACVICARIA Gibson and Bray, 1982

Plagioporinae. Body small, fusiform. Oral sucker ventrally subterminal. Ventral sucker normally larger than oral sucker, in anterior half of body. Pharynx oval to globular. Oesophagus distinct, often short, often with external gland cells. Intestinal bifurcation in mid- to posterior forebody. Caeca unbranched, end blindly in post-testicular region. Testes two, usually oblique, occasionally in tandem, in middle of hindbody, entire or slightly lobed. Cirrus sac well developed, usually extends back to position dorsal to ventral sucker, contains looped or convoluted seminal vesicle, pars prostatica, and ejaculatory duct. Genital pore sinistral, ventro-lateral, extra-caecal, at level of oesophagus. Ovary rounded, normally antero-dextral to anterior testis. Uterus pre-testicular, inter-caecal. Eggs normally few; without filaments. Vitelline follicles extending in lateral fields from about level of genital pore to posterior extremity, lateral fields confluent in post-testicular region and normally dorsally in forebody (fields at least approach one another). Excretory vesicle extending anteriorly at least to level of anterior testis, normally to level of ovary. Parasitic in intestine of marine teleosts.

Key to the species of *Macvicaria*

1 Genital pore at level of oesophagus; in Atlantic waters ....*Macvicaria soleae* (Dujardin, 1845) Gibson and Bray, 1982 (Fig. 78)
Syn.: *Plagioporus varius* (Nicoll, 1910) Price, 1934.
Sites: Intestine, pyloric caeca.
Hosts: *Glyptocephalus cynoglossus*; *Hippoglossus hippoglossus*; *Pleuronectes americanus*; *P. ferrugineus.*
Distribution: Atl.
Record: Ronald 1960.

Genital pore at level of pharynx (but see "Remarks"); in Pacific waters ....*M. isaitschikowi* (Layman, 1930) Bray, 1985
Syn.: *Plagioporus isaitschikowi* (Layman, 1930) Price, 1934.
Site: Intestine.
Host: *Oligocottus maculosus.*
Distribution: Pac.

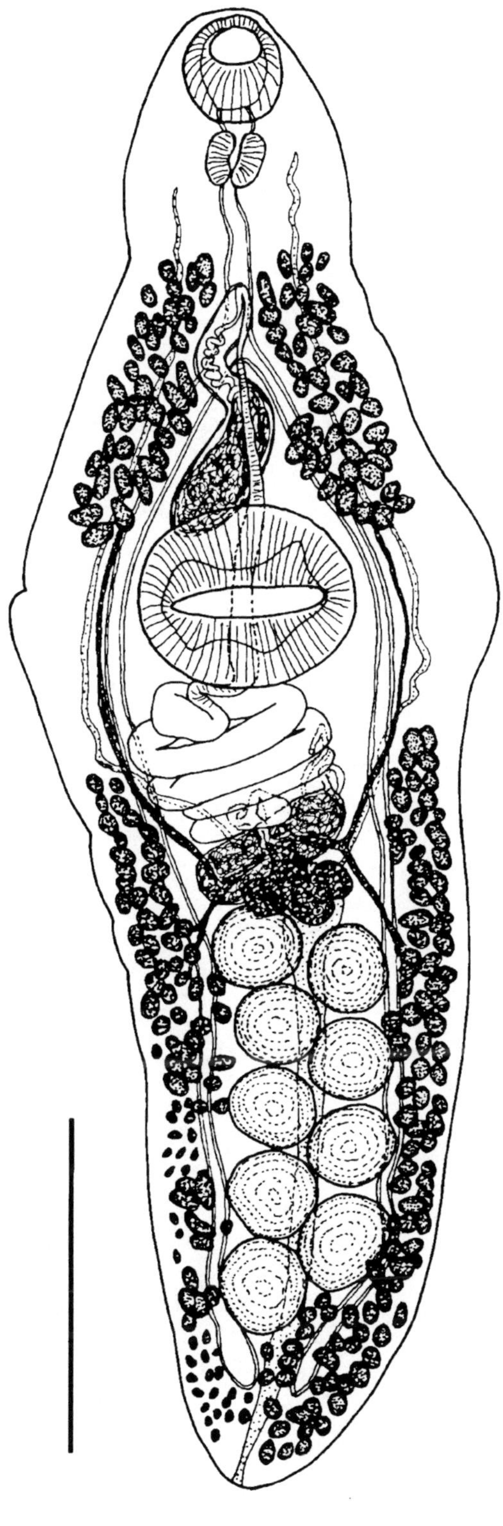

FIG. 77. *Helicometrina nimia* (after Yamaguti (1971; as *H. elongata*)). Scale bar (roughly estimated): 1 mm.

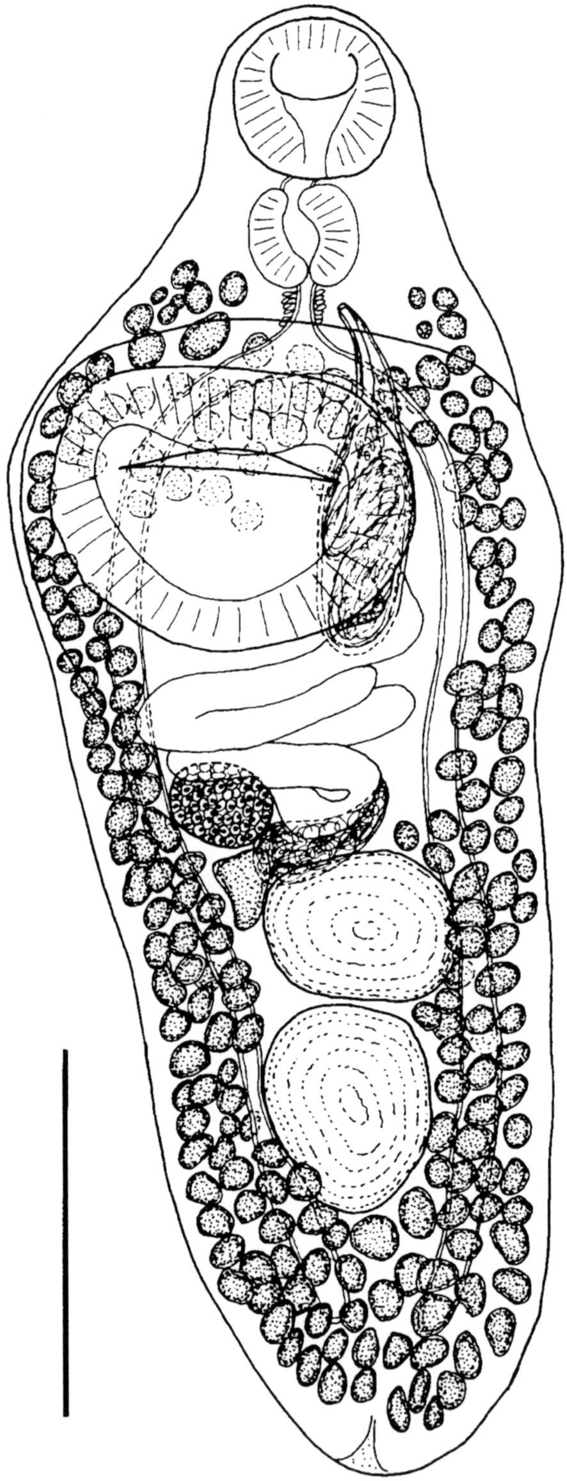

FIG. 78. *Macvicaria soleae* (after Gibson and Bray (1982)). Scale bar: 500 μm.

Record: BM(NH) collection.

Remarks: This species, based upon a single specimen of Layman (1930) from the northern Northwest Pacific, is in need of detailed study. It has been redescribed by Yamaguti (1938) in Japanese waters and reported in Californian waters and commented upon by Manter and Van Cleave (1951). One of the main problems is Layman's figure, which indicates that the genital pore is lateral to the oesophagus and not the pharynx. In the material that I have examined, which was collected by Dr. H.P. Arai off Vancouver Island, another of the main features of the worm, i.e., the gap in the lateral vitelline fields at the level of the ventral sucker, was not absolutely distinct.

NEOLEBOURIA Gibson, 1976

Plagioporinae. Body small to medium-sized, elongate-fusiform to flattened pyriform. Ventral sucker larger than oral sucker, embedded to semi-embedded, inside anterior half of body. Oesophagus usually relatively long. Intestinal bifurcation in posterior half of forebody. Caeca terminate blindly near posterior extremity or near middle of post-testicular zone. Testes two, in tandem, near middle of hindbody. Cirrus sac claviform, posterior limit varying between anterior margin of ventral sucker and well posterior to ventral sucker, containing seminal vesicle, pars prostatica, and ejaculatory duct. Seminal vesicle coiled, proximal region often saccular. Genital pore ventrally sinistro-medial at level of oesophagus. Ovary immediately pre-testicular, median or submedian, indistinctly tri-lobed. Uterine field between ovary and ventral sucker. Eggs without filaments. Vitelline follicles in lateral fields between middle of forebody and posterior extremity, occasionally with gaps lateral to ventral sucker, fields confluent or almost so in post-testicular zone and dorsally in forebody. Excretory vesicle reaching to level of ovary. In intestine of marine teleosts (especially in deeper waters).

Neolebouria tinkerbellae Thompson and Margolis, 1987 (Fig. 79)

Site: Intestine.

Hosts: *Gasterosteus aculeatus* (exp.); *Leptocottus armatus* (exp.); *Platichthys stellatus* (exp.).

Distribution: Pac.

Record: Thompson and Margolis 1987.

Remarks: Mature specimens were obtained by Thompson and Margolis (1987) in the above hosts experimentally after feeding them with metacercarial cysts removed from the decapod *Pandalus jordani* taken at a depth of 120 m off Vancouver Island.

PELLAMYZON Montgomery, 1957

Plagioporinae. Body medium-sized, elongate. Ventral sucker much larger than oral sucker, protuberant, relatively close to anterior extremity, with internal ring of thickened musculature. Oesophagus short to medium-sized. Intestinal bifurcation in posterior forebody. Caeca open through separate ani close to posterior extremity. Testes two, tandem, in middle of hindbody. Cirrus sac tubular, long, extending into hindbody where it may be looped, contains long, saccular seminal vesicle, pars prostatica, and ejaculatory duct. Genital pore ventrally sinistro-median at level of pharynx or oesophagus. Ovary tri-lobed, immediately pre-testicular. Uterine field between ovary and ventral sucker. Vitelline follicles fill hindbody laterally, except occasionally at level of testes, lateral fields confluent in post-testicular zone, dorsally to uterine field and sometimes between testes. Eggs without filaments. Excretory vesicle reaching to level of ovary. In intestine of marine teleosts (*Sebastes*).

Pellamyzon abitionis (McFarlane, 1936) Gibson and Bray, 1982 (Fig. 80)

Syn.: *Podocotyle abitionis* McFarlane, 1936; *Pellamyzon sebastodis* Montgomery, 1957.

Site: Intestine.

Hosts: *Sebastes* sp. (1); *S. maliger* (2).

Distribution: Pac.

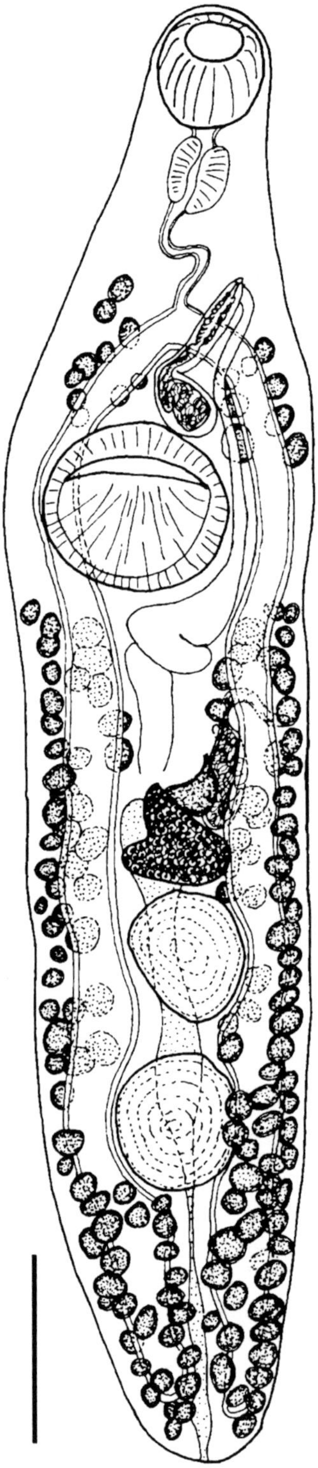

FIG. 79. *Neolebouria tinkerbellae* (after Thompson and Margolis (1987)). Scale bar: 250 μm.

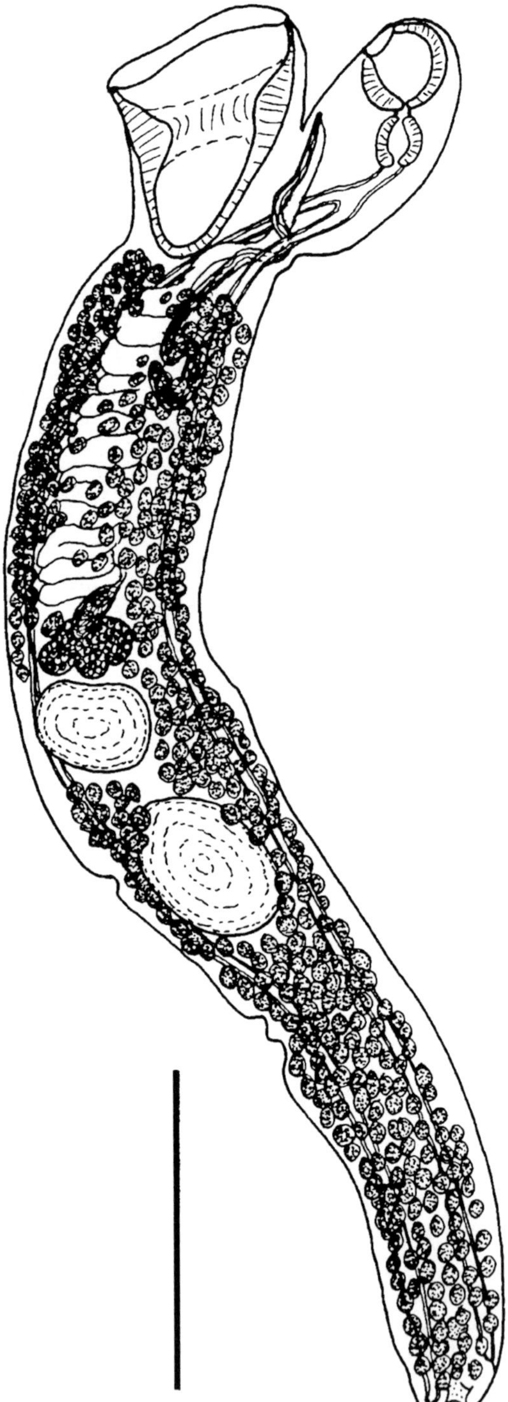

FIG. 80. *Pellamyzon abitionis* (after Gibson and Bray (1984)). Scale bar: 1 mm.

Records: 1. McFarlane 1936; 2. Gibson and Bray 1984.

Remarks: This species was also recorded by Ching (1960) from *Sebastes* sp. off Friday Harbor, Washington. It was commented upon by Gibson and Bray (1982, 1984).

PERACREADIUM Nicoll, 1909

Plagioporinae. Body small to medium-sized, elongate-oval. Ventral sucker larger than oral sucker, embedded or semi-embedded, in anterior half of body. Oesophagus short to medium-sized. Intestinal bifurcation in middle to posterior forebody. Caeca end blindly near posterior extremity. Testes two, tandem to oblique, in middle of hindbody. Cirrus sac claviform, medium-sized to large, reaching back to between level of ventral sucker and level just anterior to testes, contains coiled, tubular seminal vesicle, distinct pars prostatica, and ejaculatory duct. Genital pore mid-ventral in middle of forebody, pre-bifurcal or bifurcal. Ovary entire, dextro-median, anterior or antero-lateral to anterior testis. Uterine field usually between anterior testis and ventral sucker, may coil between ovary and anterior testis or dorsally to ventral sucker. Eggs without filaments. Vitelline follicles fill lateral fields between pharynx and posterior extremity, except in some cases at level of ventral sucker, fields confluent in post-testicular zone and in forebody. Excretory vesicle reaching to level of ovary. In intestine of marine teleosts.

Peracreadium idoneum (Nicoll, 1909) Gibson and Bray, 1982 (Fig. 81)

Syn.: *Plagioporus idoneus* (Nicoll, 1909) Price, 1934.

Site: Intestine.

Hosts: *Anarhichas lupus* (1, 2, 3); *A. minor* (1, 2).

Distribution: Atl.

Records: 1. Bray 1979; 2. Zubchenko 1980; 3. Bray 1987c.

PLAGIOPORUS Stafford, 1904

Plagioporinae. Body dorso-ventrally flattened, fusiform to elongate-oval. Body surface smooth, in one instance apparently covered with spiniform papillae. Ventral sucker larger than oral, in anterior half of body or equatorial. Pharynx well developed. Intestinal bifurcation in mid- to posterior forebody. Caeca end blindly lateral to testes or in post-testicular field. Testes two, tandem or oblique, near middle of hindbody or close to posterior extremity. Cirrus sac claviform, containing tubular to saccular seminal vesicle, pars prostatica, and ejaculatory duct. Genital pore pre-bifurcal, sinistrally submedian, usually at level of oesophagus. Ovary oval or indistinctly lobed, immediately pre-testicular or antero-lateral to anterior testis. Vitelline follicles in lateral fields, extend into forebody, may or may not extend into post-testicular field, vitelline fields may unite in post-testicular region. Uterine field between gonads and ventral sucker. Excretory vesicle short, reaching only to level of posterior testis. Parasites of freshwater teleosts.

Comments

The conception of this genus used here is that of Gibson and Bray (1982), who restricted it to freshwater forms on the grounds that morphologically similar marine forms, for which they erected the genus *Macvicaria*, had a much longer excretory vesicle. As presented above, the generic diagnosis does not include the anomalous *Plagioporus shawi* (McIntosh, 1939) (see below), although this species is included in the key below for convenience. In order to include *P. shawi* the variations "ovary distinctly lobed", "excretory vesicle extending to level of ovary", and "in anadromus fishes" would have to be incorporated.

It is very apparent that, with the exception of *P. sinitsini*, the Canadian freshwater species of this genus require restudy and redescription to confirm their validity.

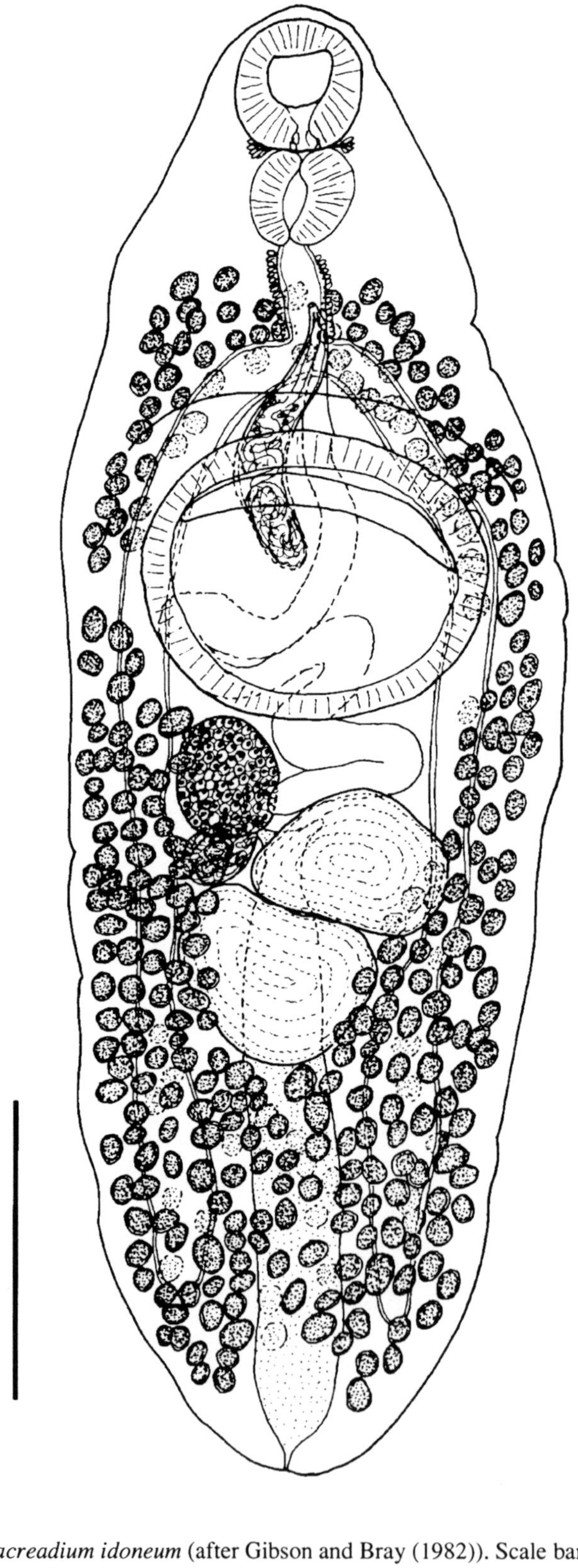

FIG. 81. *Peracreadium idoneum* (after Gibson and Bray (1982)). Scale bar: 500 µm.

Key to the species of *Plagioporus*

1 Post-testicular field very short (testes near posterior extremity), containing few or no vitelline follicles; gut caeca end laterally to testes; ovary may be indistinctly lobed; usually occurs in gall bladder 2

Post-testicular field large (testes near middle of hindbody), containing numerous vitelline follicles ... 3

2 Body surface smooth .... *Plagioporus sinitsini* Mueller, 1934 (Fig. 82)

Sites: Intestine, gall bladder.

Hosts: *Catostomus catostomus* (1); *C. commersoni* (2, 3); *Luxilis cornutus* (3, 4); *Nocomis biguttatus* (1); *Notropis hudsonius* (4); *Pimephales promelas* (3).

Distribution: Ont.

Records: 1. Bangham 1955; 2. Dechtiar 1972b; 3. Molnar *et al.* 1974; 4. Dechtiar and Christie 1988.

Body surface covered with minute spiniform papillae .... *P. serratus* Miller, 1940

Site: Gall bladder.

Host: *Hiodon tergisus.*

Distribution: Que.

Record: Miller 1940a.

Remarks: The type specimen in the BM(NH) collection is labelled "*Plagioporus spinosus*" from "whitefish": one of the less common vernacular names of the "mooneye" *H. tergisus* is the "river whitefish". This is a primitive osteoglossomorph teleost.

Miller (1940a) stated "the cuticula is characteristically extended to form conspicuous broad-based spines". Gibson and Bray (1982), after examining the type specimen in the BM(NH) collection, confirmed that "indistinct spines do appear to be present on the surface of the body", and because of this transferred the species to the family Enenteridae, a group related to the Lepocreadiidae, along with *P. niloticus* Vercammen-Grandjean, 1960 and the genus *Spinoplagioporus* Skryabin and Koval, 1958. Having re-examined the type specimen of *P. serratus,* I now consider that there is no "cuticular" element to the "spines" of this species, such as occurs in *P. niloticus* and *Spinoplagioporus.* The armature appears to be formed by projections of the tegument that form spiniform papillae. In other aspects of its morphology this worm closely resembles *P. sinitsini,* even to the extent of occurring in the gall bladder and having a slightly lobed ovary (an additional lobe is apparent in the type specimen, assuming that this is not the seminal receptacle). I should not be at all surprised if the specimen from *H. tergisus* eventually proved to be a small specimen of *P. sinitsini* and that the armature is artifactual.

3 Testes tandem; gut caeca extending into post-testicular field 4

Testes oblique; gut caeca end laterally to testes or extend into post-testicular field; sucker-ratio usually 1:<1.5 .... *P. cooperi* (Hunter and Bangham, 1932) Price, 1934

Syn.: *Allocreadium commune* of Cooper (1915) in part; *Lebouria cooperi* Hunter and Bangham, 1932.

Sites: Intestine, gall bladder.

Hosts: *Ammocrypta pellucida* (2); *Couesius plumbeus* (5); *Etheostoma nigrum* (4); *Hybopsis storeriana* (2, 3); *Luxilis cornutus* (1, 4, 5); *Notropis atherinoides* (2, 3, 5); *N. emiliae* (3); *N. hudsonius* (2, 3, 4, 5); *N. spilopterus* (3, 4); *N. stramineus* (2, 3); *N. volucellus* (2, 3); *Percina copelandi* (2); *Pimephales notatus* (3); *Rhinichthys cataractae* (2, 3).

Distribution: Ont.

Records: 1. Cooper 1915; 2. Hunter and Bangham 1932; 3. Bangham and Hunter 1939; 4. Bangham 1955; 5. Dechtiar *et al.* 1988.

Remarks: Cooper (1915) and Aliff (1977) record the egg length as being 62–68 and 67 µm, respectively. Assuming that Cooper's measurements were of this species, these are smaller than the 74–103 (av. 86) µm recorded by Hunter and Bangham (1932). Aliff also stated that the ovary in his material was 3–5 lobed, and that there was some variation in the number of

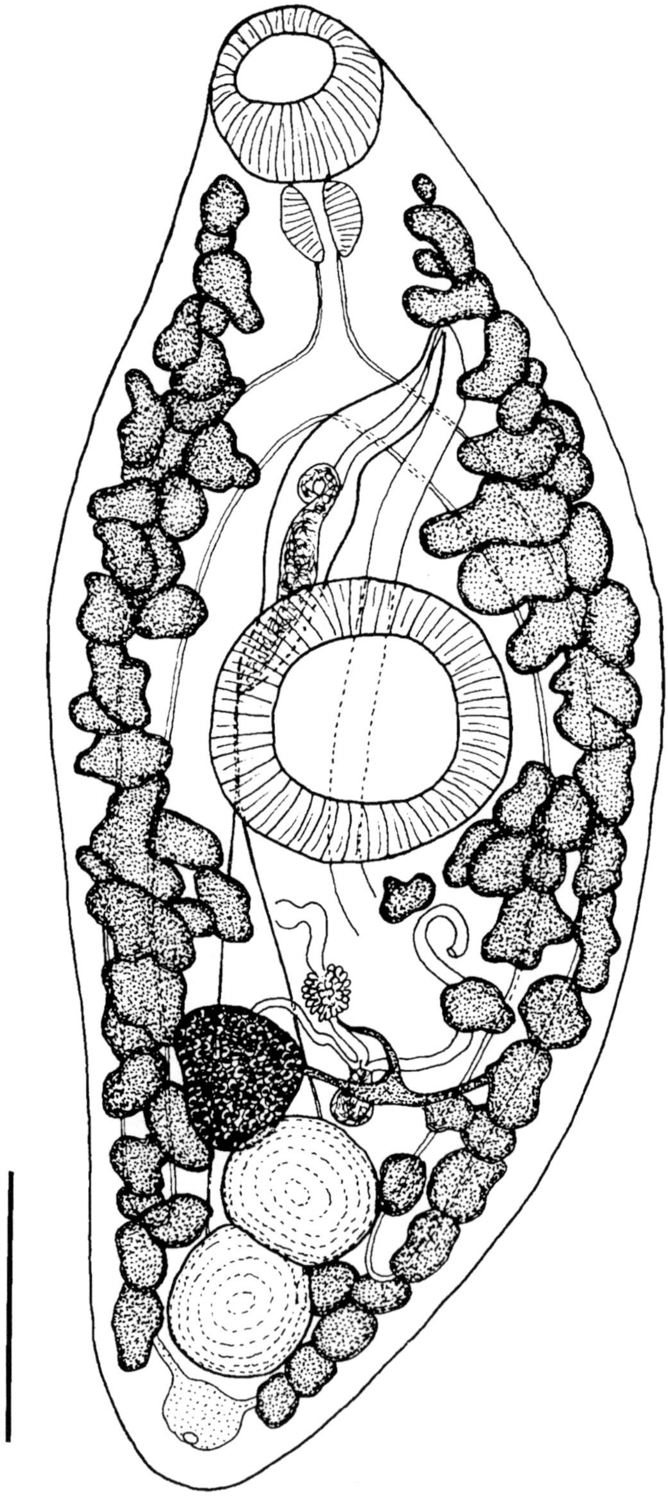

FIG. 82. *Plagioporus sinitsini* (after Dobrovolny (1939b)). Scale bar: 250 µm.

vitelline follicles in the post-testicular field and in the level of the posterior extremity of the gut caeca.

4 Excretory vesicle short (typical for *Plagioporus*), reaching to posterior testis; cirrus sac does not extend into hindbody; sucker-ratio ca.1:2; in catostomids ....*P. serotinus* Stafford, 1904

Site: Intestine.

Hosts: *Catostomus commersoni* (2); *Moxostoma macrolepidotum* (1, 2).

Distribution: Que.

Records: 1. Stafford 1904; 2. Miller 1940a.

Remarks: This species was redescribed by Miller (1940a).

Excretory vesicle long, reaching to level of ovary; cirrus sac extending into hindbody; sucker-ratio 1:1.25–1.75; in *Oncorhynchus* off Pacific coast (and in fresh water in Northwest United States)..*P. shawi* (McIntosh, 1939) Margolis, 1972

Site: Intestine.

Host: *Oncorhynchus mykiss*.

Distribution: Pac.

Records: Margolis 1990; Margolis 1992; Burgner *et al.* 1992.

Remarks: This species occurs in *Oncorhynchus* spp., especially *O. nerka*, in the rivers of Oregon, Washington, and Idaho. This is an essentially freshwater species that does not appear to occur in fresh water in Canada, but is occasionally taken in anadromous hosts in the sea.

Its exact generic position is problematical, having been associated with *Allocreadium* (see Yamaguti 1953), *Peracreadium* (see Yamaguti 1953), *Cainocreadium* (see Yamaguti 1958), *Plagioporus* (see Margolis 1972), and *Neolebouria* (see Gibson 1976). Gibson (1986) indicated that it had some similarities with *Plagioporus siliculus* Sinitzin, 1931 and two species of *Allopodocotyle*. Although morphologically this species is perhaps closer to *Neolebouria*, it is included here with *Plagioporus* because of its freshwater life cycle. It differs fundamentally from other species of *Plagioporus* in the size of its excretory vesicle, which might reflect its ability to survive in marine conditions in anadromous hosts.

PODOCOTYLE Dujardin, 1845

Plagioporinae. Body small to relatively large, elongate-oval to cylindrical. Ventral sucker larger than oral sucker, in anterior half of body, embedded to pedunculate. Oesophagus short to medium-sized. Intestinal bifurcation in middle or posterior half of forebody. Caeca terminating blindly near posterior extremity. Testes two, usually tandem, occasionally oblique, near middle of hindbody. Cirrus sac well developed, claviform, short to extending deep into anterior hindbody, contains seminal vesicle, pars prostatica, and ejaculatory duct, seminal vesicle often looped. Genital pore ventrally sinistro-median at level of oesophagus or intestinal bifurcation. Ovary tri-lobed (or at least campaniform with indication of three posterior lobes), normally pre-testicular, occasionally antero-lateral to anterior testis. Uterine field normally intercaecal between ovary and ventral sucker. Eggs without filaments. Vitelline fields normally fill all or most of hindbody laterally, except occasionally for gaps at level of gonads, fields confluent in post-testicular zone, on rare occasions fields may extend into forebody on one or both sides, but are not normally confluent. Excretory vesicle normally reaches to level of ovary. In intestine of marine or estuarine teleosts.

Comments

The identification of the species of this genus always has been, and probably always will be, difficult because of the absence of good diagnostic characters and the considerable amount of intra-specific variation that appears to occur. The following key is based upon the work of Gibson and Bray (1982) and Gibson (1986): the latter work includes a key to the known species of the genus occurring on the Pacific coast of North America.

Key to the species of *Podocotyle*

1 Cirrus sac long and usually sinuous, extending well into hindbody 2

Cirrus sac short, not normally extending much beyond posterior margin of ventral sucker 7

2 Seminal vesicle normally straight or occasionally with small anterior loop; vitelline fields often interrupted laterally to testes and usually laterally to ovary ....*Podocotyle radifistuli* (Acena, 1941) Gibson and Bray, 1984

Syn.: *Dideutosaccus radifistuli* Acena, 1941; *Opecoelina radifistuli* (Acena, 1941) Yamaguti, 1958; *O. pharynmagna* Annereaux, 1943.

Site: Intestine.

Hosts: *Leptocottus armatus* (1), *Blepsias cirrhosus* (1); *Sebastes nebulosus* (2).

Distribution: Pac.

Records: 1. Gibson and Bray 1984; 2. Holmes 1990.

Remarks: This species also occurs in *Sebastes elongatus* off Washington State (Gibson and Bray, 1984).

Main bulk of seminal vesicle in posterior part of cirrus sac; large loop in seminal vesicle normally present, often extending back close to posterior extremity of cirrus sac, but loop occasionally small or absent; cirrus sac usually reaches near to middle of uterine field or beyond; vitelline fields may be interrupted at level of testes, but not normally at level of ovary 3

3 Slender worms with protuberant ventral sucker; vitelline fields normally interrupted at level of testes 4

Lanceolate worms; vitelline fields not normally interrupted at level of testes 6

4 Large worms of up to 8 mm in length (occasionally larger) 5

Worms of up to 4 mm in length; eggs 70–85 μm in length; Pacific coast ....*P. sinusacca* Ching, 1960 (Fig. 83)

Syn.: *Neopodocotyloides sinusaccus* (Ching, 1960) Pritchard, 1966; *Podocotyle* sp. of Arai (1967a) in part; *P. reflexa* (Creplin, 1825) of McFarlane (1936).

Site: Intestine.

Hosts: *Blepsias cirrhosus* (2, 3); *Gasterosteus aculeatus* (2, 3); *Hemilepidotus hemilepidotus* (2, 3); *Hexagrammos decagrammus* (4); *H. stelleri* (2, 3, 4); *Leptocottus armatus* (1, 2, 3, 4); *Pleuronectes vetulus* (2, 3); *Syngnathus leptorhynchus* (1, 2, 3).

Distribution: Pac.

Records: 1. McFarlane 1936; 2. Arai 1967a; 3. Arai 1969b; 4. Gibson and Bray 1984.

Remarks: This species may prove to be a synonym of *P. theragrae* (see Gibson and Bray 1984; Gibson 1986); but see also "Remarks" on *P. theragrae*.

5 Eggs 80–100 μm; proximal saccular part of seminal vesicle consistently small, with distal loop reaching back almost to posterior end of cirrus sac; Pacific coast ....*P. theragrae* (Lloyd, 1938) Gibson and Bray, 1984

Syn.: *Opecoelina theragrae* Lloyd, 1938; *Podocotyle* sp. of Arai (1967a) in part; *Neopodocotyloides sinusaccus* (Ching, 1960) of Arai (1969b) in part.

Site: Intestine.

Hosts: *Ophidion elongatus* (3); *Sebastes alutus* (3); *S. caurinus* (3); *S. maliger* (1, 2); *Theragra chalcogramma* (1, 2, 3).

Distribution: Pac.

Records: 1. Arai 1967a; 2. Arai 1969b; 3. Gibson and Bray 1984.

Remarks: It is very apparent that *P. theragrae, P. sinusacca*, and *P. reflexa* are very similar. Gibson (1986) was of the opinion that the former two may prove to be synonymous. With regard to their validity as distinct from *P. reflexa*, the solution of this difficult problem will require a study of material from both Pacific and Atlantic waters that has been collected and

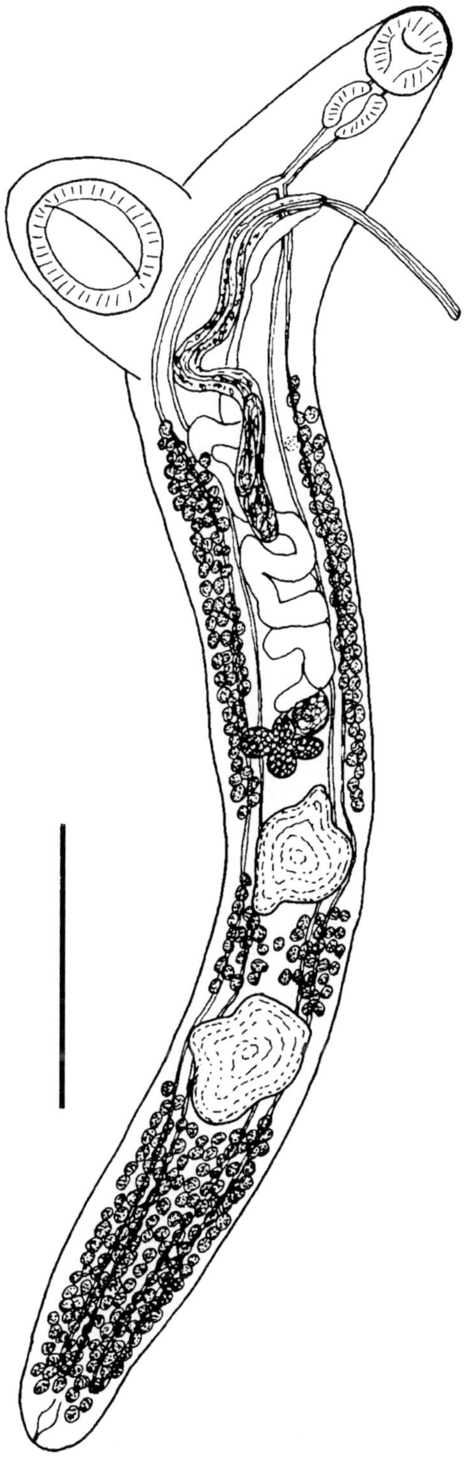

FIG. 83. *Podocotyle sinusacca* (after Gibson & Bray (1984)). Scale bar: 500 μm.

fixed in the same manner: the majority of Pacific material that I have seen was collected from frozen hosts.

Eggs 70–95 µm; proximal saccular part of seminal vesicle variable in size but often long, with distal loop reaching a variable distance from posterior end of cirrus sac; Atlantic coast ....
....*P. reflexa* (Creplin, 1925) Odhner, 1905

Syn.: *P. olssoni* Odhner, 1905; *P. simplex* (Rudolphi, 1809) of Stafford (1904, 1907) in part.

Site: Intestine.

Hosts: *Alosa sapidissima* (16); *Gadus morhua* (7, 9); *Hemitripterus americanus* (1, 18); *Hippoglossus hippoglossus* (8, 15); *Lepidion eques* (18); *Merluccius bilinearis* (13); *Microgadus tomcod* (3); *Oncorhynchus gorbuscha* (Pacific salmon introduced into Newfoundland waters) (10, 18); ? *Pleuronectes ferrugineus* (4); *Pollachius virens* (12, 13); *Reinhardtius hippoglossoides* (17); *Sebastes fasciatus* (11, 14); *S. marinus* (6, 11); *S. mentella* (11); *Tautoglabrus adspersus* (5); *Urophycis chuss* (?1,?2, 13, 18); *U. musicki* (1, 13).

Distribution: Atl, Nfld.

Records: 1. Stafford 1904 (Atl); 2. Stafford 1907 (Atl); 3. Heller 1949 (Atl); 4. Ronald 1960 (Atl); 5. Sekhar and Threlfall 1970a (Atl); 6. Noble 1973 (Atl); 7. Umnova 1979 (Atl); 8. Zubchenko 1980 (Atl); 9. Appy and Burt 1982 (Atl); 10. Margolis 1982 (Nfld); 11. Bourgeois and Ni 1984; 12. Scott 1985a (Atl): 13. Scott 1987 (Atl); 14. Scott 1988 (Atl); 15. Scott and Bray 1989 (Atl); 16. Hogans *et al.* 1993 (Atl); 17. 18.BM(NH) collection.

Remarks: McFarlane's (1936) records of *P. reflexa* from the Pacific are treated as *P. sinusacca*; but the validity of *P. sinusacca* and *P. theragrae*, as distinct from *P. reflexa*, is commented upon under *P. theragrae*.

6 Cirrus usually protruded, often swollen proximally; Pacific coast....................*P. enophrysi* Park, 1937

Syn.: *P. blennicottusi* Park, 1937; *P. pacifica* Park, 1937; *Podocotyle* sp. of Arai (1967a) in part.

Sites: Intestine, pyloric caeca.

Hosts: *Aulorhynchus flavidus* (4); *Myoxocephalus polyacanthocephalus* (2); *Oligocottus maculosus* (3, exper.); *Platichthys stellatus* (4); *Syngnathus leptorhynchus* (2); *Triglops pingeli* (1, 2, 4)

Distribution: Pac.

Records: 1. Arai 1967a; 2. Arai 1969b; 3. Ching 1979; 4. Gibson 1986.

Remarks: Ching (1979) deduced experimentally that the molluscan host is *Lacuna mormorata* and the crustacean host is *Hyale plumulosa*.

Cirrus not normally protruded, not swollen proximally; especially common in brackish water (estuarine) or migratory fishes; Atlantic coast ....*P. angulata* Dujardin, 1845

Syn.: *Podocotyle atomon* (Rud., 1802) of various authors (in part); *P. simplex* (Rud., 1809) of Stafford (1904, 1907) in part; *P. staffordi* Miller, 1941.

Site: Intestine.

Hosts: *Gasterosteus aculeatus* (1, 2, 3); *Oncorhynchus gorbuscha* (Pacific salmon introduced into Newfoundland waters) (4, 8); *Reinhardtius hippoglossoides* (7); ?*Salmo salar* (1, 2); *Salvelinus fontinalis* (5, 6).

Distribution: Atl, NB-b, NB, Nfld.

Records: 1. Stafford 1904 (Atl); 2. Stafford 1907 (Atl); 3. Miller 1941a (Atl); 4. Margolis 1982 (Nfld); 5. Frimeth 1987a (NB-b, NB); 6. Frimeth 1987b (Atl, NB-b); 7. Arthur and Albert 1994 (Atl); 8. BM(NH) collection (Nfld).

Remarks: The taxonomy of this species was discussed by Gibson and Bray (1982).

7 Sucker-ratio 1:>2; vitelline fields often, but not always, interrupted laterally to testes; ovary campaniform or indistinctly trilobate; Pacific coast, in *Sebastes*..........................*P. araii* Gibson, 1986

Syn.: *Podocotyle* sp. of Arai (1967a, 1969b); Sekarak and Arai (1973, 1977).

Sites: Intestine, pyloric caeca.

Hosts: *Sebastes aleutianus* (4, 5); *S. alutus* (1, 2, 3, 4, 5); ? *S. babcocki* (4); *S. borealis* (5); *S. brevispinis* (4, 5); ? *S. diploproa* (4); ? *S. elongatus* (4); ? *S. flavidus* (4); ? *S. goodei* (4); ? *S. helvomaculatus* (4); ? *S. maliger* (4); ? *S. paucispinus* (4); ? *S. pinniger*; ? *S. proriger* (4); ? *S. reedi* (4); ? *S. ruberrimus* (4); ? *S. variegatus* (4); ? *S. wilsoni* (4); ? *S. zacentrus* (4).

Distribution: Pac.

Records: 1. Arai 1967a; 2. Arai 1969b; 3. Sekerak and Arai 1973: 4. Sekerak and Arai 1977; 5. Gibson 1986.

Sucker-ratio 1:<2; vitelline follicles not normally interrupted laterally to testes 8

8 Proximal saccular part of seminal vesicle normally bipartite, with median constriction or bend; Pacific coast ....*P. gibbonsia* Johnson, 1949

Syn.: ?*P. atomon* (Rud., 1802) of Arai (1967a, 1969b) (in part).

Site: Intestine.

Host: *Platichthys stellatus*.

Distribution: Pac.

Records: ?Arai 1967a; ?Arai 1969b; Gibson 1986.

Proximal saccular part of seminal vesicle without constriction or bend; Atlantic coast; especially common in littoral and rock-pool fishes ....*P. atomon* (Rudolphi, 1802) Dujardin, 1845

Syn.: *Distomum simplex* (Rud., 1809) of Stafford (1907) (in part); *Sinistroporus simplex* (Rud., 1809) of Stafford (1904) (in part) and Cooper (1915).

Sites: Intestine, pyloric caeca (stomach).

Hosts: *Anguilla rostrata* (9); *Apeltes quadracus* (9); *Gadus morhua* (4, 20, 24); *Gasterosteus aculeatus* (7, 10, 15); *G. wheatlandi* (11); *Hemitripterus americanus* (20); *Hippoglossoides platessoides* (6, 16, 22, 23); *Hippoglossus hippoglossus* (6, 27); *Myoxocephalus octodecimspinosus* (20); *M. quadricornis* (13); *M. scorpius* (1, 2, 4, 21); *Pholis gunnellus* (3); *Pleuronectes americanus* (3, 4, 6, 18, 23, 25); *P. ferrugineus* (6, 17, 23); *P. putnami* (6); *Pungitius pungitius* (14); *Raja ocellatus* (5, accidental); *Salvelinus fontinalis* (14); *Scomber scombrus* (1, 2); *Sebastes marinus* (1, 2, 19); *Tautoglabrus adspersus* (12); *Urophycis chuss* (1, 2, 26).

Distribution: Atl, Lab-b, Lab, Nfld-b, Nfld, Que-b, Que.

Records: 1. Stafford 1904 (Atl); 2. Stafford 1907 (Atl); 3. Cooper 1915 (Atl); 4. Heller 1949 (Atl); 5. Myers 1959 (Atl); 6. Ronald 1960 (Atl); 7. Hanek and Threlfall 1969a (Nfld-b); 8. Hanek and Threlfall 1970a (Nfld-b); 9. Hanek and Threlfall 1970b (Lab, Nfld-b); 10. Hanek and Threlfall 1970d (Atl, Lab-b, Lab, Nfld-b, Nfld); 11. Hanek and Threlfall 1971 (Nfld); 12. Sekhar and Threlfall 1970a (Atl); 13. Threlfall and Hanek 1971 (Lab-b); 14. Hicks and Threlfall 1973 (Lab); 15. Hanek and Molnar 1974 (Que-b, Que); 16. Scott 1975a (Atl); 17. Scott 1975c (Atl); 18. Scott 1976 (Atl); 19. Gaevskaya and Umnova 1977 (Atl); 20. Linkletter *et al.* 1977 (Atl); 21. Bray 1979 (Atl); 22. Umnova 1979 (Atl); 23. Scott 1982 (Atl): 24. Appy and Burt 1982 (Atl); 25. Scott 1985b (Atl); 26. Scott 1987 (Atl); 27. Scott and Bray 1989.

Remarks: Not all of the above records are likely to have been valid. There has been great confusion between this species and *P. reflexa* and, especially, *P. angulata* (=*P. staffordi*) (see Gibson and Bray 1982). The latter species is especially common in migratory fishes and in brackish water, whereas *P. atomon* is especially common in the intertidal region. Those records from migratory fishes in fresh or brackish water are, therefore, especially suspect.

Subfamily STENAKRINAE Yamaguti, 1970

Opecoelidae. Body small to medium-sized, oval to fusiform. Ventral sucker embedded. Caeca end blindly. Testes two, tandem to symmetrical, usually close to posterior extremity. Functional cirrus sac present, contains seminal vesicle, pars prostatica, and ejaculatory duct. External seminal vesicle absent. Ovary usually pre-testicular, occasionally lateral or antero-lateral to anterior testis. Canalicular seminal receptacle absent. Uterine seminal receptacle present. Uterine field between gonads and ventral sucker. Eggs large, numerous, without filaments. Vitelline fields extend into forebody, but absent from or at least not confluent in post-testicular field. In intestine (? and stomach) of marine teleosts.

Key to the genera of the Stenakrinae

1 Ovary distinctly lobed .... *Stenakron*

Ovary oval 2

2 Oral sucker larger than ventral sucker; body often distinctly narrower in caudal (testicular) zone; excretory vesicle short, reaching to level of posterior testis .... *Anisorchis*

Ventral sucker larger than oral sucker; body usually fusiform; excretory vesicle long, reaching to level of posterior margin of ventral sucker .... *Caudotestis*

STENAKRON Stafford, 1904

Stenakrinae. Body small to medium-sized, broad spindle-shaped, posterior region often narrow. Ventral sucker larger than oral, in middle or just anterior to middle of body, rounded. Pharynx subglobular. Oesophagus relatively long. Intestinal bifurcation in posterior forebody. Caeca end blindly at level of ovary or anterior region of testes. Testes large, symmetrical to diagonal, in middle or posterior half of hindbody. Cirrus sac elongate-claviform, reaches back to level of ventral sucker or even into hindbody, contains elongate-saccular seminal vesicle, tubular pars prostatica, and ejaculatory duct. Genital atrium small. Genital pore mid-ventral, pre-bifurcal in middle of forebody. Ovary normally tri-lobed, but may apparently be entire or four-lobed, dextral, anterior to dextral testis. Uterine seminal receptacle often indistinct. Uterine field between testes and middle of ventral sucker. Vitelline follicles in lateral fields, confluent dorsally in forebody and sometimes at level of ovary. Excretory vesicle reaching to level of testes. In intestine of marine teleosts.

Stenakron vetustum Stafford, 1904 (Fig. 84)

Site: Intestine.

Hosts: *Careproctus reinhardti* (8); *Gadus morhua* (11); *Glyptocephalus cynoglossus* (6, 9, 12); *Hemitripterus americanus* (1, 8); *Hippoglossoides platessoides* (8, 9, 11, 12, 14); *Hippoglossus hippoglossus* (1, 8, 9, 17); *Lumpenus lampretaeformis* (8); *Lycodes vahli* (8); *Pleuronectes americanus* (7, 12, 13); *P. ferrugineus* (2, 5, 8, 12); *Reinhardtius hippoglossoides* (8, 10, 16, 17, 18, 19, 20, 21); *Urophycis musicki* (15).

Distribution: Atl, EArc.

Records: 1. Stafford 1904 (Atl); 2. Ronald 1960 (Atl); 3. Scott 1975a (Atl); 4. Scott 1975b (Atl); 5. Scott 1975c (Atl); 6. Scott 1975d (Atl); 7. Scott 1976 (Atl); 8. Bray 1979 (Atl); 9. Zubchenko 1980 (Atl); 10. Reimer 1981 (EArc); 11. Appy and Burt 1982 (Atl); 12. Scott 1982 (Atl); 13. Scott 1985b (Atl); 14. Zubchenko 1985a (Atl); 15. Scott 1987 (Atl); 16. Wierzbicka 1988 (Atl); 17. Scott and Bray 1989 (Atl); 18. Wierzbicka 1991a (Atl); 19. Wierzbicka 1991b (Atl); 20. Krzykawski and Wierzbicka 1992 (Atl); 21. Arthur and Albert 1994 (Atl).

Remarks: Bray (1979) also described a specimen that he called *Stenakron* sp. innom. from the intestine of *Triglops murrayi* on the Grand Banks. Bray considered that this worm may be a teratological specimen of *S. vetustum*, *S. skrjabini* (Issaichikov, 1928), or even, but less likely, *Eurycreadium vitellosum* Manter, 1934.

Redkozubova (1978) listed *Stenakron* sp. as occurring in several flatfishes in Canadian waters. This is almost certain to have been *S. vetustum*.

ANISORCHIS Polyanski, 1955

Stenakrinae. Body medium-sized, elongate-oval with narrow posterior region. Oral sucker larger than ventral sucker, latter in anterior half of body. Oesophagus well developed. Intestinal bifurcation near middle of forebody. Caeca terminate blindly lateral to anterior testis. Testes in tandem, within narrow caudal region of body close to posterior extremity. Cirrus sac long, cylindrical, just reaching into hindbody, contains

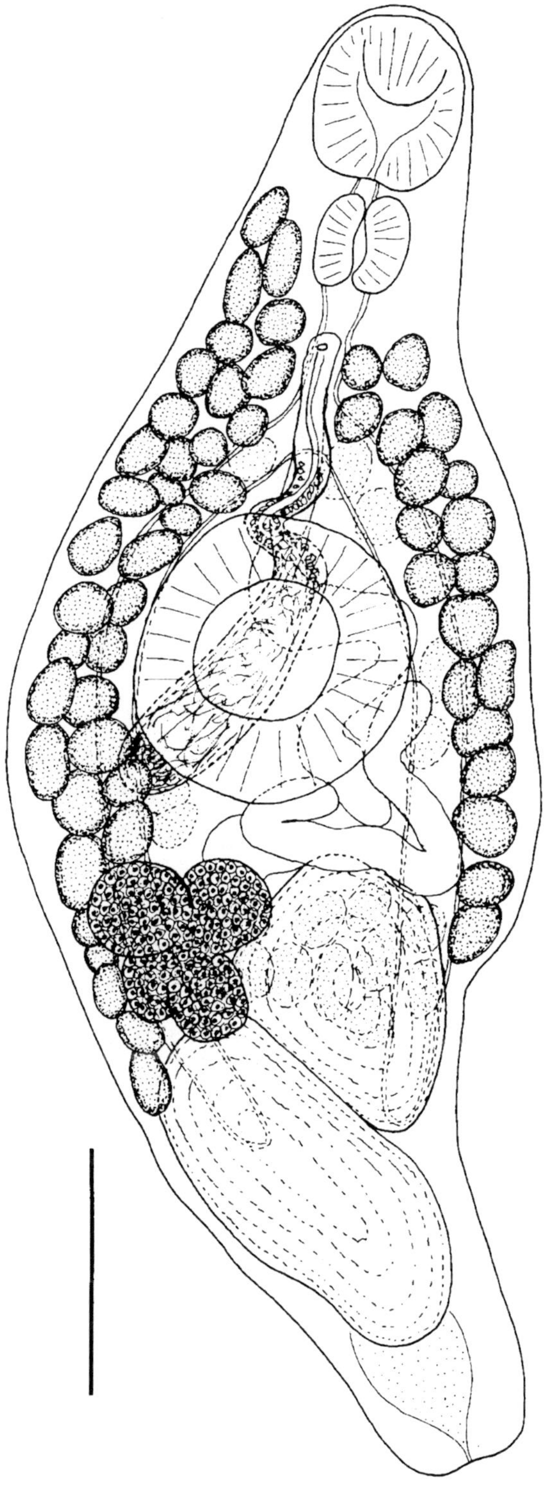

FIG. 84. *Stenakron vetustum* ex. *Hippoglossus hippoglossus*, northeast Atlantic. Original. Scale bar: 250 µm.

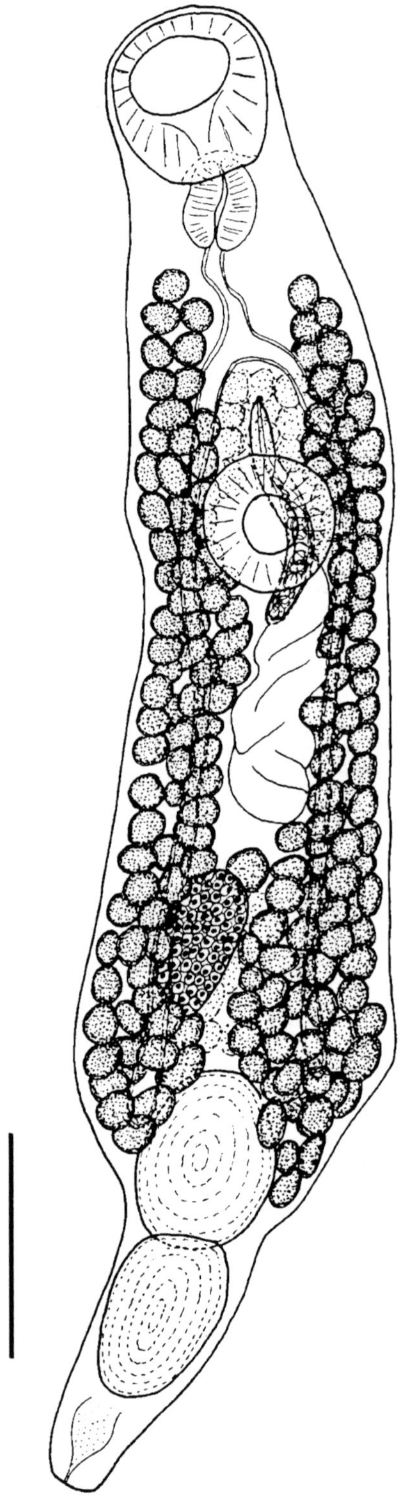

FIG. 85. *Anisorchis opisthorchis* (after Bray (1979)). Scale bar: 500 µm.

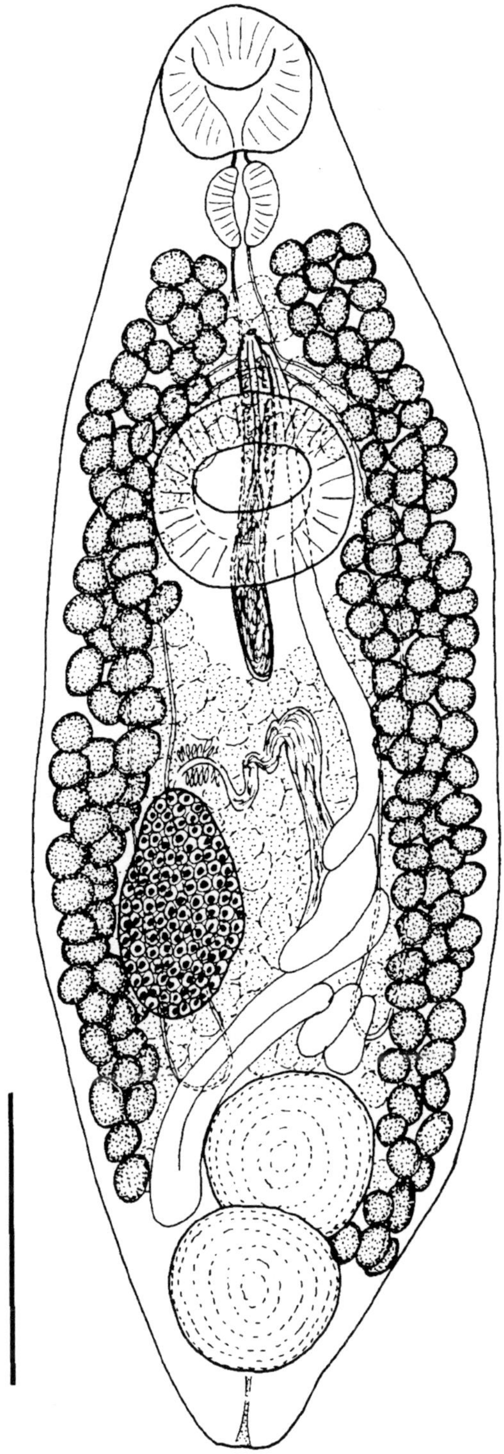

FIG. 86. *Caudotestis nicolli* (after Bray (1979)). Scale bar: 500 µm.

long, saccular seminal vesicle and tubular pars prostatica/ejaculatory duct (the two cannot be distinguished). Genital pore ventro-median, post-bifurcal in forebody. Ovary elongate-oval, short distance antero-dextral to anterior testis. Uterine field between ovary and ventral sucker. Vitelline follicles mainly in lateral fields between level of pharynx and anterior testis, fields united dorsally at level of genital pore and ovary. Excretory vesicle short, reaching to level of posterior testis. In intestine of marine teleosts.

Anisorchis opisthorchis Polyanski, 1955 (Fig. 85)

Sites: Intestine, pyloric caeca (stomach).
Hosts: *Artediellus uncinatus* (3); *Hemilepidotus hemilepidotus* (1, 2); *Hexagrammos lagocephalus* (1, 2); *H. stelleri* (1, 2); *Leptagonus decagonus* (3).
Distribution: Pac, Atl.
Records: 1. Arai 1967a (Pac); 2. Arai 1969b (Pac); 3. Bray 1979 (Atl).

CAUDOTESTIS Issaichikov, 1928 (sensu Bray, 1979)

Stenakrinae. Body small, fusiform. Oral sucker almost terminal. Ventral sucker slightly larger than oral sucker, in anterior third of body, rounded. Prepharynx very small. Oesophagus well developed. Intestinal bifurcation in posterior forebody. Caeca terminate blindly close to anterior margin of anterior testis. Testes tandem to slightly oblique, at posterior extremity. Cirrus sac elongate, cylindrical, reaching just posterior to ventral sucker, contains saccular seminal vesicle, small pars prostatica, and long, muscular ejaculatory duct. Genital pore ventro-median in forebody. Ovary pyriform to oval, antero-dextral to testes. Uterine field extending to level of posterior testis. Eggs thin-shelled. Vitelline follicles in lateral fields, confluent dorsally at level of ventral sucker and in hindbody, overlapping caeca ventrally, from level of pharynx to middle of posterior testis. Excretory vesicle reaching to level of ventral sucker. In intestine (? and stomach) of marine teleosts.

Caudotestis nicolli Issaichikov, 1928 (Fig. 86)

Syn.: *Plagioporus nicolli* (Issaichikov, 1928) Price, 1934.
Site: Intestine.
Hosts: *Anarhichas lupus* (1, 2); *Triglops murrayi* (1).
Distribution: Atl.
Records: 1. Bray 1979; 2. Bray 1987c.

SUPERFAMILY LEPOCREADIOIDEA ODHNER, 1905

Small to medium-sized worms, body shape variable, but usually elongate and dorso-ventrally flattened. Tegument spinose, occasionally with row(s) of enlarged circum-oral spines. Eye-spot material often present. Oral and ventral suckers well developed, oral sucker ventrally subterminal, ventral sucker normally in anterior half of body. Prepharynx present, short to long. Pharynx well developed. Oesophagus short to long, pseudoesophagus occasionally present. Intestinal bifurcation normally in forebody. Caeca end blindly close to posterior extremity, form uroproct or rarely open via ani. Testes two, in posterior half of body, usually tandem, occasionally oblique, rarely symmetrical. External seminal vesicle present or absent. Cirrus sac usually present, occasionally absent; when present containing internal seminal vesicle, pars prostatica, and ejaculatory duct; latter duct may be armed. Genital atrium small to long and tubular. Genital pore usually median or submedian, occasionally sublateral, normally in forebody, usually immediately anterior, to ventral sucker, rarely lateral to ventral sucker. Ovary median or submedian, occasionally lateral, normally pre-testicular near middle of hindbody, rarely lateral to anterior testis. Canalicular or uterine seminal receptacle present. Laurer's canal present. Uterine field normally between gonads and ventral sucker, occasionally extends further posteriorly. Metraterm distinct or indistinct, sometimes armed. Eggs medium-sized to large, operculate, few to numerous. Vitellarium follicular, usually widely distributed in hindbody, occasionally more restricted, occasionally extends into forebody, usually confluent in post-testicular zone and occasionally more anteriorly. Excretory pore terminal, excretory vesicle I-shaped, short to long. Mainly in intestine of marine teleosts, but occasionally in euryhaline and freshwater teleosts and chondrosteans.

Comments

Features of "larval" morphology and the life history may well be of use as diagnostic characters of this group; but recent work by Køie (1985a,b) has indicated that the lepocreadiid *Lepidapedon elongatum* (see below) does not have the typical lepocreadioid opthalmotrichocercous cercaria, while its congener *L. rachion* (see below) does. Køie (1985b) used this argument to suggest that *L. rachion* is more closely related to *Opechona* than to *L. elongatum*, dismissing adult characters as if they were, *a priori*, of no significance (Lumb *et al.* (1993) have shown that *L. elongatum* and *L. rachion* are more closely related than either is to *Opechona*). On the contrary, such plasticity in cercarial morphology, and especially, in the form of the cercarial tail, casts doubt upon the use of this feature as a phylogenetic indicator (see Gibson 1987; Bray 1988). Since "larval" characters are of no value in keying down adult specimens, the following diagnoses and keys are based entirely upon adult features.

One of the main features of this group is the presence of spination on the body surface: this is, however, easily lost in poorly fixed material or in worms, such as those removed from frozen hosts or guts, which are allowed to die prior to fixation. This fact should be borne in mind when identifying the species of this and related groups.

KEY TO THE FAMILIES OF THE LEPOCREADIOIDEA

1 Uterine seminal receptacle present (in the case of *Neophasis* sectioning is necessary to determine this) .... Acanthocolpidae

Canalicular seminal receptacle present 2

2 Cirrus sac absent .... Homalometridae

Cirrus sac present 3

3 External seminal vesicle present; ejaculatory duct and metraterm unarmed..................Lepocreadiidae

External seminal vesicle absent; ejaculatory duct and metraterm armed with large spines...........
....Deropristidae

Family ACANTHOCOLPIDAE Lühe, 1906

Lepocreadioidea. Body oval to elongate and dorso-ventrally flattened. Tegument spinose, two, three, or more rows of alternating, enlarged circum-oral spines may be present. Eye-spot material may be present. Oral sucker subglobular, ventrally subterminal. Ventral sucker in anterior half of body. Prepharynx present, short to long. Pharynx oval. Oesophagus short to long, pseudoesophagus absent. Intestinal bifurcation in forebody. Caeca end blindly near posterior extremity or unite with excretory vesicle to form uroproct. Testes two, in hindbody, usually in tandem, occasionally oblique. External seminal vesicle absent. Cirrus sac present, contains internal seminal vesicle, pars prostatica, and ejaculatory duct (latter may be armed with small spines). Genital atrium (hermaphroditic duct) tubular, short to long. Genital pore median or submedian, immediately anterior to ventral sucker. Ovary median or submedian, in hindbody, pre-testicular or occasionally lateral to anterior testis. Canalicular seminal receptacle absent. Laurer's canal and uterine seminal receptacle present. Uterine field in pre-ovarian or pre-testicular hindbody. Metraterm distinct or indistinct, unarmed. Eggs large, few to numerous, operculate. Vitellarium follicular, lateral fields extend from posterior extremity to level of uterine field or occasionally to pharynx, fields confluent in post-testicular zone and occasionally dorsally to ventral sucker. Excretory pore terminal, vesicle I-shaped (see "Comments"). In intestine of marine teleosts.

Comments

The systematic position of the Acanthocolpidae has long been a matter of contention (see Stunkard 1961; Peters 1961). Nevertheless, in adult morphology they are clearly close to the lepocreadioids. Earlier work on cercarial morphology and the life cycle that suggested affinities with the Echinostomatoidea is based upon insufficient evidence or questionable interpretation. In addition, Bray and Gibson (1991a) argued that the new suborder Acanthocolpata of Brooks *et al.* (1985) could not be substantiated. Following the hypothesis that it is the adult (marital) generation, possessing the most reliable characters that can be used to assess phylogenetic characters in the Digenea (Gibson 1987), the Acanthocolpidae are treated here as a family within the Lepocreadioidea.

Because there are only two genera in Canadian waters and because of unresolved problems within the family, especially with regard to the inclusion of *Neophasis*, I have not subdivided the Acanthocolpidae into subfamilies.

Many authors have considered the excretory vesicle to be Y-shaped. This often depends upon one's interpretation of the shape of the proximal extremity as it enlarges to give off the main collecting ducts on each side. Bray and Gibson (1991a) considered that it should be interpreted as I-shaped.

Key to the genera of the Acanthocolpidae

1 Rings of enlarged, circum-oral spines present; body elongate ....*Stephanostomum*

Rings of enlarged, circum-oral spines absent; body oval..*Neophasis*

NEOPHASIS Stafford, 1904

Acanthocolpidae. Body oval, fusiform or lanceolate. Tegument spinose, enlarged circum-oral spines absent. Eye-spot pigment normally present. Oral sucker spherical, ventrally subterminal. Ventral sucker in anterior half of body. Prepharynx short. Pharynx large. Oesophagus short. Caeca end blindly near posterior extremity. Testes two, normally oblique, occasionally symmetrical, large, adjacent in posterior half of body. Cirrus sac large, extending into hindbody, claviform, containing oval or bipartite internal seminal vesicle, pars prostatica, and short, unarmed ejaculatory duct; may project into genital atrium forming small papilla. Genital atrium short. Genital pore median immediately anterior to ventral sucker. Ovary irregularly oval, dextral, usually lateral to anterior testis and anterior to and contiguous with posterior testis. Canalicular seminal receptacle absent. Laurer's canal and uterine seminal receptacle present. Uterine field between gonads and ventral sucker. Metraterm muscular, enveloped by gland cells, lined with irregular tegumentary filaments. Eggs large, few, operculate. Vitellarium follicular, follicles large, lateral fields extend from posterior extremity into forebody, fields confluent in post-testicular zone and at level of ventral sucker. Excretory vesicle I-shaped, reaches to level of testes. In intestine of marine teleosts.

Comments

Neophasis, although it differs in shape, vitelline distribution and gonadal arrangement from other acanthocolpids, probably fits into the Acanthocolpidae because of the presence of a uterine seminal receptacle: this is discussed in more detail elsewhere (Bray and Gibson 1991a). In fact, the nature of the seminal receptacle in this genus can be rather difficult to detect. For example, Bray (1979), using material from Canadian waters, described it as being of the canalicular type. His recent examination of this material has, however, convinced him that his earlier observation was in error (Bray and Gibson 1991a).

A revision, with a key and descriptions, of North Atlantic species of this genus was given by Bray and Gibson (1991a). The latter key is followed below.

Key to the species of *Neophasis*

1 Length <750 μm; average egg length <95 μm; parasitic in *Anarhichas* spp 2

Length up to 1620 μm; average egg length >95 μm; mainly in cottids and zoarcids 3

2 Mean sucker-ratio 1:<1; average egg length >90 μm...........*Neophasis pusilla* Stafford, 1904 (Fig. 87)

Sites: Intestine, urinary bladder, gall bladder.
Host: *Anarhichas lupus*.
Distribution: Atl.
Records: Stafford 1904 (see also Miller 1941a); Bray 1979; Bray 1987c; Bray and Gibson 1991a.
Remarks: Zhukov (1960) considered this species a synonym of *N. oculata*; but this was not recognised by Brinkmann (1975), Bray (1979), or Bray and Gibson (1991a).

Mean sucker-ratio 1:>1; average egg length <90 μm; (? a Northeast Atlantic species) ....*N. anarrhichae* (Nicoll, 1909) Bray, 1987

Syn.: *Acanthopsolus anarrhichae* Nicoll, 1909; *Neophasis lageniformis* (Lebour, 1910) Miller, 1941.
Site: Intestine.
Hosts: *Anarhichas lupus*; *A. minor*.
Distribution: Atl.
Record: Zubchenko 1980.
Remarks: According to Bray and Gibson (1991a), this species normally occurs in the Northeast Atlantic. Zubchenko's records may, therefore, have been *N. pusilla*.

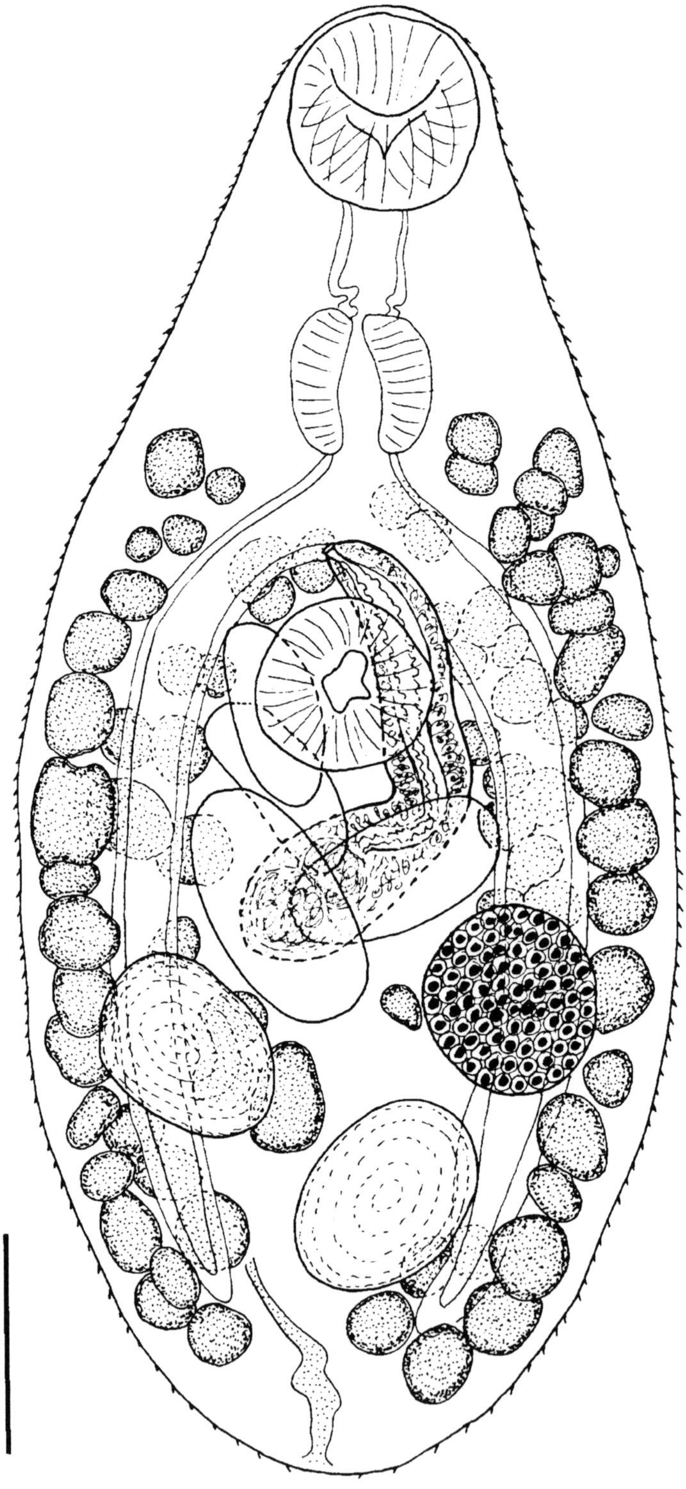

FIG. 87. *Neophasis pusilla* (after Bray and Gibson (1991a)). Scale bar: 100 µm.

3 Mean sucker-ratio 1:>1; average longitudinal overlap of testes 50–100%, average lateral overlap of testes 26–32% (relates to degree of obliqueness of testes; see Bray and Gibson 1991a) ....*N. oculata* (Levinsen, 1881) Miller, 1941

Site: Intestine.
Host: *Lycodes vahli* (1, 2).
Distribution: Atl.
Records: Bray 1979; Bray and Gibson 1991a.

Mean sucker-ratio 1:<1; average longitudinal overlap of testes 38%, average lateral overlap of testes 69% (i.e., testes are more tandem and less symmetrical than in *N. oculata*)........................ ....*N. burti* Bray and Gibson, 1991

Syn.: *Neophasis oculata* (Levinsen, 1881) of Bray (1979) in part; (?) *Neophasis* sp. of Appy and Burt (1982).
Site: Intestine.
Hosts: *Gadus morhua* (?2); *Hippoglossoides platessoides* (2); *Myoxocephalus octodecimspinosus* (1, 3).
Distribution: Atl.
Records: 1. Bray 1979; 2. Appy and Burt 1982; 3. Bray and Gibson 1991a.
Remarks: This species is discussed in detail by Bray and Gibson (1991a). Appy and Burt's (1982) specimens were immature, and in the case of *H. platessoides* they were from the mesenteries. It is worth noting that Arthur and Albert (1994) recorded immature specimens and metacercariae of *Neophasis* sp. from *Reinhardtius hippoglossoides*.

STEPHANOSTOMUM Looss, 1899

Acanthocolpidae. Body elongate to tubular. Tegument spinose, with two (sometimes three) rows of enlarged, alternating circum-oral spines. Eye-spot pigment may be present. Oral sucker ventrally subterminal. Ventral sucker well inside anterior half of body. Prepharynx long. Pharynx oval. Oesophagus short to long. Caeca terminate blindly near posterior extremity or open into excretory vesicle and form uroproct. Testes tandem in posterior half of body. Cirrus sac elongate-claviform, reaching into hindbody, contains saccular seminal vesicle, pars prostatica, and unarmed ejaculatory duct. Genital atrium (hermaphroditic duct) short to long. Ovary pre-testicular, median or submedian. Canalicular seminal receptacle absent. Laurer's canal and uterine seminal receptacle present. Uterine field between ovary and ventral sucker. Eggs large, numerous. Vitellarium follicular, lateral fields confined to hindbody, confluent in post-testicular zone. Excretory vesicle I-shaped, reaching to about level of posterior margin of ovary. In intestine of marine teleosts.

Comments

This genus is in dire need of revision. Some species may prove not to be acanthocolpids, since the nature of the seminal receptacle has not always been described.

One of the most obvious characters used to distinguish the species is the number of circum-oral spines. Although this feature is still used by most workers, its value has been questioned by Wolfgang (1955a,b) and Køie (1984). Wolfgang, working with material from Canadian waters, found that the number of spines in *S. baccatum*, supposed to have been 56, ranged from 44 to 56, and cast doubt upon its value as a taxonomic criterion. Køie suggested that the cod forms, *S. pristis* (said to have 2 × 18 spines) and *S. caducum* (Looss, 1901) (said to have 2 × 24 spines) were synonyms on the basis of her finding spine numbers ranging from 2 × 18 to 2 × 26, although she stated that most had 2 × 23–25. This contradicted her earlier (1978) work in which she found that all of her material, referred to as *S. caducum*, had 2 × 24–25 oral spines, and stated that "there is no doubt *S. caducum* is a valid species". I prefer to use the number of oral spines as a valid criterion for the following reasons: (1) all of the well-fixed material of *S. baccatum* that I have examined from both sides of the North Atlantic has a spine number in the region of 54–56 (2 × 27–28): this includes material from *Eutrigla gurnardus* off the Faroes, which indicates that *S. triglae* (Lebour, 1908) is a synonym of this species; (2) all of the material from the cod *Gadus morhua* in the BM(NH) collection has spine numbers in the region of 2 × 18 and 2 × 24 (more prevalent) with no intermediates, indicating that two species are present; and (3) it is a well-known fact that in lepocreadioids, and other groups of digeneans with oral spines, these spines are easily lost in material in poor condition: this may account for the intermediates.

One feature used by Manter and Van Cleave (1951) in their key to the species of this genus was the anterior extent of the vitelline fields. In agreement with Wolfgang (1955b), I find this a somewhat variable feature in these worms.

The use of oral spines as a diagnostic feature means that metacercariae and immature adults can be recognized. The ability to identify immature specimens in accidental hosts may give a false impression of the host specificity of these parasites, because fully developed mature adults occur in a narrower host-range than usually indicated below.

The presence or absence of a uroproct is a feature that may prove to be useful for splitting the genus.

Key to species of *Stephanostomum*

1 On Atlantic coast 2

On Pacific coast 3

2 Number of circum-oral spines in region of 56 (2 rows of 28) ....
.... *Stephanostomum baccatum* (Nicoll, 1907) Manter, 1934

(A) Encysted in tissues (metacercaria)

Syn.: *Stephanochasmus histrix* (Duj.) of Stafford (1904, 1907); unidentified metacercaria of Huntsman (1918).

Sites: Skin, fins, gills, musculature.

Hosts: *Glyptocephalus cynoglossus* (4, 7, 8); *Hemitripterus americanus* (5, 6); *Hippoglossoides platessoides* (3, 4, 6, 7, 8, 10); *Hippoglossus hippoglossus* (7); *Lophius americanus* (1, accidental); *Pleuronectes americanus* (1, 2, 4, 5, 7, 9, 12); *P. ferrugineus* (4, 7); *P. putnami* (4, 7); *Reinhardtius hippoglossoides* (11); *Scophthalmus aquosus* (4); Unspecified fishes (6).

Distribution: Atl.

Records: 1. Stafford 1904; 2. Stafford 1907; 3. Huntsman 1918; 4. Wolfgang 1954a; 5. Wolfgang 1954b; 6. Wolfgang 1955a; 7. Ronald 1960; 8. Umnova 1979; 9. Scott 1985b; 10. Morrison *et al.* 1986; 11. Arthur and Albert 1994; 12. BM(NH) collection.

(B) Free in gut (adult)

Syn.: *Stephanochasmus sobrinus* (Lev.) of Stafford (1904, 1907); *Stephanostomum* sp. of Miller (1941a).

Sites: Pyloric caeca, intestine, rectum.

Hosts: *Cryptacanthodes maculatus* (1, 2); *Gadus morhua* (9); *Hemitripterus americanus* (1, 2, 3, 4, 6, 7, 11); *Hippoglossus hippoglossus* (3, 5, 8); *Lycodes* sp. (1, 2); *Myoxocephalus octodecemspinosus* (7); *M. scorpius* (3); *Pleuronectes americanus* (5, 10); *P. ferrugineus* (5); Unspecified flounders (4).

Distribution: Atl.

Records: 1. Stafford 1904; 2. Stafford 1907; 3. Wolfgang 1954b; 4. Wolfgang 1955b; 5. Ronald 1960; 6. Linkletter *et al.*, 1977; 7. Bray 1979; 8. Zubchenko 1980; 9. Appy and Burt 1982; 10. Scott 1985b; 11. BM(NH) collection.

Remarks: This species appears to be especially common in cottids, but it can become ovigerous in several groups of hosts. Among the pleuronectids it matures only in *Hippoglossus hippoglossus* (see Ronald 1960).

Appy and Burt's (1982) record from cod was of immature specimens. It appears that the cod forms (see above), *S. pristis* (Deslongchamps, 1824) and *S. caducum* (Looss, 1901), do not occur on the western side of the North Atlantic. This may be related to the absence of the intermediate hosts (see Linkletter *et al.* 1977), *Natica alderi* and gobiid teleosts, in the case of *S. caducum* (see Køie 1978). *S. pristis* and *S. caducum* are easily recognizable by the interruption in the lateral vitelline fields as the level of the testes.

Future work may show that *S. sobrinus* (Levinsen, 1881) is the oldest valid name for this species. It may also show that *S. dentatum* (Linton, 1900) is a synonym of this species. It occurs in New England waters, is morphologically very similar, and has an identical number of oral spines (54 according to Linton 1940, and Stunkard 1961); but it is supposed to have much smaller eggs (45–60 µm, according to Linton 1940). I have found the eggs of *S. baccatum* in the BM(NH) collection, at least those which are not deformed and lying at the correct angle, to be consistently in the region of 90–100 µm, and yet Wolfgang (1955a) indicated that they ranged from 75–90 µm and some of the Pacific records list egg measurements of up to about 120 µm.

Specimens of Wolfgang and Bray are in the BM(NH) collection.

Number of circum-oral spines in region of 42 (2 rows of 21).......*S. tenue* (Linton, 1898) Linton, 1934

(A) Encysted in tissues (metacercaria)

Sites: Pericardium, spleen, gill arch, viscera, body cavity.

Hosts: *Apeltes quadracus* (1); *Fundulus heteroclitus* (3); *Salvelinus fontinalis* (1, 2).

Distribution: Atl, NB, NB-b, PEI-b.

Records: 1. Frimeth 1987a (NB, NB-b); 2. Frimeth 1987b (Atl, NB-b); 3. McGladdery *et al.* 1990 (PEI-b).

Remarks: I have examined Frimeth's specimens from *S. fontinalis* in the BM(NH) collection labelled as *Stephanostomum* sp. metacercariae and can confirm the oral spine numbers; one specimen clearly has 42 with no obvious gaps for missing spines.

(B) Free in gut (adult)

Site: Posterior intestine.

Hosts: *Anguilla rostrata* (3); *Morone saxatilis* (1); *Oncorhynchus mykiss* (3); *Salvelinus fontinalis* (2).

Distribution: NB, NB-b, PEI-b.

Records: 1. Hogans 1984 (NB); 2. Frimeth 1987a (NB, NB-b); 3. McGladdery *et al.* 1990 (PEI-b).

Remarks: This is a species that occurs off the coast of New England. Wolfgang (1955a) examined its known hosts in Canadian waters and failed to find it. A redescription of the adult is required. Its life history was described by Martin (1939).

Linton's (1940) material of this species from *Hemitripterus americanus* off the coast of Massachusetts is clearly misdetermined and belongs to *S. baccatum*.

3 Two rows of circum-oral spines (numbering 30–60)..4

Three rows of small circum-oral spines (numbering >100)..McFarlane, 1936

Syn.: *Stephanostomum* sp. of Arai (1967a) in part.

Site: Intestine.

Host: *Ophiodon elongatus*.

Distribution: Pac.

Records: McFarlane 1936; Arai 1967a; Arai 1969b.

Remarks: Zhukov (1960) claimed to have recorded this species from *Scomber japonicus* in Peter the Great Bay. This appears to have been a misdetermination of the lepocreadiid now known as *Cephalolepidapedon saba* Yamaguti, 1970 (see Bray and Gibson 1990). It is evident that a detailed study of fresh material is required to confirm that *S. tristephanum* does not belong to this lepocreadiid genus.

4 More than 40 circum-oral spines (usually 54), especially common in cottids ....
....*S. baccatum* (Nicoll, 1907) Manter, 1934 (Fig. 88)

Syn.: *Stephanostomum* sp. of Arai (1967a) in part; *S. dentatum* (Linton, 1900) Linton, 1940 of Arai (1969b) and Sekerak and Arai (1977).

Site: Intestine.

Hosts: *Dasycottus setiger* (1, 2); *Sebastes babcocki* (3); *S. borealis* (3).

Distribution: Pac.

Records: 1. Arai 1967a; 2. Arai 1969b; 3. Sekerak and Arai 1977.

Remarks: This species was also recorded from *Dasycottus setiger* off Japan by Machida (1984). In the same host off Japan, Yamaguti (1934) described *S. japonicum*, which Machida has considered a synonym of *S. baccatum* despite the fact that Yamaguti only recorded 46 oral spines (Machida considered the range to be 43–54). Zhukov (1960) found it in a variety of hosts in Peter the Great Bay and described it as having 54 oral spines. Zhukov's figure is almost identical to Stunkard's (1961) figure of *S. dentatum* (Linton, 1900) from *Paralichthys dentatus* off New England. Morphologically similar worms from *Paralichthys californicus* have been referred to *S. dentatum* off the Californian coast by several authors, including Manter and Van Cleave (1951). The British Columbian determinations under the name *S. dentatum* stem from the latter work. I have kept the Pacific records of *S. baccatum* separate from the Atlantic in case future work shows that the Pacific form is distinct or composed of more than one form. This is suggested by the egg size of the forms from *Paralichthys dentatus*, which is much smaller (61–68 µm) than that normally recorded for *S. baccatum* (see above).

Less than 40 circum-oral spines (usually 34) ....*S. californicum* Manter and Van Cleave, 1951

Syn.: *Stephanostomum casum* of McFarlane (1934, 1936) and Arai (1969b).

Site: Intestine.

Hosts: *Anoplopoma fimbria* (5); *Artedius harringtoni* (3); *Sebastes caurinus* (3, 4, 7); *S. maliger* (4); *S. nebulosus* (6); *Sebastes* sp. (1, 2).

Distribution: Pac.

Records: 1. McFarlane 1934; 2. McFarlane 1936; 3. Arai 1969b; 4. Sekerak and Arai 1977; 5. Kabata *et al.* 1988; 6. Holmes 1990; 7. BM(NH) Collection.

Remarks: This species, originally described from *Umbrina roncador* off California, is morphologically very similar to *S. bicoronatum* (Stossich, 1883) from *Umbrina cirrosa* in the Mediterranean. Unfortunately, the type specimens of the latter in the BM(NH) collection are not in good condition and the BM(NH) specimen of *S. californicum* from *Sebastes caurinus* is immature.

It is worth noting that Arthur (1984) and Stanley *et al.* (1992) recorded *Stephanostomum* sp. metacercariae in the fins of *Theragra chalcogramma* and the skin of *Sebastes flavidus*, respectively, off the coast of British Columbia. They have also been noted in *Atherestes stomias* by Kabata and Whitaker (1984).

Family DEROPRISTIDAE Cable and Hunninen, 1942

Lepocreadioidea. Body elongate, slightly dorso-ventrally flattened, medium-sized. Tegument heavily spined, spines well developed, acuminate (occasionally deformed), circum-oral ring of larger spines occasionally present. Eye-spot material may be present. Oral sucker well developed, ventrally subterminal. Ventral sucker often smaller than oral, in anterior third of body. Prepharynx short. Pharynx well developed. Oesophagus short to medium-length. Intestinal bifurcation usually in anterior half of forebody, or almost so. Caeca terminate blindly close to posterior extremity. Testes two, entire, in tandem or slightly oblique, separated or contiguous, in posterior third of body. External seminal vesicle absent. Cirrus sac well developed, large, claviform, extending into or entirely within hindbody, containing bipartite, saccular seminal vesicle, indistinct pars prostatica, and large, broad ejaculatory duct lined with large, flexible spines. Genital atrium tubular, short to very long. Genital pore median (or submedian), ventral, immediately anterior to ventral sucker. Ovary entire, median to submedian, pre-testicular, posterior to middle or near middle of body. Laurer's canal and canalicular seminal receptacle present. Uterine field large, extends

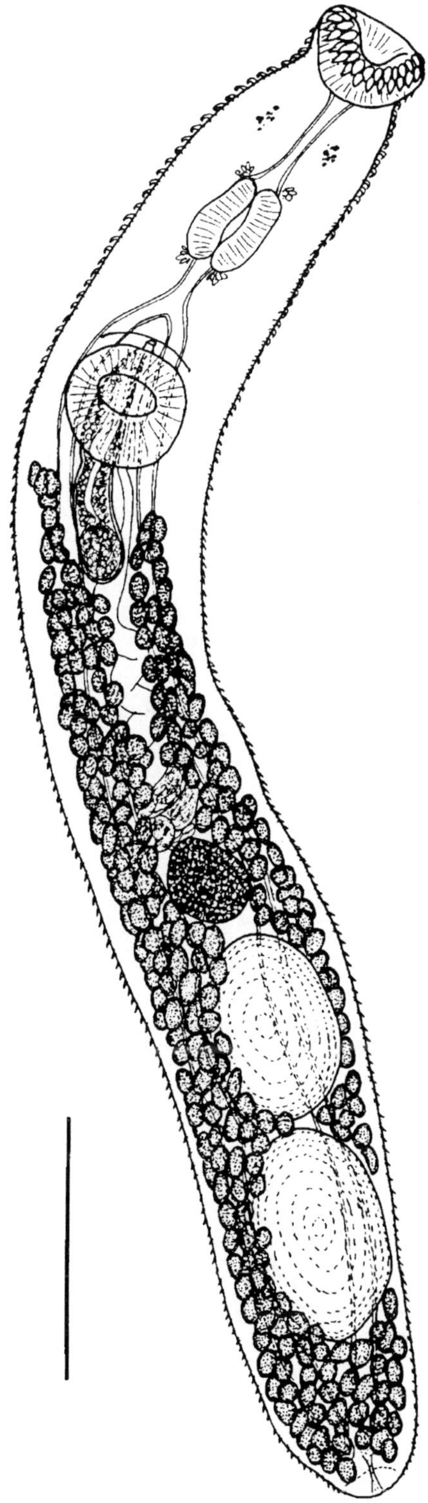

FIG. 88. *Stephanostomum baccatum* from *Hemitripterus americanus*, Canada. Original. Scale bar: 500 µm.

from about level of metraterm to level of testes or occasionally to posterior extremity; may reach extra-caecally, especially in posterior half of body, may extend between gonads. Metraterm large, distinct, claviform, lined with large, flexible spines, at level of cirrus sac. Eggs medium-sized, operculate, usually very numerous. Vitellarium follicular, lateral fields between about levels of middle of cirrus sac and testes, absent from post-testicular zone, may or may not be confluent medially. Excretory vesicle I-shaped, short, reaching to or almost to level of posterior testis, pore terminal. Parasitic mainly in spiral intestine of chondrosteans and occasionally the intestine of teleosts in euryhaline conditions.

Key to the genera of the Deropristidae

1 Lateral expansion of forebody with enlarged marginal spines present (in well-fixed material); genital atrium long, such that cirrus sac is entirely within hindbody (only when cirrus not formed); in sturgeons and eels (occasionally other migratory teleosts)................................*Deropristis*

Lateral expansion of forebody with enlarged marginal spines absent; genital atrium short, such that cirrus sac overlaps ventral sucker; in sturgeons ....*Skrjabinopsolus*

DEROPRISTIS Odhner, 1902

Deropristidae. Body elongate, forebody concave ventrally and expanded laterally. Tegument spinose, spines on margin of lateral expansions of forebody and dorsal to pharynx enlarged. Circum-oral ring of spines absent. Testes close to posterior extremity, tandem, contiguous or almost so. Cirrus sac large, in anterior hindbody, seminal vesicle bipartite, ejaculatory duct large, broad, lined with large flexible spines. Genital atrium tubular, long, usually extending into hindbody, but short when cirrus formed. Ovary posterior to middle of body, separated from anterior testis by loops of uterus. Uterine field large, extending between about level of seminal vesicle and anterior testis. Metraterm large, broad, lined with large, flexible spines. Vitellarium in lateral fields, anterior limit at about level of seminal vesicle, posterior limit varies from halfway between ovary and testis to posterior margin of anterior testis. In spiral valve of chondrosteans and intestine of migratory teleosts (especially eels) in euryhaline conditions.

Comments

It is clear that there has in the past been some difficulty in distinguishing the two species of this genus on morphological grounds. The features put forward by Odhner (1902) and Ward (1938) require confirmation. I believe that at present the most practical action, pending a detailed study, is to treat the specimens occurring in sturgeons as *D. hispida* and those occurring in eels (and occasionally other teleosts) as *D. inflata*, since this is supported by the vast majority of previous determinations of material from these hosts and by aspects of their distribution, e.g., *D. hispida* appears to be common in sturgeons in Russia, whereas *D. inflata* is unrecorded in eels in Russia. Having said this, however, the small amount of material available to me in the BM(NH) collection indicates differences in the size of the eggs and body spines. I have included these features in the key below in square brackets. The material that I examined, most of which is in poor condition, is as follows: *D. inflata* ex. *Anguilla anguilla* from off Denmark and England; *D. hispida* ex. *Acipenser sturio* from an unknown locality (this material is probably about 150 years old, having been purchased from Prof. C.T.E. von Siebold in 1851).

Key to the species of *Deropristis*

1 Parasitic in intestine of eels (or other teleosts); (eggs ca. 43–55 μm, in length (43–49 according to Ward 1938)); body spines in region of ventral sucker 11–20 μm in length, 3–4 μm in breadth: see "Comments" above) ....*Deropristis inflata* (Molin, 1859) Odhner, 1902 (Fig. 89)

Site: Intestine.
Host: *Anguilla rostrata*.
Distribution: Atl.
Records: 1. Stafford 1904; 2. Stafford 1907.
Remarks: Stafford's determination was confirmed by Miller (1941a).

Parasitic in spiral intestine of sturgeons; (eggs ca. 38–43 μm in length (36–45 according to Ward 1938)); body spines in region of ventral sucker 23–27 μm in length, ca. 9 μm in breadth: see "Comments" above) ....*D. hispida* (Abildgaard in Rudolphi, 1819) Odhner, 1902

Syn.: *Deropristis inflata* (Molin) of Linkletter *et al.* (1977).
Site: Intestine.
Host: *Acipenser oxyrhynchus*.
Distribution: Atl, NB-b.
Records: 1. Linkletter *et al.* 1977 (Atl); 2. Appy and Dadswell 1978 (NB-b).

SKRJABINOPSOLUS Ivanov in Ivanov and Murygin, 1937

Deropristidae. Body elongate, forebody without lateral expansions or ventral concavity. Tegument spinose, lacking enlarged spines in particular regions of body. Circum-oral ring of enlarged spines absent. Testes tandem to oblique, contiguous to widely separated by loops of uterus, in posterior third of body, hind testis near posterior extremity. Cirrus sac large, extending into hindbody, seminal vesicle bipartite, ejaculatory duct broad, lined with large spines. Genital atrium short. Ovary near middle of hindbody or just posterior to it, well separated from anterior testis by loops of uterus. Uterine field large, extending between about level of seminal vesicle and posterior testis, sometimes to posterior extremity. Metraterm broad, armed with spines. Vitellarium in lateral fields, anterior limit at level of seminal vesicle (or just posterior to it), posterior limit about level of anterior testis (or just anterior to it). In spiral intestine of chondrosteans.

Skrjabinopsolus manteri (Cable, 1952) Cable, 1955 (Fig. 90)

Syn.: *Deropristis hispidus* (Abild.) of Stafford (1904).
Site: Intestine.
Host: *Acipenser fulvescens*.
Distribution: ?Que, Ont.
Records: 1. Stafford 1904 (?Que); 2. Anthony 1974 (Ont); 3. Dechtiar and Lawrie 1988 (Ont); 4. Dechtiar *et al.* 1988 (Ont); 5. Dechtiar and Christie 1988 (Ont).
Remarks: Ward (1938) indicated that Stafford's (1904) record of *Deropristis hispidus* was a misdetermination, and stated that he would report on the identity of Stafford's material at a later date. Unfortunately, he did not do so. It is clear, however, that Stafford's description conforms so well with *S. manteri* as to leave little doubt as to its identity.

Family HOMALOMETRIDAE Cable and Hunninen, 1942

Lepocreadioidea. Body oval to elongate, dorso-ventrally flattened. Tegument at least partly spinous. Oral sucker ventrally subterminal, subglobular or rarely with pair of lateral projections (papillae). Ventral sucker in anterior half of body, well developed, occasionally with modified margins. Prepharynx present, short to long. Pharynx well developed. Oesophagus short to long. Intestinal bifurcation in forebody. Caeca terminate blindly near posterior extremity. Testes two, tandem, oblique, or occasionally symmetrical, in hindbody. cirrus sac absent. Naked seminal vesicle present; saccular to tubular. Pars prostatica well to

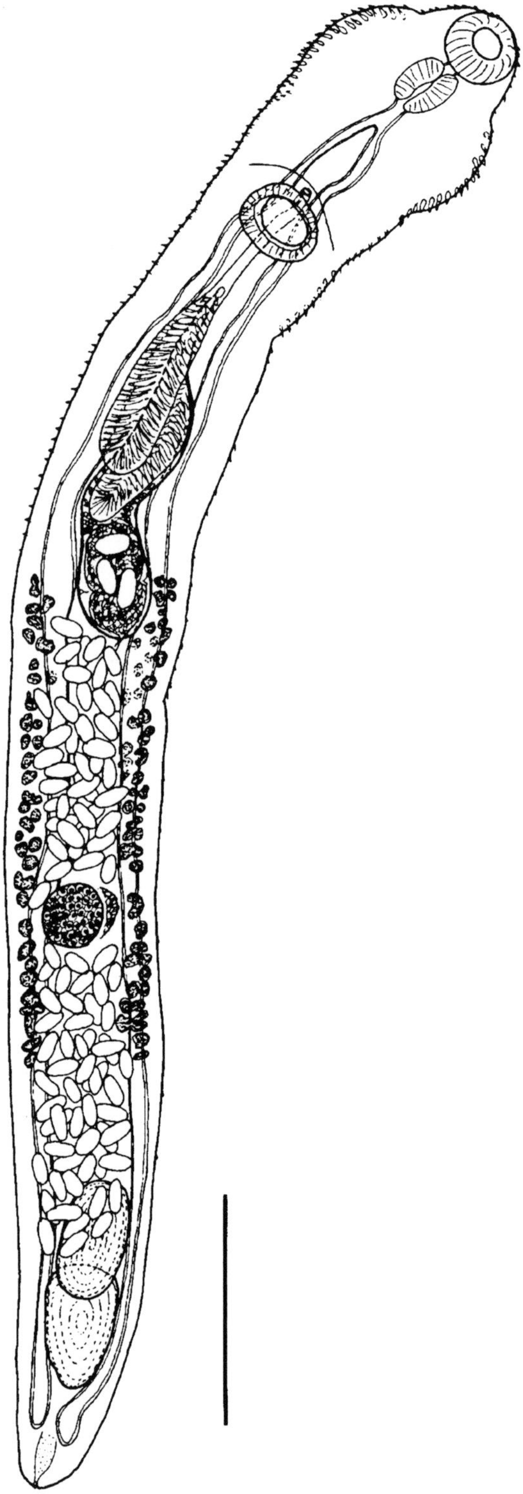

FIG. 89. *Deropristis inflata* (after Cable and Hunninen (1942)). Scale bar: 250 µm.

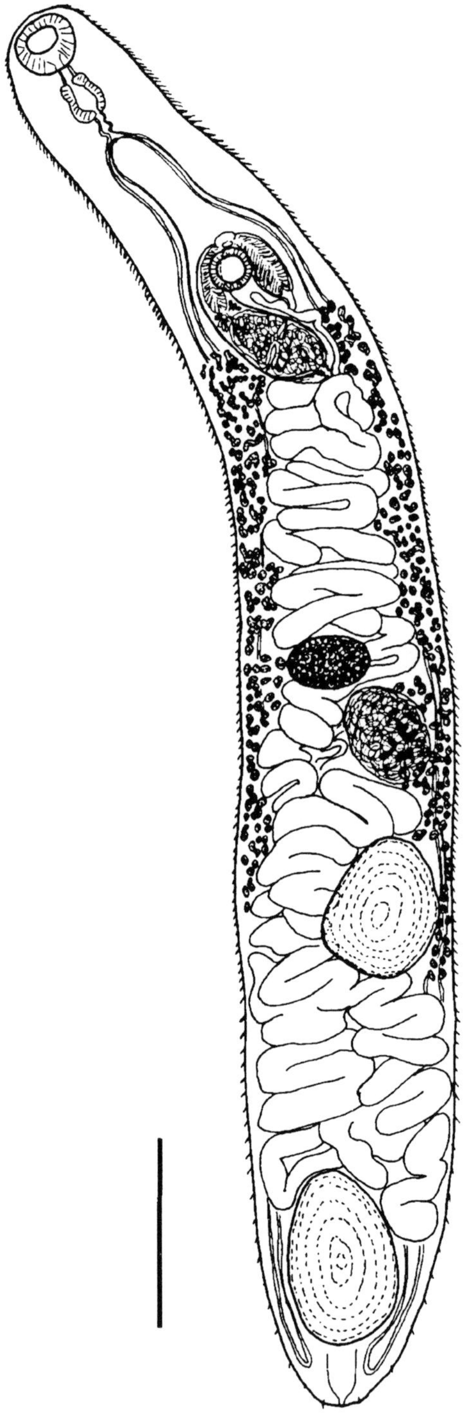

FIG. 90. *Skrjabinopsolus manteri* (after Cable (1952: as *Pristotrema manteri*)). Scale bar: 1 mm.

poorly developed. Ejaculatory duct opens into tubular genital atrium (sometimes called hermaphroditic duct). Genital pore normally immediately anterior to ventral sucker, occasionally immediately posterior to ventral sucker. Ovary entire, submedian, between ventral sucker and testes. Canalicular seminal receptacle and Laurer's canal present. Uterine field between testes and ventral sucker. Eggs usually few, large, operculate. Vitellarium follicular, fields lateral in hindbody, normally confluent in post-testicular zone; occasionally extend into forebody, where they may be confluent. Excretory pore terminal; vesicle I-shaped, short to long. Parasitic in intestine of marine, brackish, and freshwater teleosts.

Key to the genera of the Homalometridae

1 Ventral sucker as large or larger than oral sucker; testes tandem to oblique; lateral vitelline fields extend anteriorly to testes ....*Homalometron*

Ventral sucker smaller than oral sucker; testes symmetrical or almost so; vitelline field post-testicular ....*Microcreadium*

HOMALOMETRON Stafford, 1904

Homalometridae. Body oval to elongate, dorso-ventrally flattened. Tegument spinose, at least anteriorly. Oral sucker ventrally subterminal, subglobular. Ventral sucker similar in size to or larger than oral, in anterior half of body. Prepharynx short to long. Pharynx and oesophagus well developed. Caeca end blindly close to posterior extremity. Testes tandem to oblique, entire, in posterior half of body. Cirrus sac absent. Seminal vesicle naked, saccular, immediately posterior or postero-dorsal to ventral sucker. Pars prostatica often indistinct, tubular to vesicular. Ejaculatory duct unites with metraterm to form tubular genital atrium. Genital pore median, close to anterior margin of ventral sucker. Ovary entire, submedian, between anterior testis and ventral sucker. Uterine field between anterior testis and ventral sucker. Eggs few, large. Vitellarium follicular, lateral fields occur throughout hindbody, confluent or almost so in post-testicular zone. Excretory vesicle I-shaped, reaching to — or almost to — level of posterior testis. Normally in intestine of marine, brackish, or freshwater teleosts, rarely in chondrosteans.

Comments

A brief investigation of features previously used to differentiate *Homalometron armatum* and *H. pallidum* (see Miller 1959) suggests that none of them hold up. Nevertheless, the different distribution of the two forms and their reported molluscan first intermediate hosts indicate that they are distinct. It is clear, therefore, that a detailed analysis using material in good condition is required to elucidate valid morphological criteria. One feature that might prove useful is the degree of spination. *H. armatum* is described as being heavily spined as far back as the posterior testis, whereas in *H. pallidum* the spines are small, rather indistinct and, in the one mounted specimen in the BM(NH) collection with spines, restricted to the forebody. Great care must, however, be exercised in using this feature only in live-fixed material, since the spines in both species are easily lost.

Key to the species of *Homalometron*

1 (See "Comments" above) Parasitic in freshwater fishes; main host *Aplodinotus grunniens* ....*Homalometron armatum* (MacCallum, 1895) Manter, 1947 (Fig. 91)

Syn.: *Distomum isoporum* var. *armatum* MacCallum, 1895; *Anallocreadium armatum* (MacCallum, 1895) Simer, 1929; *A. pearsei* Hunter and Bangham, 1932.

Site: Intestine.

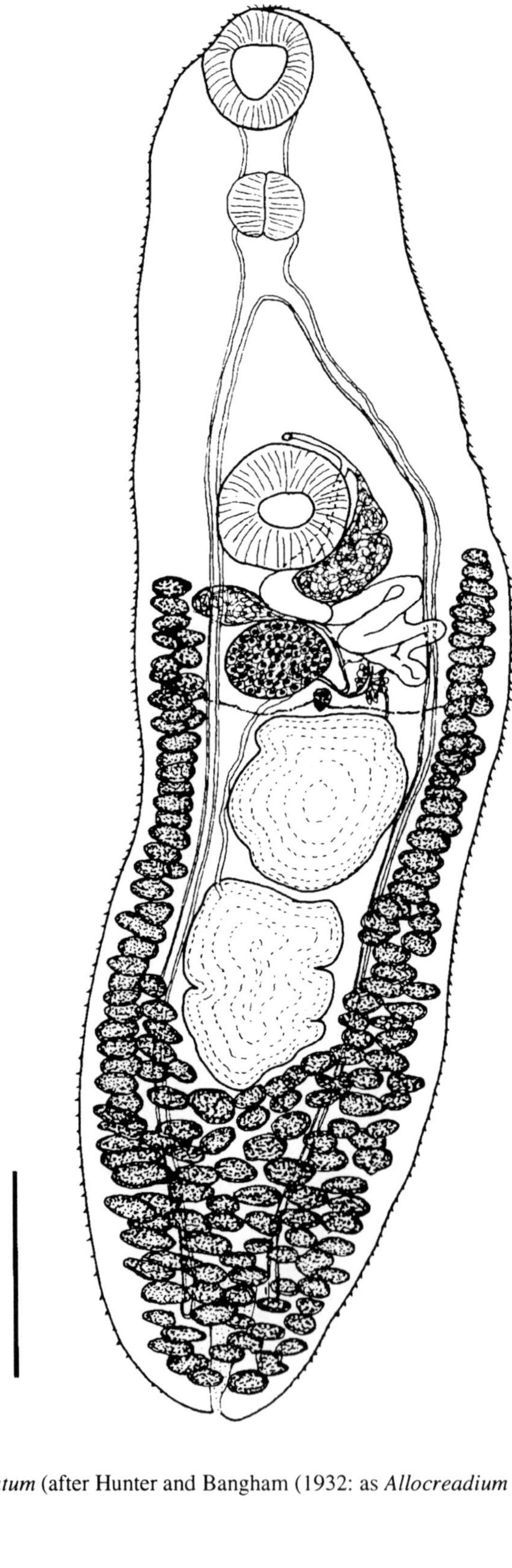

FIG. 91. *Homalometron armatum* (after Hunter and Bangham (1932: as *Allocreadium pearsei*)). Scale bar: 500 µm.

Hosts: *Acipenser fulvescens* (1); *Ameiurus nebulosus* (4); *Aplodinotus grunniens* (1, 2, 3, 5, 6, 7); *Lepomis gibbosus* (1, 4).
Distribution: Ont, Que.
Records: 1. MacCallum 1895 (Ont); 2. Hunter and Bangham 1932 (Ont); 3. Bangham and Hunter 1939 (Ont); 4. Miller 1940a (Que); 5. Anthony 1982 (Ont); 6. Dechtiar and Nepszy 1988 (Ont); 7. Dechtiar and Christie 1988 (Ont).

Parasitic in marine, brackish, and coastal freshwater fishes near the coast; main host *Fundulus heteroclitus* .... *H. pallidum* Stafford, 1904
Sites: Intestine, ?stomach.
Hosts: *Fundulus heteroclitus* (1, 2, 3, 5); *Morone saxatilis* (4).
Distribution: Atl, Nfld, NS, NB.
Records: 1. Stafford 1904 (Atl); 2. Fantham and Porter 1948 (NS); 3. Dickinson and Threlfall 1975 (Nfld); 4. Hogans 1984 (NB); 5. BM(NH) collection (NB).
Remarks: This is a species that has been found in marine, brackish, and fresh waters on the Atlantic coast of the United States, especially in killifishes. It has also been found in a wide range of other hosts, although Stunkard (1964) suggested that some of these records may be misdeterminations or accidental infections acquired with prey.
Stafford's material was briefly described by Miller (1941a).

MICROCREADIUM Simer, 1929

Homalometridae. Body very small, oval, stout. Tegument spinose. Oral sucker large, ventro-terminal. Ventral sucker considerably smaller than oral sucker, weakly muscled, well inside anterior half of body. Prepharynx very short. Pharynx well developed. Oesophagus short to apparently absent. Caeca long, pass back laterally and end blindly close to posterior extremity. Testes entire, symmetrical or almost so, mainly inter-caecal, in about middle of body. Cirrus sac absent. Naked seminal vesicle dorso-sinistral to ventral sucker, oval to sub-globular. Pars prostatica tubular, indistinct. Ejaculatory duct short, unites with uterus to form short, tubular genital atrium (hermaphroditic duct, according to Yamaguti 1971). Genital pore median, close to anterior margin of ventral sucker. Ovary large, entire, dextrally postero-dorsal to ventral sucker and between right testis and ventral sucker. Uterus coiled (Yamaguti 1971) — not coiled (Simer 1929), between left testis and ventral sucker. Eggs very few, large. Vitellarium follicular, restricted to and fills post-testicular field. Excretory vesicle ?I-shaped, reaching to testes. In intestine of freshwater fishes (*Aplodinotus*).

Microcreadium parvum Simer, 1929 (Fig. 92)
Host: *Aplodinotus grunniens*.
Distribution: Ont.
Record: Anthony 1982.

Family LEPOCREADIIDAE Odhner, 1905

Lepocreadioidea. Shape variable, but usually elongate and dorso-ventrally flattened. Tegument spinose, occasionally with one, two or more rows of circum-oral spines. Eye-spot material may be present. Oral sucker usually subglobular, occasionally modified in shape, ventrally subterminal. Prepharynx present, sometimes very long. Pharynx present. Oesophagus short to long, pseudoesophagus occasionally present. Intestinal bifurcation normally in forebody. Caeca normally reaching close to posterior extremity, usually blind, occasionally opening via uroproct or ani. Testes two (see "Comment"), in hindbody, usually in tandem, occasionally oblique, or symmetrical. External seminal vesicle present, may be surrounded by gland cells that may be delimited by membrane. Cirrus sac present, contains internal seminal vesicle, pars prostatica, and ejaculatory duct. Genital atrium usually small and indistinct. Genital pore median, submedian or

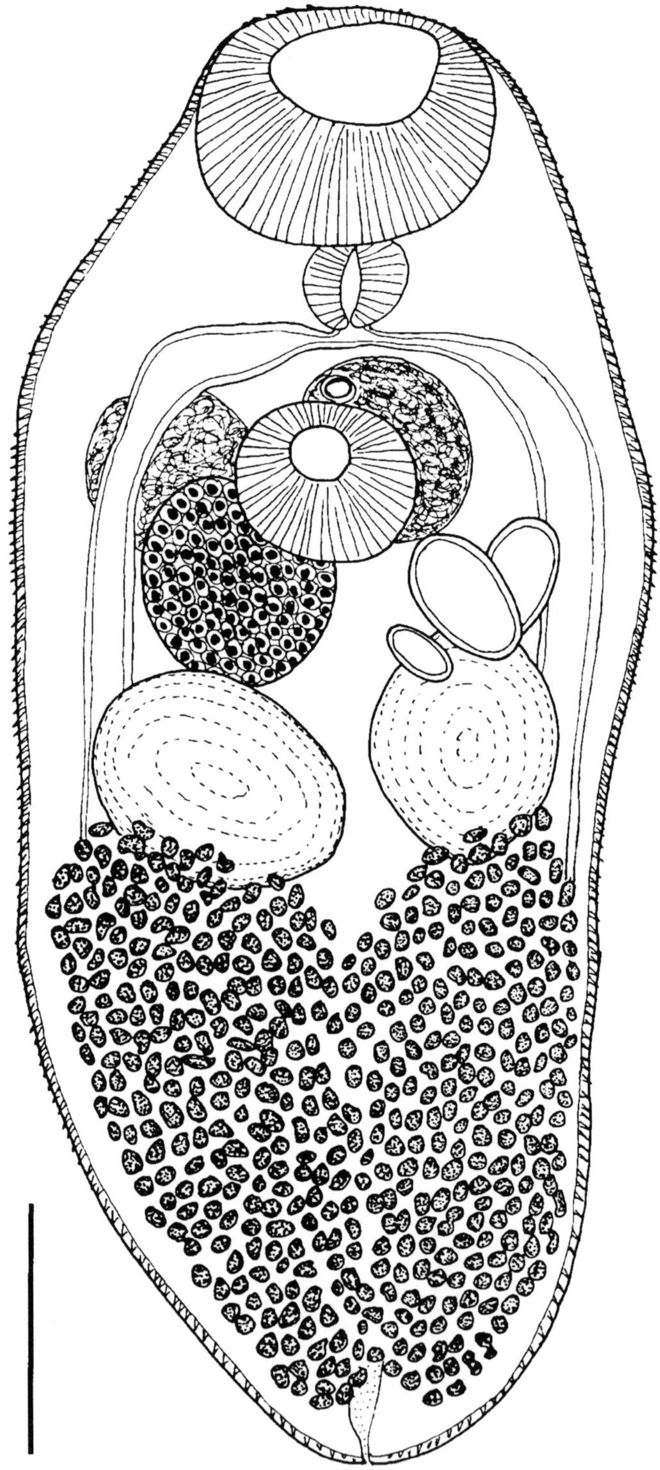

FIG. 92. *Microcreadium parvum* (after Simer (1929)). Scale bar: 200 µm.

marginal, rarely (?) dorsal, normally anterior to ventral sucker, occasionally at level of ventral sucker. Ovary median or submedian, pre-testicular in hindbody. Canalicular seminal receptacle and Laurer's canal present. Uterine field normally between ovary or anterior testis and ventral sucker, occasionally reaches further into hindbody. Eggs usually large, numerous, operculate. Vitellarium follicular, widely distributed in hindbody, occasionally extending into forebody, lateral fields normally confluent in post-testicular zone. Excretory vesicle I-shaped, pore terminal. Parasitic in intestine of mainly marine teleosts.

Comments

This family is in dire need of major revision, since the conception of Yamaguti (1971) is too broad and includes a mix of numerous forms that are not necessarily closely related. The basis of the above definition is restricted to those subfamilies that conform to Gibson and Bray's (1982) conception of the Lepocreadiidae, i.e., forms with a spinous body, possessing an external seminal vesicle and a cirrus sac containing an internal seminal vesicle and pars prostatica. In addition, features of the Folliorchiinae Yamaguti, 1958 *sensu* Yamaguti (1971) — forms with follicular testes — have not been included, because the composition and systematic position of this subfamily requires investigation.

In view of the above, full definitions of the subfamilies of the Lepocreadiidae occurring in Canadian fishes are not given, because the constituents of these subfamilies are at the present time uncertain. Many of the features used to differentiate the various subfamilies, such as the nature or presence of gland cells surrounding the external seminal vesicle, do not hold up. In terms of the Canadian forms, it has been suggested that the nature of the male duct within the cirrus sac is the most reliable diagnostic feature (Bray and Gibson 1989).

Key to the subfamilies of the Lepocreadiidae

1 Cirrus sac contains subglobular thin-walled internal seminal vesicle, vesicular pars prostatica, and long ejaculatory duct .... Lepocreadiinae

Cirrus sac contains narrow, thick-walled internal seminal vesicle, small oval to globular pars prostatica, and short ejaculatory duct .... Lepidapedinae

Subfamily LEPOCREADIINAE Odhner, 1905

Lepocreadiidae. Cirrus sac containing subglobular, thin-walled internal seminal vesicle, vesicular pars prostatica, and long ejaculatory duct.

Key to genera of Lepocreadiinae

1 Enlarged oral spines absent 2

Two ventrally interrupted rows of enlarged spines present around oral sucker (in well-fixed material); uroproct absent; vitelline fields extend into forebody; pseudoesophagus present; excretory vesicle restricted to hindbody; parasitic in intestine of *Mola* .... *Dihemistephanus*

2 Uroproct absent; vitelline fields usually extend into forebody; pseudoesophagus absent; excretory vesicle may or may not extend into forebody .... *Lepocreadium*

Uroproct present; vitelline fields confined to hindbody; pseudoesophagus present; excretory vesicle extends into forebody .... *Opechona*

DIHEMISTEPHANUS Looss, 1901

Lepocreadiinae. Body elongate-oval to elongate (forebody may be very attenuated in poorly preserved specimens). Tegument spinous, spines relatively large, acuminate; enlarged oral spines present in two ventrally interrupted rows (spines often lost in poorly preserved specimens). Oral sucker subglobular. Prepharynx long. Pharynx well developed. Oesophagus short to long (may be very short in contracted specimens); pseudoesophagus present. Intestinal bifurcation well anterior to ventral sucker. Caeca terminate blindly close to posterior extremity. Testes tandem, contiguous, usually entire, occasionally lobed, in posterior third of body. External seminal vesicle usually distinct (occasionally obscured by eggs), elongate, saccular, surrounded by free gland cells. Cirrus sac claviform, may be sinuous distally, thin-walled, extends into hindbody. Internal seminal vesicle saccular, subglobular, thin-walled. Pars prostatica distinct, subglobular, vesicular, with thick lining of anuclear blebs. Ejaculatory duct muscular, long, thick-walled, looped, sinuous or straight. Genital atrium indistinct. Genital pore sinistrally submedian, just anterior to ventral sucker, overlaid by anterior lobe (with unusual construction when viewed in sections). Ovary pre-testicular, irregularly oval, entire. Laurer's canal and canalicular seminal receptacle present. Uterus pre-testicular (usually pre-ovarian) in anterior hindbody. Metraterm distinct, muscular. Vitellarium follicular, lateral fields extend into posterior forebody, confluent in post-testicular zone and usually in forebody. Excretory pore terminal, vesicle tubular, reaching to about level of ovary. In intestine of *Mola*.

Dihemistephanus lydiae (Stossich, 1896) Looss, 1901 (Fig. 93)

Syn.: *Distomum fragile* Linton, 1900; *Stenocollum fragile* (Linton, 1900) Stafford, 1904; *Dihemistephanus fragilis* (Linton, 1900) Yamaguti, 1971.

Site: Intestine.

Host: *Mola mola*.

Distribution: Atl.

Records: Stafford 1904; Threlfall 1967.

Remarks: Miller (1941a) confirmed Stafford's determination. Cooper (1915) also claimed to have recorded this species from *Gadus morhua*, but this determination is almost certainly erroneous; it is much more likely to be a species of *Lepidapedon*.

I have followed Dollfus (1960) in considering *Dihemistephanus fragilis* a synonym of *D. lydiae*. Dollfus showed how badly preserved specimens of *D. lydiae* resemble descriptions of *D. fragilis*. Yamaguti (1971) did not agree and recognized *D. fragilis* based on differences in the number of oral spines: in view of the fact that these are easily lost and often difficult to count, this requires confirmation. As demonstrated by the accacoeliids, the trematode fauna of *Mola mola* is similar on both sides of the Atlantic.

LEPOCREADIUM Stossich, 1903

Lepocreadiinae. Body rounded to fusiform, dorso-ventrally flattened. Tegument spinose, spines small. Oral sucker well developed. Ventral sucker relatively small, often smaller than oral sucker, in anterior half of body. Prepharynx short to relatively long. Pharynx well developed. Oesophagus short; pseudoesophagus absent. Intestinal bifurcation in posterior half of forebody. Caeca terminate blindly close to posterior extremity. Testes tandem, oblique, or almost symmetrical, in mid-hindbody, entire or indistinctly lobed. External seminal vesicle present, occasionally obscured, saccular, not surrounded by gland cells. Cirrus sac claviform, extends dorsally or sinistro-dorsally to ventral sucker, often into hindbody. Internal seminal vesicle subglobular. Pars prostatica normally slightly vesicular. Ejaculatory duct long. Genital atrium small. Genital pore sinistrally submedian between ventral sucker and intestinal bifurcation. Ovary pre-testicular, usually dextro-submedian, entire or lobed. Laurer's canal and canalicular seminal receptacle present. Uterus pre-testicular, with few to many relatively large eggs. Vitellarium follicular, lateral fields usually, but not always, extend into forebody, fields confluent in post-testicular zone and occasionally in

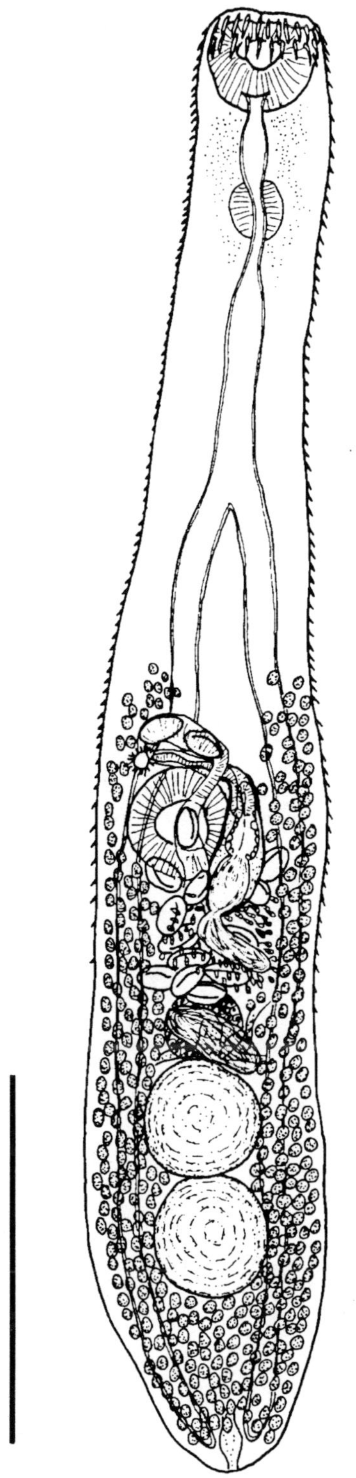

FIG. 93. *Dihemistephanus fragilis* (after Yamaguti (1971)). Scale bar (roughly estimated): 250 µm.

forebody. Excretory pore terminal, vesicle I-shaped, may or may not extend into forebody. In intestine of marine (occasionally euryhaline) teleosts.

Comments

This genus is in dire need of revision and, as conceived at the present time, may well represent a pigeon-hole for a variety of not always closely related taxa. The key to the genus provided by Edwards and Nahhas (1968) is of little use because of the problems within the genus.

Lepocreadium setiferoides (Miller and Northup, 1926) Martin, 1938 (Fig. 94)

Site: Intestine.

Host: *Morone saxatilis.*

Distribution: NB.

Record: Hogans 1984.

Remarks: This record requires confirmation as there are at least two other species of the genus occurring in marine teleosts off New England (see Stunkard 1980), i.e., *L. areolatum* (Linton, 1900) and *L. trullaforme* Linton, 1940. There has been considerable confusion between these species, some of which may be conspecific. *L. areolatum* has been previously recorded from another anadromous species of *Morone.*

The life cycle of *L. setiferoides* was studied by Martin (1938) and Stunkard (1972).

OPECHONA Looss, 1907

Lepocreadiinae. Body oval to elongate-oval. Tegument spinous; enlarged oral spines absent. Oral sucker ventrally subterminal, large and distinctly infundibuliform to small and indistinctly infundibuliform. Prepharynx distinct. Pharynx oval. Oesophagus short to long, distinct; pseudoesophagus present, short to long. Intestinal bifurcation usually in posterior half of forebody, often close to ventral sucker. Caeca reach to posterior extremity, uroproct present. Testes tandem, oval, smooth or lobed, in mid- to posterior hindbody. External seminal vesicle distinct, saccular, few associated free gland cells may be present. Cirrus sac claviform, thick-walled. Internal seminal vesicle saccular, subglobular, thin-walled. Pars prostatica distinct, oval, vesicular. Ejaculatory duct muscular, long, sinuous, wide. Genital atrium distinct. Genital pore close to ventral sucker, usually antero-sinistral. Ovary lobed to entire, pre-testicular. Canalicular seminal receptacle and Laurer's canal present. Uterine field pre-ovarian in anterior hindbody. Eggs large, numerous. Metraterm usually distinct. Vitellarium follicular, lateral fields confined to hindbody or reaching into forebody. Excretory pore terminal, vesicle long, reaching into forebody. In intestine of marine teleosts.

Comments

This genus was reviewed by Bray and Gibson (1990).

Key to the species of *Opechona*

1 Body elongate; vitellarium reaches into forebody; ventral sucker slightly larger than oral sucker; genital pore at level of anterior margin of ventral sucker .... *Opechona alaskensis* Ward and Fillingham, 1934

Sites: Intestine, pyloric caeca.

Hosts: *Sebastes alutus* (1, 2); *S. ciliatus* (2); *S. flavidus* (2); *S. maliger* (2); *S. pinniger* (2); *S. zacentrus* (2).

Distribution: Pac.

Records: 1. Sekerak and Arai 1973; 2. Sekerak and Arai 1977.

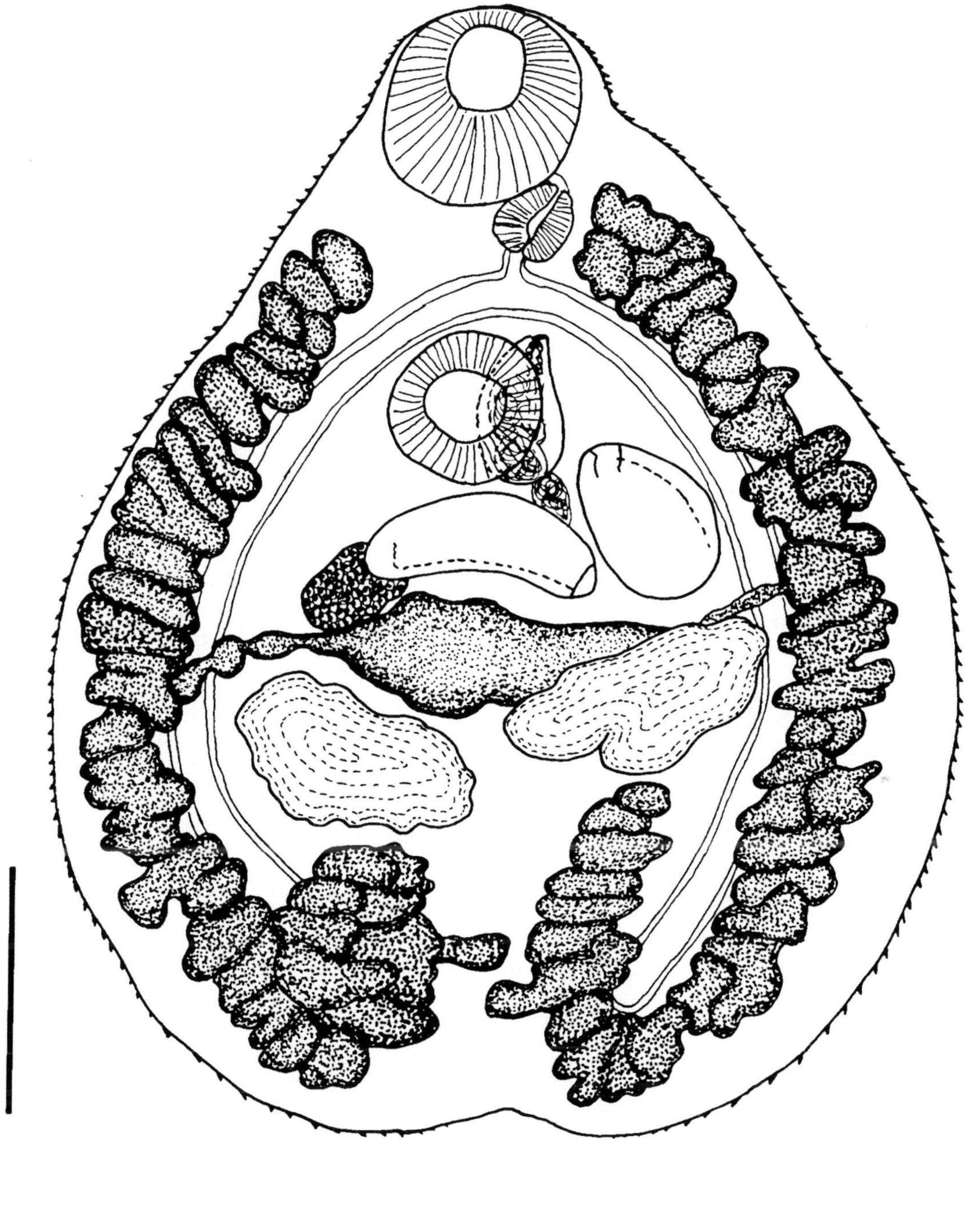

FIG. 94. *Lepocreadium setiferoides* (modified after Martin (1938)). The external seminal vesicle, a feature of the family omitted from the original illustration, has been added for completion, though it may not necessarily represent the size and shape of that actually occurring in this species. Scale bar: 100 µm.

Body very elongate, narrow; anterior limit of vitellarium posterior to ventral sucker; oral sucker larger than ventral sucker; genital pore at level of posterior margin of ventral sucker .... *O. occidentalis* Montgomery, 1957 (Fig. 95)

Syn.: *Pharyngora bacillaris* of McFarlane (1936).
Sites: Intestine, pyloric caeca.
Hosts: *Sebastes alutus* (3); *S. caurinus* (2, 3, 4); *S. ciliatus* (3); *S. elongatus* (2, 3); *S. flavidus* (3, 6, 7); *S. maliger* (3); *S. nebulosus* (3, 5); *Sebastes* sp. (1); *Syngnathus leptorhynchus* (2).
Distribution: Pac.
Records: 1. McFarlane 1936; 2. Arai 1969b; 3. Sekerak and Arai 1977; 4. Bray and Gibson 1990; 5. Holmes 1990; 6. Lee *et al.* 1990; 7. Stanley *et al.* 1992.

Subfamily LEPIDAPEDINAE Yamaguti, 1958

Lepocreadiidae. Cirrus sac containing narrow thick-walled internal seminal vesicle, small oval to globular pars prostatica, and short ejaculatory duct.

Comments

A full key to the genera of this subfamily was given by Bray and Gibson (1991b).

Key to the genera of the Lepidapedinae

1 Gland cells surrounding external seminal vesicle delimited by membrane; uroproct absent; pseudoesophagus present or absent 2

Gland cells associated with external seminal vesicle free in parenchyma; pseudoesophagus absent....3

2 Body without lateral lappets bearing marginal sucker-like structure .... *Lepidapedon*

Anterior part of body with lateral lappets bearing marginal sucker-like structure .... *Gibsonia*

3 Caeca end blindly .... *Neolepidapedon*

Uroproct present .... *Paralepidapedon*

GIBSONIA Gaevskaya and Rodyuk, 1988

Lepidapedinae. Body elongate to elongate-oval, but with broadly triangular anterior body with lateral lappets bearing sucker-like structure, dorso-ventrally flattened. Forebody distinctly spinose in well-fixed material. Oral sucker subglobular, ventrally subterminal. Ventral sucker in anterior half of body embedded. Eye-spot material absent. Prepharynx relatively short. Pharynx oval. Oesophagus of medium length; pseudoesophagus absent. Intestinal bifurcation in mid- to posterior forebody. Caeca terminate blindly close to posterior extremity. Testes entire to slightly indented, tandem, contiguous or slightly separated, just posterior to middle of hindbody. External seminal vesicle claviform, broad proximally, convoluted, surrounded by gland cells enclosed in enveloping membrane that is continuous with wall of cirrus sac. Cirrus sac small, muscular, pyriform, antero-dorsal to ventral sucker, contains internal seminal vesicle and short pars prostatica, both surrounded by gland cells and layer of connective tissue, plus muscular ejaculatory duct. Genital atrium small. Genital pore immediately anterior to ventral sucker, medial or sinistrally submedial. Ovary oval or irregularly oval, immediately pre-testicular, medial. Canalicular seminal receptacle present, Laurer's canal not observed. Uterus coils inter-caecally between ovary and ventral sucker. Metraterm present, narrow. Eggs relatively large, numerous. Vitellarium follicular, lateral fields extend throughout hind-

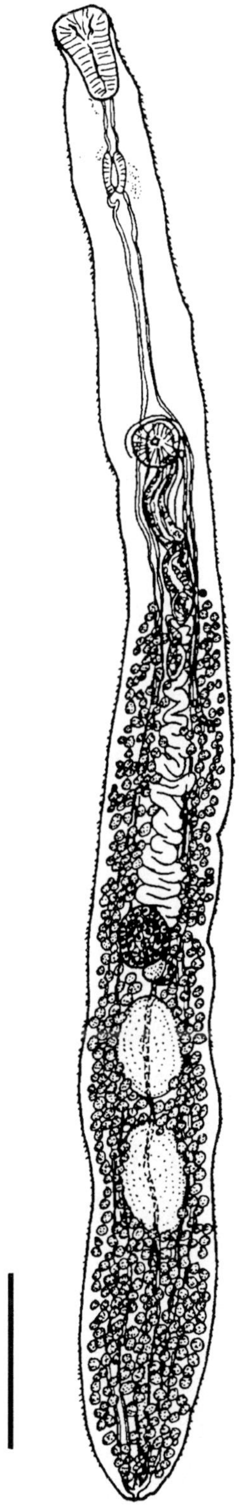

FIG. 95. *Opechona occidentalis* (after Bray and Gibson (1990)). Scale bar: 500 µm.

body and to level of about anterior margin of ventral sucker, fields confluent only in post-testicular region. Excretory pore terminal, vesicle I-shaped, reaching to about inter-testicular level. In intestine of deep-sea teleosts (Macrouridae).

Gibsonia borealis Campbell, 1992 (Fig. 96)

Site: (Intestine).
Host: *Macrourus berglax*.
Distribution: Atl.
Record: Campbell 1992.

LEPIDAPEDON Stafford, 1904

Lepidapedinae. Body elongate to elongate-oval, dorso-ventrally flattened. Tegument spinous. Oral sucker subglobular, ventrally subterminal. Ventral sucker in anterior half of body, embedded or slightly protuberant. Eye-spot material absent. Prepharynx distinct, usually long. Pharynx oval. Oesophagus short to long; pseudoesophagus present or absent. Intestinal bifurcation in mid- to posterior forebody. Caeca terminate blindly close to posterior extremity. Testes entire, tandem, contiguous or separated, near middle of hindbody. External seminal vesicle tubular to elongate-saccular, surrounded by gland cells enclosed in enveloping membrane that is continuous with the wall of the cirrus sac. Cirrus sac muscular, oval to fusiform. Internal seminal vesicle with narrow, straight lumen. Pars prostatica small, vesicular. Male duct within cirrus sac with very thick, muscular wall, which surrounds internal seminal vesicle and pars prostatica, muscular wall fills majority of cirrus sac, thin layer of prostatic cells and connective tissue lie between muscular walls of cirrus sac and male duct. Ejaculatory duct short, narrow, cirrus may be present in genital atrium. Genital atrium small. Genital pore sinistrally submedian, just anterior to ventral sucker. Ovary oval to irregularly lobed, median or submedian, pre-testicular, contiguous with or slightly separated from anterior testis. Canalicular seminal receptacle and Laurer's canal present. Uterine coils between ovary and ventral sucker. Metraterm present. Eggs large, numerous. Vitellarium follicular, widely distributed in hindbody, anterior limit of vitelline field may extend into forebody. Excretory pore terminal, vesicle I-shaped, short, often not reaching posterior testis, not extending anterior to gonads. In intestine of marine teleosts.

Comments

The work of Bray and des Clers (1992), mentioned below in relation to *Lepidapedon elongatum*, indicated that this genus was in need of detailed study and revision. Such a revision is presented in Bray and Gibson (1995).

Since the morphology of the Canadian forms of *Lepidapedon* is very variable and the diagnostic characters often difficult to elucidate in contracted specimens, I have included several distinctive features in the key.

Key to the species of *Lepidapedon*

1 Body very elongate; pseudoesophagus present (not always clearly distinguishable in poorly fixed material); oesophagus (including pseudoesophagus) long, normally at least as long as pharynx; ventral sucker similar in size to oral sucker (may be larger or smaller); ventral sucker larger than pharynx; vitellarium does not reach as far forward as ventral sucker..2

Body elongate-oval; pseudoesophagus absent; oesophagus short, normally shorter than pharynx; oral sucker much larger than ventral sucker; ventral sucker of similar length to pharynx; vitelline

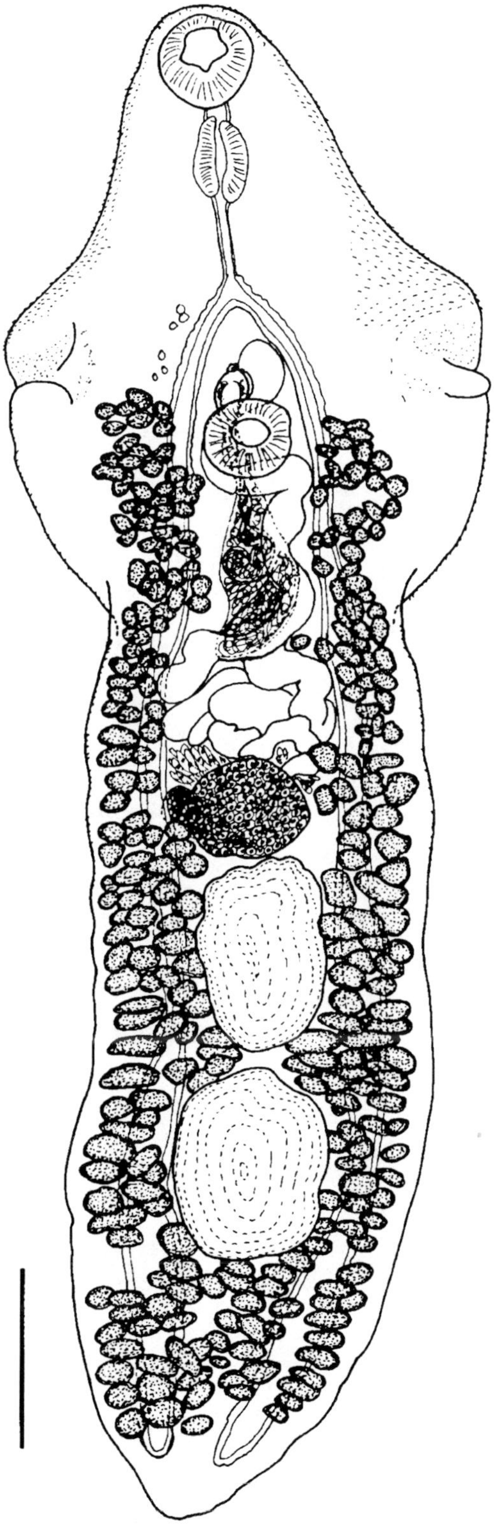

FIG. 96. *Gibsonia borealis* (after Campbell (1992)). Scale bar: 500 μm.

field reaches at least to level of ventral sucker.. ....*Lepidapedon rachion* (Cobbold, 1858) Stafford, 1904 (Fig. 97)

Syn.: *Distoma rachion* Cobbold, 1858.

Site: Intestine.

Hosts: *Gadus morhua* (5, 7, 8); *Hippoglossus hippoglossus* (9); *Lycodes vahli* (7); *Melanogrammus aeglefinus* (1, 2, 3, 6, 10); *Scophthalmus aquosus* (4).

Distribution: Atl.

Records: 1. Stafford 1904; 2. Stafford 1907; 3. Kuitunen-Ekbaum 1937; 4. Ronald 1960; 5. Linkletter *et al.* 1977; 6. Umnova 1979; 7. Bray 1979; 8. Appy and Burt 1982; 9. Scott and Bray 1989; 10. BM(NH) collection.

Remarks: Bray and Gibson (1995) consider it likely that the single worm from *Lycodes vahli* is not *L. rachion* (*sensu stricto*).

2 Atlantic coast ....*L. elongatum* (Lebour, 1908) Nicoll, 1910

Syn.: *Lepidapedon microcotyleum* Odhner in Dollfus, 1953.

Sites: Intestine, pyloric caeca.

Hosts: *Gadus morhua* (2, 3, 5, 7); *Lepidion eques* (3, 7); *Macrourus berglax* (4); *Merluccius albidus* (6); *Tautogolabrus adspersus* (1); *Urophycis chesteri* (6); *U. chuss* (6); *U. musicki* (6).

Distribution: Atl.

Records: 1. Sekhar and Threlfall 1970a; 2. Linkletter *et al.* 1977; 3. Bray 1979; 4. Zubchenko 1981a; 5. Appy and Burt 1982; 6. Scott 1987; 7. BM(NH) collection.

Remarks: The work of Bray and des Clers (1992), who used multivariate analysis of metrical features, demonstrated that *L. elongatum*, as currently conceived, is a species complex. This work was carried out on material from five species of gadiform fishes from the Northeast Atlantic and indicated that each fish had its own species of *Lepidapedon*. In the present work, it is felt most appropriate to refer all records from the Atlantic to *L. elongatum* (*sensu lato*), pending a thorough study of the group. It is worth noting that Bray and des Clers (1992) considered *L. elongatum* (*sensu stricto*) to occur only in *Gadus morhua*. This was followed by Bray and Gibson (1995), who considered specimens from other hosts as "elongatum-group (*sensu lato*)".

According to Thulin (1971) and Køie (1985a), *L. microcotyleum* is a synonym of *L. elongatum*.

Although a pseudoesophagus is normally present in *L. elongatum*, in deep-sea species of this genus from elsewhere it can be a variable or transient feature.

Pacific coast ....*L. gadi* (Yamaguti, 1934) Yamaguti, 1938

Syn.: *Lepidapedon* sp. of Arai (1967a); *L. microcotyleum* Odhner in Dollfus, 1953 of Arai (1967a,1969b).

Site: Intestine.

Host: *Theragra chalcogramma*.

Distribution: Pac.

Records: Arai 1967a; Arai 1969b; Arthur 1984; Kabata and Whitaker 1984.

Remarks: In view of Bray and des Clers' (1992) observations on *L. elongatum*, it is felt sensible to retain the Pacific form as a distinct species, but Bray and Gibson (1995) considered all specimens not from *Gadus macrocephalus* as "elongatum-group (*sensu lato*)". They stated that *L. gadi* is "morphologically indistinguishable from *L. elongatum* and is retained solely on its geographical separation".

NEOLEPIDAPEDON Manter, 1954

Lepidapedinae. Body oval to elongate-oval, flattened dorso-ventrally. Tegument spinous. Oral sucker subglobular, ventrally subterminal. Ventral sucker in anterior half of body, may be on low protuberance. Eyespot material absent. Prepharynx distinct, short to long. Pharynx oval. Oesophagus distinct, short to long, pseudoesophagus absent. Intestinal bifurcation in mid- to posterior forebody. Caeca long, reach close to posterior extremity, terminate blindly. Testes oval to lobed, tandem (occasionally oblique), contiguous or

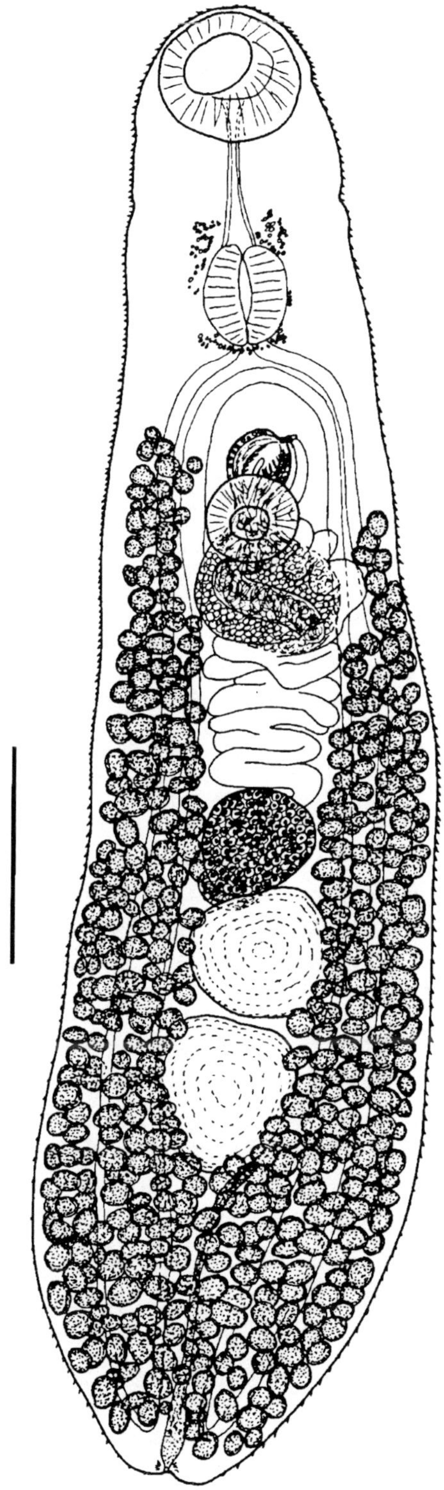

FIG. 97. *Lepidapedon rachion* (after Bray and Gibson (1995)). Scale bar: 500 μm.

separated, in mid-hindbody. External seminal vesicle tubular, convoluted, with associated gland cells surrounding whole or part of length, gland cells not enclosed in enveloping membrane. Cirrus sac muscular, oval to pyriform. Internal seminal vesicle with narrow, straight lumen. Pars prostatica vesicular, round to oval. Male duct inside cirrus sac has extremely thick muscular wall, which surrounds internal seminal vesicle and pars prostatica, muscular wall fills majority of cirrus sac; thin to very thin layer of prostatic gland cells and connective tissue lie between muscular walls of cirrus sac and male duct. Ejaculatory duct short, narrow. Cirrus short, conical, permanent, within genital atrium. Genital atrium small. Genital pore median to submedian, just anterior to ventral sucker. Ovary oval to lobed, pre-testicular, adjacent to or slightly separated from anterior testis. Canalicular seminal receptacle and Laurer's canal present. Uterus coiled between ovary and ventral sucker. Metraterm present, often with glandular sheath. Eggs large, numerous. Vitellarium follicular, widely distributed in hindbody, occasionally reaching into forebody. Excretory pore terminal, vesicle I-shaped; reaching to anterior testis or ovary. In intestine of marine teleosts.

Comments

This genus was reviewed by Bray and Gibson (1989).

Neolepidapedon pugetensis (Acena, 1947) Yamaguti, 1971 (Fig. 98)

Sites: Intestine, pyloric caeca.
Hosts: *Sebastes caurinus* (1); *S. nebulosus* (1, 2).
Distribution: Pac.
Records: 1. Sekerak and Arai 1977; 2. Holmes 1990.
Remarks: Ching (1961) considered this species to be a synonym of *Lepidapedon elongatum*, but this was rejected by Bray and Gibson (1989).

PARALEPIDAPEDON Shimazu and Shimura, 1984

Lepidapedinae. Body oval to elongate-oval, dorso-ventrally flattened. Tegument spinous. Oral sucker subglobular, ventrally subterminal. Ventral sucker in anterior part of body, may be on low protuberance. Eyespot pigment absent. Prepharynx usually long. Pharynx oval. Oesophagus short to long; pseudoesophagus absent. Intestinal bifurcation in mid- to posterior forebody. Caeca long, forming uroproct. Testes oval to lobed, tandem, contiguous or separated, in mid-hindbody. External seminal vesicle tubular, convoluted, with associated gland cells surrounding whole or part of length; enveloping membrane around gland cells seen only in young specimens, if at all. Cirrus sac muscular; oval to pyriform. Internal seminal vesicle with narrow, straight lumen. Pars prostatica vesicular, round to oval. Male duct inside cirrus sac has extremely thick, muscular wall, which surrounds internal seminal vesicle and pars prostatica, muscular wall fills majority of cirrus sac; thin to very thin layer of gland cells and connective tissue lie between muscular walls of cirrus sac and male duct. Ejaculatory duct short. Permanent cirrus short, conical. Genital atrium small. Genital pore median to submedian, just anterior to ventral sucker. Ovary oval to lobed, pre-testicular, adjacent to or slightly separated from anterior testis. Canalicular seminal receptacle and Laurer's canal present. Uterus between ovary and ventral sucker. Metraterm present, often with glandular sheath. Eggs large, numerous. Vitellarium follicular, confined to hindbody. Excretory pore terminal, vesicle I-shaped, reaching to testes or ovary. In intestine of marine teleosts.

Comments

This genus was reviewed by Bray and Gibson (1988).

Paralepidapedon sebastici (Yamaguti, 1938) Shimazu and Shimura, 1984 (Fig. 99)

Syn.: *Neolepidapedon sebastici* (Yamaguti, 1938) Manter, 1954.

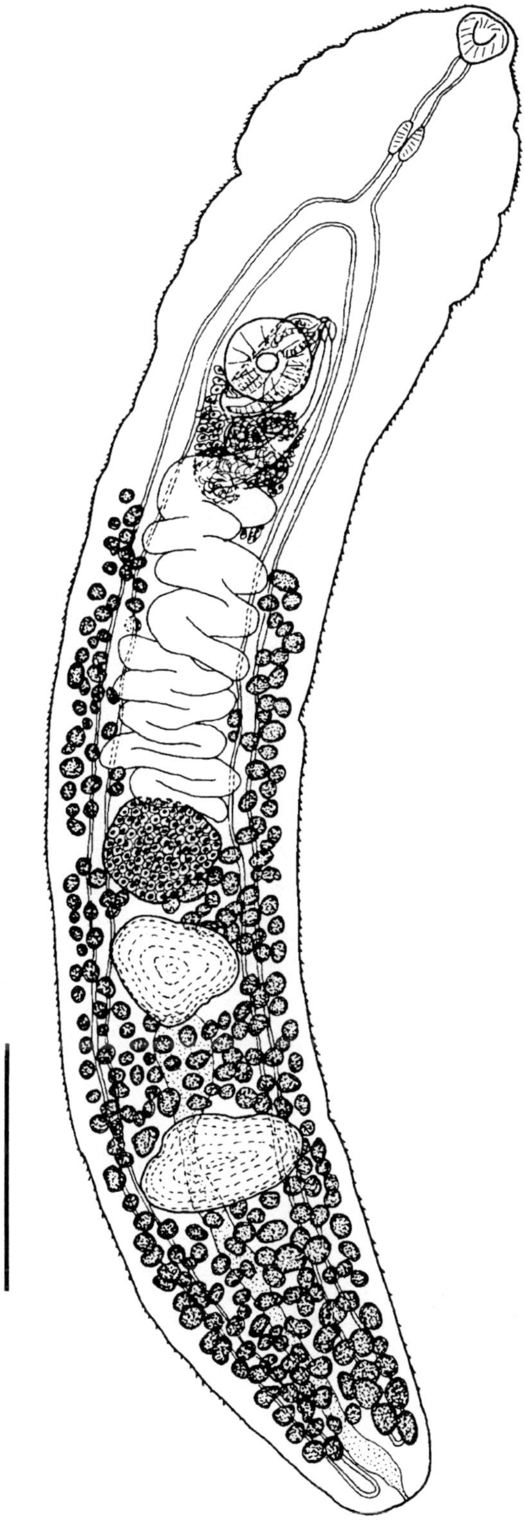

FIG. 98. *Neolepidapedon pugetensis* (after Bray and Gibson (1989)). Scale bar: 250 µm.

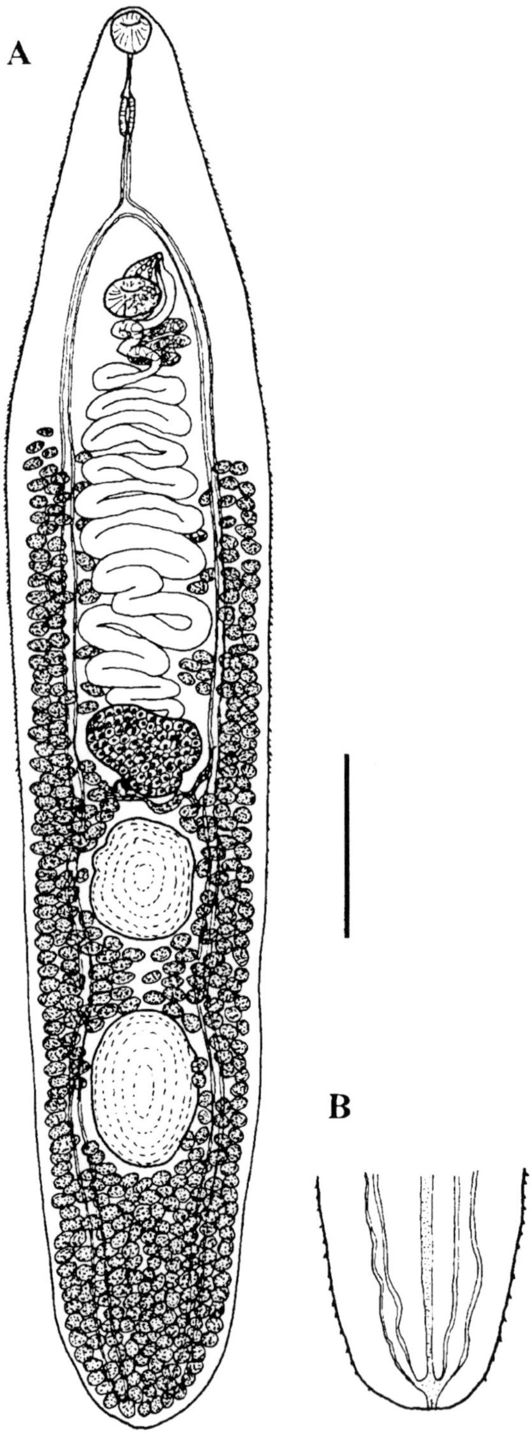

FIG. 99. A. *Paralepidapedon sebastici* (after Yamaguti (1938)). Scale bar: 500 µm. B. Posterior extremity of *P. smithi* (after Bray and Gibson (1988)) with vitelline follicles omitted to show uroproct.

Sites: Intestine, pyloric caeca.
Hosts: *Sebastes aleutianus*; *S. alutus*; *S. babcocki*; *S. borealis*; *S. helvomaculatus*; *S. paucispinus*; *S. pinniger*; *S. ruberrimus*; *S. zacentris*.
Distribution: Pac.
Record: Sekerak and Arai 1977.

SUPERFAMILY OPISTHORCHIOIDEA LOOSS, 1899

Plagiorchiida. Present in fish as adult and/or metacercaria.

Adult. Body usually small, oval to elongate. Tegument spinose. May be oculate. Suckers normally present. Ventral sucker occasionally atrophied or submedian, often forming part of acetabulo-genital complex with genital atrium. Prepharynx, pharynx, and oesophagus usually present. Intestinal bifurcation in forebody. Caeca normally two, long to short, usually blind. Testes normally two, in hindbody. Cirrus sac normally absent. Seminal vesicle, pars prostatica, and ejaculatory duct present. Genital atrium variable, often with gonotyl(s). Ovary usually pre-testicular. Canalicular seminal receptacle and Laurer's canal usually present. Uterine field usually pre-testicular, occasionally reaching to posterior extremity. Eggs small, operculate, numerous. Vitellarium normally follicular, in lateral fields. Excretory pore terminal, vesicle usually V- or Y-shaped. In gut or biliary system of piscivorous vertebrates.

Metacercaria. Encysted. Cyst globular to elongate-oval, with two-layered wall, outer layer (*host-capsule*) often thick and fibrous, of host-origin; inner layer (*parasite-cyst*) often resilient and transparent, of parasite origin. Body of worm small, oval to elongate. Tegument spinose. Suckers usually well developed, ventral sucker in middle or anterior half of body, occasionally small or submedian, sometimes associated with genital atrium forming ventro-genital complex. Prepharynx present. Pharynx well developed. Oesophagus short to long. Intestinal bifurcation in forebody. Caeca terminate blindly, usually long, reaching close to posterior extremity, occasionally short. Genital primordia present or absent. Gonotyl associated with genital atrium present or absent. Excretory vesicle I-, Y-, or V-shaped. Present in tissues, especially skin, fins, and muscles of freshwater and marine teleosts, occasionally in holosteans and amphibians.

Comments

It is not possible to produce a complete key to the opisthorchioid metacercariae in Canadian fishes for several reasons. Some metacercariae have not been described or figured, others have been identified not on morphological grounds but as a result of experimental infections of definitive hosts and in others the critical diagnostic characters used to separate taxa are based upon adult morphology and are not always developed in the metacercaria. The following keys are, therefore, where they exist, only of limited value and will enable the determination of only certain taxa.

PARTIAL KEY TO THE FAMILIES OF THE OPISTHORCHIOIDEA

It may not be possible to key certain metacercariae down if the details of the ventro-genital complex and/or gonotyl are not clear or have not yet developed.

1 Adult forms from gut....................Cryptogonimidae

Encysted metacercarial forms from tissue 2

2 In freshwater fishes; ventral sucker not associated with genital atrium to form ventro-genital complex; gonotyl not present....................3

In freshwater or marine fishes; ventral sucker may be associated with genital atrium to form ventro-genital complex; gonotyl may be present....................Heterophyidae, Cryptogonimidae

3 Developing uterus pre-testicular....................Opisthorchiidae

Developing uterus extends close to posterior extremity ....Cryptogonimidae

Family CRYPTOGONIMIDAE Ward, 1917

Opisthorchioidea. Body generally small, oval to elongate. Body surface armed with spines. Circum-oral ring of larger spines present or absent. Commonly oculate. Oral sucker terminal to ventrally subterminal. Ventral sucker usually median, occasionally submedian, in middle or anterior half of body, very often smaller than oral sucker, embedded, sometimes opening into acetabulo-genital cavity or surrounded by muscular fold. Muscular pad occasionally present postero-ventral to ventral sucker. Prepharynx present, short to long. Pharynx well developed. Oesophagus apparently absent to long. Intestinal bifurcation in middle or posterior forebody. Caeca two, blind, short to long. Testes normally two, occasionally single or numerous, entire, in hindbody. Cirrus sac absent (? rarely present). Naked seminal vesicle, pars prostatica, and ejaculatory duct present. Genital atrium present, normally tubular, short to deep. Structure referred to as gonotyl present or absent, associated with anterior margin of genital atrium or acetabulo-genital cavity, may be muscular and/or glandular, may form digitiform process. Genital pore immediately anterior to ventral sucker, often sandwiched between ventral sucker and gonotyl, often opens into common cavity with ventral sucker (acetabulo-genital cavity). Ovary entire to deeply multi-lobed, lobes occasionally appear isolated, often large, median or submedian, in hindbody; pre- or inter-testicular. Canalicular seminal receptacle and Laurer's canal present. Vitellarium normally follicular, rarely compact, in small to extensive lateral fields in anterior hindbody and/or forebody, follicles may form transverse digitiform processes, fields may be confluent dorsally. Uterus long, coiled, mainly in hindbody, reaching close to posterior extremity. Eggs numerous, usually small, operculate, occasionally with polar spine or filament. Excretory pore terminal, vesicle Y-shaped (occasionally apparently V-shaped), usually large, occasionally small, arms usually wide and extending to well into forebody. Adults in gut and occasionally other organs of freshwater and marine fishes, occasionally in amphibians and reptiles, metacercariae encysted in tissues of fishes.

Comments

This is a group in dire need of revision and the value of certain diagnostic characters requires investigation. For example, a gonotyl is present or absent in different species of the genus *Neochasmus* and yet this is the only feature that distinguishes the genera *Cryptogonimus* and *Caecincola.*

Yamaguti (1971) accepted 15 subfamilies, many based upon dubious features. If one followed Yamaguti, the seven Canadian genera in this family, some of which are clearly closely related, would appear in five different subfamilies. No subfamilial designations are made in this work, pending a revision of the group.

It is worth mentioning that cryptogonimids occur in fishes both as adults and metacercariae. Not all of the metacercariae have been described, but they often appear to be quite well developed and thus many of the features used in the following keys, even aspects of the reproductive system, may be apparent.

Key to the genera of the Cryptogonimidae

1 Circum-oral ring of spines present 6

Circum-oral ring of spines absent (note that spines are easily lost in poorly fixed specimens and material removed from frozen hosts) 2

2 Ventral sucker dextro-median; testes symmetrical, close to posterior extremity of body; lateral vitelline fields extensive, reaching almost to posterior extremity; normally in ictalurids....*Acetodextra*

Ventral sucker median; lateral vitelline fields not normally extending into posterior third of body 3

3 Oral sucker similar in size to ventral sucker; lateral vitelline fields pre-ovarian; ovary well anterior to testes (gonads often degenerate in adults) .... *Centrovarium*

Oral sucker much larger then ventral sucker, usually terminal 4

4 In marine fishes; gonotyl absent; uterus composed of many heavily coiled loops .... *Metadena*

In freshwater fishes (mainly centrarchids); uterus a single loosely coiled loop 5

5 Gonotyl absent .... *Caecincola*

Gonotyl present... *Cryptogonimus*

6 Ovary about opposite middle of vitelline fields .... *Neochasmus*

Ovary opposite posterior extremity of vitelline fields; in percichthyids (*Morone*) ..*Allacanthochasmus*

ACETODEXTRA Pearse, 1924

Cryptogonimidae. Body medium-sized, elongate-oval to fusiform. Circum-oral ring of spines absent. Oral sucker ventrally subterminal, small. Ventral sucker dextro-median, small, in anterior half of body. Caeca long, reaching to posterior extremity. Testes two, elongate, symmetrical, at posterior extremity. Seminal vesicle tubular to saccular, extending between ovary and ventral sucker. Genital atrium small. Genital pore opens into acetabulo-genital cavity immediately anterior to ventral sucker. Gonotyl present as muscular (? and glandular) thickening on anterior margin of acetabulo-genital cavity, may be protruded. Ovary multi-lobed, large, median, just posterior to middle of body. Uterine field fills most available inter-caecal space in hindbody. Vitelline fields lateral to full length of caeca except for anterior extremity. Excretory vesicle small, tubular, arms not reaching into anterior half of body. In gonads and swimbladder of ictalurid fresh-water teleosts in North America.

Comments

This is a monotypic genus that parasitizes ictalurids, especially females.

Acetodextra amiuri (Stafford, 1900) Pearse, 1924 (Fig. 100)

Syn.: *Monostomum amiuri* Stafford, 1900.
Sites: Gonads (espec. ovary); swimbladder.
Hosts: *Ameiurus nebulosus* (1, 2, 4, 5, 7, 8, 10); *Ictalurus punctatus* (3, 9); *Noturus flavus* (6); *N. gyrinus* (3).
Distribution: Ont, Que.
Records: 1. Stafford 1900 (?locality); 2. Stafford 1904 (Que); 3. Bangham and Hunter 1939 (Ont); 4. Lyster 1939 (Que); 5. Bangham 1955 (Ont); 6. Dechtiar 1972a (Ont); 7. Fréchette *et al.* 1978 (Que); 8. Dechtiar *et al.* 1988 (Ont); 9. Dechtiar and Nepszy 1988 (Ont); 10. Dechtiar and Christie 1988 (Ont).
Remarks: This species was redescribed by Arnold (1934) and Perkins (1956), and Perkins commented upon some aspects of the life history.

ALLACANTHOCHASMUS Van Cleave, 1922

Cryptogonimidae. Body small, elongate-oval to elongate. Circum-oral ring of spines absent. Oral sucker relatively large, almost terminal. Ventral sucker median, much smaller than oral sucker, in middle of body or just anterior. Caeca long, reaching to middle of post-testicular field or slightly beyond. Testes two, symmetrical or oblique, occasionally almost tandem, near middle of hindbody. Seminal vesicle elongate-

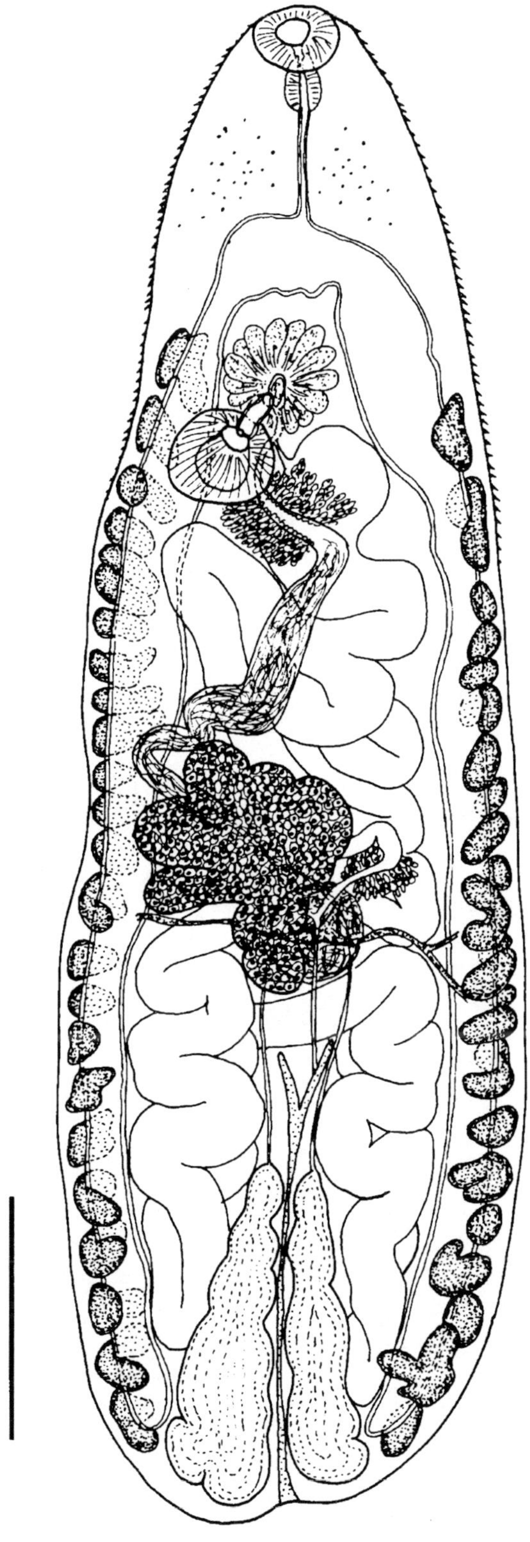

FIG. 100. *Acetodextra amiuri* (after Yamaguti (1971)). Scale bar (roughly estimated): 500 µm.

saccular, in posterior forebody or extending dorso-laterally to ventral sucker into anterior hindbody. Genital atrium tubular, short. Genital pore immediately anterior to ventral sucker, opens into acetabulo-genital cavity. Gonotyl present as muscular region immediately anterior to acetabulo-genital cavity, may form papilliform process. Ovary multi-lobed, sometimes very wide, occupying almost full width of body, just posterior to middle of body. Uterus occupying most of post-ovarian region of body. Lateral vitelline fields extending between level of gonads and a level within posterior half of forebody, fields may be confluent dorsally within forebody. Excretory vesicle large, arms wide, reaching to level of pharynx. In intestine of Nearctic percichthyid freshwater teleosts (*Morone*).

Key to the species of *Allacanthochasmus*

1 Body stout; prepharynx and oesophagus short or apparently absent; transverse loops of uterus fill posterior third of body; oral spines gently curved ....*Allacanthochasmus varius* Van Cleave, 1922

Site: Digestive tract.

Host: *Morone chrysops.*

Distribution: Ont.

Records: 1. Bangham and Hunter 1939; 2. Bangham 1955; 3. Anthony 1984; 4. Dechtiar *et al.* 1988; 5. Dechtiar and Nepszy 1988; 6. Dechtiar and Christie 1988.

A. varius Van Cleave, 1922 (metacercaria)

Site: Mesenteries.

Host: *Labidesthes sicculus.*

Distribution: Ont.

Record: Bangham and Hunter 1939.

Body slender; prepharynx and oesophagus well developed; no distinct lateral loops of uterus in posterior third of body; tip of oral spines sharply curved..*A. artus* Mueller and Van Cleave, 1932 (Fig. 101)

Site: Intestine.

Host: *Morone chrysops.*

Distribution: Ont.

Records: 1. Bangham 1955; 2. Anthony 1984; 3. Dechtiar *et al.* 1988; 4. Dechtiar and Nepszy 1988; 5. Dechtiar and Christie 1988.

Remarks: Some of the differences between this species and *A. varius* given by Mueller and Van Cleave (1932) might relate to the state of contraction of the worms. Nevertheless, although both species have been recorded on numerous occasions, nobody appears to have suggested that they might be synonymous.

CAECINCOLA Marshall and Gilbert, 1905

Cryptogonimidae. Body very small, oval to fusiform. Circum-oral ring of spines absent. Oral sucker large, terminal, often funnel-shaped. Ventral sucker median, much smaller than oral sucker, usually anterior to middle of body, occasionally in middle. Caeca short, terminating at about level of ovary. Testes two, relatively large, oblique, just posterior to caecal extremities, near middle of hindbody. Seminal vesicle saccular, bipartite, extending posteriorly into anterior hindbody. Genital atrium tubular, deep. Genital pore median, immediately anterior to ventral sucker. Gonotyl absent. Ovary tri-lobed, median or submedian, anterior to or ventral to testes, close to middle of body or more posterior. Uterus with single loop, loosely coiled, reaching close to posterior extremity. Eggs with or without anopercular spine or process. Lateral vitelline fields with restricted distribution, entirely in forebody or between level of pharynx and gonads, fields not confluent, follicles may form into transverse digitiform processes. Excretory vesicle large, arms wide, reaching to about level of pharynx. In intestine of Nearctic centrarchid freshwater teleosts.

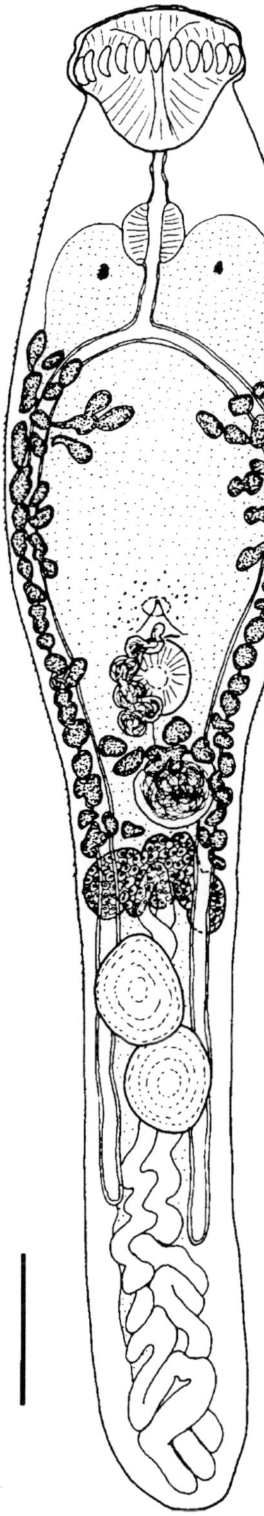

FIG. 101. *Allacanthochasmus artus* (after Mueller and Van Cleave (1932)). Scale bar: 100 µm.

Comments

It is clear that this genus is morphologically very close to *Cryptogonimus* Osborn, 1903, differing only in the absence of a gonotyl. Both occur in centrarchids and it seems very possible that confusion may have occurred between species of these two genera.

Caecincola parvulus Marshall and Gilbert, 1905 (Fig. 102)

Site: Intestine.

Hosts: *Micropterus dolomieui* (2); *M. salmoides* (1).

Distribution: Ont.

Records: 1. Bangham 1955; 2. Dechtiar 1972b.

Remarks: This minute worm appears to be a parasite mainly of *Micropterus* spp., especially *M. salmoides*. The descriptions by Marshall and Gilbert (1905), Van Cleave and Mueller (1934), Mueller (1934), Lundahl (1941), and Amin (1982b) exhibit considerable variation in ovarian shape, position of the gonads, and both the shape and size of the vitelline follicles, and the extent of the vitelline field. Two other species of the genus, *C. wakullata* Premvati, 1967 and *C. latostoma* Greer and Corkum, 1979, have also been described from *M. salmoides*, but from more southern regions of the United States. Since many of the features said to differentiate these species appear to be within the limits of variation described by other authors for *C. parvulus*, the validity of these two species requires confirmation.

The life history of this species was described by Lundahl (1941). Szalai and Dick (1991) recorded an encysted metacercaria of *Caecincola* sp. in the digestive tract of *Perca flavescens* in Manitoba.

CENTROVARIUM Stafford, 1904

Cryptogonimidae. Body small, elongate-oval or slightly pyriform, wider anteriorly or posteriorly. Circum-oral ring of spines absent. Oral sucker ventrally subterminal, relatively small. Ventral sucker median, similar in size to or slightly larger than oral sucker at about anterior third of body. Caeca short, terminating at about level of ovary. Details of reproductive system visible in metacercaria, but often degenerate in adult. Testes two, symmetrical to oblique, widely separated by uterus, in middle of hindbody or just posterior. Seminal vesicle long, saccular, tapering distally, posterior extremity dorsal to ventral sucker or extending into anterior hindbody, passes anteriorly almost to intestinal bifurcation prior to looping back towards genital pore. Genital atrium short, tubular. Genital pore on anterior margin of ventral sucker. Gonotyl absent. Ovary large, multi-lobed, median, in about middle of body, anterior to testes. Uterine coils filling available space lateral to ovary and in post-ovarian region. Lateral vitelline fields extending between about level of intestinal bifurcation and posterior margin of ovary (in metacercaria), often reduced in extent and number of follicles in adult, fields may be confluent dorsally. Excretory vesicle large, arms wide, reaching to level of pharynx. In intestine of Nearctic freshwater teleosts.

Comments

This is a monotypic genus.

Centrovarium lobotes (MacCallum, 1895) Stafford, 1904 (Fig. 103)

Syn.: *Distomum lobotes* MacCallum, 1895.

Sites: Intestine, pyloric caeca.

Hosts: *Ambloplites rupestris* (3, 4); *Ameiurus nebulosus* (4, 6); *Anguilla rostrata* (1); *Esox americanus vermiculatus* (5, 7); *E. lucius* (2, 4, 8, 14, 15, 16, 18); *Micropterus dolomieui* (5); *Notropis hudsonius* (14); *Perca flavescens* (1, 9, 13, 17, 19); *Percopsis omiscomaycus* (5); *Stizostedion canadense* (5, 6, 9); *S. vitreum glaucum* (5); *S. vitreum vitreum* (1, 2, 4, 5, 10, 11, 12, 13, 14, 18).

Distribution: Man, Ont, Que.

Records: 1. MacCallum 1895 (Ont); 2. Stafford 1904 (Que?); 3. Cooper 1915 (Ont); 4. Lyster 1939 (Que): 5. Bangham and Hunter 1939 (Ont); 6. Bangham 1955 (Ont); 7. Crossman 1962 (Ont);

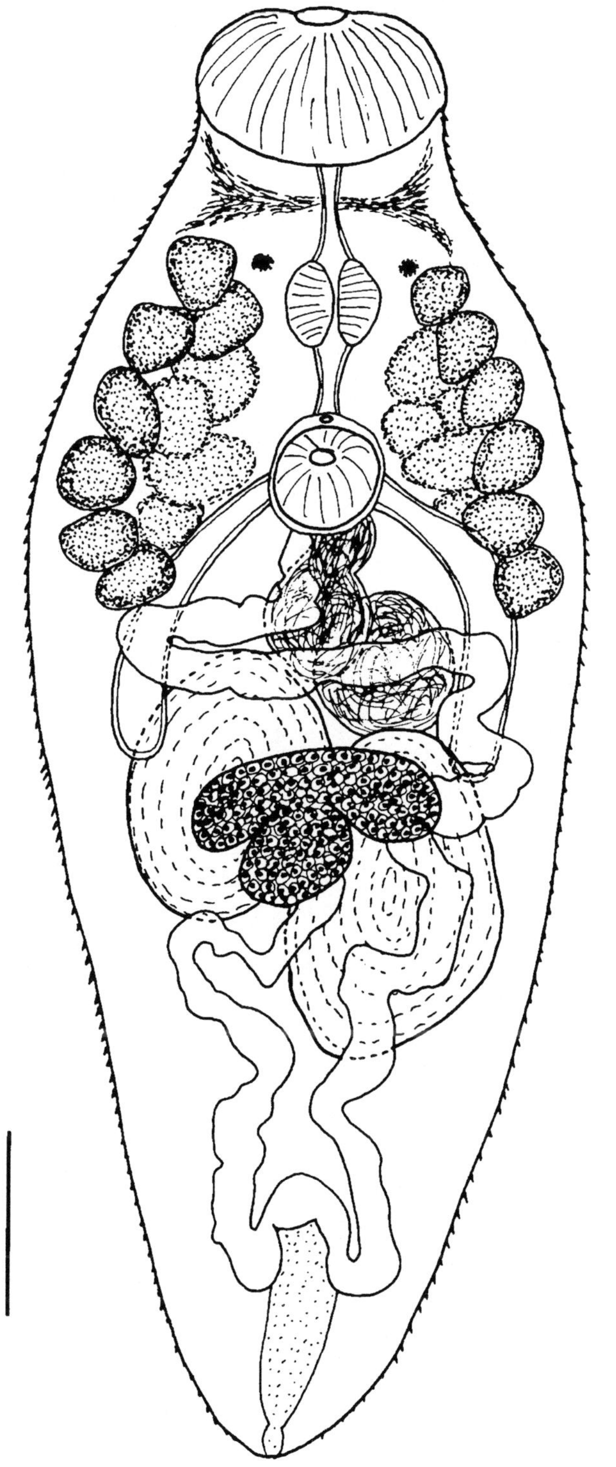

FIG. 102. *Caecincola parvula* (after Lundahl (1941)). Scale bar: 50 µm.

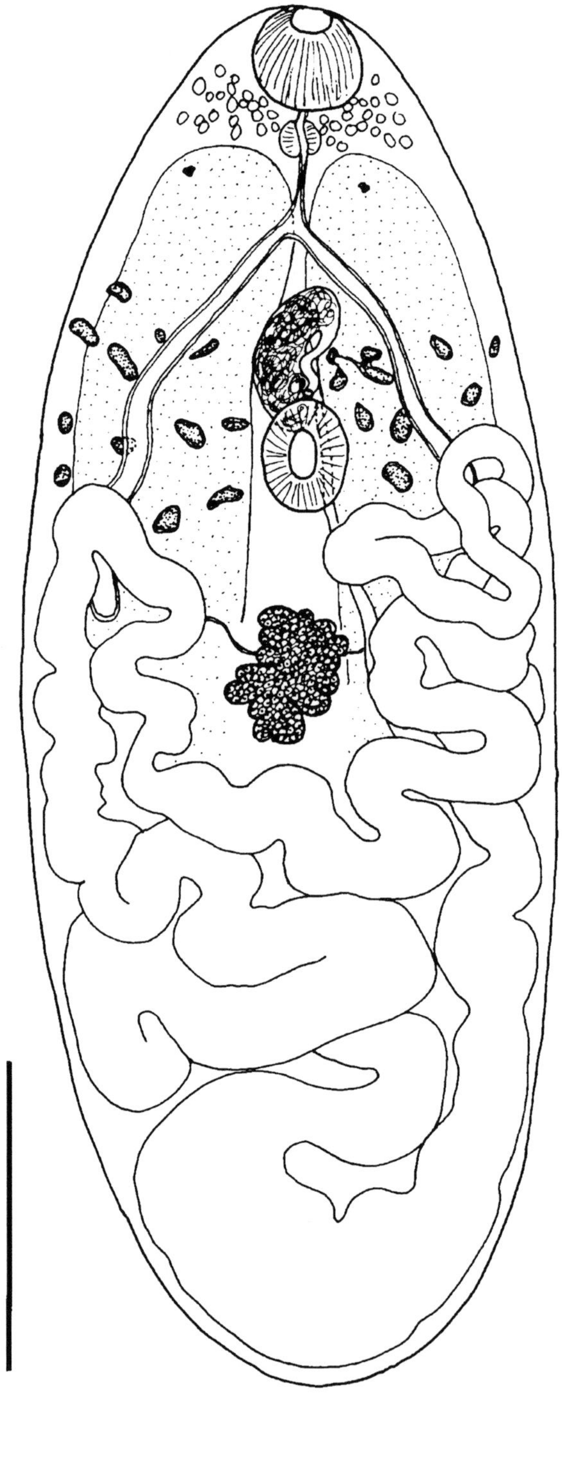

FIG. 103. *Centrovarium lobotes* (after Van Cleave and Mueller (1934)). Scale bar: 500 µm.

8. Dechtiar 1972a (Ont); 9. Dechtiar 1972b (Ont); 10. Anthony 1976 (Ont); 11. Stewart-Hay in Lubinsky 1976 (Man); 12. Dickson in Lubinsky 1976 (Man); 13. Anthony 1978b (Ont); 14. Stewart-Hay, Dickson and Watson, in Lubinsky and Loch 1979 (Man); 15. Watson and Dick 1980 (Man); 16. Anthony 1983 (Ont); 17. Dechtiar *et al.* 1988 (Ont); 18. Dechtiar and Nepszy 1988 (Ont); 19. Szalai and Dick 1991 (Man).

Remarks: In adult specimens the gonads and vitellarium, especially the testes, are often degenerate. They are best seen in the metacercariae, which, according to Van Cleave and Mueller (1934), occur commonly in *Percopsis omiscomaycus* (see also below).

C. lobotes (MacCallum, 1895) Stafford, 1904 (metacercaria) (Fig. 104)

Site: (Musculature).

Hosts: *Notropis hudsonius* (2, 4, 5); *N. volucellus* (1); *Percopsis omiscomaycus* (5, 6); *Pimephales notatus* (1, 5); *Stizostedion vitreum vitreum* (3).

Distribution: Ont.

Records: 1. Bangham 1955; 2. Dechtiar 1972a; 3. Anthony 1984; 4. Dechtiar and Lawrie 1988; 5. Dechtiar *et al.* 1988; 6. Dechtiar and Christie 1988.

CRYPTOGONIMUS Osborn, 1903

Cryptogonimidae. Body very small to small, fusiform to elongate-oval. Circum-oral ring of spines absent. Oral sucker large, terminal, often funnel-shaped. Ventral sucker median, in anterior half of body, much smaller than oral sucker, may be enclosed in fold of body wall. Caeca short to long, extending to level between ovary and posterior extremity. Testes two, oblique, just posterior to middle of hindbody. Seminal vesicle elongate, saccular, often appearing bipartite, extending back into anterior hindbody. Genital atrium deep, tubular. Genital pore median, sandwiched between ventral sucker and gonotyl, opening into acetabulo-genital sinus. Gonotyl a muscular pad, may resemble sucker. Ovary median or submedian, apparently entire, tri-lobed or irregularly multi-lobed, just posterior to middle of body, anterior to testes. Uterus a single loop, loosely coiled, reaching close to posterior extremity. Eggs with or without anopercular spine or process. Lateral vitelline fields with restricted distribution, anterior limit between pharynx and ventral sucker, posterior limit between level just anterior to ovary and level of anterior testis, fields not confluent, follicles may form into transverse digitiform processes. Excretory vesicle large, arms wide, reaching to level of pharynx. In intestine of Nearctic centrarchid freshwater teleosts.

Cryptogonimus chili Osborn, 1903 (Fig. 105)

Sites: Intestine, pyloric caeca, stomach.

Hosts: *Ambloplites rupestris* (1, 2, 3, 4, 6, 7, 8); *Micropterus dolomieui* (2, 3, 4, 5, 6, 8, 9); *M. salmoides* (3); *Perca flavescens* (3).

Distribution: Ont, Que.

Records: 1. Stafford 1905 (?locality); 2. Cooper 1915 (Ont); 3. Bangham and Hunter 1939 (Ont); 4. Miller 1940a (Que); 5. Fantham and Porter 1948 (Que); 6. Bangham 1955 (Ont); 7. Dechtiar 1972b (Ont); 8. Dechtiar *et al.* 1988 (Ont); 9. Dechtiar and Nepszy 1988 (Ont).

Remarks: This species appears to be especially common in *Ambloplites rupestris* and *Micropterus dolomieui*. Van Cleave and Mueller (1934) considered that the former was the most important host. A similar species, *C. spinovum* Greer and Corkum, 1979, occurs in *M. salmoides* and *M. punctulatus* in Louisiana, but has much longer caeca.

C. chili Osborn, 1903 (metacercaria)

Site: Musculature.

Hosts: *Ambloplites rupestris*; *Micropterus dolomieui*; *Notropis hudsonius*; "sunfish" (?*Lepomis gibbosus* or reference to *M. dolomieui*).

Distribution: Ont.

Record: Cooper 1915.

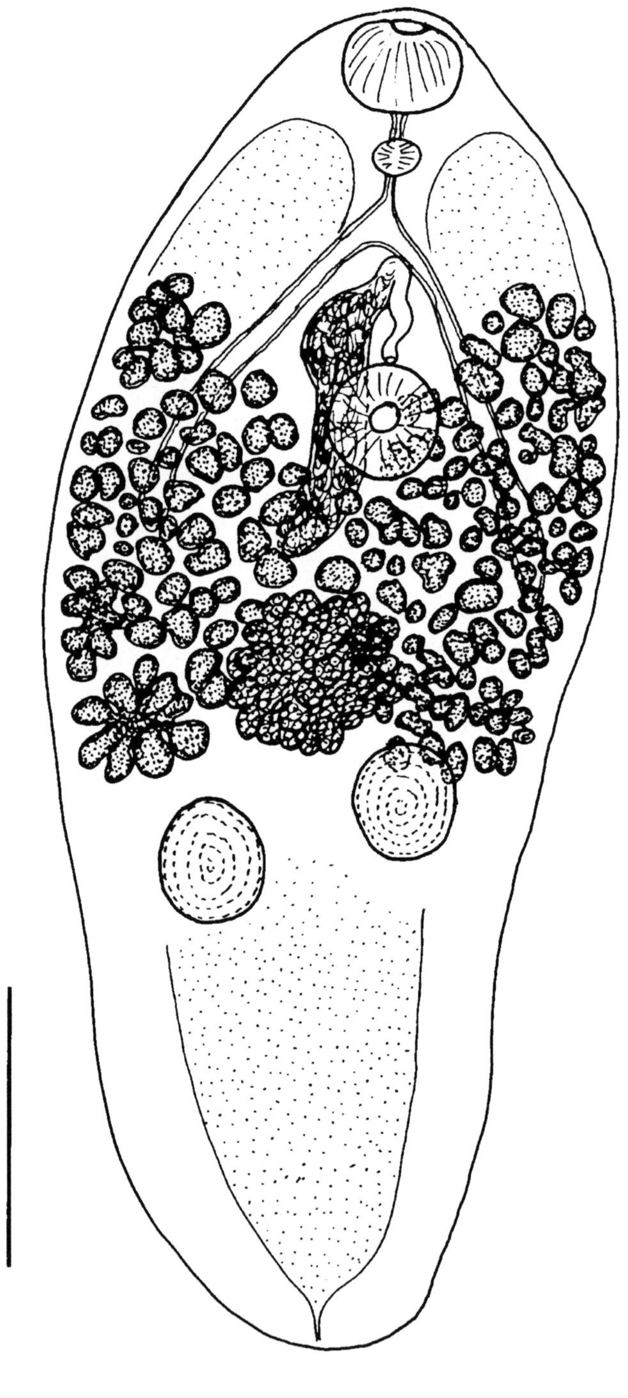

FIG. 104. *Centrovarium lobotes* metacercaria (after Van Cleave and Mueller (1934)). Scale bar: 500 μm.

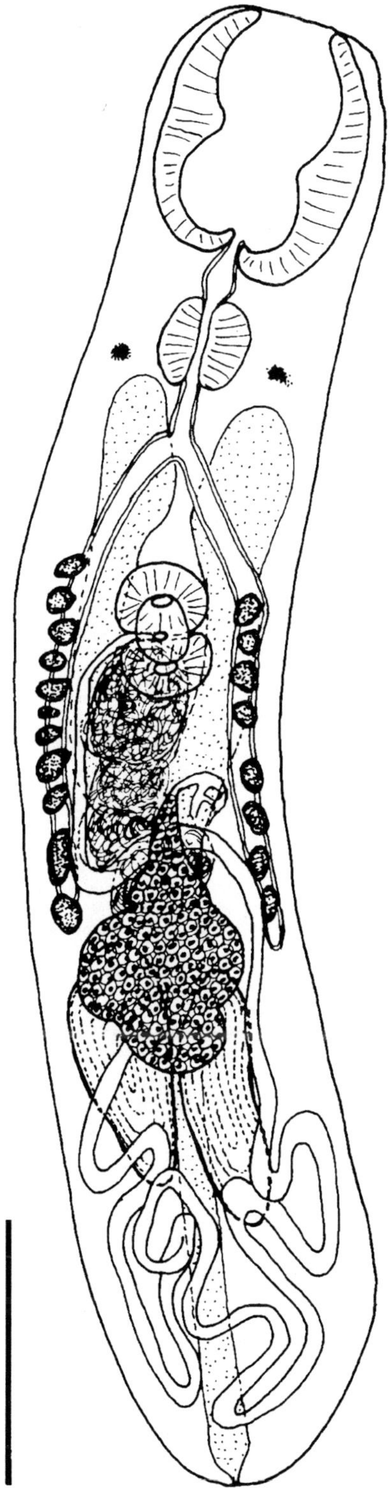

FIG. 105. *Cryptogonimus chili* (modified after Osborn (1910)). Body spines omitted; coils of uterus simplified. Scale bar (roughly estimated): 200 μm.

C. diaphanus (Stafford, 1904) Miller, 1941

Syn.: *Protenteron diaphanum* Stafford, 1904.
Site: Intestine.
Host: *Ambloplites rupestris*.
Distribution: Que?
Record: Stafford 1904.
Remarks: This species is known from only a small number of specimens that have been inadequately described by both Stafford (1904) and Miller (1941a). Because they come from the same host as *C. chili*, this species may prove to be a synonym; but Stafford (1904) indicated that a live worm measured 1.54 mm and Miller (1941a) stated that a mounted worm measured 1.9 mm, which is twice the size of records of *C. chili* (Osborn's, 1903, maximum size of 9.3 mm is presumably a lapsus for 0.93 mm).

METADENA Linton, 1910

Cryptogonimidae. Body minute to small, oval to fusiform. Circum-oral ring of spines absent. Oral sucker small to large, ventrally subterminal to almost terminal. Ventral sucker median, small, usually smaller than oral sucker, in anterior half of body. Caeca long, terminating between middle of post-testicular field and posterior extremity. Testes two, often large, usually symmetrical, occasionally oblique, separated by coils of uterus or ovary, just posterior to middle of body. Seminal vesicle elongate-saccular, often bipartite, anterior part may be in forebody or hindbody, posterior part usually in or extends into hindbody. Genital atrium short, tubular. Genital pore immediately anterior to ventral sucker. Gonotyl absent. Ovary large, irregularly multi-lobed, median, near middle of body, often inter-testicular. Uterus filling much of hindbody, especially in post-testicular field. Lateral vitelline fields generally between level of pharynx and testes, fields may be confluent dorsally. Excretory vesicle large, arms broad, reaching to level of pharynx. In intestine of marine teleosts.

Metadena sp. (Fig. 106)

Site: Intestine.
Host: *Tautogolabrus adspersus*.
Distribution: Atl.
Record: Sekhar and Threlfall 1970a.
Remarks: Sekhar and Threlfall (1970a) indicated that their two specimens were morphologically somewhat similar to *M. crassulata* Linton, 1910, but had much larger eggs. This labrid would appear a reasonable host for a species of *Metadena*, as most species occur in marine perciforms, but the record seems rather far north, because most records are from the West Indies and the southern states of the United States. *M. crassulata* is used to illustrate this genus.

NEOCHASMUS Van Cleave and Mueller, 1932

Cryptogonimidae. Body small, elongate-oval. Circum-oral ring of larger spines present. Oral sucker relatively large, terminal. Ventral sucker median, smaller than oral sucker, deep within acetabulo-genital sinus. Large circular muscular pad present or absent postero-ventral to ventral sucker. Caeca reaching back to middle of hindbody or slightly beyond. Testes two, subsymmetrical, in about middle of body to middle of hindbody. Seminal vesicle long, saccular, winding, reaching back into hindbody, sometimes as far as anterior margin of ovary. Genital atrium tubular. Genital pore opens into acetabulo-genital cavity immediately anterior to ventral sucker. Bi-lobed gonotyl present or absent anterior to genital pore (some authors have considered circular muscular pad to be vestige of gonotyl, despite its posterior position in relation to aperture of acetabulo-genital cavity). Ovary large, deeply multi-lobed, lobes often appearing to be isolated follicles; extends across almost full width of body ventrally, in middle or second quarter of body. Uterus filling post-testicular field. Lateral vitelline fields occur in region between ventral sucker and posterior margin of testes, fields may be confluent dorsally, follicles may form digitiform processes. Excretory

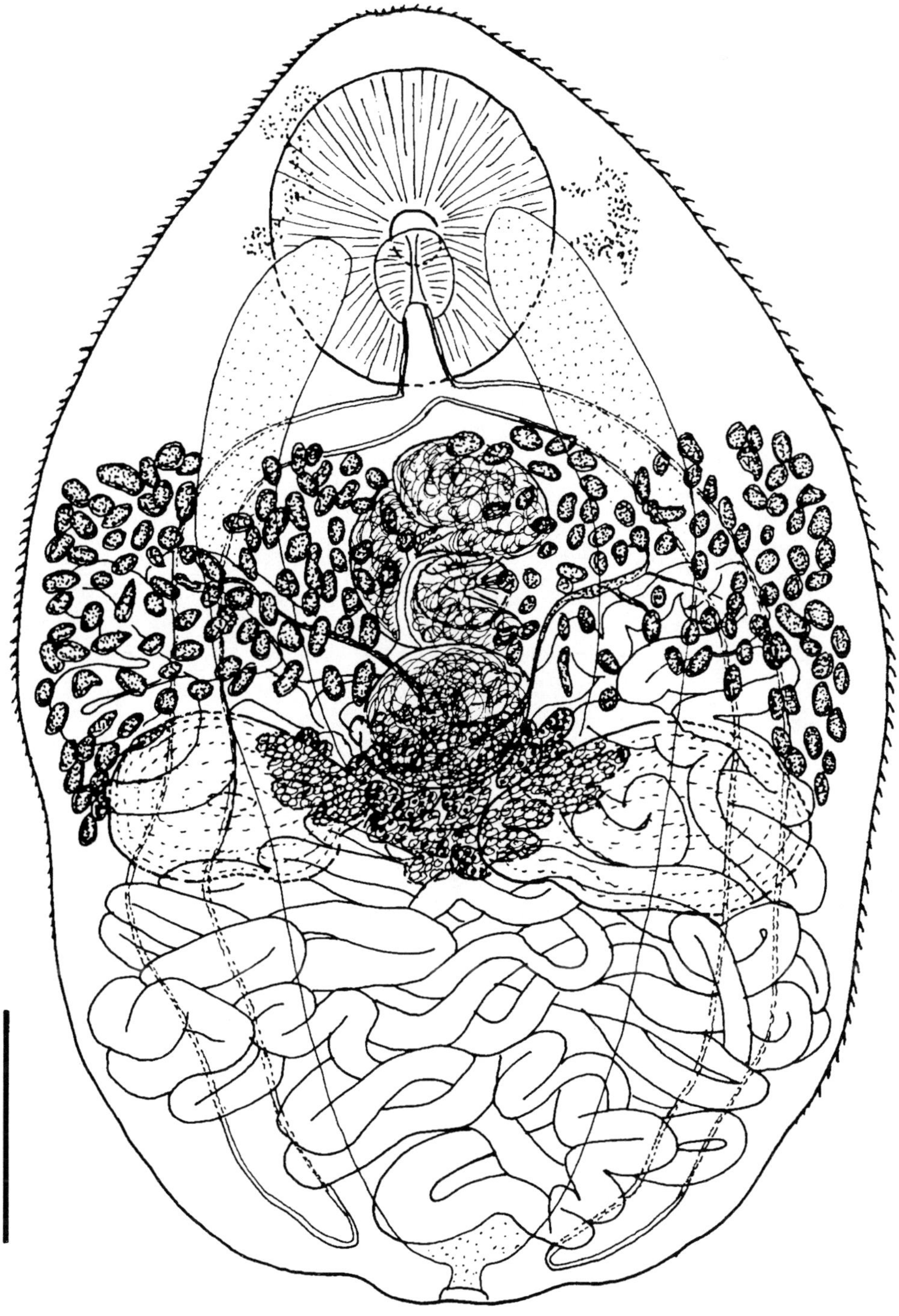

FIG. 106. *Metadena crassulata* (after Yamaguti (1971)) , dorsal view. Scale bar (roughly estimated): 250 µm.

vesicle large, arms reaching to level of pharynx. In intestine of freshwater and ?marine teleosts and ?reptiles (?accidental).

Neochasmus umbellus Van Cleave and Mueller, 1932 (Fig. 107)

Site: Intestine.

Hosts: *Micropterus dolomieui* (1, 2, 3); *Morone chrysops* (4, 5).

Distribution: Ont.

Records: 1. Bangham and Hunter 1939; 2. Dechtiar 1972b; 3. Anthony 1985; 4. Dechtiar and Nepszy 1988; 5. Dechtiar and Christie 1988.

Remarks: This species has also been recorded from amphibians of the genus *Necturus*.

Family HETEROPHYIDAE Leiper, 1909

Opisthorchioidea. Present in fishes as encysted metacercaria. Cyst bi-layered, with thin inner layer of parasite origin and thick outer layer of host origin; outer layer fibrous, occasionally partly calcified. Body oval to elongate. Tegument spined. Oral and ventral suckers present. Ventral sucker occasionally small or submedian, often associated with genital atrium forming ventro-genital complex (not always clearly developed in metacercaria). Prepharynx, pharynx, and oesophagus present. Intestinal bifurcation in forebody. Caeca short to long, terminate blindly. Genital primordia often developed in hindbody. Gonotyl(s) (muscular structure associated with genital atrium) may or may not be developed, armed or unarmed. Excretory vesicle I-shaped, often saccular and interpreted as V- or Y-shaped. On skin and fins and in tissues of teleosts (occasionally holosteans). Adults in piscivorous birds and mammals.

Comments

When present on the skin and fins these metacercariae (e.g., *Cryptocotyle* spp. and *Apophallus imperator*) may elicit the deposition of melanin, producing the condition known as "blackspot" (Miller 1940b). *A. brevis* in the tissue of perch may form large, partly calcified cysts and is known as the "sand-grain grub".

Features of the ventro-genital complex and gonads are sometimes not visible in specimens fixed under pressure. It is best to examine unflattened whole mounts.

Key to the genera of the Heterophyidae

1 Armed gonotyl* developed (in marine teleosts of the Pacific coast)............................*Galactosomum*

Gonotyl unarmed or not developed..2

2 Gonotyl not yet developed (in freshwater fishes)..*Apophallus*

Gonotyl muscular but difficult to discern (in marine teleosts of Atlantic coast, occasionally in euryhaline or freshwater fishes in coastal regions)...*Cryptocotyle*

* According to Prof. John Pearson (pers. comm.), it is not actually the gonotyl that is armed but parts of the ventral sucker that contribute to the ventro-genital complex.

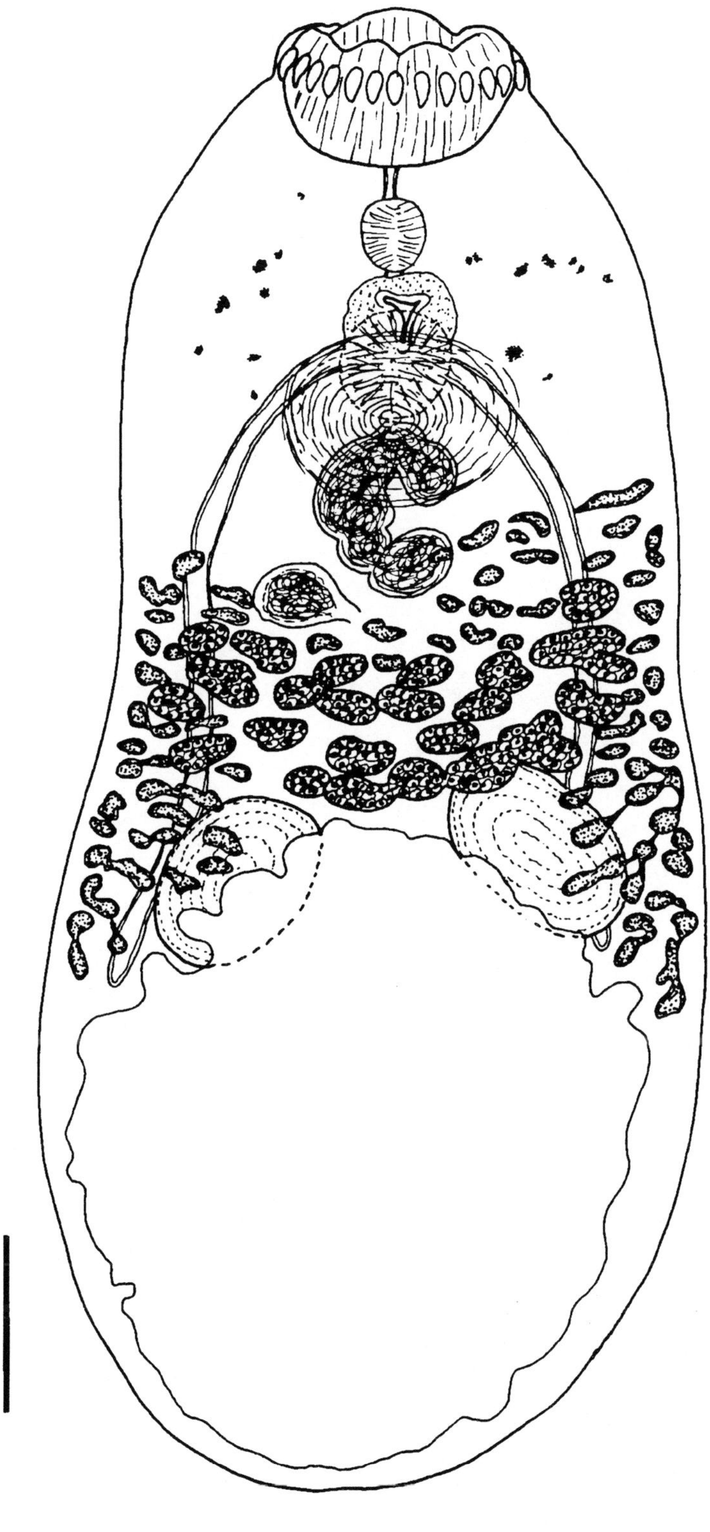

FIG. 107. *Neochasmus umbellus* (after Van Cleave and Mueller (1932)). Body spines omitted. Scale bar: 100 µm.

APOPHALLUS Lühe, 1909

Heterophyidae. Outer layer of cyst may be calcified. Body elongate. Ventral sucker near middle of body. Oesophagus long. Caeca long, terminating near posterior extremity. Gonads developed or apparently not. Gonotyl not yet developed.

Comments

There has obviously been some confusion in the identification of metacercariae of this genus. It seems likely that some of the records from atypical hosts are erroneous.

Key to the species of *Apophallus*

1 Outer host-cyst large (often more than 1 mm in diameter), often with calcified layer; excysted metacercaria 680–1100 µm in length; gonads well developed; usuallyin *Perca flavescens* ....*Apophallus brevis* Ransom, 1920 (metacercaria) (Fig. 108)

Syn.: *Apophallus americanus* Van Cleave and Mueller, 1932; *A. itascaensis* Warren, 1953; *Distomum* sp. larva of Cooper (1915).

Sites: Skin, fins, gills, musculature.

Hosts: *Etheostoma exile* (5, 20); *E. nigrum* (5, 7, 20); *Gasterosteus aculeatus* (8); *Lepomis gibbosus* (9); *Micropterus dolomieui* (21); *Notemigonus crysoleucas* (20); *Notropis hudsonius* (16); *Perca flavescens* (1, 4, 5, 6, 7, 11, 12, 13, 14, 15, 16, 17, 18, 19, 20, 21, 22 (exp), 23); *Prosopium cylindraceum* (7); *Salmo salar* (2, 3, 10); *Salmo trutta* (2); *Salvelinus fontinalis* (2).

Distribution: Lab, Man, NB-b, NB, NS, Nfld, Ont, Que, Sask.

Records: 1. Cooper 1915 (Ont); 2. Sandeman and Pippy 1967 (Nfld); 3. Pippy 1969 (Lab, NB-b, NB, Nfld); 4. Dechtiar 1972a (Ont); 5. Dechtiar 1972b (Ont); 6. Sinclair 1972a (Ont); 7. Sinclair 1972b (Ont, Man, Sask); 8. Hanek and Molnar 1974 (Que); 9. Cone and Anderson 1977 (Ont); 10. Pippy 1980 (Lab, Nfld, NB); 11. Pike and Burt 1981 (NB); 12. Pike and Burt 1982 (NB); 13. Pike and Burt 1983 (NB); 14. Pike, Thompson, and Burt 1983 (NB); 15. Poole and Dick 1983 (Man); 16. Poole and Dick 1985 (Man); 17. Dechtiar and Lawrie 1988 (Ont); 18. Dechtiar *et al.* 1988 (Ont); 19. Dechtiar and Nepszy 1988 (Ont); 20. Dechtiar and Christie 1988 (Ont); 21. Dechtiar *et al.* 1989 (Ont); 22. Taylor *et al.* 1993 (NS); 23. BM(NH) collection (NB).

Remarks: This species, according to Sinclair (1972a,b), is specific to *Perca flavescens*. Sinclair considered Cooper's material of *Distomum* sp. to belong to this species and both Miller's (1941b) and Cameron's (1945) material of "*A. brevis*" to be *A. imperator*.

Host-cyst thick and fibrous but not calcified; excysted metacercariae normally <600 µm in length 2

2 Gonads poorly differentiated; usually in *Salvelinus fontinalis* ....*A. imperator* Lyster, 1940 (metacercaria)

Syn.: *Apophallus brevis* of Miller (1941b, 1942) and Cameron (1945); *Apophallus* sp. of Lyster (1940a).

Sites: Skin, fins.

Hosts: *Salmo salar* (10); *Salvelinus fontinalis* (1, 2, 3, 4, 5, 6, 7, 8, 9, 10).

Distribution: NB, NS, Ont, Que, Nfld.

Records: 1. Lyster 1940a (Que); 2. Lyster 1940b (Que); 3. Miller 1941b (Que); 4. Miller 1942 (Que); 5. Cameron 1945 (locality not specified); 6. Sinclair 1972a (Ont, Que); 7. Hare and Frantsi 1974 (NB, NS); 8. Cone and Ryan 1984 (Nfld); 9. Dechtiar *et al.* 1989 (Ont); 10. Marcogliese and Cone 1991b (Nfld).

Remarks: This species is one that causes "blackspot" disease. Sinclair (1972a) considered this species to be specific to *Salvelinus fontinalis*. Material referred to as "blackspot" by Choquette (1948) may also belong to this species, as may the *Apophallus* sp. metacercariae

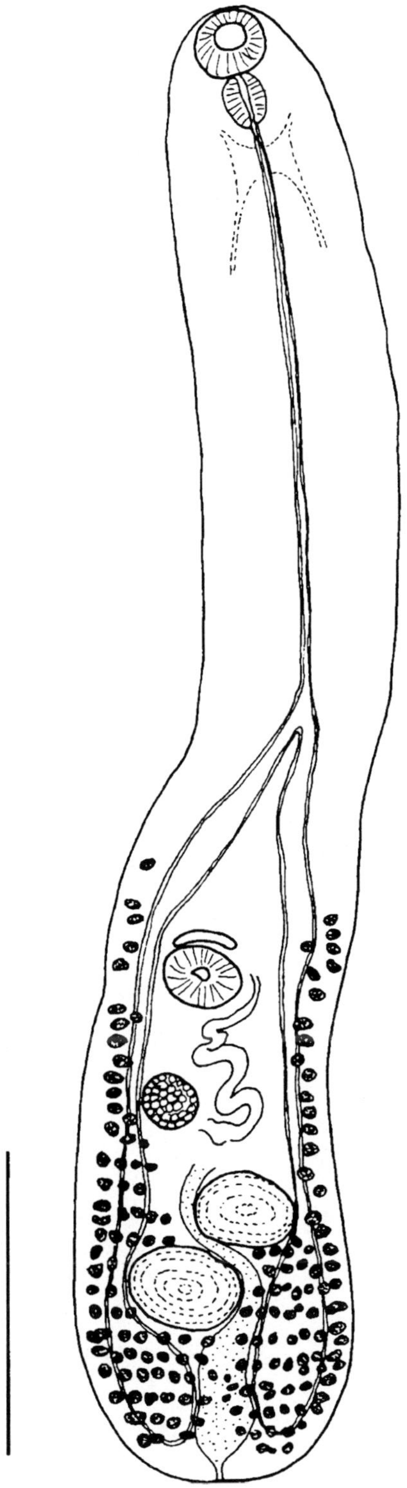

FIG. 108. *Apophallus brevis* metacercaria (after Van Cleave and Mueller (1932)). Body spines omitted. Scale bar: 200 μm.

reported by Bangham and Venard (1946) from *S. fontinalis* and *S. namaycush* in Ontario. It is also possible that other records from salmonoids may be related.

Gonads well differentiated; mainly in centrarchids and ictalurids ....*A. venustus* (Ransom, 1920) Cameron, 1936 (metacercaria)

Site: Musculature.

Hosts: *Ameiurus nebulosus* (2, 4); *Amia calva* (3, 4); *Catostomus commersoni* (2, 4); *Cyprinus carpio* (2, 4); *Esox lucius* (2, 4); *Ictalurus punctatus* (3, 4); *Lepisosteus osseus* (2, 4); *Lepomis gibbosus* (2, 4); *Luxilis cornutus* (2, 4); *Micropterus dolomieui* (2, 4); *Morone americana* (6); *Moxostoma macrolepidotum* (4); *Perca flavescens* (2, 4, 5, 7); *Stizostedion vitreum vitreum* (2, 4); Unspecified fish (1).

Distribution: Ont, Que.

Records: 1. Cameron 1936 (Que); 2. Cameron 1937a (Que); 3. Cameron 1937b (Que); 4. Cameron 1945 (locality not specified); 5. Tedla and Fernando 1969a (Ont); 6. Tedla and Fernando 1969b (Ont); 7. Tedla and Fernando 1972 (Ont).

Remarks: According to Cameron (1945) the main hosts are species of *Ictalurus*, *Micropterus*, and *Lepomis*.

Other records that may also be *Apophallus* are:

Apophallus sp. (metacercaria)

Site: Skin.

Host: *Perca flavescens*.

Distribution: Ont.

Record: Anthony 1978b.

Heterophyes sp. (?metacercaria)

Site: Not specified.

Host: *Hiodon alosoides*.

Distribution: Man.

Record: Kennedy and Sprules 1967.

Remarks: This is presumably an error, because adults of *Heterophyes* do not occur in North America (Margolis and Arthur, 1979).

Heterophyidae gen. spp. (metacercariae)

Sites: Skin, gills, musculature.

Hosts: *Cottus cognatus* (2); *Lepomis gibbosus* (3); *Salvelinus fontinalis*.

Distribution: Ont, Nfld, YT.

Records: 1. Frost 1940 (Nfld); 2. Arthur *et al.* 1976 (YT); 3. Cone and Anderson 1977 (Ont).

CRYPTOCOTYLE Lühe, 1899

Heterophyidae. Body round to elongate, may be concave. Oesophagus short to relatively long. Caeca terminate near posterior extremity. Gonads developed. Muscular gonotyl developed but often apparently difficult to discern.

Cryptocotyle lingua (Creplin, 1825) Fischoeder, 1903 (metacercaria) (Fig. 109)

Syn.: *Tocotrema lingua* (Creplin, 1825); *Dermocystis ctenolabri* Stafford, 1905.

Sites: Skin, gills.

Hosts: *Clupea harengus* (6, 10); *Gadus morhua* (9, 12); *Glyptocephalus cynoglossus* (5); *Hemitripterus americanus* (2); *Hippoglossoides platessoides* (5); *Hippoglossus hippoglossus* (5); *Melanogrammus aeglefinus* (12); *Osmerus mordax* (8); *Pleuronectes americanus* (3, 5, 11); *P. ferrugineus* (5); *P. putnami* (5); *Pollachius virens* (12); *Salvelinus fontinalis* (13, 14); *Scomber scombrus* (12); *Scophthalmus aquosus* (5); *Tautoglabrus adspersus* (1, 7); Unspecified flounder (4).

Distribution: Atl, NB, NB-b, Nfld.

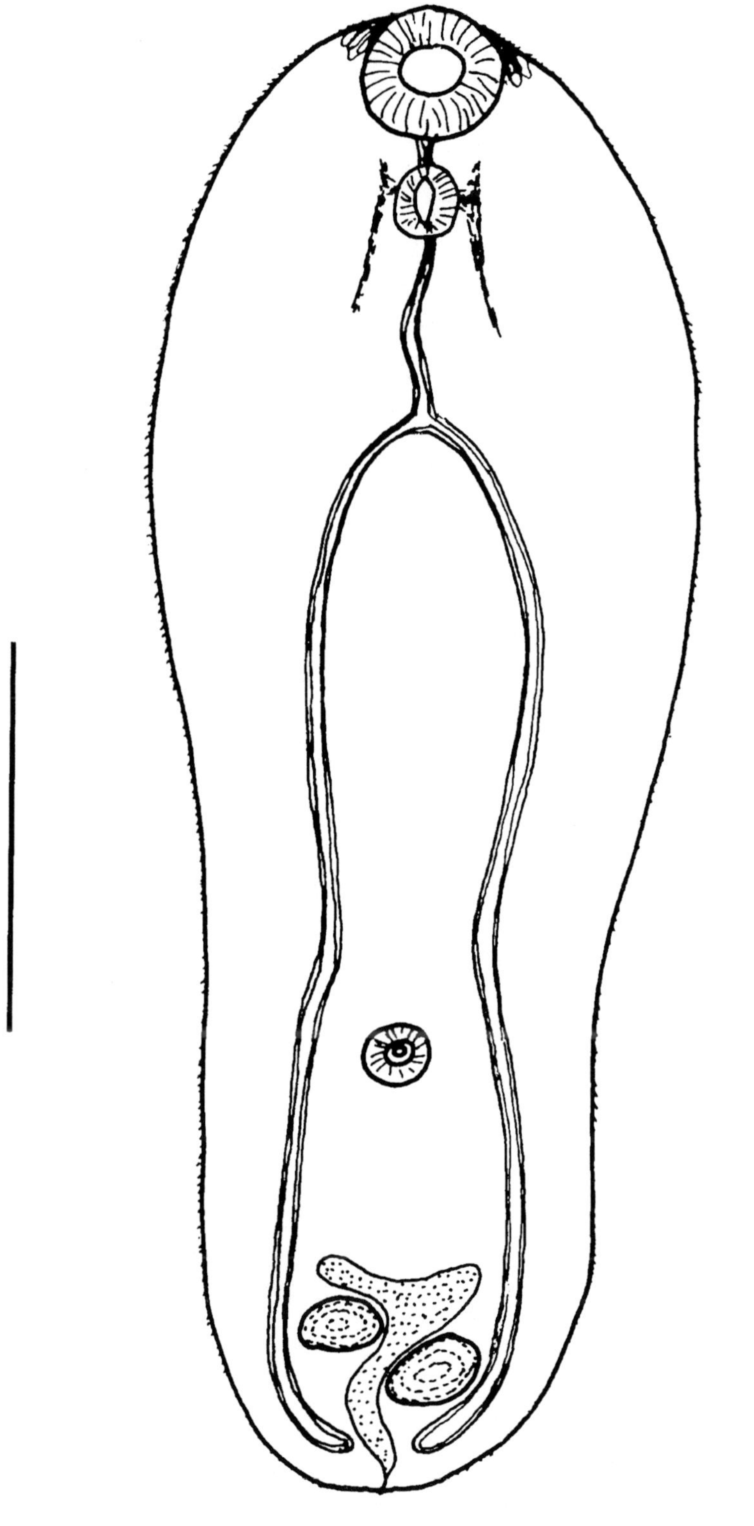

FIG. 109. *Cryptocotyle lingua* metacercaria (modified after Stunkard (1930)). Scale bar (estimated): 100 µm.

Records: 1. Stafford 1905 (Atl); 2. Cooper 1915 (Atl); 3. Wolfgang 1954b (Atl); 4. Wolfgang 1955a (Atl); 5. Ronald 1960 (Atl); 6. Sindermann 1965 (Atl); 7. Sekhar and Threlfall 1970b (Atl); 8. Threlfall 1981 (Nfld); 9. Appy and Burt 1982 (Atl); 10. McGladdery and Burt 1985 (Atl); 11. Scott 1985a (Atl); 12. Morrison *et al.* 1986 (Atl); 13. Frimeth 1987a (NB, NB-b): 14. Frimeth 1987b (Atl, NB-b).

Remarks: This species is one which causes "blackspot" disease. A second common Holarctic species, *C. concava* (Creplin, 1825), may occur in Canadian waters. Although most common in coastal regions, the latter has been recorded in freshwater fishes in the United States. Its metacercaria is more oval and concave in shape than that of *C. lingua*, and the developing testes tend to be symmetrical rather than oblique.

GALACTOSOMUM Looss, 1899

Heterophyidae. Body elongate. Prepharynx often long, oesophagus short. Caeca terminate close to posterior extremity. Gonads developed. Armed gonotyl present.

Comments

According to Prof. John Pearson (pers. comm.), it is not actually the gonotyl that is armed but parts of the ventral sucker that form part of the ventro-genital complex.

Key to the species of *Galactosomum*

1 Ventro-genital complex in anterior two-fifths of body; excretory vesicle reaching to level of posterior testis; ventral sucker larger than oral sucker .... *Galactosomum humbargari* Park, 1936 (metacercaria)

Syn.: *Galactosomum* sp. of Arai (1967a,1969b) in part.

Sites: Intestinal wall, gill arch.

Host: *Cymatogaster aggregata*.

Distribution: Pac.

Records: Arai 1967a; Arai 1969b; BM(NH) collection.

Remarks: Arai's material from *Cymatogaster aggregata* is assumed to be conspecific with that in the BM(NH) collection. Pearson *et al.* (1978) indicated that it was not *G. phalacrocoracis*. Arai (1967a,b) also listed *Galactosomum* sp. from *Hexagrammos lagocephalus*. Pearson *et al.* (1978) did not consider this material as being *G. phalacrocoracis*.

Ventro-genital complex close to middle of body; excretory vesicle reaching to level of ovary; ventral sucker smaller than oral sucker .... *G. phalacrocoracis* Yamaguti, 1939 (metacercaria) (Fig. 110)

Syn.: *Galactosomum* sp. of Arai (1967a, 1969b) in part.

Sites: Viscera, pharynx, gill arches.

Hosts: *Ammodytes hexapterus* (1, 2); *Clupea pallasi* (4, 5); *Oncorhynchus gorbuscha* (3); *O. keta* (1, 2).

Distribution: Pac.

Records: 1. Arai 1967a; 2. Arai 1969b; 3. Pearson, Margolis, and Boyce 1978; 4. Arthur and Arai 1980a; 5. Arthur and Arai 1980b.

Family OPISTHORCHIIDAE Looss, 1899

Opisthorchioidea. Present in fish as encysted metacercaria. Cyst relatively small, globular to oval, with two-layered wall: outer layer of host origin, often thick and readily digested by enzymes, inner layer of parasite origin, often thin, transparent, and resistant to enzymes. Metacercarial body small, elongate-oval. Tegument spined. Suckers well developed. Prepharynx present. Pharynx well developed. Oesophagus short

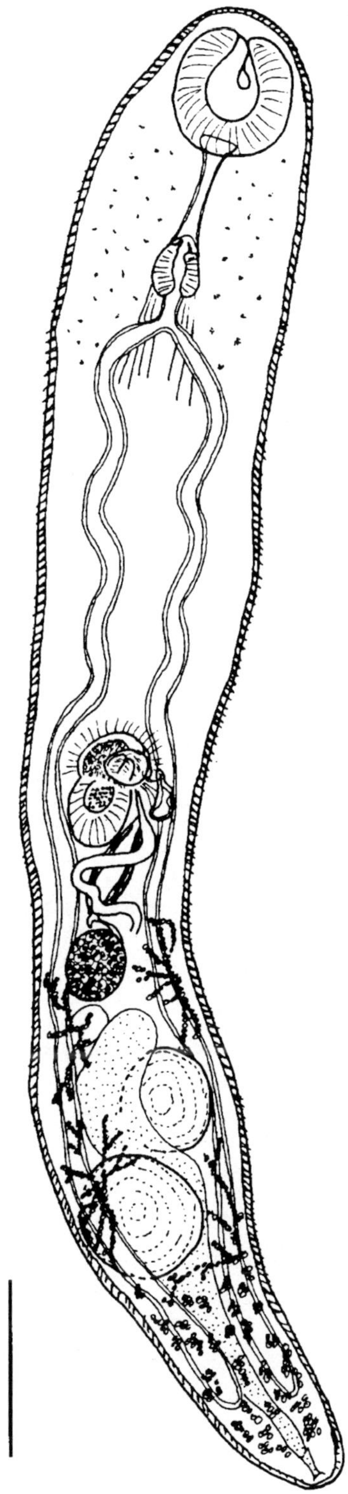

FIG. 110. *Galactosomum phalacrocoracis* metacercaria (after Pearson *at al.* (1978)). Scale bar: 200 µm.

to long. Intestinal bifurcation usually in middle of forebody or more anterior. Caeca long, terminating blindly close to posterior extremity. Primordia of gonads present or not in hindbody. Gonotyl absent. Excretory vesicle Y-shaped, stem large, saccular, arms narrow, tubular. Present in tissues, especially muscles or fins, of freshwater fishes. Adults in piscivorous mammals and birds.

Comments

Because only one species of opisthorchiid metacercaria recorded from Canadian fishes has been described in any detail, it is not possible to produce a key to these forms. It is worth noting, however, that there does appear to be a size difference (see below) between cysts of *Amphimerus elongatus* and *Metorchis conjunctus*, and that metacercariae of the former species differ from many *Metorchis* spp. metacercariae (and adults of *Parametorchis* spp. and *A. pseudofelineus*) in that the oesophagus is relatively long. There may be differences in the site of infection for *A. elongatus* and *M. conjunctus* (see below), but the observations are somewhat conflicting.

AMPHIMERUS Barker, 1911

Opisthorchiidae. Cysts small. Body of metacercaria elongate-oval to elongate. Pharynx well developed. Oesophagus long (in *A. elongatus*). In muscles and fins of freshwater teleosts.

Amphimerus elongatus Gower, 1938 (metacercaria) (Fig. 111)

Sites: Fins (mainly), musculature.
Host: *Catostomus commersoni.*
Distribution: Que.
Record: Cameron 1944.
Remarks: Cameron (1944) maintained that this species could be distinguished from *Metorchis conjunctus* by the facts that the metacercariae were smaller (<200 μm in diameter, according to Wallace (1939), as opposed to about twice that size), and that they tended to occur in the fins and adjacent tissue (plus the head) as opposed to the musculature. However, according to Wallace (1939) the metacercariae of this species occur mainly in the muscles, with fewer in the fins. Font (1991) indicated that they are in the fins and musculature. It seems likely, therefore, that there may be some variation, depending upon the host and level of infection, although there is always the possibility that other opisthorchiid species may be involved.

A. pseudofelineus (Ward, 1901) Barker, 1911 (metacercaria)

Syn.: *Opisthorchis pseudofelineus* Ward, 1901.
Site: Musculature.
Hosts: *Catostomus commersoni* (2); Unspecified fish (1).
Distribution: Man.
Records: 1. Parnell 1934 (unspecified locality); 2. Evans 1963 (Man).
Remarks: Evans' determination was confirmed by feeding experiments. There appears to be no description of the metacercaria of this species.

METORCHIS Looss, 1899

Opisthorchiidae. Cyst small to medium-sized. Body of metacercaria elongate-oval. Pharynx small to well developed. Oesophagus short. In musculature of freshwater fishes.

Metorchis conjunctus (Cobbold, 1860) Looss, 1899 (metacercaria) (Fig. 112)

Site: Musculature.
Hosts: *Catostomus catostomus* (3, 5); *C. commersoni* (1, 2, 4, 5); *Perca flavescens* (4); *Salvelinus fontinalis* (5); *Semotilus corporalis* (5).

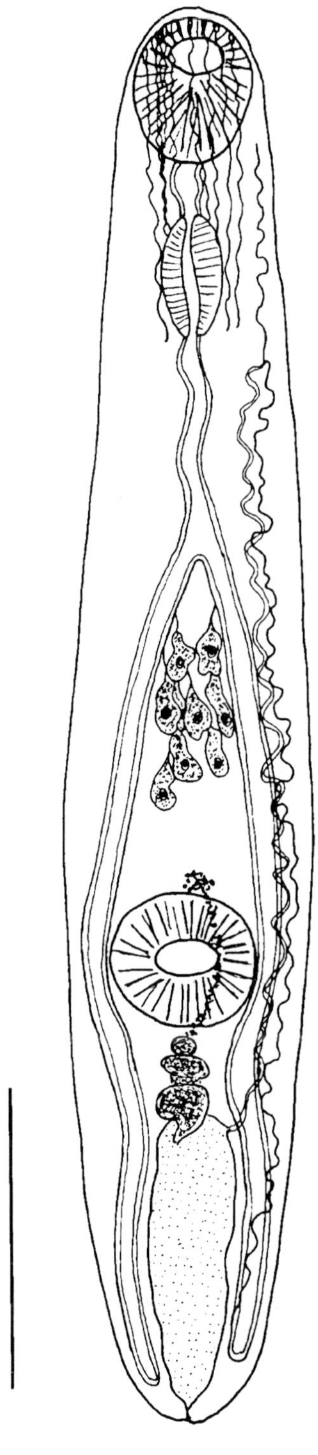

FIG. 111. *Amphimerus elongatus* metacercaria (after Wallace (1939)): spines on body surface not shown. Scale bar: 100 µm.

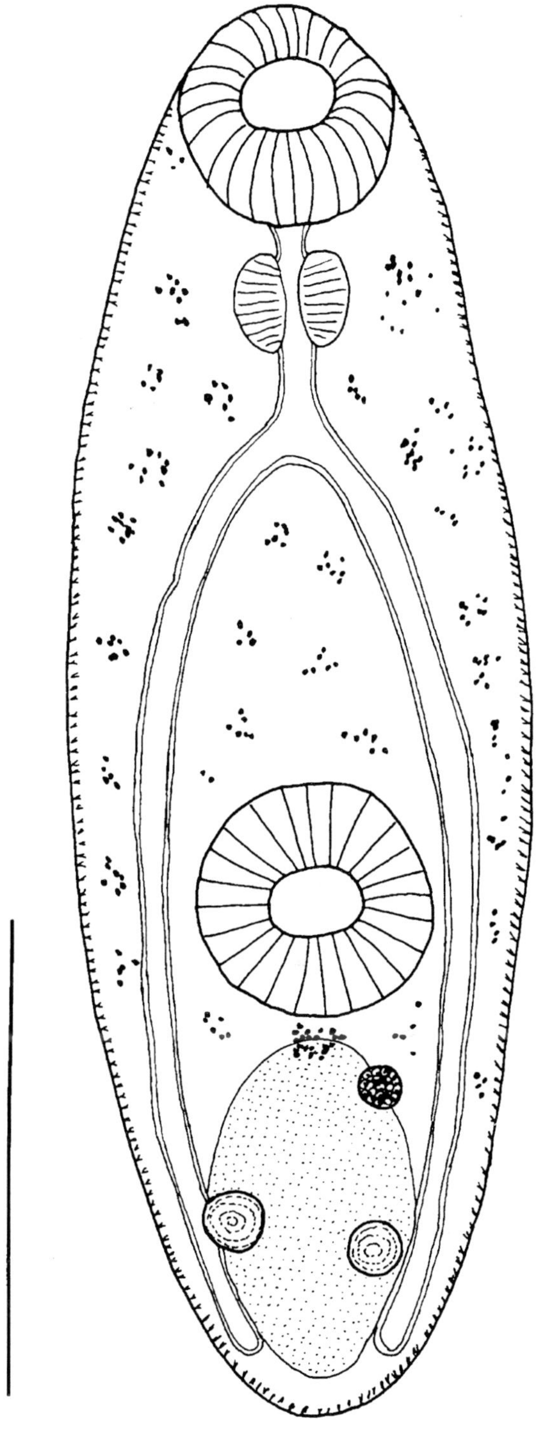

FIG. 112. *Metorchis xanthosomus* (Creplin, 1846) metacercaria (after Razmashkin (1978)). Scale bar (roughly estimated): 100 µm.

Distribution: Man, Que, Sask.

Records: 1. Cameron 1944 (Que); 2. Cameron 1945 (unspecified locality); 3. Rawson 1960 (Sask); 4. Evans (1963 MSc thesis) cited in Lubinsky 1976, Lubinsky and Loch 1979, and Watson 1981 (Man, Que).

Remarks: According to Watson (1981), *C. commersoni* is the main host. This author also questioned Rawson's record from *C. catostomus*. The identification of Cameron's (1944) material was confirmed by experimental infections. Watson also confirmed that these metacercariae occur mainly in the musculature. The metacercaria of *M. xanthosomus* is used to illustrate this genus.

PARAMETORCHIS Skrjabin, 1913

Opisthorchiidae. Metacercaria not described. ?Encysted in the tissues of freshwater fishes.

Comments

Several species of *Parametorchis* have been reported from Canadian mammals. For example, Kennedy (1983) recorded *P. intermedius* Price, 1939 and *P. canadensis* Price, 1939, but their metacercariae have not yet been found.

Parametorchis complexus (Stiles and Hassal, 1894) Skrjabin, 1913 (metacercaria)

Syn.: *Parametorchis noveboracensis* Hung, 1926, *P. manitobensis* Allan and Wardle, 1934.

Site: Not specified.

Host: Unspecified fish.

Distribution: Prairie provinces.

Record: Parnell 1934.

Remarks: Cameron (1940, 1944) considered *P. noveboracensis*, *P. manitobensis*, *P. canadensis* Price, 1929, and *P. intermedius* Price, 1929 all synonyms of *Metorchis conjunctus*. However, Kennedy (1983) disagreed and considered the former two synonyms of *P. complexus* and the latter two as valid species.

SUPERFAMILY PLAGIORCHIOIDEA LÜHE, 1901

Plagiorchiida. Normally present in fish as adult, occasionally as metacercaria (see Troglotrematidae). Body usually small, occasionally larger, spherical to elongate. Tegument usually armed, may be unarmed especially when not parasitic in gut. Suckers normally present. Ventral sucker rarely reduced or absent, usually in anterior half of body, occasionally near middle. Prepharynx and pharynx may be absent. Oesophagus present. Intestinal bifurcation normally in forebody. Caeca two, short to long, blind. Testes normally two, occasionally many, normally in hindbody, symmetrical, oblique, or tandem. Cirrus sac absent or more often present. Seminal vesicle, pars prostatica, and ejaculatory duct internal when cirrus sac present. Genital atrium large to absent. Genital pore variable in position between postero-lateral to ventral sucker and antero-dorsal to oral sucker. Ovary normally in hindbody, pre-testicular. Laurer's canal and canalicular (most cases) or uterine seminal receptacle normally present. Uterus coiled throughout much of hindbody. Eggs numerous. Vitellarium usually follicular, occasionally reduced to two symmetrical masses. Excretory vesicle V-, Y-, or I-shaped; pore terminal. Parasitic in intestine, occasionally urinary system, of vertebrates.

Comments

This superfamily contains a mix of families that tend to be similar in morphology and/or life history and is often used as a repository for groups of uncertain affinities.

KEY TO THE FAMILIES OF THE PLAGIORCHIOIDEA

1 Encysted in tissues as metacercariae (NB. metacercariae of other families, especially microphallids, are occasionally reported, presumably as temporary accidental infections in the gut of some teleosts); in anadromous salmonids off coast of British Columbia and in salmonids and other freshwater teleosts in Northwest States of United States .... Troglotrematidae

As adults; occasionally as excysted metacercariae or juveniles 2

2 Parasitic in urinary system; body unarmed; vitellarium in form of two symmetrical masses; cirrus sac absent; pharynx absent .... Gorgoderidae

Parasitic in intestine; body armed (spines may be lost in poorly preserved material); vitellarium follicular; cirrus sac present or absent; pharynx present 3

3 Caeca long, reaching into post-testicular field; uterus extending most of length of body; genital pore median or submedian immediately anterior to ventral sucker; usually develops into ovigerous adults .... Macroderoididae

Caeca short, usually ending anterior to testes, never extending into post-testicular field; uterine field, if developed, restricted to post-caecal region; genital pore sinistro-lateral to ventral sucker; usually immature in fishes .... Microphallidae

Family GORGODERIDAE Looss, 1899

Plagiorchioidea. Body large to small, elongate to discoidal, hindbody often much wider than forebody. Body surface unarmed. Oral sucker terminal or subterminal, subglobular. Ventral sucker subglobular, semi-embedded, normally in anterior half of body and at junction of narrow and wide parts of body when these are differentiated. Prepharynx and pharynx present only in Anaporrhutinae (=Probolitrematinae). Oesophagus short to long. Intestinal bifurcation in forebody. Caeca long, terminating blindly in posterior region of hindbody. Testes two to numerous, entire to deeply lobed. Usually near middle of hindbody. Cirrus sac absent. Seminal vesicle small, saccate, naked, normally in posterior forebody. Pars prostatica and ejaculatory duct present. Genital atrium small. Genital pore mid-ventral in forebody, often closely post-bifurcal. Ovary entire or lobed, pre-testicular or at level of anterior testis. Blind seminal receptacle or uterine seminal receptacle plus Laurer's canal present. Uterus usually extending throughout hindbody, occasionally restricted to inter-caecal field. Eggs numerous, relatively small. Vitellarium two symmetrical or almost symmetrical masses, entire, lobed or branched, in anterior hindbody. Excretory pore terminal, vesicle tubular, I-shaped, occasionally Y-shaped. Parasitic in urinary bladder and ureters of teleosts, amphibians, and reptiles (one genus in gall bladder of latter) or body cavity of elasmobranchs (Anaporrhutinae).

Subfamily PHYLLODISTOMINAE Nybelin, 1926

Gorgoderidae. Forebody tapered; hindbody lanceolate, spatulate, discoidal, or foliate. Ventral sucker well developed, at junction of narrow forebody and wider hindbody. Testes two entire to lobed masses or follicular, near middle or just inside anterior half of hindbody. Ovary entire or lobed, usually but not always pretesticular or lateral to anterior testis. Uterine seminal receptacle and Laurer's canal present. Uterus occupies most of hindbody, occasionally confined to inter-caecal field. Vitellarium two masses, symmetrical or subsymmetrical, immediately posterior to ventral sucker. Excretory vesicle I-shaped. Parasitic in urinary bladder and ureters of teleosts and amphibians.

PHYLLODISTOMUM Braun, 1899

Phyllodistominae. Body divided at level of ventral sucker into two regions; forebody tapered, often narrow; hindbody wider, lanceolate, spatulate, discoidal, or foliate. Testes two entire or lobed masses. Vitelline masses entire or lobed. Parasitic in urinary bladder and ureters of teleosts and amphibians.

Comments

The species of the genus *Phyllodistomum* are in dire need of revision; but such studies, as have been carried out by Bakke (1984, 1985) and Bakke and Bailey (1987) for *P. umblae* (Fabricius, 1780), must await detailed redescriptions of the various forms attributed to the genus. After several years of studying forms mainly from salmonoids, Bakke (pers. comm.) is of the opinion that "no clear way of distinguishing adults of the genus has yet been found". *Phyllodistomum* spp. are known for their considerable variation in features, such as egg size, sucker ratio, body shape, etc., considered as diagnostic in other genera. Many of these features are severely affected by fixation techniques, flattening, and the condition of the worm at fixation (see Bakke 1988). Bakke and Bailey's recent (1987) detailed study of Canadian species from salmonoids solves some of the problems, but still leaves unanswered questions concerning some of the nominal species from this group of hosts. Bakke and Bailey's work demonstrates that the situation in North America cannot be looked at in isolation, since, as in the case of the nominal species from salmonoids, they may be conspecific with European forms. This poses further problems, as the situation in Europe is confused. Even the type species of the genus is not well known.

Earlier keys to North American species of *Phyllodistomum* by Pearse (1924), Holl (1929), van Cleave and Mueller (1934), and Lewis (1935) are uncritical and tend to distinguish the nominal species, many of which are poorly known and inadequately described, by features that tend to be susceptible to treatment during or prior to fixation or to ontogenetic variation. As it would seem pointless to provide another rather impractical key, I have taken rather drastic action based upon the following assumptions:

(1) Strong evidence from species such as *P. umblae* and *P. megalorchis* Nybelin, 1926 and indications from other species in European waters suggests that some degree of host group specificity (stenoxeny) is involved.

(2) Work on *P. umblae* indicates that species of the European fauna must be taken into consideration.

(3) A drastic reduction in the number of nominal species would not result in a loss of ecological data, since many of the species and many of the records are very questionable.

(4) The synonymy of questionable species would mean that future workers wishing to recognize them would have to justify their action.

Although this type of action would not be appropriate in most instances, I feel that the following key, which is based upon host specificity and gross morphology, does create some sort of order out of the current chaos until major redescriptions and a complete revision, possibly along the lines indicated by Cribb (1987a,b), are available.

Key to the species of *Phyllodistomum*

1 Body lanceolate, with maximum width of hindbody in unflattened specimens usually less than twice maximum width of forebody; parasites of salmonoids ....*Phyllodistomum umblae* (Fabricius, 1780) Bakke, 1982 (Fig. 113)

Syn.: *P. conostomum* (Olsson, 1876) Looss, 1902; *P. lachancei* Choquette, 1947; *P. limnosa* Sandeman and Pippy, 1967; *P. coregoni* of Leong and Holmes (1974, 1981).

Sites: Ureters, urinary bladder.

Hosts: *Coregonus clupeaformis* (5, 6, 9); *Oncorhynchus mykiss* (12, 13); *O. nerka* (11, 12, 14, 15, 17); *Salmo salar* (3, 4, 7, 8); *Salvelinus alpinus* (5, 10, 16); *S. fontinalis* (1, 2, 3, 5, 7, 13, 18, 19); *S. namaycush* (7); *Sebastes* spp. (see "Remarks").

Distribution: BC, Lab, Nfld, Que, Ont, Alta.

Records: 1. Choquette 1947 (Que); 2. Choquette 1948 (Que); 3. Sandeman and Pippy 1967 (Nfld); 4. Pippy 1969 (Nfld); 5. Hicks and Threlfall 1973 (Lab); 6. Leong and Holmes 1974 (Alta); 7. Chinniah and Threlfall 1978 (Lab); 8. Pippy 1980 (Nfld); 9. Leong and Holmes 1981 (Alta); 10. Curtis 1984 (Que); 11. Bailey and Margolis 1987 (BC); 12. Bakke and Bailey 1987 (BC); 13. Dechtiar *et al.* 1988 (Ont); 14. Bailey, Margolis, and Groot 1988 (BC); 15. Bailey, Margolis, and Workman 1989 (BC); 16. Bouillon and Dempson 1989 (Lab); 17. Groot *et al.* 1989 (BC); 18. Wright *et al.* 1989 (Que); 19. Albert and Curtis 1991 (Que).

Remarks: Preliminary work on the nominal species from salmonoids in Canadian waters, including the examination of the types of *P. lachancei* in the BM(NH) collection, convinced me that *P. lachancei* and *P. limnosa* are conspecific, as indicated by Chinniah and Threlfall (1978). Bakke and Bailey's (1987) detailed work on this group confirmed these as synonyms of the Palaearctic species *P. umblae*, although they did pose a slight question against *P. lachancei*, considering it as "probably synonymous". In my preliminary study I was also inclined to consider *P. coregoni* synonymous with *P. lachancei* and *P. limnosa*; but Bakke and Bailey refrained from this synonymy because of the anterior position of the testes described for *P. coregoni* and the strange arrangement of the papillae on the suckers in Dechtiar's (1966) description. Bakke (pers. comm.) is of the opinion that the papillae arrangement on the ventral sucker of *P. coregoni* differs considerably from *P. umblae*, and yet that of *P. umblae* is identical to other species of *Phyllodistomum* and the unrelated gorgoderid *Gorgoderina vitelliloba* from amphibians. It is clear, therefore, that a detailed study of *P. coregoni* from type host and locality is required. In the meantime, I have treated it as a *species inquirenda* and listed its records below. I have examined Leong and Holmes' (1974, 1981) specimens

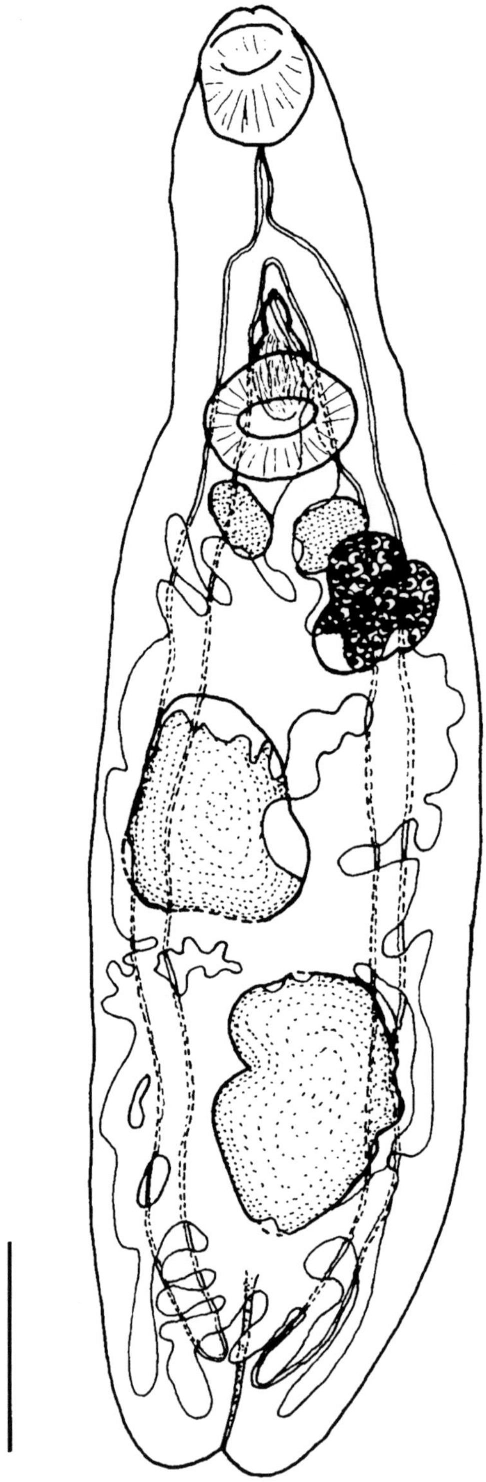

FIG. 113. *Phyllodistomum umblae* (after Bakke (1984)). Scale bar: 500 μm.

of *P. coregoni* and consider them to be flattened specimens of *P. umblae* that appear to have been removed from frozen hosts.

Additional records of specimens identified as *Phyllodistomum* sp. have been reported from *Sebastes* spp. off the coast of British Columbia. These are four specimens from *Sebastes caurinus*, one specimen from *S. diploproa*, and one specimen from *S. ruberrimus*, site not given, all immature, by Sekerak and Arai (1977). In addition, through the kindness of Dr. J.C. Holmes, I have been able to examine three mature specimens that he collected from *S. nebulosus* (University of Alberta parasite collection 11277-1279) (see Holmes 1990). The latter specimens key down reasonably well to *P. umblae*. In view of this, and their site, I consider it best to consider them as accidental parasites of *Sebastes* spp. acquired from migratory salmonoids in their diet, rather than an endemic parasite of these marine hosts.

There are also what appear to be non-original records of this species in *Salvelinus* spp. from Quebec/Labrador by Curtis (1988) and, as *P. limnosa*, in *S. fontinalis* from Labrador and Newfoundland by Marcogliese and Cone (1991a).

Possibly identical species:

P. coregoni Dechtiar, 1966, sp. inq.
Sites: Ureters, urinary bladder.
Hosts: *Coregonus clupeaformis* (1, 2, 3, 4, 5); *Prosopium cylindraceum* (2).
Distribution: Ont.
Records: 1. Dechtiar 1966 (Ont); 2. Chinniah and Threlfall 1978 (Lab); 3. Dechtiar and Lawrie 1988 (Ont); 4. Dechtiar *et al.* 1988 (Ont); 5. Dechtiar and Christie 1988 (Ont).

Body discoidal, spatulate, or foliate, with maximum width of hindbody usually more than twice maximum width of forebody 2

2 Uterine fields usually restricted in inter-caecal fields even in fully developed worms; parasitic in catostomids .... *P. lysteri* Miller, 1940
Site: Ureters.
Hosts: *Carpiodes cyprinus* (7); *Catostomus catostomus* (6); *C. commersoni* (1, 2, 3, 4, 6, 7, 8, 9); *Moxostoma macrolepidotus* (6).
Distribution: Ont, Que.
Records: 1. Miller 1940a (Que); 2. Dechtiar 1972a (Ont); 3. Dechtiar 1972b (Ont); 4. Molnar *et al.* 1974 (Ont); 5. Dechtiar and Lawrie 1988 (Ont); 6. Dechtiar *et al.* 1988 (Ont); 7. Dechtiar and Nepszy 1988 (Ont); 8. Dechtiar and Christie 1988 (Ont); 9. Dechtiar *et al.* 1989 (Ont).
Remarks: The distribution of the uterus may not be a good criterion, because it may be affected by pressure, fixation technique, or ontogenetic development. I can, however, confirm that the uterus is inter-caecal in the holotype and paratype (labelled cotype) in the BM(NH) collection.

Uterine fields reach extra-caecally in full-developed worms 3

3 Hindbody widest in middle or anterior region, with shape of irregular triangle or maple leaf; parasitic in percoids .... *P. superbum* Stafford, 1904
Syn.: (?) *P. fausti* Pearse, 1924; *P. pearsei* Holl, 1929; *P. lohrenzi* (Loewen, 1935) Bhalerao, 1937.
Sites: Ureters, urinary bladder.
Hosts: *Ambloplites rupestris* (7); *Aplodinotus grunniens* (6); *Micropterus dolomieui* (2, 5); *Perca flavescens* (1, 3, 4, 5); *Stizostedion canadense* (3); *S. vitreum vitreum* (3).
Distribution: Ont, Que.
Records: 1. Stafford 1904 (Que?); 2. Fantham and Porter 1948 (Que); 3. Dechtiar 1972b (Ont); 4. Dechtiar and Lawrie 1988 (Ont); 5. Dechtiar *et al.* 1988 (Ont); 6. Dechtiar and Nepszy 1988 (Ont); 7. Dechtiar and Christie 1988 (Ont).
Remarks: All the above records are listed under *P. superbum*. The descriptions of this species appear morphologically very close to a Czechoslovakian specimen in the BM(NH) collection, labelled *P. angulatum* Linstow, 1907, from a percid. It is clear that the relationship between *P. superbum* and *P. angulatum* requires investigation. Its similarity with the other European species from percids, *P. pseudofolium* Nybelin, 1926, should also be investigated.

Some of the centrarchid records are listed under *P. lohrenzi*. Detailed study is required to determine whether the forms from centrarchids are different from those from percids.

Other records listed under *P. superbum*:

Hosts: *Ameiurus nebulosus* (3); *Catostomus commersoni* (4); *Esox lucius* (2, 4); *E. niger* (4); *Salvelinus fontinalis* (4).

Distribution: Ont, Que.

Records: 1. Stafford 1904 (Que?); 2. Lyster 1939 (Que); 3. Bangham and Hunter 1939 (Ont); 4. Fantham and Porter 1948 (Que).

Remarks: The records from salmonoids, ictalurids, and catostomids may be *P. umblae*, *P. staffordi*, and *P. lysteri*, respectively. Due to their carnivorous nature, it may well be that esocids can harbour several species, although Lyster's (1939) description does show some resemblance to *P. superbum*. It should not be forgotten, however, that the European cyprinid form *P. folium* (Olfers, 1816) occurs regularly in *Esox*.

Hindbody widest in middle, with discoidal outline..4

4 Parasitic in ictalurids..*P. staffordi* Pearse, 1924

Syn.: *P. folium* (Olfers, 1816) (in part) of Stafford (1902); *P. superbum* Stafford, 1904 (in part); (?) *P. carolini* Holl, 1929; *P. lacustri* of Dechtiar (1972a) and Dechtiar and Nepszy (1988); (?) *P. hunteri* (Arnold, 1934).

Sites: Ureters, urinary bladder.

Hosts: *Ameiurus nebulosus* (1, 2, 3, 4, 5, 6, 7, 8, 9, 11); *Ictalurus melas* (5); *I. punctatus* (10).

Distribution: BC, Ont, Que.

Records: 1. Stafford 1902 (?); 2. Stafford 1904 (?Que); 3. Bangham 1941 (Ont); 4. Bangham 1955 (Ont); 5. Bangham and Adams 1954; 6. Dechtiar 1972a (Ont); 7. Dechtiar 1972b (Ont); 8. Dechtiar and Lawrie 1988 (Ont); 9. Dechtiar *et al.* 1988 (Ont); 10. Dechtiar and Nepszy 1988 (Ont); 11. Dechtiar and Christie 1988 (Ont).

Remarks: Numerous morphologically similar nominal species have been described from ictalurids in North America. Despite the minor morphological differences used to justify the erection of these forms, I consider it most prudent to consider many of these as synonyms or probable synonyms of *P. staffordi*. Future work may show that *P. lacustri* (Loewen, 1929) and *P. caudatum* Steelman, 1938 may be added to the list.

Other records from esocoids listed under *P. folium*, *P. staffordi*, and *P. brevicaecum* Steen, 1938 that appear to resemble *P. staffordi*, although no detailed description exists.

Hosts: *Esox lucius* (1, 3); *E. masquinongy* (2); *Umbra limi* (4).

Distribution: Que, Ont.

Records: 1. Stafford 1904 (?Que); 2. Lyster 1939 (Que); 3. Miller 1941a (Stafford's, 1904, material); 4. Dechtiar *et al.* 1988 (Ont).

Remarks: See "Remarks" on esocid records of *P. superbum*. Lyster's figure resembles *P. staffordi* more than *P. superbum*. *P. brevicaecum*, which Dechtiar *et al.* recorded from *Umbra limi*, requires confirmation. This species resembles *P. staffordi* and its so-called short caeca are not especially short when one considers them in relation to the posterior margin of the testes.

Parasitic in cottids ....*P. simile* Nybelin, 1926

Syn.: *P. undulans* Steen, 1938.

Sites: Ureter, urinary bladder.

Host: *Cottus bairdi*.

Distribution: Ont.

Records: Dechtiar 1972b; Dechtiar *et al.* 1988.

Remarks: The description of *P. undulans* appears morphologically very similar to the syntype specimens of the European species *P. simile* in the BM(NH) collection. Because both occur in cottids, I have considered them synonymous. In gross morphology, *P. undulans* is also similar to *P. staffordi*.

Parasitic in cyprinids ....*P. semotili* Fischthal, 1942

Syn.: (?) *P. notropidus* Fischthal, 1942.

Site: Ureters.

Hosts: *Luxilis cornutus*, *Semotilus atromaculatus*.

Distribution: Ont.

Records: Dechtiar *et al.* 1989.

Remarks: There is very little information on the variability of the nominal species of *Phyllodistomum* from North American cyprinids. Little can be done to clarify the position until a detailed study has been undertaken. This would need to take into account other forms recorded from North American cyprinids, such as *P. nocomis* Fischthal, 1942, those from European cyprinids and species, such as *P. lysteri* and *P. staffordi*, from the related cypriniform and siluriform groups, the catostomids and ictalurids.

Family LECITHODENDRIIDAE Lühe, 1901

Fantham and Porter (1948) referred to a single specimen of "*Pleurogenes* sp." from the intestine of *Salvelinus fontinalis* in Quebec. In agreement with Hoffman (1967) and Margolis and Arthur (1979), this is most likely to have been an accidental infection. *Pleurogenes* spp. are parasites of amphibians but, according to Prudhoe and Bray (1982), this genus does not occur in North America. The worm may, therefore, have been a young specimen of the related *Loxogenoides bicolor* (Krull, 1933), which occurs in North American frogs.

Family MACRODEROIDIDAE McMullen, 1937

Plagiorchioidea. Body oval to elongate. Tegument spinose, slightly enlarged spines on oral sucker occasionally present. Suckers well developed, large or small, often similar in size, oral sucker ventro-terminal, ventral sucker in anterior half of body. Prepharynx present, short to long. Pharynx well developed. Oesophagus short to long. Intestinal bifurcation near mid-forebody. Caeca terminate blindly between level of testes and posterior extremity. Testes two, tandem, oblique, or occasionally symmetrical, in middle or posterior half of body. Cirrus sac present, small to large, oval to claviform, containing bipartite, occasionally saccular, seminal vesicle, pars prostatica, and unarmed ejaculatory duct. Genital atrium insignificant. Genital pore median or submedian, immediately anterior to ventral sucker or at least post-bifurcal. Ovary entire or lobed, submedian or median, between ventral sucker and testes, rarely dorsal to ventral sucker. Canalicular seminal receptacle or uterine seminal receptacle present. Laurer's canal present. Uterine field extending full length of hindbody, often present between testes. Eggs usually numerous, small to medium-sized, operculate. Vitellarium follicular, lateral fields vary in extent, but distribution often limited, sometimes extending into forebody, sometimes extending into post-testicular zone, rarely extending full length of body or confluent in post-testicular zone. Excretory pore terminal, vesicle I-shaped (may appear Y-shaped in juveniles), short to long, tubular or saccular. In intestine of freshwater teleosts and holosteans, reptiles, and occasionally in invertebrates.

Comments

This is a family that is in dire need of revision. Based upon cercarial morphology, the group is clearly related to the plagiorchiids, differing in adult morphology only in the shape of the excretory vesicle that is I-shaped rather than Y-shaped. This feature, however, is not always straightforward, because in some young macroderoidids it may be Y-shaped, and the enlargement of the proximal extremity of the vesicle, as it gives rise to the main collecting ducts, has been interpreted as giving the vesicle a Y-shape. In adult morphology the worms are morphologically very similar to some lepocreadioid groups, but can be distinguished in the majority of instances by the posterior extent of the uterus (the exception being the deropristid genus *Pristicola* Cable, 1952 from acipenserids in the United States).

In view of the above, I have refrained from any comment with regard to subfamilies; but the above definition excludes the Walliniinae, a group often linked with the macroderoidids even though they possess an unarmed tegument.

Key to the genera of the Macroderoididae

1 Minute, pyriform to oval worms with symmetrical testes..*Vietosoma*

Small to medium-sized, oval to elongate worms with tandem or oblique testes 2

2 Excretory vesicle long, extending to about level of ovary. ....*Alloglossidium*

Excretory vesicle short (in mature worms), not reaching middle of posterior testis ....*Macroderoides*

ALLOGLOSSIDIUM Simer, 1929

Syn.: *Parastiotrema* Miller, 1940.

Macroderoididae. Body fusiform to elongate. Tegument spinose. Oral sucker ventro-terminal. Ventral sucker well inside anterior half of body. Prepharynx short to relatively long. Pharynx well developed. Oesophagus short to relatively long. Caeca terminating blindly in post-testicular zone, usually in anterior half of zone. Testes tandem to oblique, in middle or posterior half of body. Cirrus sac oval to claviform, often small, may extend into hindbody, contains bipartite seminal vesicle, pars prostatica, and unarmed ejaculatory duct. Genital pore median or submedian, immediately anterior to ventral sucker. Ovary between anterior testis and ventral sucker, entire to lobed. Canalicular seminal receptacle absent. Uterine seminal receptacle and Laurer's canal present. Uterine field extends full length of hindbody, mainly intercaecal. Eggs numerous, operculate, relatively small. Vitellarium follicular, in lateral fields between limits of testicular zone and oral sucker. Excretory pore terminal, vesicle I-shaped, tubular, reaching to about level of ovary. In intestine of ictalurid freshwater teleosts and hirudinids, occasionally in antennary glands of freshwater decapod crustaceans.

Comments

A phylogenetic analysis of this genus has been carried out recently by Carney and Brooks (1991).

I can see no reason for recognizing *Parastiotrema* Miller, 1940 as a distinct genus (see below under *A. corti*). Other species of *Alloglossidium* occur in Canadian waters in invertebrates but not in fishes (see Carney and Brooks 1991).

Key to the species of *Alloglossidium*

1 Vitelline fields extend into forebody; ovary indistinctly lobed ....
..................................*Alloglossidium corti* (Lamont, 1921) Van Cleave and Mueller, 1934 (Fig. 114)

Syn.: *Parastiotrema ottawanensis* Miller, 1940.
Site: Intestine.
Hosts: *Ameiurus nebulosus* (3, 5); *Ictalurus punctatus* (2, 4, 7, 8, 9); *Noturus flavidus* (9); *N. gyrinus* (1, 6, 9).
Distribution: BC, Ont, Que.

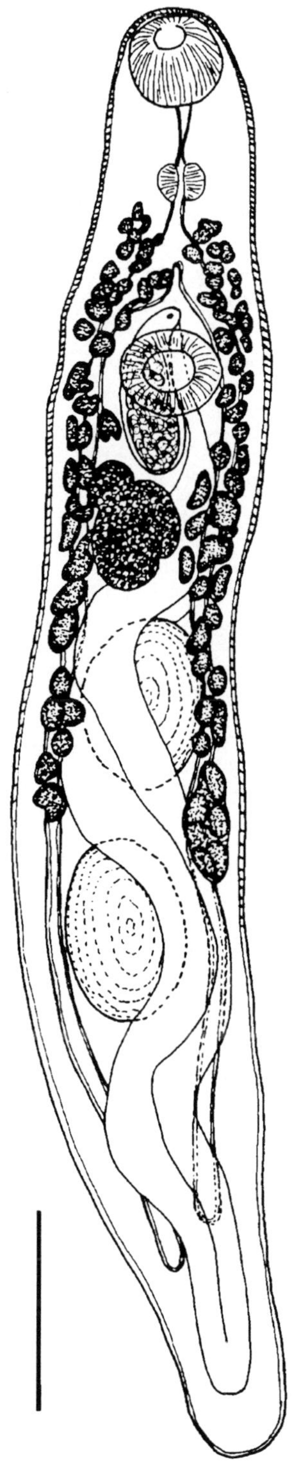

FIG. 114. *Alloglossidium corti* (after Van Cleave and Mueller (1934)). Scale bar: 300 μm.

Records: 1. Bangham and Hunter 1939 (Ont); 2. Miller 1940a (Que); 3. Bangham 1941 (Ont); 4. Bangham 1955 (Ont); 5. Bangham and Adams 1954 (BC); 6. Dechtiar 1972b (Ont); 7. Dechtiar *et al.* 1988 (Ont); 8. Dechtiar and Nepszy 1988 (Ont); 9. Dechtiar and Christie 1988 (Ont).

Remarks: I have listed *Parastiotrema ottawanensis*, described by Miller (1940a) from *Ictalurus punctatus* in the Ottawa River, as a synonym of *A. corti*. This action was taken because (1) this species has never subsequently been recorded and was originally described from two obviously contracted specimens; (2) as indicated above, I consider *Parastiotrema* a synonym of *Alloglossidium*; (3) *A. corti* is the normal species of *Alloglossidium* that occurs in *I. punctatus* and Miller (1940a) did record this species from this host in the same locality; and (4) taking into account the contracted nature of the worms, the distribution of the vitellarium and the shape of the ovary are closer to Miller's figures of *A. corti* than to his figure of *A. gemimum*.

I have examined the holotype of *P. ottawanensis* in the BM(NH) collection and can confirm that the ovary is more distinctly tri-lobed than indicated in Miller's figure of the same worm and the caeca extend further posteriorly than he figured. The large size of the cirrus sac in this worm suggests that it is perhaps more likely to have been a species of *Macroderoides* than *Alloglossidium*. The species of *Macroderoides* that is morphologically most similar is *M. flavus*, but the body shape, size, vitelline distribution, arrangement of the testes, etc., are all inconsistent with this species. In fact the cirrus sac of *A. corti* is longer than those of its congeners and it is possible that the apparent size of the cirrus sac in the type specimen of *P. ottawanensis* is because it is a young, contracted worm.

Vitelline fields not extending into forebody; ovary entire ....
....*A. gemimum* (Mueller, 1930) Van Cleave and Mueller, 1932

Syn.: *Glossidium gemimum* (Mueller, 1930) Yamaguti, 1953.

Site: Intestine.

Hosts: *Ameiurus nebulosus* (1, 2, 3, 4, 5, 6, 7, 8); *Noturus flavus* (8).

Distribution: Ont, Que.

Records: 1. Miller 1940a (Que); 2. Bangham 1941 (Ont); 3. Bangham 1955 (Ont); 4. Bangham and Venard 1946 (Ont); 5. Dechtiar 1972b (Ont); 6. Molnar *et al.* 1974 (Ont); 7. Fréchette *et al.* 1978 (Que); 8. Dechtiar and Christie 1988 (Ont).

Remarks: The systematic position of this species was discussed by Timmers (1979) and Vande Vusse (1980). *A. nebulosus* appears to be the main host.

MACRODEROIDES Pearse, 1924

Macroderoididae. Body fusiform to elongate. Tegument spinose. Oral sucker ventro-terminal. Ventral sucker relatively small, in anterior half of body. Prepharynx short to relatively long. Pharynx small. Oesophagus short to relatively long. Caeca terminate blindly in posterior half of post-testicular zone. Testes tandem to oblique, occasionally almost symmetrical, in posterior half of body. Cirrus sac claviform, large, extends into hindbody, contains bipartite seminal vesicle, well-developed pars prostatica, and muscular, unarmed ejaculatory duct. Genital pore median or submedian, immediately anterior to ventral sucker. Ovary entire or lobed, median or submedian, usually about halfway between ventral sucker and anterior testis. Canalicular seminal receptacle (?) absent (see "Comments" below). Uterine seminal receptacle and Laurer's canal reported. Uterine field reached full length of hindbody, mainly inter-caecal. Eggs numerous, small to medium-sized, operculate. Vitellarium follicular, lateral fields extend between level of ventral sucker or just posterior and level between anterior testis and posterior extremity. Excretory pore terminal, vesicle I-shaped, tubular or saccular, short, not reaching posterior testis (may be Y-shaped in juvenile worms). In intestine or holosteans and teleosts in fresh water.

Comments

This genus is in need of revision and the validity of some species requires confirmation. As can be judged by the key presented below, some of the diagnostic criteria are questionable. There is a clear need for information on host specificity, host-induced variation, and fixation artifacts. The problem is exacerbated by the poor condition of some of the type and voucher material in the USNM collection, which is in a variety of states of preservation and contraction.

The presence or absence of a canalicular seminal receptacle in *M. spiniferus* requires confirmation. Yamaguti (1971) figured this organ, and in some whole mounts that I have seen, including that figured by Yamaguti, a small sac which may be the seminal receptacle is visible; but, even after seeing the sections in the USNM collection, I am not yet entirely convinced that this does not represent a saccular loop of the uterus.

Key to the species of *Macroderoides*

1 Medium-sized worms of up to 3 mm in length; body elongate; length at least six times breadth (depending upon fixation and condition); testes in tandem, usually well separated by loops of uterus; cirrus sac long, extending deep into hindbody; normally parasitic in holosteans 2

Small worms of less than 1.5 mm in length; body fusiform to elongate; length less than six times breadth; testes oblique, slightly separated by loops of uterus; cirrus sac long or short, depending upon degree of eversion of cirrus; parasitic in holosteans or teleosts 3

2 Fully formed eggs 40–46 µm in length; ovary entire or indistinctly to clearly lobed; usually in *Lepisosteus*...*Macroderoides spiniferus* Pearse, 1924

Site: Gut.

Host: *Lepisosteus osseus*.

Distribution: Ont.

Record: Bangham and Hunter 1939.

Remarks: The work of Simer (1929) and Leigh (1958) indicates that this species may have a lobed ovary, although in material that I have seen the ovary is entire to indistinctly lobed. This species has been recorded in *Amia* in the United States (I have seen material of Ward from *A. calva* in Michigan in the USNM Collection), although *Lepisosteus* spp. appear to be its main hosts.

Pearse (1924) claimed to have found this species in *Ictalurus* spp. and there is a Canadian record of *Macroderoides* sp. from *Ictalurus punctatus* by Bangham and Hunter (1939): it seems perhaps more likely that, at least the latter record, refers to a species of *Alloglossidium*.

Fully formed eggs normally 20–30 µm in length; ovary distinctly 3- to 4-lobed; in *Amia* ....*M. trilobatus* Taylor, 1978

Site: Intestine.

Host: *Amia calva*.

Locality: Ont.

Record: BM(NH) collection.

Remarks: This record is represented by a single specimen in poor condition from Lake Ontario. This species is known only from *Amia calva*, the previous records being from Georgia (Taylor 1978) and, as *M. spiniferus*, Wisconsin (Amin 1982b). Amin claimed that the eggs of his specimens were 22–54 µm in length, but in his voucher specimens in the USNM collection I found fully formed eggs to be in the range of 22–30 µm.

3 Forebody narrower than hindbody; cirrus sac usually long and extending deep into hindbody; in *Esox* ....*M. flavus* Van Cleave and Mueller, 1932 (Fig. 115)

Site: Intestine.

Hosts: *Esox americanus vermiculatus* (2); *E. niger* (1).

Distribution: Ont, Que.

Records: 1. Fantham and Porter 1948 (Que); 2. Crossman 1962 (Ont).

Remarks: Van Cleave and Mueller (1934) suggested that this species is very specific to *E. niger* in Lake Oneida, New York, where it did not occur in *E. lucius*.

Forebody similar in width to hindbody; cirrus sac may be short when cirrus is formed; in *Amia* ....*M. typicus* (Winfield, 1929) Van Cleave and Mueller, 1932

Site: Gut.

Host: *Amia calva.*

Distribution: Ont.

Records: Bangham and Hunter 1939; Bangham 1955; Dechtiar and Christie 1988.

VIETOSOMA Van Cleave and Mueller, 1932

Macroderoididae. Body minute, pyriform to oval, dorso-ventrally flattened, outline may be indented or lobed posteriorly. Tegument spinose. Oral sucker ventro-terminal, significantly larger than ventral sucker. Ventral sucker small, inside anterior half of body. Prepharynx short to similar in length to pharynx. Pharynx relatively small. Oesophagus short to apparently absent. Caeca end blindly just posterior to testes. Testes symmetrical, entire or slightly indented, just inside posterior half of body. Cirrus sac oval, small, antero-dorsal or dorso-lateral to ventral sucker. Genital pore median, immediately anterior to ventral sucker. Ovary large, median, postero-dorsal to ventral sucker, distinctly lobate, dorso-ventrally flattened. Canalicular seminal receptacle absent. Uterine seminal receptacle present. Uterine field extends full length of hindbody, inter-caecal, present between testes, main bulk in post-testicular zone. Eggs not especially numerous, small. Vitellarium follicular, in lateral fields from about level of pharynx to about level of caecal extremities. Excretory pore terminal, vesicle I-shaped (or Y-shaped — see "Comments" below). In intestine of ictalurid freshwater teleosts.

Comments

Until more is known of the biology of these worms it is probably best to retain this genus in the Macroderoididae, although they differ in certain aspects of their morphology from most members of this family. It is worth noting a certain morphological similarity with some species of the genus *Glypthelmins* Stafford, 1904, the species of which occur in amphibians.

Van Cleave and Mueller (1932) gave the shape of the excretory vesicle as Y-shaped in young specimens, but stated that it acquires a number of posteriorly oriented accessory branches in older worms. They figured the excretory vesicle of adult worms, showing a long stem with numerous sub-branches anteriorly and posteriorly. Later (1934), they described it as "a long median unbranched stem..., anterior end of bladder square, or with very slight indication of forking" and "the excretory bladder is not plainly Y-shaped...". In the type specimens from the USNM that I have examined, the excretory vesicle, even in juvenile specimens, appears to be I-shaped. Lyster (1939) described a single specimen from Canadian waters in which he illustrated and stated that the cirrus sac reached back to the level of the testes, and he suggested that the ovary might be a single, small mass and situated in the posterior forebody. These observations do not conform with the type material and require confirmation.

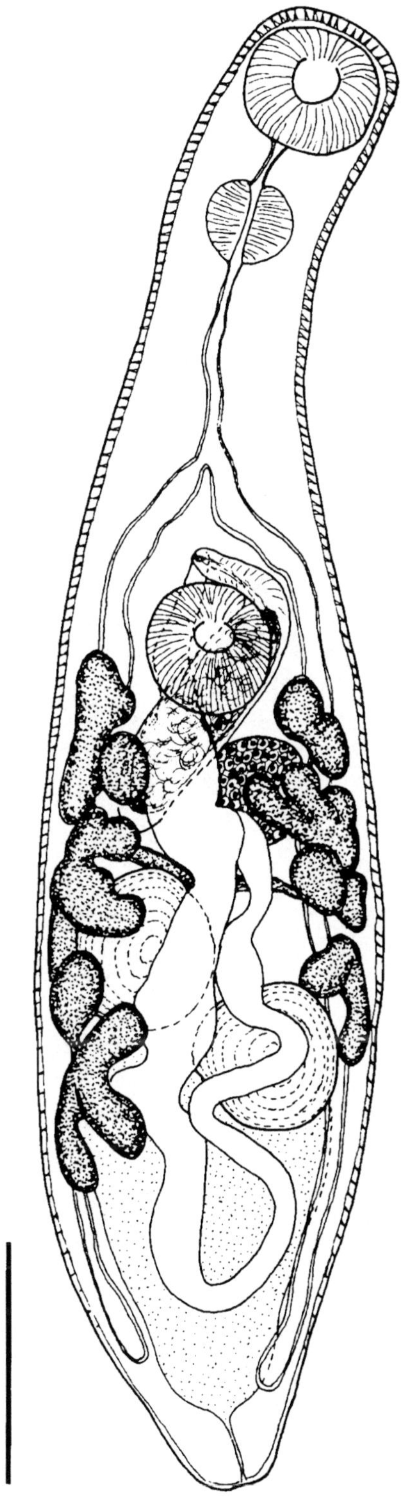

FIG. 115. *Macroderoides flavus* (after Van Cleave and Mueller (1932)). Scale bar: 100 µm.

Vietosoma parvum Van Cleave and Mueller, 1932 (Fig. 116)

Site: Gut.

Hosts: *Ameiurus nebulosus* (3, 4); *Ictalurus punctatus* (1, 2).

Distribution: Ont, Que.

Records: 1. Lyster 1939 (Que); 2. Bangham and Hunter 1939 (Ont); 3. Bangham 1941 (Ont); 4. Bangham and Venard 1946 (Ont).

Remarks: Van Cleave and Mueller (1932) and Becker and Cloutman (1975) have recorded this species in very large numbers (hundreds) in *I. punctatus* in the United States. Van Cleave and Mueller (1932) have suggested that it occurs only accidentally in *A. nebulosus*.

Family MICROPHALLIDAE Ward, 1901

Plagiorchioidea. Body very small, oval, pyriform, linguiform, or elongate. Tegument spinous. Oral sucker well developed, ventrally subterminal. Ventral sucker normally well developed, occasionally doubled or apparently absent, just anterior or more often near or just posterior to middle of body. Prepharynx and pharynx normally present, small to medium-sized. Oesophagus short to long. Intestinal bifurcation in forebody. Caeca blind, divergent, short to medium-length, not extending posteriorly to testes. Testes two, symmetrical, post-ovarian in hindbody. Cirrus sac present or absent, often large when present, claviform, rectilinear or arcuate, situated transversely between intestinal bifurcation and ventral sucker, containing seminal vesicle, pars prostatica, and ejaculatory duct, may not enclose terminal copulatory organ (in this form Deblock (1971) refers to it as a "poche vésiculo-prostatique"). When cirrus sac absent, seminal vesicle, pars prostatica, and ejaculatory duct free in parenchyma. Copulatory organ of variable form, armed or unarmed, eversible cirrus, protrusible, permanent papilla, or permanent atrial complex. Genital atrium small or large, simple or diverticulate. Genital pore usually lateral to ventral sucker, on opposite side to ovary. Ovary usually dextral, posterior or postero-lateral to ventral sucker and pre-testicular, rarely between testes. Laurer's canal present or absent. Canalicular or oviductal seminal receptacle normally absent, reduced if present; uterine seminal receptacle may be present. Uterus generally post-caecal, reaching close to posterior extremity, occasionally reaching into forebody or even pre-caecally. Metraterm present, variable in form. Eggs numerous, small, operculate. Vitellarium follicular, small number of large follicles or bands or groups of numerous small follicles, arrangement variable, but often grouped in lateral fields in hindbody or less often in forebody, rarely confluent posteriorly. Excretory pore terminal, vesicle V- or Y-shaped. Adults in gut of birds, mammals, reptiles, amphibians, and fishes.

Comments

Members of this group are not normally parasitic in Canadian fishes. I suspect that all records are due to the accidental ingestion of metacercariae and that records of ovigerous forms are due to metacercarial progenesis (see "Remarks" on *Microphallus opacus*). Other unidentified records in Canadian waters are *Microphallus* sp. of Sekhar and Threlfall (1970a) from the intestine and stomach of the marine fish *Tautoglabrus adsperus* off the Atlantic coast and "Microphallidae gen. sp." of Bangham (1941) from an unspecified site in *Lota lota*, *Micropterus dolomieui*, and *Perca flavescens* in Ontario.

The definitions of the microphallids used here are based upon the works of Deblock (1971, 1973).

Key to the subfamilies of the Microphallidae

1 Cirrus sac absent; copulatory organ a protrusible male papilla or atrial organ (this feature may not be visible in immature specimens from fish)..Microphallinae

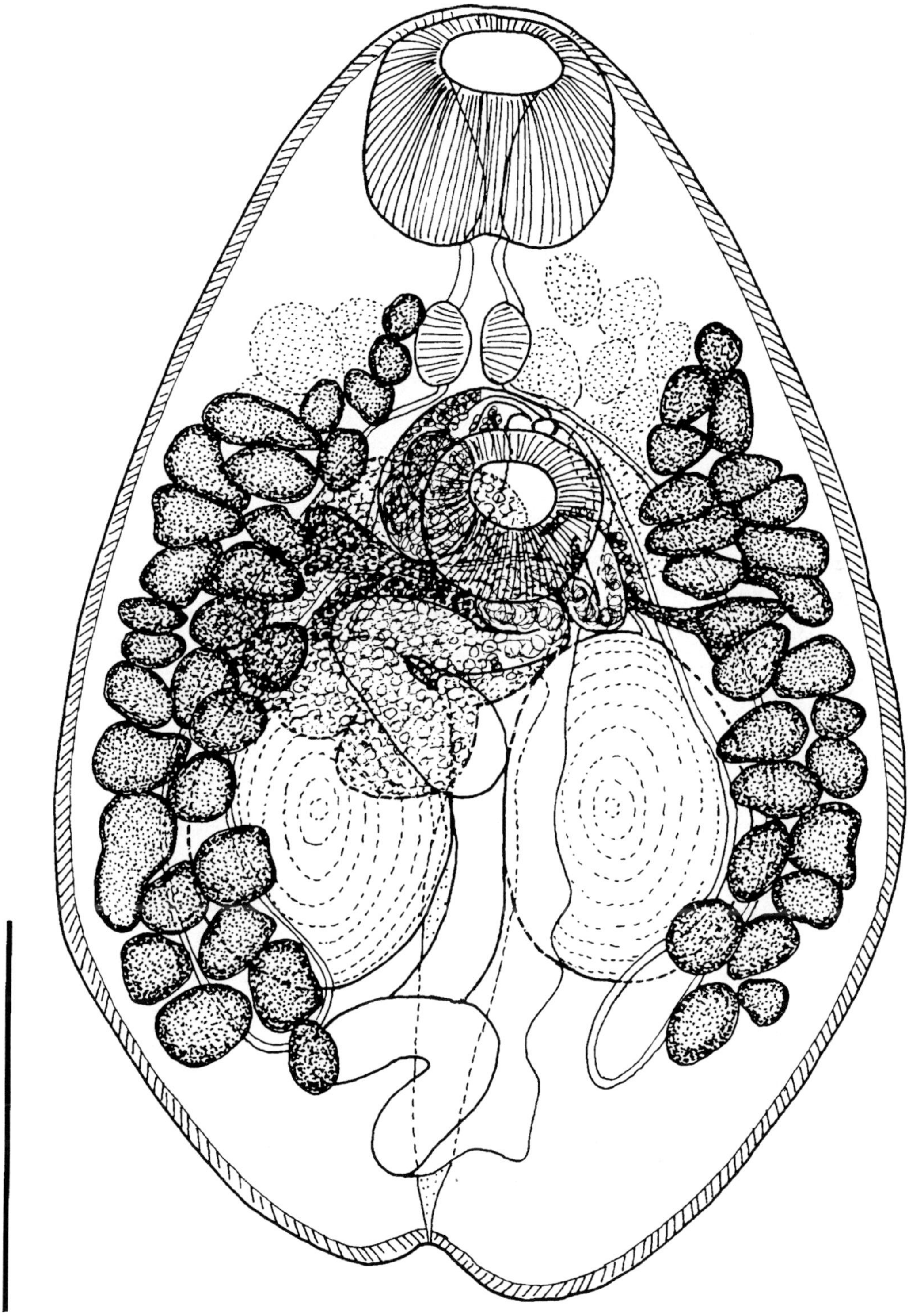

FIG. 116. *Vietosoma parva* (after Yamaguti (1971)). Scale bar (roughly estimated): 100 µm.

Cirrus sac present; copulatory organ an eversible ejaculatory duct (cirrus) (this feature may not be visible in immature specimens from fish) .... Maritrematinae

Subfamily MICROPHALLINAE Ward, 1901

Microphallidae. Caeca usually short, exceptionally absent. Cirrus sac absent, seminal vesicle and pars prostatica free in parenchyma. Copulatory organ in form of atrial organ or protrusible male papilla. Genital atrium usually simple and often poorly developed, sometimes diverticulate and large, usually thin-walled, occasionally thick-walled. Genital pore usually sinistral to ventral sucker. Metraterm often well differentiated, sometimes hypertrophied. Vitelline follicles few, large, grouped laterally in testicular or post-testicular region, may be diffuse in old specimens. Excretory vesicle V- to Y-shaped. In gut of birds, mammals, reptiles, amphibians, and fishes.

MICROPHALLUS Ward, 1901

Microphallinae. Body small, oval, pyriform or linguiform. Tegument spinose anteriorly. Suckers small, similar in size. Oral sucker subterminal. Ventral sucker near middle of body. Prepharynx, pharynx, and oesophagus present, latter short to medium length. Caeca short or reaching to level of ventral sucker, blind, very divergent, occasionally atrophied. Testes two, ovoid, symmetrical, post-ovarian. Cirrus sac absent. Seminal vesicle and pars prostatica free in parenchyma (prostatic gland cells may be delimited in some species). Seminal vesicle oval, close to ventral sucker and ovary. Pars prostatica small, within or close to base of male papilla. Ejaculatory duct short. Male papilla permanent and protrusible, of variable size and form. Genital atrium sinistral to ventral sucker, non-diverticulate, wall usually thin, size related to that of male papilla. Ovary dextral, at level of ventral sucker. Uterus post-caecal, surrounding testes. Metraterm always differentiated, musculature variable, postero-sinistral to genital atrium. Eggs numerous, small. Vitellarium in two lateral groups of large follicles, postero-ventral and posterior to testes. Excretory vesicle V- to Y-shaped. In gut of birds, mammals, and occasionally reptiles and fishes.

Microphallus opacus (Ward, 1894) Ward, 1901 (Fig. 117)

Syn.: *Distomum opacum* Ward, 1894.

Site: Intestine.

Hosts: *Ambloplites rupestris* (4); *Ameiurus nebulosus* (4, 7); *Amia calva* (2); *Anguilla rostrata* (1, 5); *Esox lucius* (3); *E. niger* (3); *Ictalurus punctatus* (4, 8); *Micropterus dolomieui* (2, 3, 4, 6); *M. salmoides* (4); *Perca flavescens* (2, 3).

Distribution: Ont, Que.

Records: 1. MacCallum 1895 (Ont); 2. Bangham and Hunter 1939 (Ont); 3. Fantham and Porter 1948 (Que); 4. Bangham 1955 (Ont); 5. Dechtiar 1972a (Ont); 6. Dechtiar 1972b (Ont); 7. Dechtiar *et al.* 1988 (Ont); 8. Dechtiar and Christie 1988 (Ont).

Remarks: The life history of this species was commented on by Rausch (1947) and Caveny and Etges (1971). It would appear that metacercariae of this species can be progenetic, and thus the presence of ovigerous adults in fishes may be accidental. Ovigerous specimens have been recovered from the crustacean intermediate host, fishes, reptiles, and mammals. Caveny and Etges maintain that a vertebrate host may not be necessary for the survival of this parasite.

Subfamily MARITREMATINAE Nicoll, 1907

Microphallidae. Caeca short. Cirrus sac present, distinct. Ejaculatory duct may form protrusible, digitiform cirrus. Genital atrium simple, small. Genital pore normally sinistral opposite middle of ventral sucker. Vitelline distribution variable, linear in hindbody, in pre-testicular bands or fields, in complete or incom-

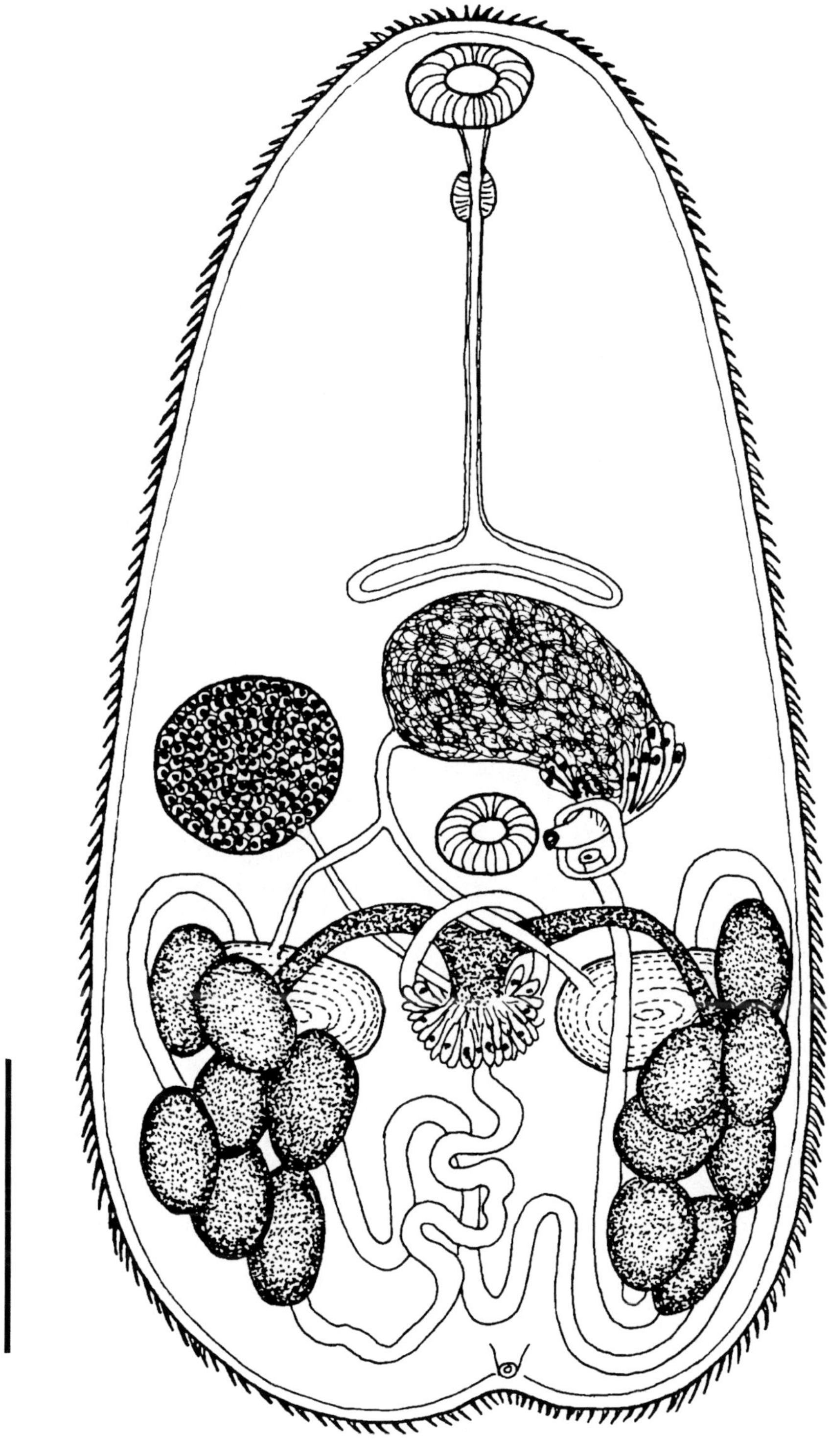

FIG. 117. *Microphallus opacus* (after Caveny and Etges (1971)). Scale bar: 500.

plete posterior ring, or combinations of these arrangements. Uterus post-caecal or with pre-caecal loops. Excretory vesicle V- to Y-shaped. In gut of birds, occasionally mammals.

Key to the genera of the Maritrematinae

1 Uterus not extending into forebody; vitellarium composed of two narrow bands of follicles that extend along anterior border of testes and then posteriorly in lateral fields to varying extent ....*Maritrema*

Loops of uterus extend on both sides of body into forebody to level of caeca; vitelline fields composed of small number of large follicles, which may appear to fuse, between testes and ovary ....*Quasimaritremopsis*

MARITREMA Nicoll, 1907

Maritrematinae. Body small, oval, pyriform or linguiform. Tegument spinose. Suckers similar in size, oral sucker subterminal, ventral sucker in middle of body or just posterior. Prepharynx variable in length. Pharynx small to medium-sized. Oesophagus short to medium in length. Caeca blind, divergent, short to long (terminate anterior or posterior to level of cirrus sac, but not reaching level of posterior margin of testes). Testes two, rounded, entire, symmetrical, post-ovarian. Cirrus sac present, claviform, transverse, curved between caeca and ventral sucker, with proximal end anterior to left testis; contains rectilinear or curved seminal vesicle that tapers distally, small pars prostatica (apparently missing in some species), and ejaculatory duct of variable length that may form cirrus. Genital pore sinistral to ventral sucker. Ovary relatively large, oval or rounded, dorsal, dorso-dextral or postero-dextral to ventral sucker. Laurer's canal and seminal receptacle (?) (present according to Rankin 1939). Uterus post-caecal, sometimes looping around testes. Metraterm present, unarmed. Eggs numerous, small. Vitellarium in two narrow, lateral bands of follicles forming peripheral ring around part or most of hindbody, bands may unite anteriorly to testes and reach back to varying degrees between level of testes and posterior extremity. Excretory pore terminal; vesicle V- to Y-shaped. In intestine of birds and occasionally mammals.

Maritrema obstipum (Van Cleave and Mueller, 1932) Mueller, 1934 (Fig. 118)

Syn.: *Maritreminoides obstipus* (Van Cleave and Mueller, 1932) Rankin, 1939; *Maritreminoides nettae* (Gower, 1938) Rankin, 1939.
Site: Gut.
Host: *Micropterus dolomieui.*
Distribution: Ont.
Record: Bangham 1941.
Remarks: Like *Quasimaritremopsis medius*, specimens of this species occurring in fishes are likely to be accidentally ingested metacercariae that have been excysted. The life history of *M. obstipum* was described by Etges (1953). See also the "Remarks" under *Q. medius.*

QUASIMARITREMOPSIS Deblock, 1973

Maritrematinae. Body small, oval, pyriform or linguiform, rounded posteriorly, conical anteriorly. Tegument spinose. Oral sucker subterminal. Ventral sucker larger than oral sucker, near middle or just anterior to middle of body. Prepharynx variable in length. Pharynx small. Oesophagus relatively long. Intestinal bifurcation in mid-forebody. Caeca short, divergent, blind, retained in forebody. Testes two, rounded, entire, symmetrical, post-ovarian. Cirrus sac present, large, claviform, transverse, curved between caeca and ventral sucker, with proximal end antero-lateral to ovary, contains simple, saccular seminal vesicle, pars prostatica, and eversible ejaculatory duct. Genital atrium small, unarmed. Genital pore immediately

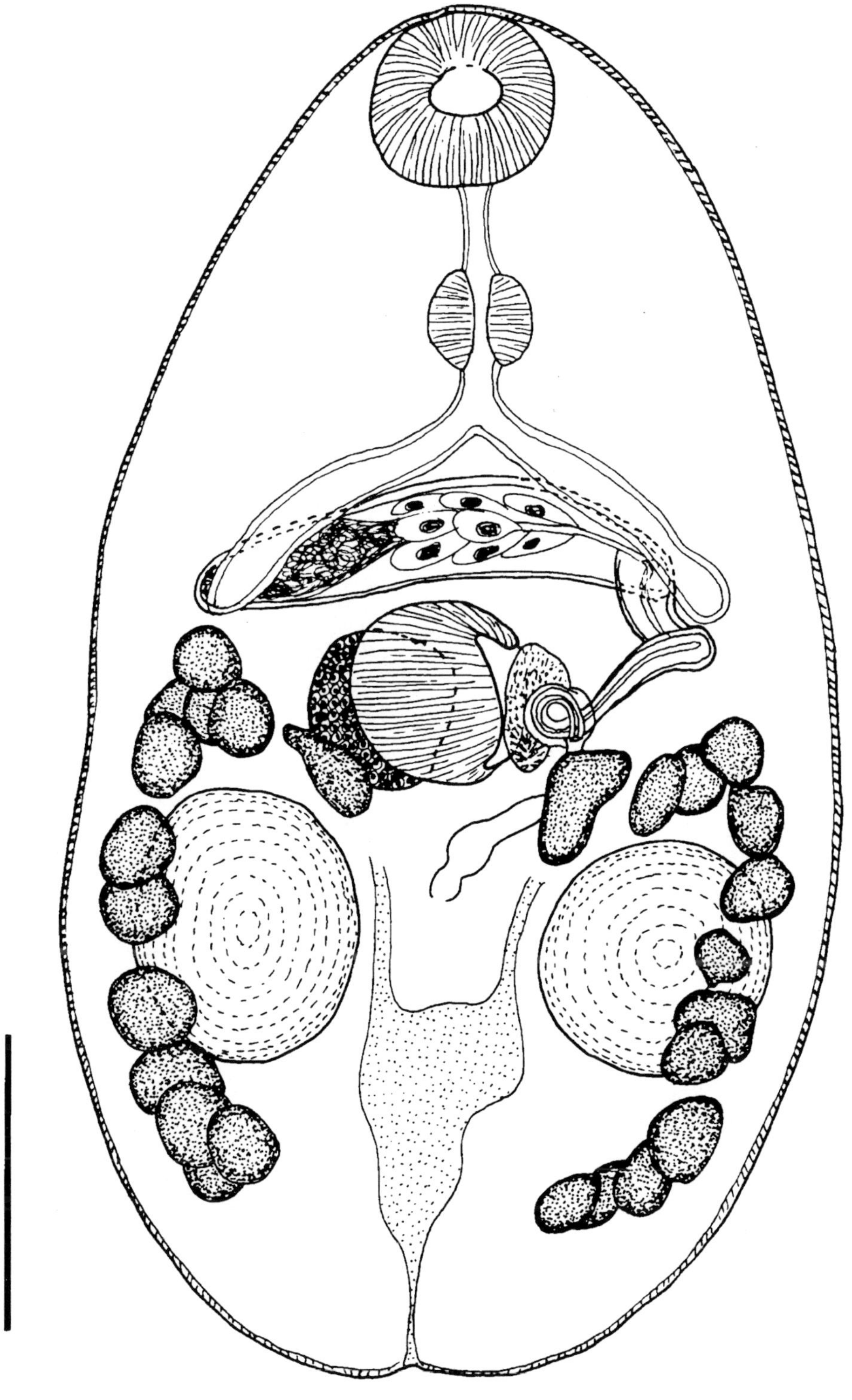

FIG. 118. *Maritrema obstipum* (modified after Van Cleave and Mueller (1932) and Deblock (1973)). Scale bar: 100 μm.

sinistral to ventral sucker. Ovary large, subspherical, dextral or postero-dextral to ventral sucker. Laurer's canal and seminal receptacle (?). Uterus surrounds testes and loops on either side of body into forebody to level of caeca. Metraterm present, unarmed. Eggs numerous, small. Vitellarium in two narrow, lateral fields of few large follicles between ovary and testes, follicles may appear to fuse to form large mass in adults. Excretory pore terminal, vesicle V- to Y-shaped. In intestine of mammals or (?) birds.

Quasimaritremopsis medius (Van Cleave and Mueller, 1932) Deblock, 1973 (Fig. 119)

Syn.: *Maritrema medium* (Van Cleave and Mueller, 1932) Mueller, 1934; *Maritreminoides obstipus* (Van Cleave and Mueller, 1932) of Margolis and Arthur (1979) in part.

Site: Gut.

Host: *Micropterus dolomieui*.

Distribution: Ont.

Record: Bangham 1941.

Remarks: This species was recorded under the name *Maritreminoides obstipus* by Margolis and Arthur (1979): they presumably followed workers, such as Chen (1956), who considered these two species to be synonymous. Deblock (1973) studied the type material of both of these species and considered that they were generically distinct and used the combinations *Quasimaritremopsis medius* and *Maritrema obstipum*. Both of these forms occur as adults in mammals and/or birds. The specimens from fishes appear to be excysted metacercariae that have been accidentally ingested with their food, although Sheldon (1938) has suggested that fish might act as paratenic hosts.

The variations in the distribution of the vitellarium outlined by Etges (1953) in his life history study of *M. obstipum* make identifications based upon this character questionable. Yet, Deblock (1973) has used this feature as one of the diagnostic characters of the genus *Quasimaritremopsis*, since most authors (e.g., Chen *et al.*, 1985) have figured the vitelline follicles as almost forming a ring around the testes in *M. obstipum* and related species.

The life history of *Q. medius*, as *Microphallus medius*, was described by Sheldon (1938); but Deblock (1973) has suggested that the form studied by Sheldon may not be conspecific with *M. medius* of Van Cleave and Mueller (1932).

Family TROGLOTREMATIDAE Odhner, 1914

Syn.: Nanophyetidae Wallace, 1935.

Plagiorchioidea. Sometimes encysted as metacercariae in tissues of fishes. Cyst wall thin or thick, double-layered. Body oval. Tegument spinose. Cercarial stylet may be retained in young worms. Suckers well developed, often similar in size. Ventral sucker in middle or anterior half of body. Prepharynx present or apparently absent. Pharynx well developed. Gonadal primordia may develop; testicular primordia, if present symmetrical in hindbody; terminal genitalia and genital pore, if visible, closely posterior to ventral sucker. Excretory vesicle saccular, filled with calcareous corpuscles in live worms. Metacercaria in tissues of fishes, crustaceans. (Parasitic as adults in intestine, body sinuses, kidney, liver, etc. of mammals.)

Comments

This family is in need of revision, especially as regards its relationships with the Paragonimidae Dollfus, 1939 and the Nanophyetidae Wallace, 1935, the latter of which is here treated as a synonym.

NANOPHYETUS Chapin, 1927

Troglotrematidae. Encysted as metacercariae in tissues of fishes. Cyst wall relatively thin, transparent. Body small, oval to pyriform, slightly flattened. Tegument spinose. Cercarial stylet may be retained in young worms. Oral sucker large. Ventral sucker fractionally smaller than oral sucker, near middle of body.

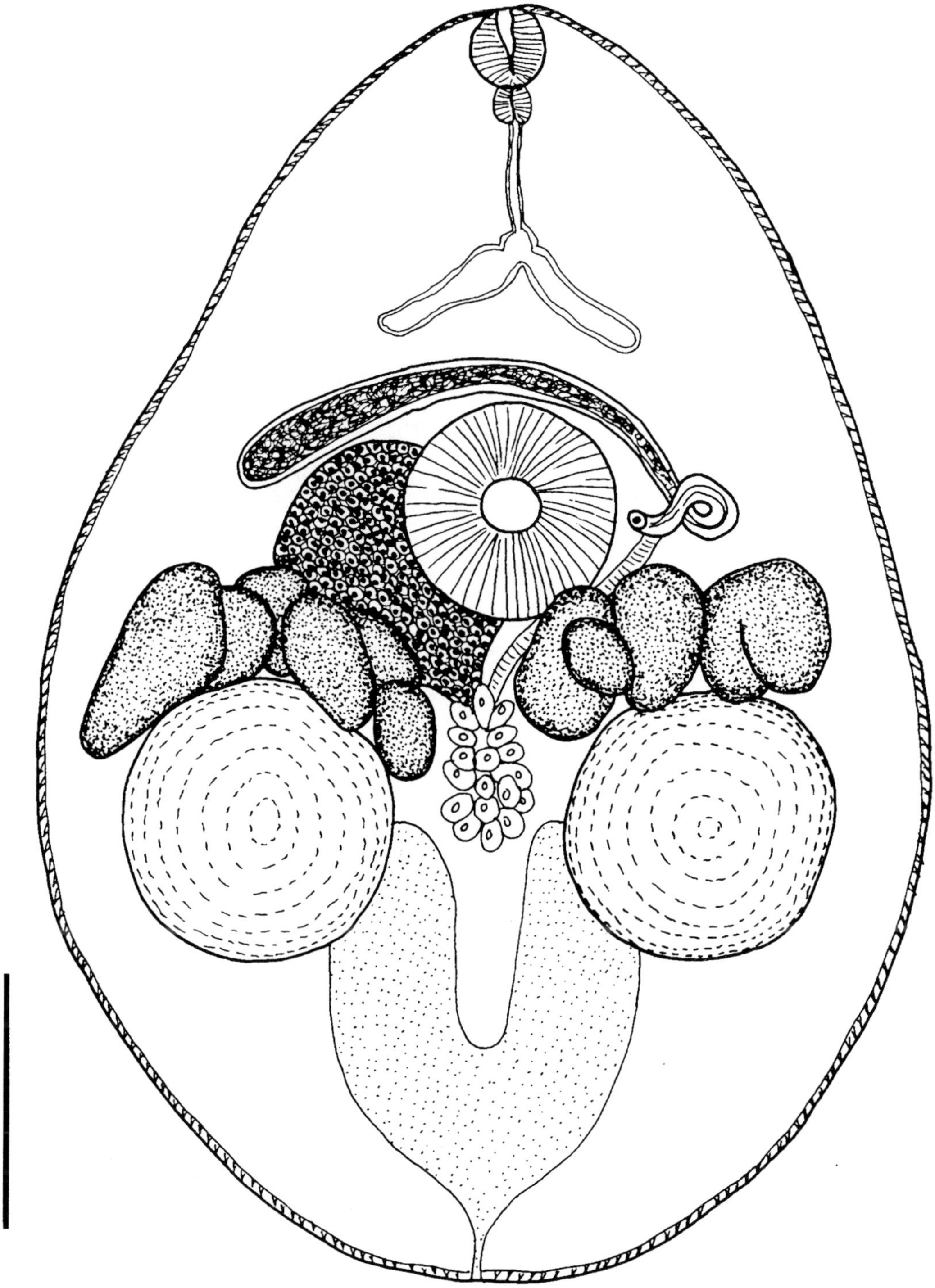

FIG. 119. *Quasimaritremopsis medius* (after Van Cleave and Mueller (1932)). Scale bar: 100 µm.

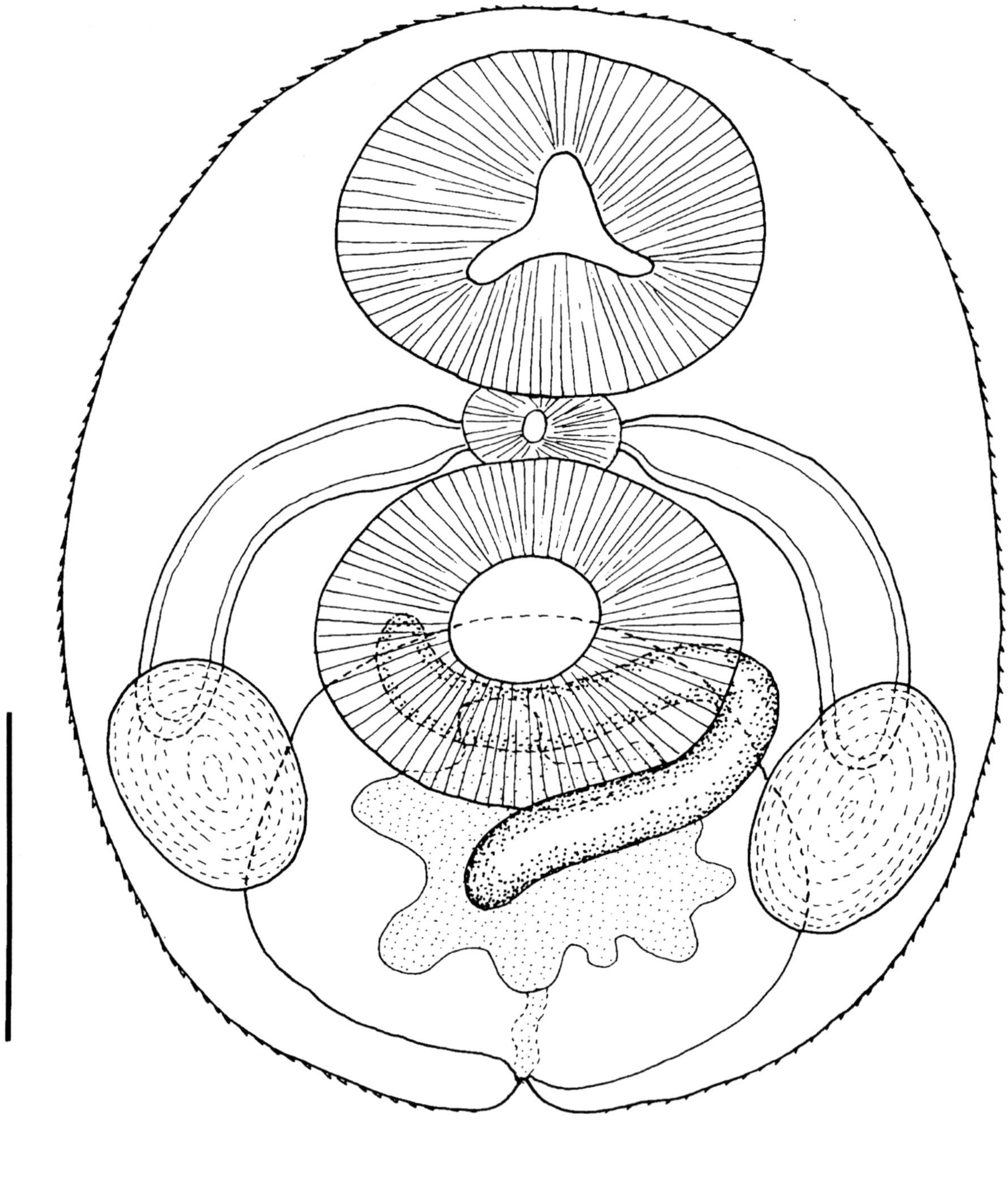

FIG. 120. *Nanophyetus salmincola* (modified after Ward and Mueller (1926: as *Distomulum oregonensis*)). Scale bar: 50 μm.

Pharynx well developed. Oesophagus short to medium-sized. Intestinal bifurcation in forebody. Caeca end blindly in anterior half of forebody at about level of testicular primordia. Primordia of testes develop symmetrically in anterior hindbody. Primordia of female reproductive system develops medially or submedially just posterior to ventral sucker. Excretory vesicle large, saccular, fills much of hindbody. (Adults in intestine of piscivorous mammals.)

Nanophyetus salmincola (Chapin, 1926) Chapin, 1927 (Fig. 120)

Site: (Kidney).

Host: *Oncorhynchus mykiss*.

Distribution: Pac.

Records: Margolis 1990; Margolis 1992; Burgner *et al.* 1992.

Remarks: This species occurs in a range of hosts, and especially salmonids, in fresh waters of the western states of the United States but not apparently in Canada. This parasite is the cause of the so-called "salmon poisoning disease" of dogs and other carnivorous mammals: for review see Dollfus (1975). The life history of this parasite was described by Bennington and Pratt (1960).

SUPERFAMILY ZOOGONOIDEA ODHNER, 1902

Plagiorchiida. Body usually small to medium-sized, cylindrical to oval. Tegument normally, but not always, armed. Oral and ventral suckers present, ventral sucker near middle of body or in anterior half, rarely in posterior half. Prepharynx and pharynx present, oesophagus usually present. Intestinal bifurcation usually in forebody, occasionally in anterior hindbody. Caeca normally two, occasionally one, blind, short to long. Testes one or more usually two, tandem, oblique, or symmetrical, usually in hindbody. Cirrus sac well developed, contains seminal vesicle, pars prostatica, and ejaculatory duct. Ejaculatory duct/cirrus may be armed. Genital atrium apparently absent to large, occasionally armed. Genital pore marginal at level of ventral sucker, sometimes more anteriorly, rarely more posteriorly, or medially in forebody. Ovary usually pre-testicular. Laurer's canal and canalicular seminal receptacle normally present. Uterus fills much of hindbody and extends into post-testicular zone. Metraterm may be hypertrophied and armed. Eggs operculate, small to medium-sized, numerous, occasionally with membranous capsule. Vitellarium in form of one or two compact masses, two acinous bunches of follicles or large lateral field of follicles, normally in hindbody, occasionally forebody. Excretory pore terminal. Excretory vesicle I-, V-, or Y-shaped. Usually in intestine, occasionally gall bladder, bile duct, or urinary bladder, of fishes.

KEY TO THE FAMILIES OF THE ZOOGONOIDEA

1 Testis single; ejaculatory duct and metraterm heavily armed .... Monorchiidae

Testes two; ejaculatory duct and metraterm unarmed (take care not to interpret filamentous structures sometimes found lining terminal genital ducts as spines) 2

2 In marine fishes; genital pore usually in forebody, occasionally lateral to ventral sucker or slightly more posterior; occasionally in organs other than gut; vitellarium composed of one or two masses or restricted distribution of follicles .... Zoogonidae

In freshwater fishes; genital pore lateral to ventral sucker; parasitic in gut; vitellarium follicular .... Lissorchiidae

Family LISSORCHIIDAE Poche, 1926

Zoogonoidea. Body small to medium-sized, elongate-oval to fusiform. Body surface covered, at least anteriorly, with spines. Suckers well developed. Ventral larger than or similar in size to oral, near middle or more often in anterior half of body. Prepharynx short. Pharynx well developed. Oesophagus short. Intestinal bifurcation in forebody. Caeca long, reaching at least to level of testes and often close to posterior extremity, blind. Testes two, tandem, in posterior half of body, entire or slightly irregularly lobed. Cirrus-sac claviform; extending postero-dorsally to ventral sucker or into anterior hindbody, contains bipartite, saccular seminal vesicle, small, often indistinct, vesicular pars prostatica, and ejaculatory duct capable of forming cirrus; latter armed or not. Genital atrium small to apparently absent. Genital pore lateral at level of ventral sucker. Ovary lobed, often tri-lobed, median or submedian, closely anterior to anterior testis. Small canalicular seminal receptacle present (?Yamaguti (1971) indicated that the seminal receptacle is absent and illustrated a uterine seminal receptacle). Uterus long, coiled, uterine field extending throughout most of hindbody, encircling gonads, often with large post-testicular zone. Metraterm present. Eggs numerous, small, operculate. Vitellarium follicular, in lateral fields with restricted distribution in hindbody, normally limited to within region between ventral sucker and post-testicular field. Excretory vesicle tubular, I-shaped. In intestine of catastomid (and occasionally other cypriniform) freshwater teleosts.

Comments

The systematic position and validity of this family is problematical. It has been linked with both the Monorchiidae and the Zoogonidae. The main problem has been a discrepancy between the life cycle of the type species of *Lissorchis* and that of other species of the genus. The most recent discussion on this topic was given by Bray (1987a), who excluded the family from the Zoogonidae.

LISSORCHIS Magath, 1917

Syn.: *Triganodistomum* Simer, 1929.

Lissorchiidae. With characters of family. Parasites of catastomids and occasionally other cypriniforms and their predators.

Comments

The genus is in need of critical revision. Smith's (1968) synonymy of *Lissorchis* and *Triganodistomum* has been accepted by most authors, but not by Yamaguti (1971), who placed them in different families. As indicated by Bray (1987a), the verification of the life history of the type species of *Lissorchis* is crucial to this question.

The validity of some species of the genus must also be in question when works such as Amin (1982b) are taken into consideration. The allometric growth of *L. attenuatus* was studied by Fischthal *et al.* (1982). The most recent key to the genus is that of Krygier and Macy (1969), but this is rather uncritical and accepts all nominal species.

In addition to the species listed below, Bangham recorded "encysted larval forms" from *Notropis hudsonius* in Ontario.

Key to the species of *Lissorchis*

1 Medium-sized worms (2–7 mm); fusiform, often attenuated posteriorly; with large number of small vitelline follicles on each side of body; caeca terminate close to posterior extremity 2

Small worms (<2 mm); elongate-oval; with small number of relatively large vitelline follicles on each side of body; gut caeca terminate well short of posterior extremity 3

2 Spines restricted to anterior half of body. In lateral fields ....*Lissorchis attenuatus* (Mueller and Van Cleave, 1932) Kryger and Macy, 1969 (Fig. 121)

Syn.: *Triganodistomum attenuatum* Mueller and Van Cleave, 1932.

Site: Intestine.

Hosts: *Carpioides cyprinus* (5, 6, 13); *Catostomus catostomus* (4, 9, 10, 11); *C. commersoni* (1, 2, 3, 7, 11, 12, 13, 14); *C. macrocheilus* (4, 9, 10); *Moxostoma anisurum* (6); *Semotilus atromaculatus* (3).

Distribution: BC, Ont, Alta.

Records: 1. Bangham 1941 (Ont); 2. Bangham 1955 (Ont); 3. Bangham and Venard 1946 (Ont); 4. Bangham and Adams 1954 (BC); 5. Dechtiar 1972a (Ont); 6. Dechtiar 1972b (Ont); 7. Molnar *et al.* 1974 (Ont); 8. Leong and Holmes 1981 (Alta); 9. Anonymous 1981 (BC); 10. Anonymous 1984 (BC); 11. Dechtiar and Lawrie 1988 (Ont); 12. Dechtiar *et al.* 1988 (Ont); 13. Dechtiar and Nepszy 1988 (Ont); 14. Dechtiar and Christie 1988 (Ont).

Remarks: This species was recently recorded from *Catostomus commersoni* in American waters of Lake Superior by Hogue *et al.* (1993).

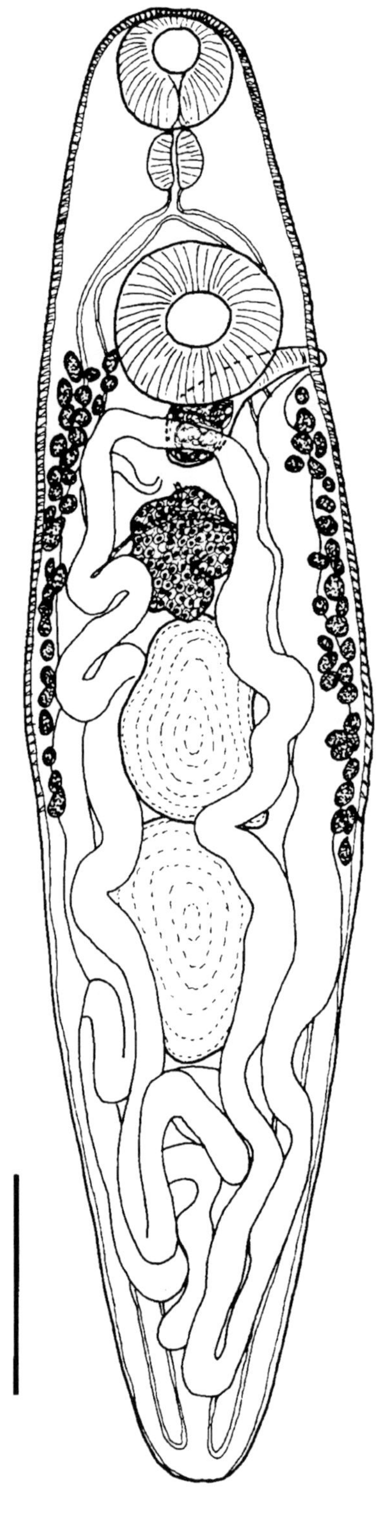

FIG. 121. *Lissorchis attenuatus* (after Van Cleave and Mueller (1932)). Scale bar: 500 µm.

Spines extend full length of body in lateral fields ....*L. gullaris* Self and Campbell, 1956

Site: Intestine.

Host: *Carpiodes cyprinus* (2); *Catostomus catostomus* (1).

Distribution: Lab, Man.

Records: 1. Threlfall and Hanek 1970a (Lab); 2. Szalai and Dick 1987 (Man).

Remarks: It is possible that this species will prove to be a synonym of *L. attenuatus*. Both of these species resemble the type species, *L. fairporti* Magath, 1917. I have examined Szalai and Dick's (1987) specimen in the Canadian National Collection, but, unfortunately, the spines have been lost.

3 Eggs >20 µm in length; post-testicular field greater in length than posterior testis ....*L. simeri* (Mueller and Van Cleave, 1932) Krygier and Macy, 1969

Syn.: *Triganodistomum simeri* Mueller and Van Cleave, 1932.

Site: (Gut).

Hosts: *Catostomus catostomus* (2); *C. commersoni* (1).

Distribution: Ont.

Records: 1. Bangham 1941; 2. Bangham 1955.

Remarks: This may prove to be a synonym of *L. mutabile* (Cort, 1919) Smith, 1968.

Eggs <20 µm in length; length of post-testicular field less than that of posterior testis ....*L. kritskyi* Barnhart and Powell, 1979

Site: Not given (assumed to be intestine).

Host: *Perca flavescens*.

Distribution: Man.

Record: Poole and Dick 1985.

Remarks: The presence of a single specimen in a perch was presumably accidental, being acquired with the fish's diet. This species is morphologically similar to *L. mutabile* (see Amin's 1982b, description) and possibly a synonym. I have examined Poole and Dick's (1985) specimen from the Canadian National Collection. It does key down to *L. kritskyi*, but the vitelline follicles are smaller, the pharynx is much larger, almost as large as the oral sucker, and the spines are lost.

Family MONORCHIIDAE Odhner, 1911

Zoogonoidea. Body small, oval to elongate. Body surface spinose. Suckers well developed, often small. Oral sucker ventrally subterminal. Ventral sucker usually in anterior half of body. Prepharynx present, short to long. Pharynx well developed, sometimes small. Oesophagus short to long. Intestinal bifurcation in forebody. Caeca short to long, blind. Testes single (often) or double, usually in hindbody. Cirrus sac present, contains saccular seminal vesicle, pars prostatica, and heavily spined ejaculatory duct that may form armed cirrus. Genital atrium present, often small, sometimes armed. Genital pore usually in forebody, ventro-median to lateral. Ovary usually pre-testicular, occasionally post-testicular. Laurer's canal present. Canalicular seminal receptacle or uterine seminal receptacle present; in latter case rudimentary seminal receptacle may also occur. Uterus occupies most of available space in hindbody. Metraterm normally well developed, may extend into distinct "terminal organ", normally lined with spines over at least part of its length. Eggs usually numerous, small, operculate, occasionally filamented. Vitellarium normally follicular, usually in form of two small symmetrical acinous bunches, occasionally two compact masses, position varies. Excretory pore terminal, excretory vesicle of variable shape, saccular, V-shaped or Y-shaped (interpretations vary). In intestine of marine (mainly) and freshwater teleosts.

Comments

In the absence of a detailed modern revision of the family and the recorded presence of only two genera in Canadian waters, there seems little point in subfamily designations. It is worth noting, however, that Mamaev (1968) listed both genera in the Monorchiinae, while Yamaguti (1971) recognized both the Lasiotocinae Yamaguti, 1958 and the Telolecithinae Yamaguti, 1958.

Key to the genera of the Monorchiidae

1 Vitelline fields close to ventral sucker; Atlantic coast, especially in *Menidia* .... *Lasiotocus*

Vitelline fields close to posterior extremity; Pacific coast, in embiotocids .... *Telolecithus*

LASIOTOCUS Looss, 1907

Monorchiidae. Body oval to elongate. Tegument spined. Eye-spots present or absent. Suckers sometimes small. Ventral sucker in anterior half of body. Prepharynx short to long. Pharynx relatively small. Oesophagus short to long. Caeca usually ending well short of posterior extremity. Testis single, often large, lobed or not, near middle or in posterior half of body. Cirrus sac claviform, usually reaches into hindbody, contains saccular seminal vesicle, pars prostatica, and heavily armed ejaculatory duct. Genital atrium small to long and tubular, unarmed. Genital pore median to medio-lateral. Ovary anterior or antero-lateral to testis, submedian, lobed or not. Uterine seminal receptacle present. Uterus fills much of hindbody. Metraterm extended into distinct "terminal organ", lined with spines distally. Eggs without filaments. Vitellarium follicular, symmetrical fields usually at about level of ovary. Excretory vesicle saccular. In intestine of marine teleosts.

Lasiotocus elongatus (Manter, 1931) Thomas, 1959 (Fig. 122)

Site: Intestine.

Host: *Menidia menidia.*

Distribution: NB-b.

Record: BM(NH) collection.

Remarks: This record is based upon a single specimen collected by Dr. R.A. Bray at Sam Orr Pond, St. Andrews, New Brunswick. It is a common parasite of *Menidia menidia* on the Atlantic coast of the United States, along with its congener *L. minor* (Manter, 1931). The life history of *L. elongatus* was described by Stunkard (1981).

TELOLECITHUS Lloyd and Guberlet, 1932

Monorchiidae. Body elongate-oval, slightly tapering anteriorly. Tegument spined. Suckers well developed, relatively large. Ventral sucker in anterior half of body. Prepharynx short. Pharynx relatively small. Oesophagus short. Caeca long, terminate near posterior extremity. Testes single, in mid-hindbody. Cirrus sac large, claviform, extends into hindbody, contains saccular seminal vesicle, muscular pars prostatica, and heavily armed ejaculatory duct. Genital atrium small, unarmed. Genital pore median, post-bifurcal. Ovary submedian, immediately pre-testicular. Uterine seminal receptacle present. Uterus filling hindbody. Metraterm extended into distinct, claviform "terminal organ", armed with spines. Eggs without filaments. Vitellarium in form of symmetrical pair of c.6–7 large follicles, post-testicular. Excretory vesicle saccular. In intestine of marine teleosts (Embiotocidae).

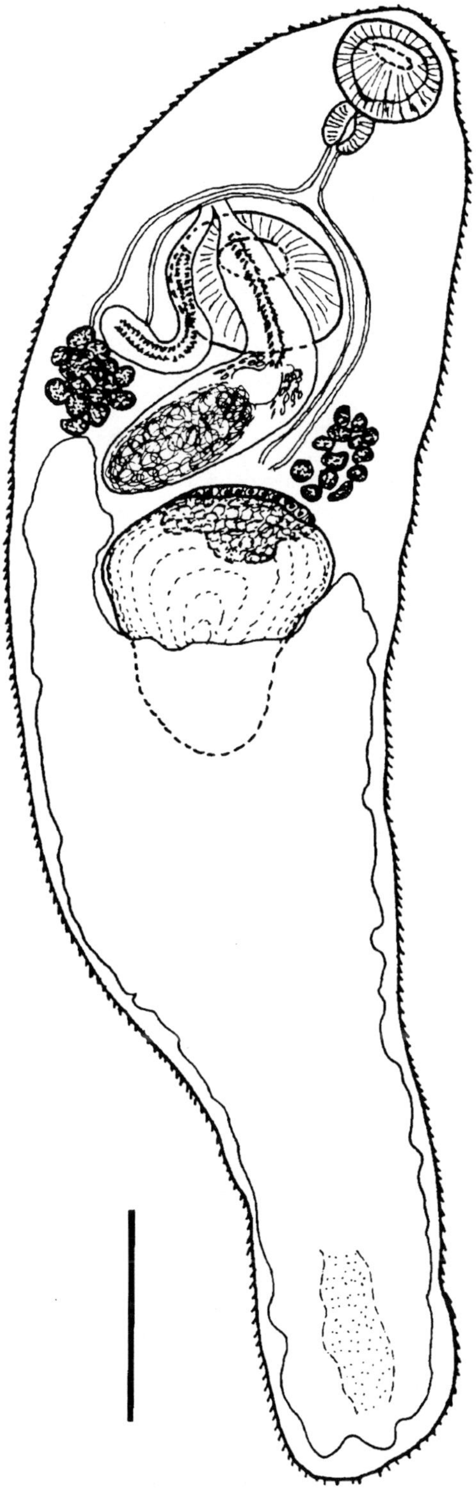

FIG. 122. *Lasiotocus elongatus* (after Manter (1931)). Scale bar: 100 µm.

Telolecithus pugetensis Lloyd and Guberlet, 1932 (Fig. 123)

Site: Posterior intestine, rectum.

Host: *Cymatogaster aggregata.*

Distribution: Pac.

Records: McFarlane 1936; Arai 1967a; Arai 1967b; Arai 1969b; Arai *et al.* 1988; BM(NH) collection.

Remarks: This is a parasite that appears to be specific to embiotocid teleosts on the western coast of North America. Its life history has been described by De Martin and Pratt (1964) and its pathogenicity to its bivalve molluscan host was studied by Kabat (1986).

Family ZOOGONIDAE Odhner, 1902

Zoogonoidea. Body small to large, elongate cylindrical to flattened spatulate. Body surface usually spinose, spines occasionally absent or lost. Oral sucker subterminal to terminal, globular to infundibuliform. Ventral sucker small to large, simple, ornamented or divided transversely, may bear muscular lips; in about middle of body. Prepharynx long to apparently absent. Pharynx globular to oval, large to small. Oesophagus long to absent. Intestinal bifurcation in forebody or anterior hindbody. Caeca normally two, but occasionally one, short and saccular to long; blind, terminating in forebody to ending close to posterior extremity. Testes two, oval to globular, lobed or entire, symmetrical, oblique, or tandem; in hindbody or posterior forebody. Cirrus sac well developed, short saccular to elongate cylindrical, straight to sharply recurved. Seminal vesicle internal, saccular, narrow sinuous tubular or bipartite. Pars prostatica usually prominent, narrow to vesicular. Ejaculatory duct muscular, may form protuberant cirrus bearing spine-like structures or bosses. Genital atrium either small and simple or large, pocketed, and muscular, surrounded by gland cells. Genital pore ventrally or occasionally dorsally submarginal but more usually marginal, usually sinistral, usually in forebody, occasionally lateral to ventral sucker or in anterior hindbody. Ovary entire to multi-lobate, in hindbody or posterior forebody, usually pre-testicular. Canalicular seminal receptacle present. Laurer's canal present. Uterus normally extends into post-gonadal field, reaching close to posterior extremity, containing numerous eggs. Eggs with thin to thick tanned, operculate capsule or occasionally with thin, membranous capsule, occasionally filamented. Vitellarium in form of small, single mass or pair of masses, or divided into follicles in one median or two lateral fields, in fore- or hindbody. Excretory pore terminal to subterminal, vesicle short saccular to long tubular, I-shaped (although occasionally reported as Y-shaped). Parasitic in alimentary tract, gall bladder, bile duct, or urinary bladder of teleosts and occasionally elasmobranchs.

Comments

Full descriptions and discussion of the Northeast Atlantic forms of this family, which include many of the Canadian Atlantic forms, were given by Bray and Gibson (1986) and the Zoogonidae was revised by Bray (1987a,b). Subsequent cladistic analyses of the group have been carried out by Brooks (1990) and Brooks and McLennan (1993). In the latter work, several of the genera recognised in this work are synonymised.

Key to the subfamilies of the Zoogonidae

1 Vitellarium one to two masses or several follicles in undivided median field; egg capsule weakly tanned or membranous .... Zoogoninae

Vitellarium in paired fields of follicles; egg capsule strongly tanned, forming shell ....Lepidophyllinae

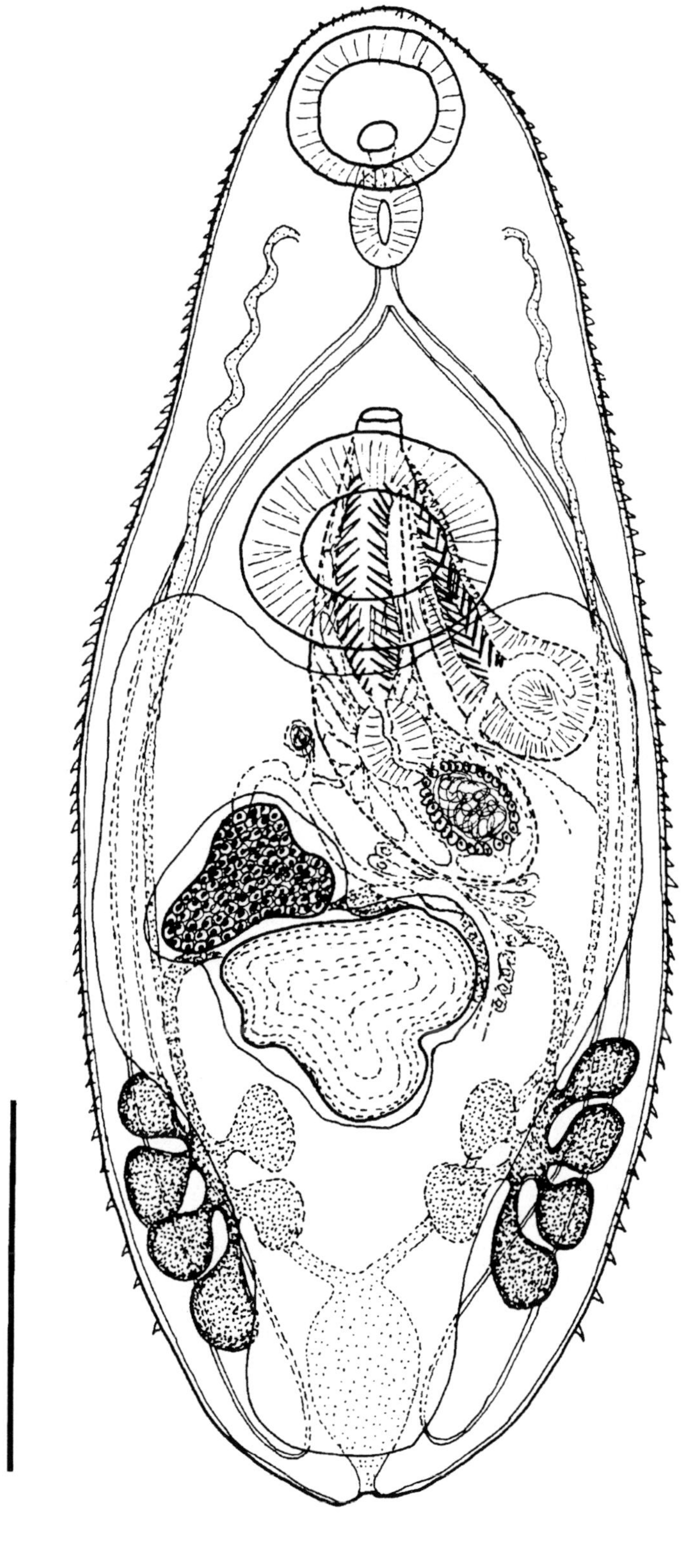

FIG. 123. *Telolecithus pugetensis* (modified after Lloyd and Guberlet (1932)). Scale bar: 200 µm.

Subfamily ZOOGONINAE Odhner, 1902

Zoogonidae. Body small, oval. Body surface usually spinous. Ventral sucker small to large, simple or bearing muscular lips. Testes entire, symmetrical to oblique, in hindbody or lateral to ventral sucker. Seminal vesicle normally bipartite, occasionally tubular. Genital atrium small, may possess atrial sac. Peri-atrial gland absent. Genital pore marginal, in forebody, at level of ventral sucker or in anterior hindbody, usually sinistral. Ovary entire, in hindbody or dorsal to ventral sucker, pre- or post-testicular. Mehlis' gland weakly developed. Uterus mainly post-gonadal. Often viviparous. Egg capsules membranous or weakly tanned. Vitellarium small, single mass, two masses or field of follicles, in median region of hindbody. Excretory pore terminal, vesicle short, saccular. In alimentary canal of marine teleosts and elasmobranchs.

Comments

A revision and keys to this subfamily were given by Bray (1987a).

Key to the genera of the Zoogoninae

1 Genital pore and intestinal bifurcation in mid-forebody 3

Genital pore and intestinal bifurcation posterior to mid-forebody 2

2 Caeca short, saccular .... *Zoogonus*

Caeca pass into posterior half of hindbody .... *Neozoogonus*

3 Vitelline mass single .... *Zoogonoides*

Vitelline masses two .... *Pseudozoogonoides*

NEOZOOGONUS Arai, 1954

Zoogoninae. Body small, oval to elongate-oval. Body surface spinous anteriorly. Oral sucker infundibuliform and terminal or globular and subterminal. Ventral sucker globular, small, similar in size or smaller than oral sucker, just in anterior half of body. Prepharynx distinct, short to long. Pharynx globular. Oesophagus long, narrow. Intestinal bifurcation in anterior hindbody. Caeca reach well past testes, almost to posterior extremity. Testes oval, symmetrical to oblique, in anterior half of hindbody. Cirrus sac claviform, curved. Seminal vesicle bipartite, saccular or (?) tubular. Genital atrium distinct, with atrial sac. Genital pore dextral or sinistral, at level of ventral sucker or in anterior hindbody. Ovary globular to oval, post-testicular. Vitellarium a single irregular mass, adjacent to and often just posterior to ovary. Uterus fills much of post-testicular field. Metraterm muscular. Viviparous. Miracidium surrounded by membranous capsule. Excretory vesicle short, saccular. In intestine of marine teleosts.

Comments

This genus was revised by Bray (1987a) and considered a synonym of *Zoogonoides* by Brooks and McLennan (1993).

Neozoogonus californicus Arai, 1954 (Fig. 124)

Site: Intestine.
Host: *Cymatogaster aggregata.*
Distribution: Pac.
Records: 1. Arai 1967a; 2. Arai 1969b.
Remarks: This species was transferred to *Zoogonoides* by Brooks and McLennan (1993).

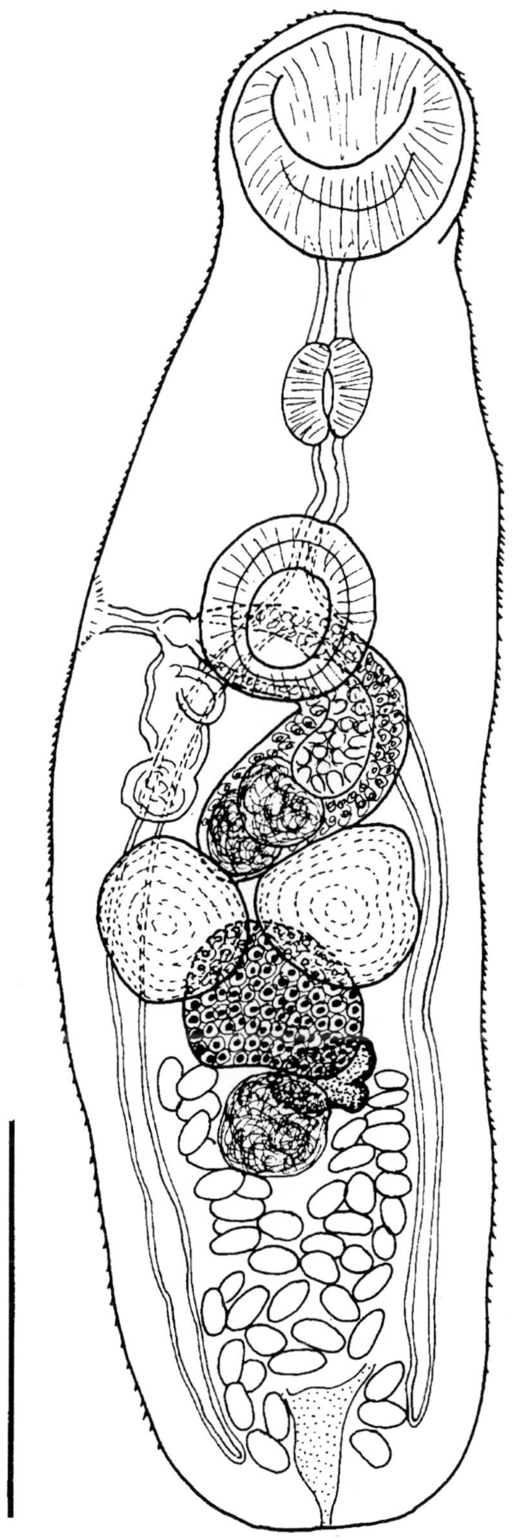

FIG. 124. *Neozoogonus californicus* (after Bray (1987a)). Scale bar: 200 µm.

PSEUDOZOOGONOIDES Zhukov, 1957

Zooogoninae. Body small, pyriform. Body surface spinous, spines in regular transverse rows. Oral sucker globular, subterminal. Ventral sucker rounded, in middle of body or more posterior, similar in size or larger than oral sucker. Prepharynx short. Pharynx globular. Oesophagus distinct. Intestinal bifurcation in forebody. Caeca narrow, extend just into hindbody. Testes symmetrical, at level of ventral sucker. Cirrus sac claviform, straight or curved. Seminal vesicle bipartite. Pars prostatica vesicular. Ejaculatory duct distinct. Genital atrium with atrial sac. Genital pore marginal, sinistral, in mid-forebody. Ovary oval; submedian; post-testicular. Vitellarium two subequal masses. Uterus fills much of hindbody. Viviparous. Miracidium surrounded by membranous or very weakly tanned capsule. Excretory pore slightly elongate saccular. In intestine and rectum of teleosts (occasionally spiral valve of elasmobranchs).

Comments

This genus was revised by Bray (1987a) and considered to be a synonym of *Zoogonoides* by Brooks and McLennan (1993).

Pseudozoogonoides subaequiporus (Odhner, 1911) Bray and Gibson, 1986 (Fig. 125)

Syn.: *Zoogonoides subaequiporus* Odhner, 1911; *Diphterostomum microacetabulum* Shulman-Albova, 1952; *Pseudozoogonoides microacetabulum* (Shulman-Albova, 1952) Zhukov, 1957; *Zoogonoides viviparus* of Scott (1975a,b); Trematoda gen. sp. of Zubchenko (1980).

Site: Intestine.

Hosts: *Anarhichas minor* (4); *Hippoglossoides platessoides* (1, 2, 3, 5, 6, 7).

Distribution: Atl.

Records: 1. Scott 1975a; 2. Scott 1975b; 3. Bray 1979; 4. Zubchenko 1980; 5. Scott 1982; 6. Zubchenko 1985a; 7. Bray and Gibson 1986.

Remarks: This species was returned to *Zoogonoides* by Brooks and McLennan (1993).

ZOOGONOIDES Odhner, 1902

Zoogoninae. Body small, pyriform to elongate-oval. Body surface spinous, spines usually restricted to forebody, occasionally also in hindbody. Oral sucker globular; subterminal. Ventral sucker in anterior half or middle of body in fully developed worms, similar in size or larger than oral sucker. Prepharynx short. Pharynx globular. Oesophagus distinct. Intestinal bifurcation in posterior forebody. Caeca reach to level of testes or just beyond. Testes oval, symmetrical to slightly oblique, just posterior to or at level of ventral sucker. Cirrus sac claviform, short to long, straight or curved. Seminal vesicle bipartite. Pars prostatica oval, vesicular. Ejaculatory duct short to long, transversely wrinkled (? spined). Genital atrium distinct. Atrial sac usually evident. Genital pore usually sinistral, lateral in mid-forebody. Ovary oval, inter-testicular but slightly posterior. Vitellarium a single subglobular to oval mass, normally between ovary and seminal receptacle. Uterus fills much of hindbody. Metraterm muscular. Viviparous. Miracidium in membranous capsule (or very thin shell). Excretory vesicle small, oval. In rectum and posterior intestine of marine teleosts.

Comments

This genus was revised by Bray (1987a).

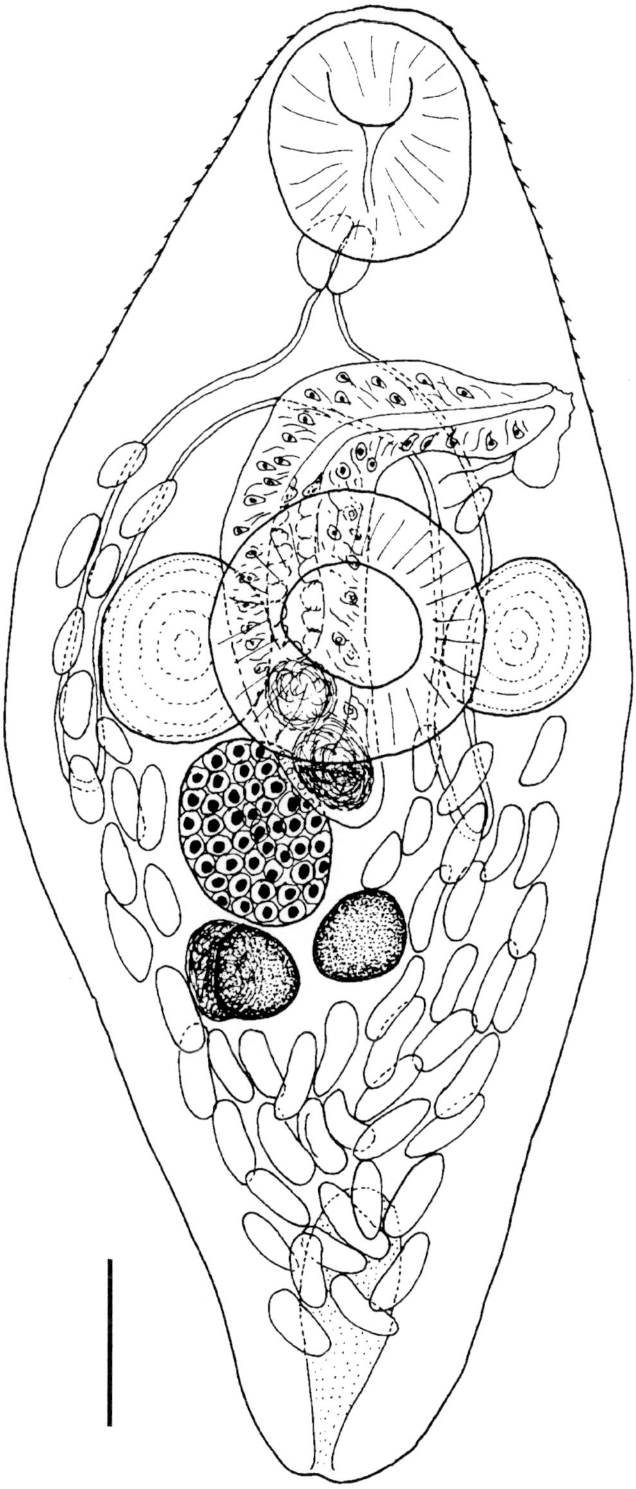

FIG. 125. *Pseudozoogonoides subaequiporus* (after Bray and Gibson (1986)). Scale bar: 100 µm.

Zoogonoides viviparus (Olsson, 1868) Odhner, 1902 (Fig. 126)

Site: ? (normally rectum or posterior intestine).

Host: *Hippoglossoides platessoides*.

Distribution: Atl.

Records: 1. Umnova 1979; 2. Zubchenko 1980.

Remarks: It is doubtful if this common Northeast Atlantic species occurs in Canadian waters. Scott's (1975a,b) records of *Z. viviparus* proved to be *Pseudozoogonoides subaequiporus* (see Bray 1979; Scott 1982), and, in the studies of *Hippoglossoides platessoides* in Canadian waters carried out by Bray (1979), Scott (1982), and Zubchenko (1985a), only *P. subaequiporus* was found. The evidence that *Z. viviparus* occurs in Canadian waters, therefore, rests upon Zubchenko's (1980) list and Umnova's (1979) abstract of a record from the Northwest Atlantic. The species was described in detail by Bray and Gibson (1986).

ZOOGONUS Looss, 1901

Zoogoninae. Body small, pyriform to elongate-oval. Body surface spinous, spines in regular transverse rows in anterior half of body. Oral sucker globular, subterminal. Ventral sucker in anterior half of body, similar in size to oral sucker. Prepharynx distinct, short to long. Pharynx oval. Oesophagus long, reaches to posterior margin of ventral sucker or into hindbody. Intestinal bifurcation dorsal to ventral sucker or in anterior hindbody. Caeca short, saccular. Testes large, symmetrical, in anterior hindbody. Cirrus sac elongate, reaches into hindbody. Seminal vesicle bipartite. Pars prostatica vesicular. Ejaculatory duct narrow. Genital atrium small. Genital pore marginal, at or about level of ventral sucker. Ovary globular to oval, median, inter-caecal, post-testicular. Vitellarium single (occasionally bi-lobed). Uterus fills hindbody posterior to gonads. Viviparous; miracidium surrounded by membranous capsule (? or very thin shell). In lower intestine and rectum of marine teleosts.

Comments

This genus was revised by Bray (1987a).

Key to the species of *Zoogonus*

1 Genital pore dextral; Pacific coast .... *Zoogonus dextrocirrus* Aldrich, 1961

Site: ?

Host: *Sebastes nebulosus*.

Distribution: Pac.

Record: Holmes 1990.

Remarks: This species has been recorded from pleuronectid, zoarcid, stichaeid, and embiotocid fishes off the coast of Oregon (Bray 1987a).

Genital pore sinistral; Atlantic coast .... *Z. lasius* (Leidy, 1891) Stunkard, 1940 (Fig. 127)

Site: Posterior intestine.

Hosts: *Menidia menidia* (1); *Salvelinus fontinalis* (2, 3).

Distribution: NB-b, NB, Atl.

Records: 1. Bray and Gibson 1986 (NB-b); 2. Frimeth 1987a (NB-b,NB); 3. Frimeth 1987b (Atl, NB-b).

Remarks: This species was commented upon by Bray and Gibson (1986), while discussing *Z. rubellus* (Olsson, 1868). Although Canadian records are from brackish and fresh water, it is likely to be a marine and brackish water species.

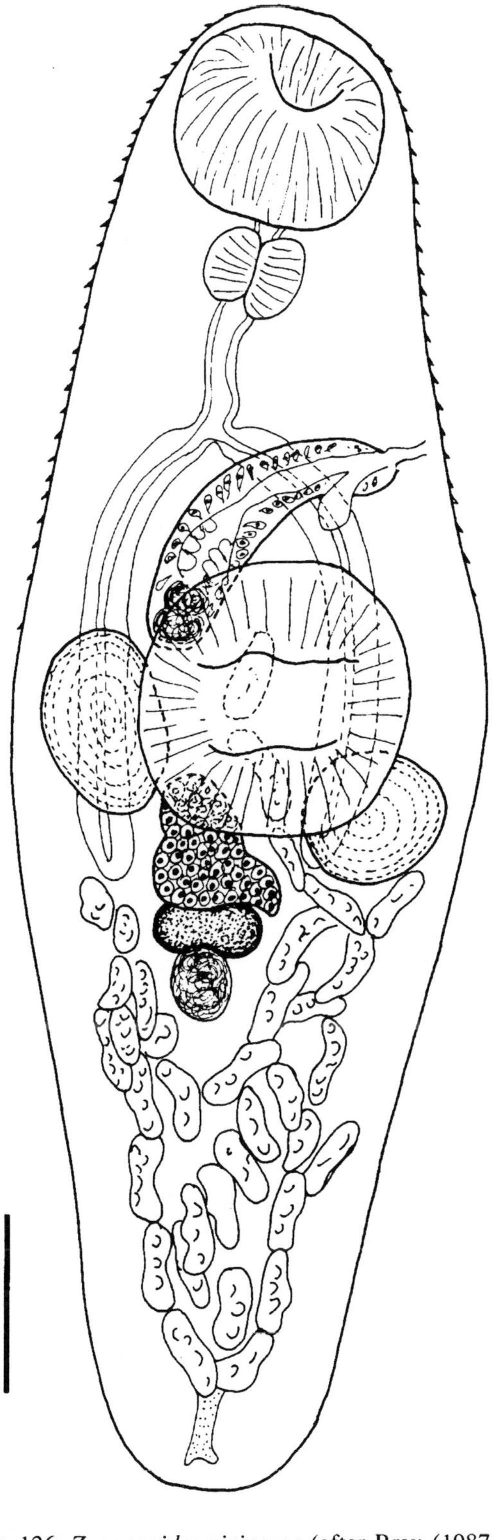

FIG. 126. *Zoogonoides viviparus* (after Bray (1987a)). Scale bar: 100 μm.

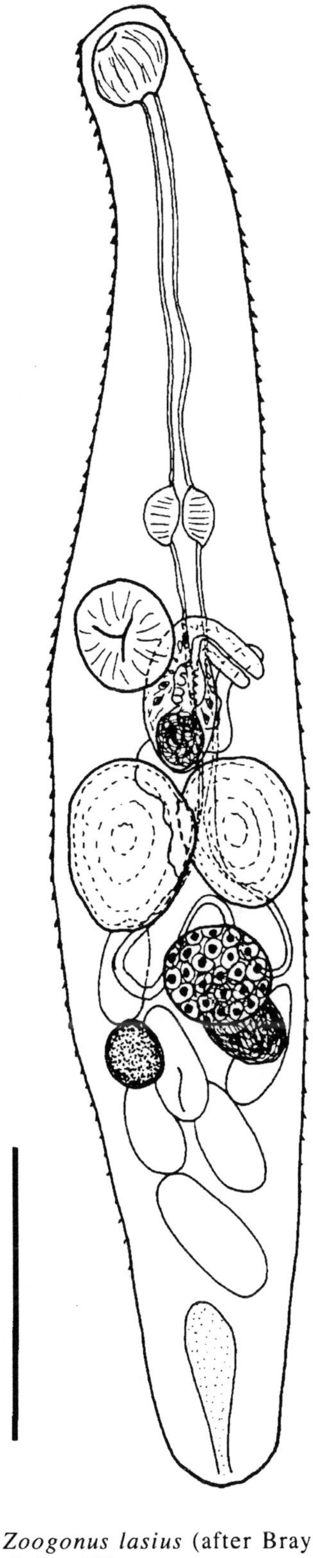

FIG. 127. *Zoogonus lasius* (after Bray and Gibson (1986)). Scale bar: 200 μm.

Subfamily LEPIDOPHYLLINAE Stossich, 1903

Zoogonidae. Body small to large, cylindrical to flattened, oval to spatulate. Body surface usually spinous. Prepharynx usually short to apparently absent. Spines occasionally present in lining of alimentary system. Intestinal bifurcation in forebody. Caeca two, blind. Testes two, in hindbody, at level of ventral sucker or occasionally in posterior forebody. Genital atrium large or small, usually simple, occasionally pocketed or surrounded by gland cells (peri-atrial gland). Genital pore in forebody, lateral. Mehlis' gland well developed. Ovary usually in anterior hindbody, occasionally overlapping ventral sucker or in posterior forebody. Uterus usually mainly post-gonadal, but occasionally partly or mainly pre-gonadal. Eggs with tanned, operculate shells. Vitellarium two symmetrical fields of relatively few distinct globular to subglobular follicles in fore- or hindbody. In alimentary tract, gall bladder, bile duct, or urinary bladder of marine and freshwater teleosts.

Comments

A revision and keys to this subfamily were given by Bray (1987b). It is worth noting that two species of *Urinatrema* Yamaguti, 1934 have been recorded in Northern Pacific waters from the urinary bladder of *Hexagrammos* spp. (see Bray 1987b).

Key to the genera of the Lepidophyllinae

1 Body flattened, spatulate; testes deeply lobed; excretory pore subterminal; in urinary bladder .... *Lepidophyllum*

Body more or less oval or rounded in cross-section; oval to pyriform or fusiform in ventral view; testes more or less entire; excretory pore terminal; not usually found in urinary bladder 2

2 Caeca short, saccular; not reaching into hindbody 3

Ceaca reach to testes or beyond 4

3 Testes not extending into forebody; ovary pre-testicular (occasionally inter-testicular); vitellarium at or near level of ventral sucker or in posterior forebody .... *Brachyenteron*

Testes overlapping ventral sucker into forebody; ovary post-testicular; vitellarium in forebody .... *Panopula*

4 Vitellarium in hindbody 5

Vitellarium in forebody .... *Deretrema*

5 Ventral sucker divided equatorially by row of papillae .... *Steganodermatoides*

Ventral sucker not divided equatorially by row of papillae .... *Steganoderma*

BRACHYENTERON Manter, 1934

Syn.: *Cypseluritrematoides* Yamaguti, 1970; *Prolateroporus* Yamaguti, 1971.

Lepidophyllinae. Body fusiform to pyriform. Body surface spinous varying to apparently aspinous. Oral sucker globular or infundibuliform. Ventral sucker in about mid-region of body, larger than or similar in size to oral sucker; may have tegumental pit on posterior lip. Prepharynx distinct. Pharynx globular, small.

Oesophagus distinct, long. Intestinal bifurcation in forebody. Caeca short, divergent, saccular, terminating in forebody or at level of ventral sucker. Testes subglobular, smooth, symmetrical, or subsymmetrical, close to or overlapping ventral sucker. Cirrus sac claviform to elongate claviform, reaching to ventral sucker or not. Seminal vesicle bipartite or elongate coiled. Pars prostatica vesicular, oval to elongate. Ejaculatory duct long, may be extruded to form cirrus. Genital pore sinistral, marginal or submarginal at level of intestinal bifurcation or pharynx. Ovary globular, overlapping ventral sucker or in forebody, pre-testicular (often dextral) in inter-testicular. Seminal receptacle large. Uterus fills most of hindbody. Metraterm thick-walled, muscular. Eggs numerous, shell tanned. Vitellarium two lateral fields of follicles in forebody or at and around level of ventral sucker, entirely posterior to cirrus sac on poral side. Excretory pore terminal, excretory vesicle elongate saccular. In intestine or gall bladder of marine teleosts.

Comments

This genus was revised by Bray (1987b).

Brachyenteron pycnorganum (Rees, 1953) Overstreet and Pritchard, 1977) (Fig. 128)

Syn.: *Steganoderma pycnorganum* Rees, 1953; *S. spinosa* Polyanski, 1955; *Yamagutia anarhichae* Brinkmann, 1956.
Site: Gall bladder (also occurs in bile duct).
Hosts: *Anarhichas lupus* (1, 3, 4); *A. minor* (1, 2, 3).
Distribution: Atl.
Records: 1. Bray 1979; 2. Zubchenko 1980; 3. Bray and Gibson 1986; 4. Bray 1987b.
Remarks: This species was described in detail by Bray and Gibson (1986).

DERETREMA Linton, 1910

Lepidophyllinae. Body oval to pyriform. Body surface with or without spines. Oral sucker globular, subterminal. Ventral sucker sessile, usually distinctly larger than oral sucker. Prepharynx short. Pharynx oval. Oesophagus usually long, occasionally short. Pharynx and/or oesophagus often spinous internally. Intestinal bifurcation in forebody. Caeca reach to testes or beyond. Testes oval, symmetrical to subsymmetrical, in anterior hindbody or occasionally at level of anterior margin of ventral sucker. Cirrus sac small, claviform. Seminal vesicle tubular, coiled saccular or bipartite. Pars prostatica weakly to well developed. Ejaculatory duct may form cirrus. Genital pore marginal, occasionally ventral, in mid- or occasionally posterior forebody. Ovary entire, oval to globular, pre- or occasionally post-testicular. Uterus mostly post-gonadal to substantially pre-gonadal. Metraterm weakly developed. Eggs numerous, shells tanned. Vitellarium in posterior half of forebody, sometimes overlapping ventral sucker, two lateral fields of relatively few follicles (6–10). Excretory pore terminal, excretory vesicle elongate saccular. In gall bladder or intestine of marine teleosts.

Comments

This genus was revised by Bray (1987b).

Deretrema cholaeum McFarlane, 1936 (Fig. 129)

Syn.: *Deretrema pooli* Annereaux, 1947.
Site: Gall bladder.
Hosts: *Sebastes caurinus* (3); *S. elongatus* (3); *S. maliger* (2, 3); *S. nebulosus* (4); *S. nigrocinctus* (3); *S. ruberrimus* (3); *Sebastes* sp. (1).
Distribution: Pac.
Records: 1. McFarlane 1936; 2. Arai 1969b; 3. Sekerak and Arai 1977; 4. Holmes 1990.
Remarks: This species was also described, as *D. pooli*, from off the coast of California by Annereaux (1947).

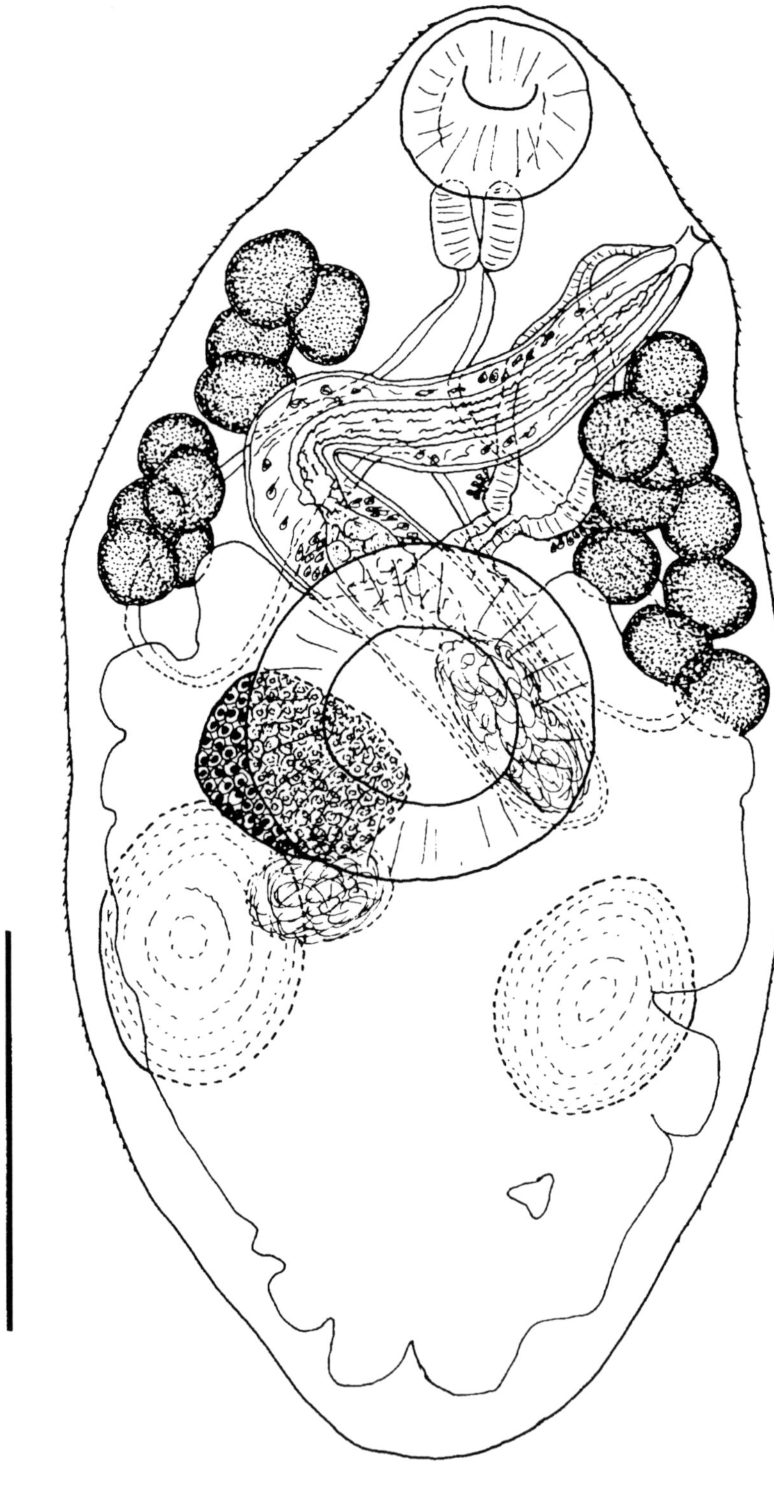

FIG. 128. *Brachyenteron pycnorganum* (after Bray (1987b)). Scale bar: 500 μm.

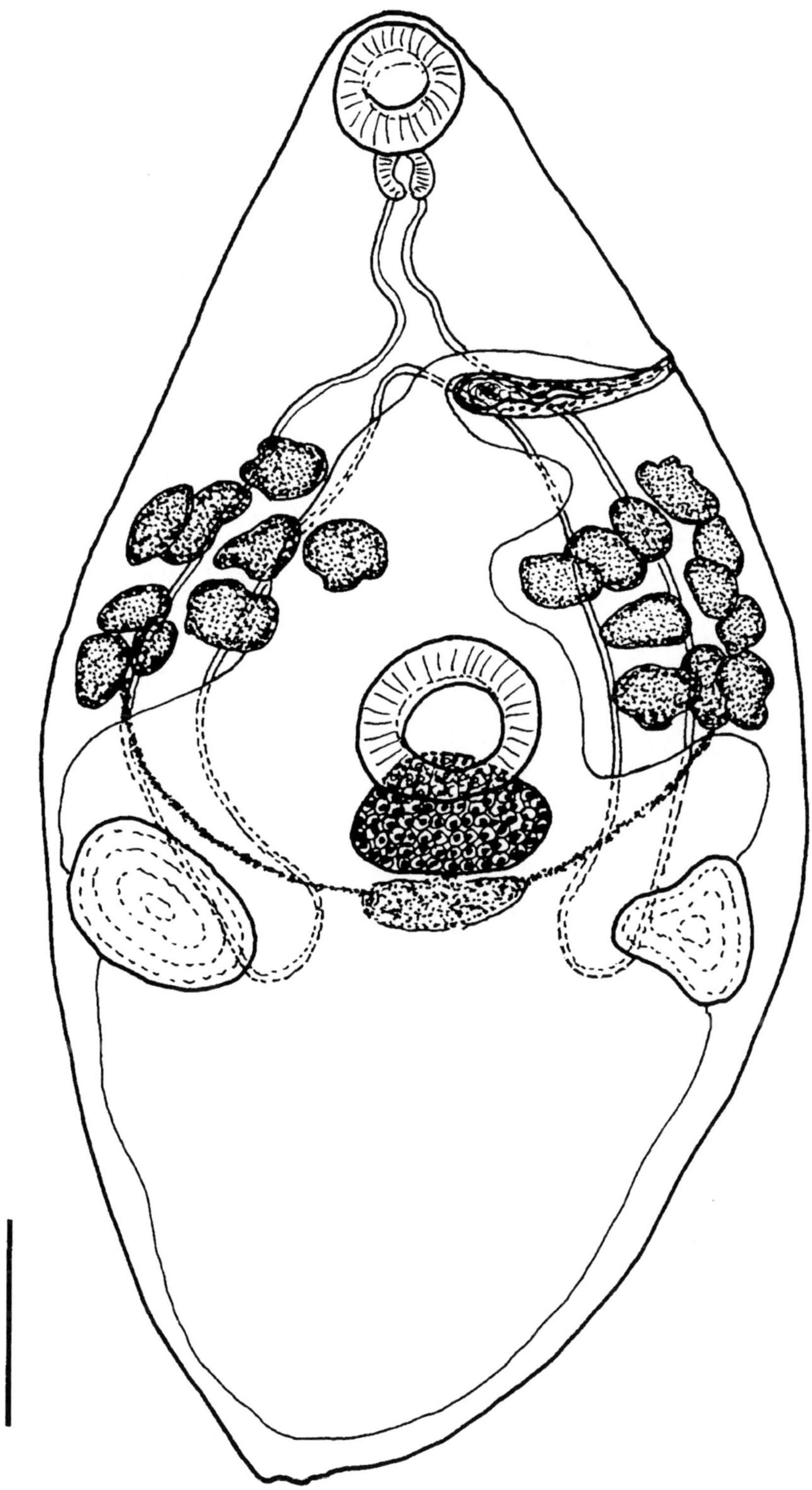

FIG. 129. *Deretrema cholaeum* (after Annereaux (1947: as *D. pooli*)). Scale bar: 500 µm.

LEPIDOPHYLLUM Odhner, 1902

Lepidophyllinae. Body broadly pyriform to ampullaceous, strongly flattened dorso-ventrally. Body surface spinous. Oral sucker small, globular. Ventral sucker small, rounded, in anterior half of body. Prepharynx short. Oesophagus distinct. Caeca reach into hindbody usually to level of vitellarium or testes. Testes lobed to deeply lobed, post-ovarian, symmetrical in mid- to anterior hindbody. Cirrus sac narrow, elongate. Seminal vesicle bipartite. Pars prostatica long. Ejaculatory duct short, junction with pars prostatica indistinct. Genital pore sinistral, dorsally submarginal or marginal, in mid- to anterior forebody. Ovary oval to slightly lobed, in anterior hindbody; submedian. Seminal receptacle distinct, may be large, post-ovarian. Uterus mainly post-testicular, may occupy bulk of this region or be surrounded by broad band of parenchyma. Metraterm muscular. Eggs tanned, filaments of egg-shell material present in uterus. Vitellarium in two groups of about 9–13 distinct globular follicles, anterior to antero-medial to testes. Excretory pore subterminal, excretory vesicle elongate, restricted to posterior hindbody. In urinary bladder of marine teleosts.

Comments

This genus was revised by Bray (1987b).

Key to the species of *Lepidophyllum*

1 Cirrus sac reaching close to posterior margin of ventral sucker; on Pacific coast ....*Lepidophyllum cameroni* Arai, 1969

Syn.: *Lepidophyllum* sp. of Arai (1967a).
Site: Urinary bladder.
Host: *Hemilepidotus hemilepidotus.*
Distribution: Pac.
Records: 1. Arai 1967a; 2. Arai 1969a,b.

Cirrus sac not overlapping or only just reaching ventral sucker; on Atlantic coast...............................2

2 Uterus always surrounded by wide band of parenchyma, with no loops reaching to body margin; ovary always posterior to ventral sucker, with distinct gap between them; ovary oval to smoothly rounded, relatively small; testes relatively small, with on average 5.6 shallow lobes; body spines relatively small; in *Anarhichas* spp ....*L. steenstrupi* Odhner, 1902 (Fig. 130)

Site: Urinary bladder.
Hosts: *Anarhichas denticulatus* (5); *A. lupus* (1, 2, 3, 4, 5, 6, 7, 8); *A. minor* (6).
Distribution: Atl.
Records: 1. Stafford 1904; 2. Stafford 1907; 3. Noble 1973; 4. Linkletter *et al.* 1977; 5. Bray 1979; 6. Zubchenko 1980; 7. Bray and Gibson 1986; 8. Bray 1987b.
Remarks: This species was described and discussed in detail by Bray and Gibson (1986).

Uterus tends to fill post-testicular region and/or loops reach to body margin; ovary often overlaps or is at least contiguous with ventral sucker; ovary relatively large, often smooth but may be distinctly lobed (lobes shallow); testes relatively large, deeply lobed with on average 7.7 lobes; body spines relatively large; in zoarcids ....*L. appyi* Bray and Gibson, 1986

Syn.: *L. steenstrupi* Odhner, 1902 of Stafford (1904, 1907) in part, Bray (1979) in part, Linkletter *et al.* (1977) in part, and Cooper (1915).
Site: Urinary bladder.
Hosts: *Lycodes reticulatus* (5, 6); *L. vahli* (5, 6); *Macrozoarces americanus* (1, 2, 3, 4, 6).
Distribution: Atl.
Records: 1. Stafford 1904; 2. Stafford 1907; 3. Cooper 1915; 4. Linkletter *et al.* 1977; 5. Bray 1979; 6. Bray and Gibson 1986.
Remarks: This species was described and discussed in detail by Bray and Gibson (1986) based mainly upon Canadian material.

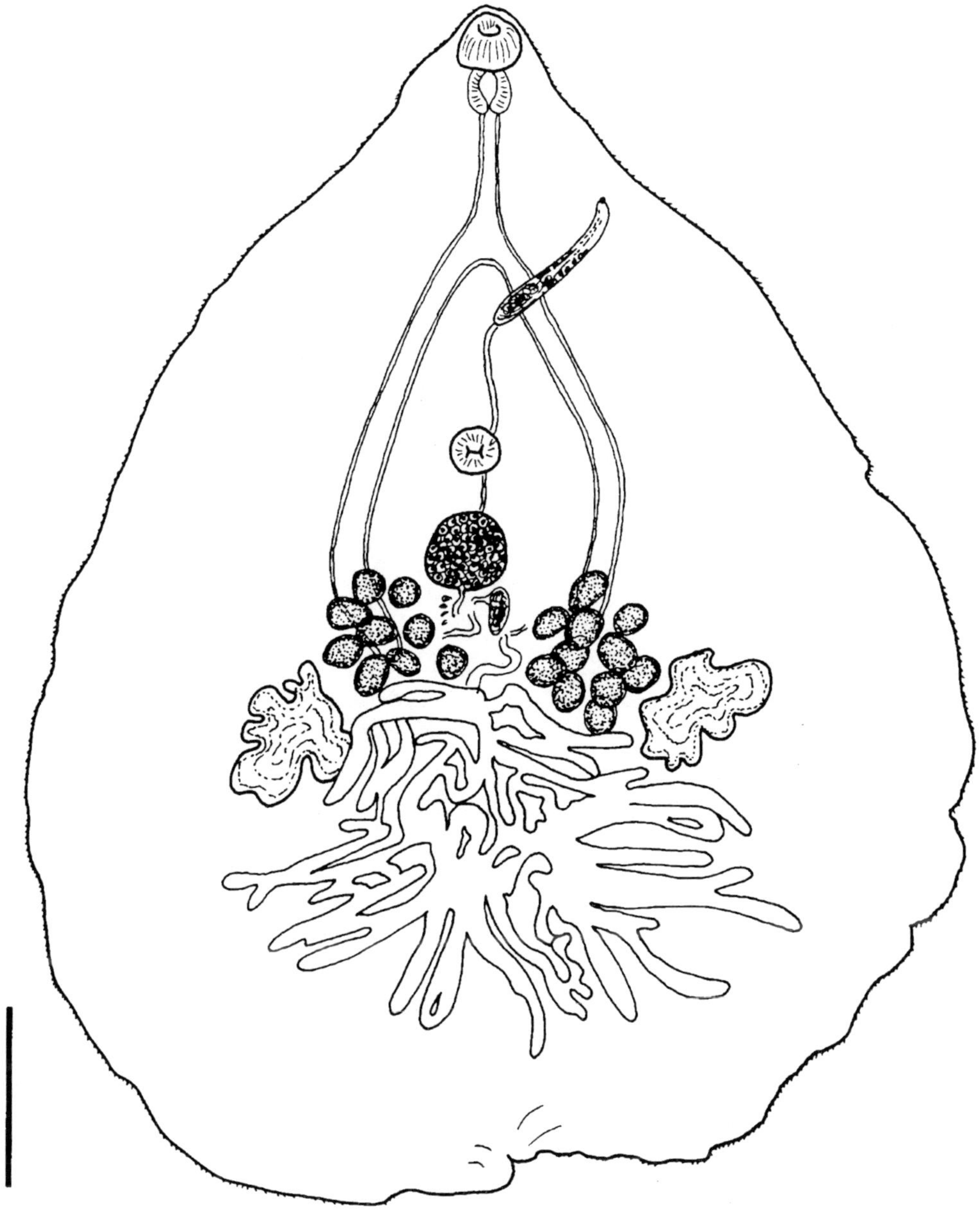

FIG. 130. *Lepidophyllum steenstrupi* (after Bray and Gibson (1986)). Scale bar: 500 µm.

PANOPULA Overstreet and Pritchard, 1977

Lepidophyllinae. Body fusiform. Entire body surface spinous. Oral sucker large, globular to infundibuliform, subterminal to terminal. Ventral sucker similar in size or larger than oral sucker, may have tegumental pit on posterior lip. Prepharynx short. Pharynx oval to globular. Oesophagus distinct. Intestinal bifurcation in forebody. Caeca short, saccular, reaching to testes or just overlapping ventral sucker. Testes symmetrical, at level of mid- to anterior region of ventral sucker or in posterior forebody. Cirrus sac large, claviform. Seminal vesicle bipartite or coiled. Pars prostatica wide, vesicular. Ejaculatory duct long, muscular. Genital atrium simple, distinct. Genital pore sinistral, lateral or sub-lateral, in mid-forebody. Ovary globular to oval, smooth to irregular, postero-dorsal to ventral sucker, post-testicular. Seminal receptacle and Mehlis' gland large. Uterus occupies much of hindbody, may extend dorsally to ventral sucker. Metraterm thick-walled, muscular. Eggs with tanned shells. Vitellarium two lateral fields of follicles, in mid- to posterior forebody, poral field may be divided by cirrus sac. Excretory pore terminal; excretory vesicle saccular. In intestine of deep-sea teleosts.

Comments

This genus was revised by Bray (1987b) and considered a synonym of *Deretrema* by Brooks and McLennan (1993).

Panopula spinosa (Zubchenko, 1978) Bray and Gibson, 1986 (Fig. 131)

Syn.: *Antorchis spinosus* Zubchenko, 1978.
Site: Intestine.
Host: *Notacanthus chemnitzii.*
Distribution: Atl.
Records: 1. Zubchenko 1978; 2. Bray and Gibson 1986.
Remarks: This species, known only from the notocanthid *Notocanthus chemnitzii*, was redescribed by Bray and Gibson (1986). It was transferred to *Deretrema* by Brooks and McLennan (1993).

STEGANODERMA Stafford, 1904

Lepidophyllinae. Body oval to elongate-oval, flattened to cylindrical. Body surface spinous. Oral sucker small, globular. Ventral sucker globular, sessile or pedunculate, similar in size to oral sucker or larger. Prepharynx short. Pharynx small to large. Oesophagus absent to long. Intestinal bifurcation in forebody. Caeca extend to between level of testes and close to posterior extremity. Testes oval, symmetrical to oblique, in anterior or mid-hindbody. Cirrus sac claviform, straight or curved. Seminal vesicle saccular to elongate, straight or coiled. Pars prostatica short to long, vesicular. Ejaculatory duct short. Genital atrium small. Genital pore sinistral, in mid- to anterior forebody, lateral to sub-lateral. Ovary globular to irregularly oval, pre-testicular, median or submedian, normally in hindbody. Mehlis' gland large. Vitellarium symmetrical fields of 8–16 regular follicles, mainly in lateral regions of anterior hindbody. Uterus in hindbody, mainly post-testicular, but may have significant proportion pre-testicular. Metraterm muscular. Eggs numerous, tanned. Excretory pore terminal; excretory vesicle saccular, short to long. In intestine of marine teleosts.

Comments

This genus was revised by Bray (1987b).

Steganoderma formosum Stafford, 1904 (Fig. 132)

Syn.: *Steganoderma messjatzevi* (Issaichikov, 1928) Yamaguti, 1934; *Deretrema messjatzevi* (Issaichikov, 1928) Manter, 1954.
Sites: Intestine, pyloric caeca.

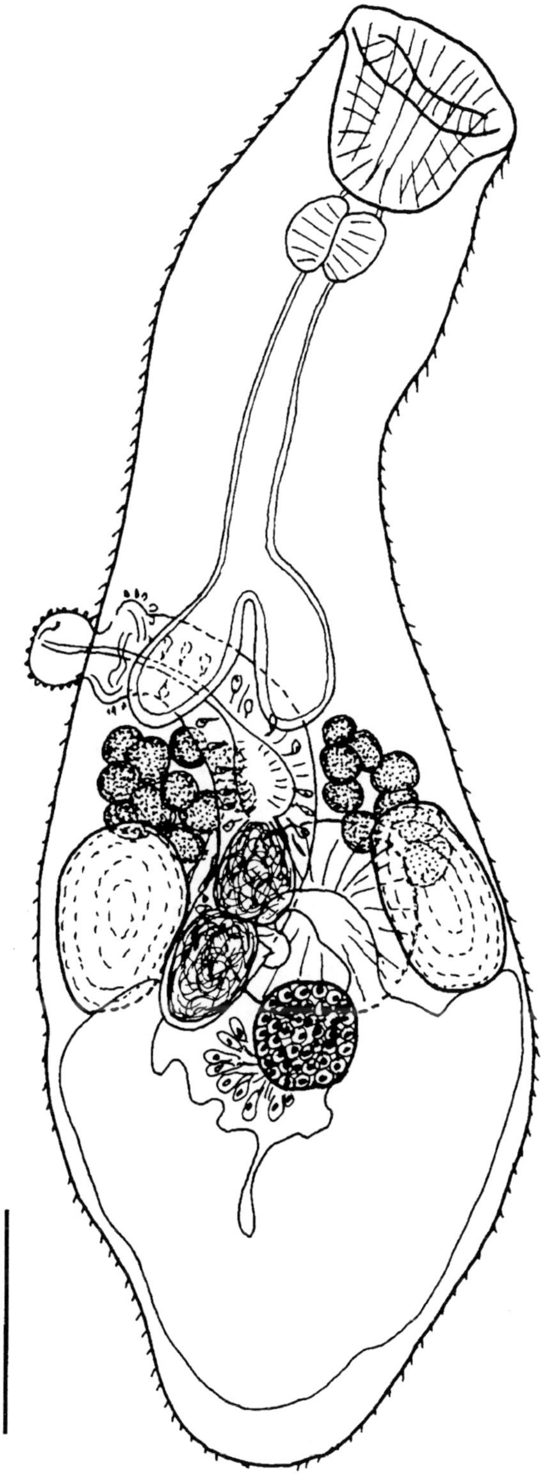

FIG. 131. *Panopula spinosa* (after Bray and Gibson (1986)). Scale bar: 500 μm.

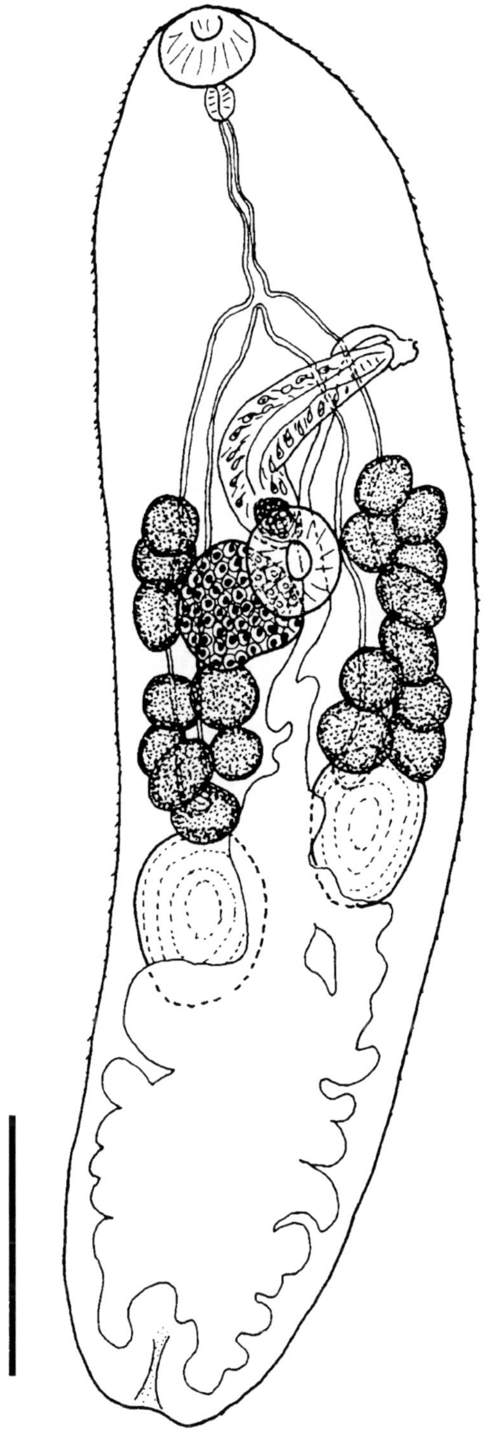

FIG. 132. *Steganoderma formosum* (after Bray (1987b)). Scale bar: 500 μm.

Hosts: *Atheresthes stomias* (14); *Gadus morhua* (12); *Hemilepidotus hemilepidotus* (5); *Hippoglossus hippoglossus* (1, 3, 9, 10, 20, 25); *Myoxocephalus octodecemspinosus* (9); *M. scorpius* (9); *Pleuronectes americanus* (7, 11, 15); *P. ferrugineus* (6, 11, 16); *Raja laevis* (accidental) (2); *Reinhardtius hippoglossoides* (19, 20, 21, 22, 23, 24); *Sebastes fasciatus* (18); *Squalus acanthias* (accidental) (2, 4); *Theragra chalcogramma* (13); *Urophycis chuss* (17); *U. musicki* (17).

Distribution: Atl, Pac.

Records: 1. Stafford 1904 (Atl); 2. Myers 1959 (Atl); 3. Ronald 1960 (Atl); 4. Threlfall 1969 (Atl); 5. Arai 1969b (Pac); 6. Scott 1975c (Atl); 7. Scott 1976 (Atl); 8. Sekerak and Arai 1977 (Pac); 9. Bray 1979 (Atl); 10. Zubchenko 1980 (Atl); 11. Scott 1982 (Atl); 12. Appy and Burt 1982 (Atl); 13. Arthur 1984 (Pac); 14. Kabata and Whitaker 1984 (Pac); 15. Scott 1985b (Atl); 16. Morrison *et al.* 1986 (Atl); 17. Scott 1987 (Atl); 18. Scott 1988 (Atl); 19. Wierzbicka 1988 (Atl); 20. Scott and Bray 1989 (Atl); 21. Wierzbicka 1991a (Atl); 22. Wierzbicka 1991b (Atl); 23. Krzykawski and Wierzbicka 1992 (Atl); 24. Arthur and Albert 1994 (Atl); 25. BM(NH) collection (Atl).

Remarks: This species was discussed and figured by Bray (1987b). Its distribution appears to be restricted to the Northwest Atlantic, Barents Sea, and the northern North Pacific Ocean. Canadian material was described by Stafford (1904), Miller (1941a), and Ronald (1960).

STEGANODERMATOIDES Parukhin and Lyadov, 1979

Syn.: *Hudsonia* Campbell, 1975, *nec* Edwards, 1923, *nec* Leroux, 1940; *Allosteganoderma* Campbell, 1983.

Lepidophyllinae. Body fusiform to elongate fusiform. Body surface bears large, prominent spines. Oral sucker globular to transversely elongate, subterminal to terminal. Ventral sucker large, in about middle of body, internal equator bears five to seven large papillae. Prepharynx distinct. Pharynx oval. Oesophagus long. Intestinal bifurcation in posterior forebody. Caeca reach to testes. Testes symmetrical to diagonal, in mid- or posterior hindbody. Cirrus sac large, curved. Seminal vesicle bipartite, coiled-tubular, or (?)saccular. Pars prostatica long, convoluted. Ejaculatory duct muscular. Genital atrium large, muscular, surrounded by gland cells (peri-atrial gland). Genital pore on left margin, posterior to middle of forebody. Ovary oval, pre-testicular, submedian. Seminal receptacle large, post-ovarian. Uterus in hindbody, mainly in pre-testicular zone. Metraterm strongly muscular with glandular sheath. Eggs numerous, tanned. Vitellarium two symmetrical lateral fields of follicles at posterior edge of ventral sucker. Excretory pore terminal; excretory vesicle saccular, reaching to testes or not. In gut (especially rectum and posterior intestine) of deep-sea teleosts.

Comments

A key to the species of this genus was given by Bray (1987b). It was considered a synonym of *Proctophantastes* Odhner, 1911 by Brooks and McLennan (1993).

Steganodermatoides agassizi (Campbell, 1975) Bray, 1985 (Fig. 133)

Syn.: *Hudsonia agassizi* Campbell, 1975; *Allosteganoderma agassizi* (Campbell, 1975) Campbell, 1983.

Site: ? (probably posterior intestine).

Hosts: *Alepocephalus agassizii*; *A. bairdii*.

Distribution: Atl.

Record: Zubchenko 1984.

Remarks: Bray and Gibson (1986) described a second Atlantic species from *Alepocephalus*, *Steganodermatoides maceri*. They recorded this species from *A. bairdii* in the Northeast Atlantic and considered that Zubchenko's (1984) record of *S. agassizi* from *A. bairdii* in the Northeast Atlantic might belong to this species. It is not known for certain, however, whether Zubchenko's undescribed Northwest Atlantic material is *S. agassizi*, originally described from the Hudson Submarine Canyon off the Atlantic coast of the United States, *S. maceri*, or a combination of the two species. Bray (1987b) has suggested that Zubchenko's Northwest Atlantic material from *A. bairdii* might

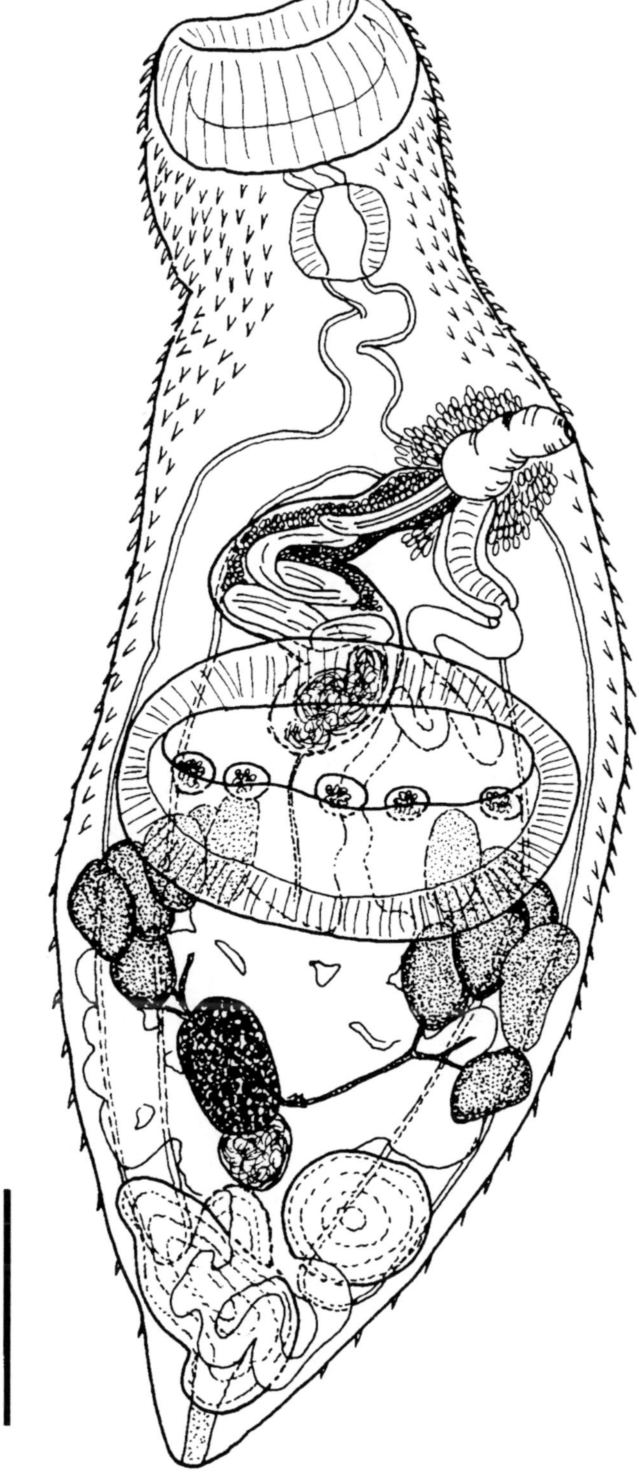

FIG. 133. *Steganodermatoides agassizi* (after Campbell (1975)). Scale bar: 500 µm.

be *S. maceri*. According to Bray and Gibson (1986), *S. maceri* differs from *S. agassizi* in having a significant amount of the uterus in the post-testicular region, with the result that the testes are relatively further forward and the excretory vesicle does not always reach the testes, and the eggs are significantly smaller (28–33 µm, as opposed to 48–55 µm). *S. agassizi* was transferred to *Proctophantastes* by Brooks and McLennan (1993).

HOST-DIGENEAN PARASITE LIST

The following list of hosts, in systematic order, includes the currently used name and recently used synonyms. The scientific names, recent synonyms, and common names are indexed in the Host Index. The parasite records normally exclude experimental infections, records of metacercariae not attributable to particular species, and in many cases records of adults identified to the generic level and not attributable to particular species. Full host-parasite lists can be found in Margolis and Arthur (1979) and McDonald and Margolis (1995).

CLASS CEPHALASPIDOMORPHI

ORDER PETROMYZONTIFORMES

Family Petromyzontidae

Lampetra richardsoni Vladykov and Follett — western brook lamprey
Ophioxenos microphagus

Petromyzon marinus Linnaeus — sea lamprey
Diplostomum huronense metacercaria
Allopodocotyle lepomis

CLASS ELASMOBRANCHII

ORDER SQUALIFORMES

Family Squalidae

Centroscyllium fabricii (Reinhardt) — black dogfish
Otodistomum cestoides

Squalus acanthias Linnaeus — spiny dogfish
Derogenes varicus (accidental)
Hemiurus levinseni (accidental)
Otodistomum sp. metacercaria
Prosorhynchus squamatus (accidental)
Steganoderma formosum (accidental)

ORDER RAJIFORMES

Family Rajidae

Bathyraja richardsoni (Garrick) — deepwater skate
Otodistomum cestoides

Raja jenseni Bigelow and Schroeder — shorttail skate
Otodistomum cestoides

Raja laevis Mitchill — barndoor skate
Hemiurus levinseni (accidental)
Otodistomum cestoides
Steganoderma formosum (accidental)

Raja ocellata Mitchill — winter skate
Podocotyle atomon (accidental)
Otodistomum cestoides

Raja radiata Donovan — thorny skate
Otodistomum cestoides

CLASS HOLOCEPHALI

ORDER CHIMAERIFORMES

Family Chimaeridae

Hydrolagus colliei (Lay and Bennett) — spotted ratfish
Rugogaster hydrolagi

Family Rhinochimaeridae

Rhinochimaera atlantica Holt and Byrne — knifenose chimaera
Lecithocladium sp. (accidental)
Opecoeloides vitellosis (accidental)

CLASS ACTINOPTERYGII

ORDER ACIPENSERIFORMES

Family Acipenseridae

Acipenser brevirostrum Le Sueur — shortnose sturgeon
Spirorchiidae gen. sp. (?accidental)

Acipenser fulvescens Rafinesque — lake sturgeon
Azygia longa
Bunodera luciopercae
Crepidostomum auriculatum
Diplostomum spathaceum metacercaria
Homalometron armatum
Skrjabinopsolus manteri

Acipenser oxyrhynchus Mitchill — Atlantic sturgeon
Derogenes varicus
Deropristis hispida

Acipenser transmontanus Richardson — white sturgeon
Crepidostomum auriculatum

ORDER LEPISOSTEIFORMES

FAMILY LEPISOSTEIDAE

Lepisosteus osseus (Linnaeus) — longnose gar
Apophallus venustus metacercaria
Diplostomum spathaceum metacercaria
Macroderoides spiniferus

ORDER AMIIFORMES

Family Amiidae

Amia calva Linnaeus — bowfin

Apophallus venustus metacercaria
Azygia longa
Crepidostomum cornutum
Diplostomum spathaceum metacercaria
Leuceruthrus micropteri
Macroderoides trilobatus
Macroderoides typicus
Microphallus opacus

ORDER OSTEOGLOSSIFORMES

Family Hiodontidae

Hiodon alosoides (Rafinesque) — goldeye

Paurorhynchus hiodontis

Hiodon tergisus Le Sueur — mooneye

Crepidostomum illinoiense
"*Heterophyes*" = (?) *Apophallus* sp. metacercaria
Plagioporus serratus
Paurorhynchus hiodontis

ORDER NOTACANTHIFORMES

Family Notacanthidae

Notacanthus chemnitzi Bloch — largescale tapirfish

Panopula spinosa

ORDER ANGUILLIFORMES

Family Anguillidae

Anguilla rostrata (Le Sueur) — American eel

Azygia longa
Bunodera luciopercae
Centrovarium lobotes
Crepidostomum brevivitellum
Crepidostomum cornutum
Derogenes varicus
Deropristis inflata
Diplostomum flexicaudum metacercaria
Diplostomum spathaceum metacercaria
Microphallus opacus
Podocotyle atomon
Stephanostomum tenue

Order Clupeiformes

Family Clupeidae

Alosa aestivalis (Mitchill) — blueback herring
- *Brachyphallus crenatus*
- *Derogenes varicus*
- *Diplostomum spathaceum* metacercaria
- *Lecithaster confusus*

Alosa pseudoharengus (Wilson) — alewife/gaspereau
- *Brachyphallus crenatus*
- *Derogenes varicus*
- *Diplostomum spathaceum* metacercaria
- *Diplostomum spathaceum indistinctum* metacercaria
- *Hemiurus levinseni*
- *Ichthyocotylurus erraticus* metacercaria
- *Lecithaster confusus*
- *Posthodiplostomum minimum* metacercaria

Alosa sapidissima (Wilson) — American shad
- *Brachyphallus crenatus*
- *Derogenes varicus*
- *Genitocotyle atlantica*
- *Hemiurus levinseni*
- *Podocotyle reflexa*

Clupea harengus Linnaeus — Atlantic herring
- *Brachyphallus crenatus*
- *Cryptocotyle lingua* metacercaria
- *Derogenes varicus*
- *Hemiurus levinseni*
- *Lecithaster confusus*
- *Lecithaster gibbosus*

Clupea pallasi Valenciennes — Pacific herring

Syn.: *Clupea harengus pallasi*
- *Brachyphallus crenatus*
- *Galactosomum phalacrocoracis* metacercaria
- *Lecithaster gibbosus*
- *Parahemiurus merus*
- *Prosorhynchoides basargini* metacercaria
- *Pronoprymna petrowi*
- *Rhipidocotyle* sp. metacercaria

Dorosoma cepedianum (Le Sueur) — gizzard shad
- *Diplostomum spathaceum* metacercaria

Order Cypriniformes

Family Cyprinidae

Acrocheilus alutaceus Agassiz and Pickering — chiselmouth
- *Posthodiplostomum minimum* metacercaria

Campostoma anomalum (Rafinesque) — central stoneroller
- *Posthodiplostomum minimum* metacercaria

Couesius plumbeus (Agassiz) lake chub
Allocreadium lobatum
Crassiphiala bulboglossa metacercaria
Diplostomum baeri bucculentum metacercaria
Diplostomum spathaceum metacercaria
Plagioporus cooperi
Posthodiplostomum minimum metacercaria

Cyprinus carpio Linnaeus common carp
Apophallus venustus metacercaria
Crepidostomum cooperi
Diplostomum flexicaudum metacercaria
Diplostomum spathaceum metacercaria

Ericymba buccata Cope silverjaw minnow
Crassiphiala bulboglossa metacercaria

Hybognathus hankinsoni Hubbs brassy minnow
Diplostomum spathaceum metacercaria
Rhipidocotyle sp. metacercaria
Ornithodiplostomum ptychocheilus metacercaria
Posthodiplostomum minimum minimum metacercaria
Uvulifer ambloplitis metacercaria

Hybopsis storeriana (Kirtland) silver chub
Plagioporus cooperi

Luxilis cornutus (Mitchill) common shiner
Syn.: *Notropis cornutus* (Mitchill)
Allocreadium lobatum
Apophallus venustus metacercaria
Bunodera sacculata
Clinostomum complanatum metacercaria
Crassiphiala bulboglossa metacercaria
Diplostomum flexicaudum metacercaria
Diplostomum spathaceum metacercaria
Neascus pyriformis metacercaria
Ornithodiplostomum ptychocheilus metacercaria
Phyllodistomum semotili
Plagioporus cooperi
Plagioporus sinitsini
Posthodiplostomum minimum metacercaria
Posthodiplostomum minimum minimum metacercaria
Tylodelphys scheuringi metacercaria
Uvulifer ambloplitis metacercaria

Margariscus margarita Cope pearl dace
Syn.: *Semotilus margarita* (Cope)
Uvulifer ambloplitis metacercaria

Mylocheilus caurinus (Richardson) peamouth
Allocreadium lobatum
Clinostomum complanatum metacercaria
Posthodiplostomum minimum metacercaria
Tylodelphys scheuringi metacercaria

Nocomis biguttatus (Kirtland) hornyhead chub

Allocreadium lobatum
Diplostomum spathaceum metacercaria
Ornithodiplostomum ptychocheilus metacercaria
Plagioporus sinitsini
Posthodiplostomum minimum metacercaria
Posthodiplostomum minimum minimum metacercaria
Uvulifer ambloplitis metacercaria

Nocomis micropogon (Cope) river chub

Posthodiplostomum minimum metacercaria

Notemigonus crysoleucas (Mitchill) golden shiner

Allocreadium lobatum
Apophallus brevis metacercaria
Clinostomum complanatum metacercaria
Crassiphiala bulboglossa metacercaria
Diplostomum spathaceum metacercaria
Echinochasmus sp. metacercaria
Plagiocirrus primus
Posthodiplostomum minimum metacercaria
Uvulifer ambloplitis metacercaria

Notropis atherinoides Rafinesque emerald shiner

Diplostomum flexicaudum metacercaria
Diplostomum spathaceum metacercaria
Leuceruthrus micropteri
Plagioporus cooperi
Posthodiplostomum minimum metacercaria
Posthodiplostomum minimum minimum metacercaria

Notropis emiliae Hay pugnose minnow

Plagioporus cooperi

Notropis heterodon (Cope) blackchin shiner

Posthodiplostomum minimum metacercaria

Notropis heterolepis Eigenmann and Eigenmann blacknose shiner

Clinostomum complanatum metacercaria
Diplostomum spathaceum metacercaria
Ornithodiplostomum ptychocheilus metacercaria
Posthodiplostomum minimum metacercaria
Uvulifer ambloplitis metacercaria

Notropis hudsonius (Clinton) spottail shiner

Allocreadium lobatum
Apophallus brevis metacercaria
Bucephalus sp. metacercaria
Centrovarium lobotes
Centrovarium lobotes metacercaria
Clinostomum complanatum metacercaria
Crassiphiala bulboglossa metacercaria
Cryptogonimus chili metacercaria
Diplostomum spathaceum metacercaria
Plagioporus cooperi

Plagioporus sinitsini
Posthodiplostomum minimum metacercaria
Posthodiplostomum minimum minimum metacercaria
Sanguinicola sp.
Uvulifer ambloplitis metacercaria

Notropis rubellus (Agassiz) — rosyface shiner
Diplostomum spathaceum metacercaria
Posthodiplostomum minimum metacercaria

Notropis spilopterus (Cope) — spotfin shiner
Plagioporus cooperi

Notropis stramineus (Cope) — sand shiner
Plagioporus cooperi

Notropis volucellus (Cope) — mimic shiner
Centrovarium lobotes metacercaria
Plagioporus cooperi
Posthodiplostomum minimum metacercaria

Phoxinus eos (Cope) — northern redbelly dace
Syn.: *Chrosomus eos* Cope
Clinostomum complanatum metacercaria
Ichthyocotylurus platycephalus metacercaria
Posthodiplostomum minimum metacercaria
Uvulifer ambloplitis metacercaria

Phoxinus neogaeus (Cope, *in* Günther) — finescale dace
Syn.: *Chrosomus neogaeus* (Cope, *in* Günther)
Clinostomum complanatum metacercaria
Diplostomum spathaceum metacercaria
Posthodiplostomum minimum metacercaria
Uvulifer ambloplitis metacercaria

Pimephales notatus (Rafinesque) — bluntnose minnow
Centrovarium lobotes metacercaria
Clinostomum complanatum metacercaria
Crassiphiala bulboglossa metacercaria
Diplostomum spathaceum metacercaria
Diplostomum spathaceum indistinctum metacercaria
Ornithodiplostomum ptychocheilus metacercaria
Plagioporus cooperi
Plagioporus sinitsini
Posthodiplostomum minimum metacercaria
Posthodiplostomum minimum minimum metacercaria
Sanguinicola sp.
Uvulifer ambloplitis metacercaria

Pimephales promelas Rafinesque — fathead minnow
Allocreadium lobatum
Clinostomum complanatum metacercaria
Crassiphiala bulboglossa metacercaria
Diplostomum spathaceum metacercaria
Ornithodiplostomum ptychocheilus metacercaria
Posthodiplostomum minimum metacercaria
Posthodiplostomum minimum minimum metacercaria

Sanguinicola sp.
Uvulifer ambloplitis metacercaria

Ptychocheilus oregonensis (Richardson) — northern squawfish
Allocreadium lobatum
Diplostomum baeri bucculentum metacercaria
Diplostomum spathaceum metacercaria
Ichthyocotylurus platycephalus metacercaria
Posthodiplostomum minimum metacercaria
Tylodelphys scheuringi metacercaria

Rhinichthys atratulus (Hermann) — blacknose dace
Clinostomum complanatum metacercaria
Neascus rhinichthysi metacercaria
Posthodiplostomum minimum metacercaria

Rhinichthys cataractae (Valenciennes) — longnose dace
Allocreadium lobatum
Crassiphiala bulboglossa metacercaria
Diplostomum spathaceum metacercaria
Neascus rhinichthysi metacercaria
Plagioporus cooperi
Posthodiplostomum minimum metacercaria
Tylodelphys scheuringi metacercaria

Richardsonius balteatus (Richardson) — redside shiner
Allocreadium lobatum
Clinostomum complanatum metacercaria
Posthodiplostomum minimum metacercaria
Tylodelphys scheuringi metacercaria

Semotilus atromaculatus (Mitchill) — creek chub
Allocreadium lobatum
Bucephalus sp. metacercaria
Clinostomum complanatum metacercaria
Crassiphiala bulboglossa metacercaria
Diplostomum spathaceum metacercaria
Lissorchis attenuatus
Neascus pyriformis metacercaria
Ornithodiplostomum ptychocheilus metacercaria
Phyllodistomum semotili
Posthodiplostomum minimum metacercaria
Posthodiplostomum minimum minimum metacercaria
Uvulifer ambloplitis metacercaria

Semotilus corporalis (Mitchill) — fallfish
Allocreadium lobatum
Clinostomum complanatum metacercaria
Crassiphiala bulboglossa metacercaria
Metorchis conjunctus metacercaria
Posthodiplostomum minimum metacercaria
Tylodelphys scheuringi metacercaria

Family Catostomidae

Carpiodes cyprinus (Le Sueur) quillback
- *Diplostomum spathaceum* metacercaria
- *Lissorchis attenuatus*
- *Lissorchis gullaris*
- *Phyllodistomum lysteri*
- *Posthodiplostomum minimum* metacercaria
- *Posthodiplostomum minimum minimum* metacercaria
- *Sanguinicola* sp.

Catostomus catostomus (Forster) longnose sucker
- *Allocreadium lobatum*
- *Bucephalus* sp. metacercaria
- *Crepidostomum farionis*
- *Diplostomum flexicaudum* metacercaria
- *Diplostomum spathaceum* metacercaria
- *Diplostomum spathaceum indistinctum* metacercaria
- *Ichthyocotylurus platycephalus* metacercaria
- *Lissorchis attenuatus*
- *Lissorchis gullaris*
- *Lissorchis simeri*
- *Metorchis conjunctus* metacercaria
- *Phyllodistomum lysteri*
- (?) *Phyllodistomum superbum*
- *Plagiocirrus* sp.
- *Plagioporus sinitsini*
- *Posthodiplostomum minimum* metacercaria
- *Sanguinicola* sp.

Catostomus commersoni (Lacépède) white sucker
- *Allocreadium lobatum*
- *Amphimerus elongatus* metacercaria
- *Amphimerus pseudofelineus* metacercaria
- *Apophallus venustus* metacercaria
- *Clinostomum complanatum* metacercaria
- *Crepidostomum cooperi*
- *Diplostomum flexicaudum* metacercaria
- *Diplostomum spathaceum* metacercaria
- *Diplostomum spathaceum indistinctum* metacercaria
- *Ichthyocotylurus erraticus* metacercaria
- *Ichthyocotylurus platycephalus* metacercaria
- *Lissorchis attenuatus*
- *Lissorchis simeri*
- *Metorchis conjunctus* metacercaria
- *Phyllodistomum lysteri*
- *Plagioporus serotinus*
- *Plagioporus sinitsini*
- *Posthodiplostomum minimum* metacercaria
- *Sanguinicola* sp.
- *Uvulifer ambloplitis* metacercaria

Catostomus macrocheilus Girard largescale sucker
- *Ichthyocotylurus platycephalus* metacercaria

Lissorchis attenuatus
Posthodiplostomum minimum metacercaria
Tylodelphys scheuringi metacercaria

Moxostoma anisurum (Rafinesque) — silver redhorse
Diplostomum flexicaudum metacercaria
Lissorchis attenuatus
Sanguinicola sp.

Moxostoma erythrurum (Rafinesque) — golden redhorse
Diplostomum flexicaudum metacercaria
Sanguinicola sp.

Moxostoma macrolepidotum (Le Sueur) — shorthead redhorse
Syn.: *Moxostoma aureolum* (Le Sueur)
Apophallus venustus metacercaria
Diplostomum flexicaudum metacercaria
Phyllodistomum lysteri
Plagioporus serotinus
Sanguinicola sp.

ORDER SILURIFORMES

Family Ictaluridae

Ameiurus nebulosus (Le Sueur) — brown bullhead
Syn.: *Ictalurus nebulosus* (Le Sueur)
Acetodextra amiuri
Alloglossidium corti
Alloglossidium gemimum
Apophallus venustus metacercaria
Azygia angusticauda
Centrovarium lobotes
Clinostomum complanatum metacercaria
Crepidostomum cornutum
Diplostomum spathaceum metacercaria
Homalometron armatum
Megalogonia ictaluri
Microphallus opacus
Phyllodistomum staffordi
(?) *Phyllodistomum superbum*
Polylecithum ictaluri
Posthodiplostomum minimum metacercaria
Rhipidocotyle sp. metacercaria
Vietosoma parvum

Ictalurus melas (Rafinesque) — black bullhead
Clinostomum complanatum metacercaria
Crepidostomum cornutum
Leuceruthrus micropteri
Megalogonia ictaluri
Phyllodistomum staffordi
Polylecithum ictaluri

Ictalurus punctatus (Rafinesque) — channel catfish

- *Acetodextra amiuri*
- *Alloglossidium corti*
- *Apophallus venustus* metacercaria
- *Azygia angusticauda*
- *Azygia longa*
- *Clinostomum complanatum* metacercaria
- *Crepidostomum cornutum*
- *Diplostomum spathaceum* metacercaria
- *Megalogonia ictaluri*
- *Microphallus opacus*
- *Phyllodistomum staffordi*
- *Vietosoma parvum*

Noturus flavus Rafinesque — stonecat

- *Acetodextra amiuri*
- *Alloglossidium corti*
- *Alloglossidium gemimum*
- *Megalogonia ictaluri*

Noturus gyrinus (Mitchill) — tadpole madtom

- *Acetodextra amiuri*
- *Alloglossidium corti*
- *Crepidostomum cornutum*
- *Diplostomum spathaceum* metacercaria
- *Megalogonia ictaluri*

Noturus miurus Jordan — brindled madtom

- *Megalogonia ictaluri*

Order Salmoniformes

Family Esocidae

Esox americanus vermiculatus Le Sueur — grass pickerel

- *Azygia angusticauda*
- *Centrovarium lobotes*
- *Crassiphiala bulboglossa* metacercaria
- *Macroderoides flavus*

Esox lucius Linnaeus — northern pike

- *Allocreadium lobatum*
- *Apophallus venustus* metacercaria
- *Azygia angusticauda*
- *Azygia longa*
- *Bunodera luciopercae*
- *Centrovarium lobotes*
- *Clinostomum complanatum* metacercaria
- *Crepidostomum cooperi*
- *Crepidostomum farionis*
- *Diplostomum spathaceum* metacercaria
- *Microphallus opacus*
- (?) *Phyllodistomum staffordi*
- (?) *Phyllodistomum superbum*
- *Posthodiplostomum minimum* metacercaria
- *Uvulifer ambloplitis* metacercaria

Esox masquinongy Mitchill	muskellunge
Azygia angusticauda	
Azygia longa	
Diplostomum flexicaudum metacercaria	
(?) *Phyllodistomum staffordi*	
Esox niger LeSueur	chain pickerel
Crepidostomum cooperi	
Macroderoides flavus	
Microphallus opacus	
(?) *Phyllodistomum superbum*	

Family Umbridae

Umbra limi (Kirtland)	central mudminnow
Bunodera eucaliae	
Crassiphiala bulboglossa metacercaria	
Creptotrema funduli	
Diplostomum spathaceum metacercaria	
(?) *Phyllodistomum staffordi*	

SUBORDER ARGENTINOIDEI

Family Argentinidae

Argentina silus (Ascanius)	Atlantic argentine
Derogenes varicus	
Hemiurus levinseni	
Lampritrema miescheri (imm.)	
Lecithophyllum botryophorum	

Family Alepocephalidae

Alepocephalus agassizii Goode and Bean	dusky slickhead
Dinosoma triangulata	
Lecithophyllum botryophorum	
Olssonium turneri	
Steganodermatoides agassizi	
Steringophorus pritchardae	
Alepocephalus bairdii Goode and Bean	manyray smoothhead
Dinosoma triangulata	
Lecithophyllum botryophorum	
Olssonium turneri	
Steganodermatoides agassizi	
Steringophorus pritchardae	

SUBORDER SALMONOIDEI

Family Osmeridae

Hypomesus pretiosus (Girard)	surf smelt
Lecithaster gibbosus	
Pronoprymna petrowi	

Mallotus villosus (Müller) capelin
Derogenes varicus
Hemiurus levinseni
Lecithaster gibbosus
Pronoprymna petrowi

Osmerus mordax (Mitchill) rainbow smelt
Brachyphallus crenatus
Cryptocotyle lingua metacercaria
Derogenes varicus
Diplostomum flexicaudum metacercaria
Diplostomum huronense metacercaria
Diplostomum spathaceum metacercaria
Diplostomum spathaceum indistinctum metacercaria
Ichthyocotylurus erraticus metacercaria
Lecithaster confusus

Spirinchus thaleichthys (Ayres) longfin smelt
Pronoprymna petrowi

Thaleichthys pacificus (Richardson) eulachon
Lecithaster gibbosus
Pronoprymna petrowi

Family Salmonidae

Coregonus artedii Le Sueur lake cisco
Crepidostomum cooperi
Crepidostomum farionis
Diplostomum spathaceum metacercaria
Diplostomum spathaceum indistinctum metacercaria
Ichthyocotylurus erraticus metacercaria

Coregonus clupeaformis (Mitchill) lake whitefish
Crepidostomum cooperi
Crepidostomum farionis
Crepidostomum metoecus
Diplostomum baeri bucculentum metacercaria
Diplostomum flexicaudum metacercaria
Diplostomum spathaceum metacercaria
Diplostomum spathaceum indistinctum metacercaria
Ichthyocotylurus erraticus metacercaria
Phyllodistomum coregoni
Phyllodistomum umblae
Tylodelphys scheuringi metacercaria

Coregonus hoyi (Gill) bloater
Crepidostomum farionis
Diplostomum spathaceum metacercaria
Diplostomum spathaceum indistinctum metacercaria
Ichthyocotylurus erraticus metacercaria

Coregonus nasus (Pallas) broad whitefish
Brachyphallus crenatus

Coregonus sardinella Valenciennes — least cisco
- *Crepidostomum farionis*
- *Diplostomum baeri bucculentum* metacercaria

Oncorhynchus clarki (Richardson) — cutthroat trout
Syn.: *Salmo clarki* Richardson
- *Clinostomum complanatum* metacercaria
- *Crepidostomum farionis*
- *Crepidostomum metoecus*

Oncorhynchus gorbuscha (Walbaum) — pink salmon
- *Brachyphallus crenatus*
- *Copiatestes filiferus*
- *Crepidostomum farionis*
- *Derogenes varicus*
- *Galactosomum phalacrocoracis* metacercaria
- *Genolinea anura*
- *Hemiurus levinseni*
- *Lecithaster gibbosus*
- *Lecithophyllum botryophorum*
- *Parahemiurus merus*
- *Podocotyle angulata*
- *Podocotyle reflexa*
- *Pronoprymna petrowi*
- *Prosorhynchoides basargini*
- *Tubulovesicula lindbergi*

Oncorhynchus keta (Walbaum) — chum salmon
- *Brachyphallus crenatus*
- *Crepidostomum farionis*
- *Galactosomum phalacrocoracis* metacercaria
- *Hemiurus levinseni*
- *Lecithaster gibbosus*
- *Lecithophyllum botryophorum*
- *Tubulovesicula lindbergi*

Oncorhynchus kisutch (Walbaum) — coho salmon
- *Brachyphallus crenatus*
- *Crepidostomum farionis*
- *Crepidostomum metoecus*
- *Derogenes varicus*
- *Diplostomum spathaceum* metacercaria
- *Hemiurus levinseni*
- *Lecithaster gibbosus*
- *Parahemiurus merus*
- *Pronoprymna petrowi*
- *Tubulovesicula lindbergi*

Oncorhynchus mykiss (Walbaum) — rainbow trout
Syn.: *Salmo gairdneri* Richardson
- *Allocreadium lobatum*
- *Bunodera luciopercae*
- *Clinostomum complanatum* metacercaria
- *Crepidostomum farionis*

Diplostomum baeri bucculentum metacercaria
Diplostomum spathaceum metacercaria
Ichthyocotylurus erraticus metacercaria
Nanophyetus salmincola metacercaria
Phyllodistomum umblae
Plagioporus shawi
Stephanostomum tenue metacercaria
Tylodelphys scheuringi metacercaria

Oncorhynchus nerka (Walbaum) — sockeye salmon
Brachyphallus crenatus
Copiatestes filiferus
Crepidostomum farionis
Derogenes varicus
Diplostomum baeri bucculentum metacercaria
Diplostomum spathaceum metacercaria
Genolinea anura
Hemiurus levinseni
Ichthyocotylurus erraticus metacercaria
Lecithaster gibbosus
Lecithophyllum botryophorum
Parahemiurus merus
Phyllodistomum umblae
Tubulovesicula lindbergi

Oncorhynchus tshawytscha (Walbaum) — chinook salmon
Brachyphallus crenatus
Crepidostomum farionis
Diplostomum baeri bucculentum metacercaria
Diplostomum spathaceum metacercaria
Genolinea anura
Lampritrema miescheri (imm.)
Lecithaster gibbosus
Tubulovesicula lindbergi

Oncorhynchus sp.
Lecithaster gibbosus

Prosopium cylindraceum (Pallas) — round whitefish
Apophallus brevis metacercaria
Bunodera luciopercae
Crepidostomum cooperi
Crepidostomum farionis
Diplostomum spathaceum metacercaria
Ichthyocotylurus erraticus metacercaria
Phyllodistomum coregoni

Prosopium williamsoni (Girard) — mountain whitefish
Allocreadium lobatum
Crepidostomum farionis
Diplostomum baeri bucculentum metacercaria
Diplostomum spathaceum metacercaria
Ichthyocotylurus platycephalus metacercaria
Tylodelphys scheuringi metacercaria

Salmo salar Linnaeus — Atlantic salmon

- *Apophallus brevis* metacercaria
- *Apophallus imperator* metacercaria
- *Azygia longa*
- *Brachyphallus crenatus*
- *Bunodera luciopercae*
- *Crepidostomum cooperi*
- *Crepidostomum farionis*
- *Derogenes varicus*
- *Diplostomum spathaceum* metacercaria
- *Hemiurus levinseni*
- *Lampritrema miescheri* (imm.)
- *Lecithaster gibbosus*
- *Phyllodistomum umblae*
- (?) *Podocotyle angulata*

Salmo trutta Linnaeus — brown trout

- *Apophallus brevis* metacercaria
- *Crepidostomum farionis*

Salvelinus alpinus (Linnaeus) — Arctic charr

- *Brachyphallus crenatus*
- *Bunodera luciopercae*
- *Crepidostomum cooperi*
- *Crepidostomum farionis*
- *Derogenes varicus*
- *Diplostomum spathaceum* metacercaria
- *Lecithaster gibbosus*
- *Phyllodistomum umblae*
- *Prosorhynchus squamatus*

Salvelinus fontinalis (Mitchill) — brook charr

- *Apophallus brevis* metacercaria
- *Apophallus imperator* metacercaria
- *Azygia angusticauda*
- *Azygia longa*
- *Brachyphallus crenatus*
- *Bunodera luciopercae*
- *Clinostomum complanatum* metacercaria
- *Crepidostomum cooperi*
- *Crepidostomum cornutum*
- *Crepidostomum farionis*
- *Crepidostomum metoecus*
- *Cryptocotyle lingua* metacercaria
- *Derogenes varicus*
- *Diplostomum spathaceum* metacercaria
- "Heterophyid" = (?) *Apophallus* sp. metacercaria
- *Ichthyocotylurus erraticus* metacercaria
- *Lecithaster gibbosus*
- *Metorchis conjunctus* metacercaria
- (?) *Phyllodistomum superbum*
- *Phyllodistomum umblae*
- *Podocotyle angulata*

Podocotyle atomon
Posthodiplostomum minimum metacercaria
Stephanostomum tenue
Stephanostomum tenue metacercaria
Zoogonus lasius

Salvelinus fontinalis (Mitchill) × *S. namaycush* (Walbaum) — splake
Crepidostomum farionis
Diplostomum spathaceum metacercaria
Diplostomum spathaceum indistinctum metacercaria
Ichthyocotylurus erraticus metacercaria

Salvelinus malma (Walbaum) — Dolly Varden
Brachyphallus crenatus
Bunodera luciopercae
Crepidostomum farionis
Diplostomum baeri bucculentum metacercaria
Diplostomum spathaceum metacercaria
Lecithaster gibbosus
Prosorhynchoides basargini
Tubulovesicula lindbergi

Salvelinus namaycush (Walbaum) — lake charr
Syn.: *Cristivomer namaycush* (Walbaum)
Azygia angusticauda
Azygia longa
Brachyphallus crenatus
Bunodera luciopercae
Crepidostomum cooperi
Crepidostomum farionis
Crepidostomum metoecus
Diplostomum baeri bucculentum metacercaria
Diplostomum spathaceum metacercaria
Phyllodistomum umblae

Thymallus arcticus (Pallas) — Arctic grayling
Allocreadium lobatum
Crepidostomum farionis
Crepidostomum metoecus
Diplostomum spathaceum metacercaria

ORDER PERCOPSIFORMES

Family Percopsidae

Percopsis omiscomaycus (Walbaum) — trout-perch
Centrovarium lobotes
Centrovarium lobotes metacercaria
Crepidostomum isostomum
Diplostomum flexicaudum metacercaria
Diplostomum spathaceum metacercaria
Ichthyocotylurus erraticus metacercaria
Ichthyocotylurus pileatus metacercaria
Ichthyocotylurus platycephalus metacercaria

Posthodiplostomum minimum metacercaria
Tylodelphys scheuringi metacercaria

Order Batrachoidiformes

Family Batrachoididae

Porichthys notatus Girard — plainfin midshipman
Tubulovesicula lindbergi

Order Gadiformes

Family Macrouridae

Coryphaenoides rupestris Gunnerus — rock grenadier
Syn.: *Macrourus rupestris* (Gunnerus)
Aporocotyle sp.
Derogenes varicus
Dolichoenterum sp. metacercaria
Glomericirrus macrouri
Gonocerca phycidis

Macrourus berglax Lacépède — roughhead grenadier
Derogenes varicus
Genolinea laticauda
Gibsonia borealis
Gonocerca phycidis
Gonocerca sp.
Hemiurus levinseni
Lecithophyllum botryophorum
Lepidapedon elongatum

Nezumia bairdi (Goode and Bean) — marlin-spike
Derogenes varicus
Genolinea laticauda
Glomericirrus macrouri

Family Moridae

Lepidion eques (Günther) — largeye lepidion
Lepidapedon elongatum
Podocotyle reflexa
Progonus muelleri
Steringophorus furciger

Family Gadidae

Boreogadus saida (Lepechin) — Arctic cod
Brachyphallus crenatus
Derogenes varicus
Hemiurus levinseni
Lecithaster gibbosus

Gadus macrocephalus Tilesius — Pacific cod
Lecithaster gibbosus
Parahemiurus merus

Gadus morhua Linnaeus — Atlantic cod

Brachyphallus crenatus
Cryptocotyle lingua metacercaria
Derogenes varicus
Gonocerca macroformis
Gonocerca phycidis
Hemiurus levinseni
Lecithaster gibbosus
Lepidapedon elongatum
Lepidapedon rachion
Neophasis burti
Otodistomum sp. metacercaria
Podocotyle atomon
Podocotyle reflexa
Prosorhynchus squamatus
Steganoderma formosum
Stenakron vetustum
Stephanostomum baccatum
Steringophorus furciger

Melanogrammus aeglefinus (Linnaeus) — haddock

Cryptocotyle lingua metacercaria
Derogenes varicus
Genolinea laticauda
Hemiurus levinseni
Lepidapedon rachion
Prosorhynchus squamatus

Microgadus tomcod (Walbaum) — Atlantic tomcod

Podocotyle reflexa

Pollachius virens (Linnaeus) — pollock

Brachyphallus crenatus
Cryptocotyle lingua metacercaria
Derogenes varicus
Genolinea laticauda
Hemiurus levinseni
Lecithaster gibbosus
Podocotyle reflexa

Theragra chalcogramma (Pallas) — walleye pollock

Aporocotyle theragrae
Brachyphallus crenatus
Copiatestes filiferus
Derogenes varicus
Hemiurus levinseni
Lecithaster gibbosus
Lecithophyllum botryophorum
Lepidapedon gadi
Parahemiurus merus
Podocotyle theragrae
Prosorhynchoides basargini metacercaria
Prosorhynchus sp. metacercaria
Rhipidocotyle sp. metacercaria

Steganoderma formosum
Stephanostomum sp. metacercaria

Family Lotidae

Brosme brosme (Ascanius) — cusk
Prosorhynchus squamatus

Lota lota (Linnaeus) — burbot
Azygia angusticauda
Azygia longa
Bunodera luciopercae
Bunodera sacculata
Crepidostomum cooperi
Crepidostomum farionis
Diplostomum baeri bucculentum metacercaria
Diplostomum spathaceum metacercaria
Diplostomum spathaceum indistinctum metacercaria
Posthodiplostomum minimum metacercaria
Tylodelphys scheuringi metacercaria

Family Phycidae

Urophycis chesteri (Goode and Bean) — longfin hake
Syn.: *Phycis chesteri* Goode and Bean
Anomalotrema koiae
Derogenes varicus
Dissosaccus laevis
Genolinea laticauda
Hemiurus levinseni
Lepidapedon elongatum
Lethadena profunda
Paraccacladium jamiesoni

Urophycis chuss (Walbaum) — red hake
Derogenes varicus
Dissosaccus laevis
Genolinea laticauda
Hemiurus levinseni
Lecithaster gibbosus
Lecithophyllum botryophorum
Lepidapedon elongatum
Otodistomum sp. metacercaria
Podocotyle atomon
Podocotyle refexa
Steganoderma formosum
Steringophorus furciger

Urophycis musicki Cohen and Lavenberg — white hake
Syn.: *Urophycis tenuis* (Mitchill)
Derogenes varicus
Dissosaccus laevis
Genolinea laticauda
Hemiurus levinseni

Lecithaster gibbosus
Lecithophyllum botryophorum
Lepidapedon elongatum
Otodistomum sp. metacercaria
Podocotyle reflexa
Progonus muelleri
Pseuodopecoelus vulgaris
Steganoderma formosum
Stenakron vetustum
Steringophorus furciger

Family Merlucciidae

Merluccius albidus (Mitchill) — offshore hake
Derogenes varicus
Hemiurus levinseni
Lecithaster gibbosus
Lepidapedon elongatum

Merluccius bilinearis (Mitchill) — silver hake
Derogenes varicus
Hemiurus levinseni
Lecithaster gibbosus
Podocotyle reflexa
Pseudopecoelus vulgaris

Merluccius productus (Ayres) — Pacific hake
Aporocotyle margolisi
Derogenes varicus
Lecithaster gibbosus
Lecithophyllum botryophorum
Parahemiurus merus

ORDER LOPHIIFORMES

Family Lophiidae

Lophius americanus Valenciennes — monkfish
Derogenes varicus
Otodistomum sp. metacercaria
Progonus muelleri
Stephanostomum baccatum metacercaria

ORDER ATHERINIFORMES

Family Atherinidae

Labidesthes sicculus (Cope) — brook silverside
Allacanthochasmus varius metacercaria
Diplostomum spathaceum metacercaria
Posthodiplostomum minimum metacercaria

Menidia menidia (Linnaeus) — Atlantic silverside
Lasiotocus elongatus
Zoogonus lasius

Order Cyprinodontiformes

Family Cyprinodontidae

Fundulus diaphanus (Le Sueur) — banded killifish
- *Clinostomum complanatum* metacercaria
- *Crassiphiala bulboglossa* metacercaria
- *Creptotrema funduli*
- *Diplostomum spathaceum* metacercaria
- *Ornithodiplostomum ptychocheilus* metacercaria
- *Posthodiplostomum minimum* metacercaria
- *Posthodiplostomum minimum minimum* metacercaria
- *Tylodelphys scheuringi* metacercaria

Fundulus heteroclitus (Linnaeus) — mummichog
- *Crepidostomum cooperi*
- *Homalometron pallidum*
- *Stephanostomum tenue* metacercaria

Order Gasterosteiformes

Family Gasterosteidae

Apeltes quadracus (Mitchill) — fourspine stickleback
- *Podocotyle atomon*
- *Stephanostomum tenue* metacercaria

Aulorhynchus flavidus Gill — tube-snout
- *Lecithaster gibbosus*
- *Podocotyle endophrysi*

Culaea inconstans (Kirtland) — brook stickleback

Syn.: *Eucalia inconstans* Kirtland
- *Apatemon gracilis* metacercaria
- *Bunodera eucaliae*
- *Culaeatrema inconstans*
- *Diplostomum scudderi* metacercaria
- *Diplostomum spathaceum* metacercaria
- *Posthodiplostomum minimum* metacercaria

Gasterosteus aculeatus Linnaeus — threespine stickleback
- *Apatemon gracilis* metacercaria
- *Apophallus brevis* metacercaria
- *Brachyphallus crenatus*
- *Bunodera eucaliae*
- *Bunodera luciopercae*
- *Bunodera mediovitellata*
- *Crepidostomum cooperi*
- *Crepidostomum farionis*
- *Derogenes varicus*
- *Diplostomum spathaceum* metacercaria
- *Diplostomum scudderi* metacercaria
- *Lecithaster gibbosus*
- *Neolebouria tinkerbellae* (exper.)
- *Podocotyle angulata*

Podocotyle atomon
Podocotyle sinusacca
Posthodiplostomum minimum metacercaria

Gasterosteus wheatlandi Putnam — blackspotted stickleback
Podocotyle atomon

Pungitius pungitius (Linnaeus) — ninespine stickleback
Apatemon gracilis metacercaria
Brachyphallus crenatus
Bunodera eucaliae
Bunodera luciopercae
Derogenes varicus
Diplostomum spathaceum metacercaria
Ichthyocotylurus erraticus metacercaria
Podocotyle atomon

Order Syngnathiformes

Family Syngnathidae

Syngnathus leptorhynchus (Girard) — bay pipefish
Syn.: *Syngnathus griseolineatus* Ayres
Lecithaster gibbosus
Opechona occidentalis
Parahemiurus merus
Podocotyle endophrysi
Podocotyle sinusacca
Tubulovesicula lindbergi

Order Scorpaeniformes

Suborder Scorpaenoidei

Family Scorpaenidae

Sebastes aleutianus (Jordan and Evermann) — rougheye rockfish
Derogenes varicus
Helicometra sebastis
Lecithophyllum botryophorum
Paralepidapedon sebastici
Podocotyle araii
Prosorhynchus sp.
Psettarium sebastodorum

Sebastes alutus (Gilbert) — Pacific ocean perch
Brachyphallus crenatus
Copiatestes filiferus
Derogenes varicus
Fellodistomum sebastodis
Helicometra sebastis
Lecithaster gibbosus
Lecithophyllum botryophorum
Opechona alaskensis

Opechona occidentalis
Paralepidapedon sebastici
Podocotyle araii
Podocotyle theragrae
Prosorhynchus sp.
Prosorhynchus sp. metacercaria
Psettarium sebastodorum
Tubulovesicula lindbergi

Sebastes aurora (Gilbert) — aurora rockfish
Psettarium sebastodorum

Sebastes babcocki (Thompson) — redbanded rockfish
Derogenes varicus
Paralepidapedon sebastici
(?) *Podocotyle araii*
Prosorhynchus sp.
Psettarium sebastodorum
Stephanostomum baccatum

Sebastes borealis Barsukov — shortraker rockfish
Derogenes varicus
Lecithophyllum botryophorum
Paralepidapedon sebastici
Podocotyle araii
Prosorhynchus sp.
Stephanostomum baccatum
Tubulovesicula lindbergi

Sebastes brevispinis (Bean) — silvergray rockfish
Syn.: *Sebastodes brevispinis* (Bean)
Copiatestes filiferus
Derogenes varicus
Podocotyle araii
Prosorhynchus sp.
Psettarium sebastodorum
Tubulovesicula lindbergi

Sebastes caurinus Richardson — copper rockfish
Aporocotyle macfarlani
Deretrema cholaeum
Derogenes varicus
Helicometra sebastis
Lecithaster gibbosus
Lecithochirium exodicum
Opechona occidentalis
Neolepidapedon pugetensis
Parahemiurus merus
Phyllodistomum umblae (accidental)
Podocotyle theragrae
Psettarium sebastodorum
Stephanostomum californicum
Tubulovesicula lindbergi

Sebastes ciliatus (Tilesius) — dusky rockfish
- *Opechona alaskensis*
- *Opechona occidentalis*

Sebastes crameri (Jordan) — darkblotched rockfish
- *Derogenes varicus*
- *Lecithophyllum botryophorum*
- *Prosorhynchus* sp.
- *Psettarium sebastodorum*
- *Tubulovesicula lindbergi*

Sebastes diploproa (Gilbert) — splitnose rockfish
- *Derogenes varicus*
- *Helicometra sebastis*
- *Phyllodistomum umblae* (accidental)
- (?) *Podocotyle araii*
- *Prosorhynchus* sp.
- *Psettarium sebastodorum*

Sebastes elongatus Ayers — greenstriped rockfish
- *Deretrema cholaeum*
- *Derogenes varicus*
- *Fellodistomum sebastodis*
- *Lecithochirium exodicum*
- *Opechona occidentalis*
- *Podocotyle araii*
- *Prosorhynchus* sp.

Sebastes entomelas (Jordan and Gilbert) — widow rockfish
- *Derogenes varicus*
- *Prosorhynchus* sp.

Sebastes fasciatus Storer — Acadian redfish
- *Anomalotrema koiae*
- *Derogenes varicus*
- *Hemiurus levinseni*
- *Lecithaster gibbosus*
- *Lecithophyllum botryophorum*
- *Podocotyle reflexa*
- *Steganoderma formosum*

Sebastes flavidus (Ayres) — yellowtail rockfish
- *Aporocotyle macfarlani*
- *Copiatestes filiferus*
- *Derogenes varicus*
- *Helicometra sebastis*
- *Lecithaster gibbosus*
- *Lecithochirium exodicum*
- *Lecithophyllum botryophorum*
- *Opechona alaskensis*
- *Opechona occidentalis*
- (?) *Podocotyle araii*
- *Prosorhynchus* sp.
- *Prosorhynchus* sp. metacercaria
- *Psettarium sebastodorum*
- *Stephanostomum* sp. metacercaria

Sebastes goodei (Eigenmann and Eigenmann) chilipepper
Derogenes varicus
(?) *Podocotyle araii*

Sebastes helvomaculatus Ayres rosethorn rockfish
Derogenes varicus
Helicometra sebastis
Paralepidapedon sebastici
(?) *Podocotyle araii*
Psettarium sebastodorum

Sebastes maliger (Jordan and Gilbert) quillback rockfish
Aporocotyle macfarlani
Deretrema cholaeum
Derogenes varicus
Helicometra sebastis
Opechona alaskensis
Pellamyzon abitionis
(?) *Podocotyle araii*
Podocotyle theragrae
Psettarium sebastodorum
Stephanostomum californicum
Tubulovesicula lindbergi

Sebastes marinus (Linnaeus) golden redfish
Syn.: *Sebastes norvegicus* (Ascanius)
Anomalotrema koiae
Brachyphallus crenatus
Derogenes varicus
Hemiurus levinseni
Lecithaster gibbosus
Lecithophyllum botryophorum
Podocotyle atomon
Podocotyle reflexa

Sebastes melanops Girard black rockfish
Aporocotyle macfarlani
Prosorhynchus sp.
Prosorhynchus sp. metacercaria
Psettarium sebastodorum

Sebastes mentella Travin deepwater redfish
Anomalotrema koiae
Derogenes varicus
Hemiurus levinseni
Lecithaster gibbosus
Lecithophyllum botryophorum
Podocotyle reflexa

Sebastes nebulosus Ayres China rockfish
Deretrema cholaeum
Derogenes varicus
Genolinea anura
Helicometra sebastis
Helicometrina nimia

Hemiurus levinseni
Lecithaster gibbosus
Lecithochirium exodicum
Neolepidapedon pugetensis
Opechona occidentalis
Parahemiurus merus
Phyllodistomum umblae (accidental)
Podocotyle radfistuli
Prosorhynchus sp.
Stephanostomum californicum
Tubulovesicula lindbergi
Zoogonus dextrocirrus

Sebastes nigrocinctus Ayres — tiger rockfish
Deretrema cholaeum
Helicometra sebastis
Psettarium sebastodorum

Sebastes paucispinis Ayres — bocaccio
Derogenes varicus
Lecithochirium exodicum
Lecithophyllum botryophorum
Paralepidapedon sebastici
(?) *Podocotyle araii*
Prosorhynchus sp.
Psettarium sebastodorum
Tubulovesicula lindbergi

Sebastes pinniger (Gill) — canary rockfish
Aporocotyle macfarlani
Copiatestes filiferus
Derogenes varicus
Lecithaster gibbosus
Opechona alaskensis
Paralepidapedon sebastici
(?) *Podocotyle araii*
Prosorhynchus sp.
Psettarium sebastodorum
Tubulovesicula lindbergi

Sebastes proriger (Jordan and Gilbert) — redstripe rockfish
Copiatestes filiferus
Derogenes varicus
Lecithaster gibbosus
(?) *Podocotyle araii*
Psettarium sebastodorum

Sebastes reedi (Westrheim and Tsuyuki) — yellowmouth rockfish
Derogenes varicus
Helicometra sebastis
Lecithophyllum botryophorum
(?) *Podocotyle araii*
Prosorhynchus sp.
Psettarium sebastodorum

Sebastes ruberrimus (Cramer) — yelloweye rockfish

Deretrema cholaeum
Derogenes varicus
Fellodistomum sebastodis
Helicometra sebastis
Paralepidapedon sebastici
Phyllodistomum umblae (accidental)
(?) *Podocotyle araii*
Prosorhynchus sp.

Sebastes variegatus Quast — harlequin rockfish

Derogenes varicus
Helicometra sebastis
Lecithaster gibbosus
(?) *Podocotyle araii*
Psettarium sebastodorum

Sebastes wilsoni (Gilbert) — pygmy rockfish

(?) *Podocotyle araii*

Sebastes zacentrus (Gilbert) — sharpchin rockfish

Derogenes varicus
Fellodistomum sebastodis
Helicometra sebastis
Lecithaster gibbosus
Opechona alaskensis
Paralepidapedon sebastici
(?) *Podocotyle araii*
Prosorhynchus sp.

Psettarium sebastodorum

Sebastes sp.
Aporocotyle macfarlani
Deretrema cholaeum
Opechona occidentalis
Pellamyzon abitionis
Prosorhynchus sp. metacercaria
Stephanostomum californicum

Suborder Anoplopomatoidei

Family Anoplopomatidae

Anoplopoma fimbria (Pallas) — sablefish

Brachyphallus crenatus
Derogenes varicus
Didymozoidae gen. sp.
Lecithaster gibbosus
Lecithochirium exodicum
Parahemiurus merus
Pseudopecoelus vulgaris
Stephanostomum californicum
Steringophorus brevis

Suborder Hexagrammoidei

Family Hexagrammidae

Hexagrammos decagrammus (Pallas)	kelp greenling
Genolinea anura	
Lecithaster gibbosus	
Podocotyle sinusacca	
Hexagrammos lagocephalus (Pallas)	rock greenling
Anisorchis opisthorchis	
Genolinea anura	
Lecithaster gibbosus	
Lecithophyllum botryophorum	
Hexagrammos stelleri Tilesius	whitespotted greenling
Anisorchis opisthorchis	
Lecithaster gibbosus	
Lecithophyllum botryophorum	
Parahemiurus merus	
Podocotyle sinusacca	
Ophiodon elongatus Girard	lingcod
Lecithaster gibbosus	
Lecithochirium exodicum	
Parahemiurus merus	
Podocotyle theragrae	
Prosorhynchus apertus	
Rhipidocotyle elongata	
Stephanostomum tristephanum	
Tubulovesicula lindbergi	

Suborder Cottoidea

Family Cottidae

Artediellus uncinatus (Reinhardt)	snowflake hookear
Anisorchis opisthorchis	
Progonus muelleri	
Artedius fenestralis Jordan and Gilbert	padded sculpin
Genolinea anura	
Artedius harringtoni Jordan and Gilbert	scalyhead sculpin
Stephanostomum californicum	
Blepsias cirrhosus (Pallas)	silverspotted sculpin
Genolinea anura	
Lecithaster gibbosus	
Podocotyle radfistuli	
Podocotyle sinusacca	
Cottus asper Richardson	prickly sculpin
Crepidostomum isostomum	
Diplostomum baeri bucculentum metacercaria	
Diplostomum spathaceum metacercaria	
Posthodiplostomum minimum metacercaria	
Tylodelphys scheuringi metacercaria	

Cottus bairdi Girard | mottled sculpin
Diplostomum spathaceum metacercaria
Phyllodistomum simile

Cottus cognatus Richardson | slimy sculpin
"Heterophyid" = (?) *Apophallus* sp. metacercaria
Crepidostomum cooperi
Crepidostomum farionis
Diplostomum spathaceum metacercaria

Cottus ricei (Nelson) | spoonhead sculpin
Diplostomum spathaceum metacercaria

Dasycottus setiger Bean | spinyhead sculpin
Lecithaster gibbosus
Lecithophyllum botryophorum
Stephanostomum baccatum

Hemilepidotus hemilepidotus (Tilesius) | red Irish lord
Anisorchis opisthorchis
Anomalotrema putjatini
Derogenes varicus
Genolinea anura
Lecithaster gibbosus
Lepidophyllum cameroni
Podocotyle sinusacca
Steganoderma formosum

Hemitripterus americanus (Gmelin) | sea raven
Cryptocotyle lingua metacercaria
Derogenes varicus
Genolinea laticauda
Hemiurus levinseni
Podocotyle atomon
Podocotyle reflexa
Prosorhynchus squamatus
Stenakron vetustum
Stephanostomum baccatum

Leptocottus armatus Girard | Pacific staghorn sculpin
Neolebouria tinkerbellae (exper.)
Podocotyle radfistuli
Podocotyle sinusacca
Rhipidocotyle elongata
Tubulovesicula lindbergi

Myoxocephalus octodecemspinosus (Mitchill) | longhorn sculpin
Derogenes varicus
Genolinea laticauda
Hemiurus levinseni
Neophasis burti
Podocotyle atomon
Prosorhynchus sp.
Steganoderma formosum
Stephanostomum baccatum
Steringophorus furciger
Tubulovesicula lindbergi

Myoxocephalus polyacanthocephalus (Pallas) — great sculpin
- *Genolinea anura*
- *Podocotyle endophrysi*

Myoxocephalus quadricornis (Linnaeus) — fourhorn sculpin
- *Podocotyle atomon*

Myoxocephalus scorpius (Linnaeus) — shorthorn sculpin
- *Derogenes varicus*
- *Podocotyle atomon*
- *Progonus muelleri*
- *Prosorhynchus squamatus*
- *Steganoderma formosum*
- *Stephanostomum baccatum*

Nautichthys oculofasciatus (Girard) — sailfin sculpin
- *Genolinea anura*
- *Lecithaster gibbosus*

Oligocottus maculosus Girard — tidepool sculpin
- *Genolinea anura*
- *Lecithaster gibbosus*
- *Macvicaria isaitschikowi*
- *Podocotyle endophrysi* (exper.)
- *Tubulovesicula lindbergi*

Scorpaenichthys marmoratus (Ayres) — cabezon
- *Genolinea anura*
- *Prosorhynchus scalpellus*
- *Tubulovesicula lindbergi*

Triglops murrayi Günther — moustache sculpin
- *Caudotestis nicolli*
- *Derogenes varicus*
- *Helicometra plovmornini*
- *Lecithaster gibbosus*
- *Progonus muelleri*

Triglops pingeli Reinhardt — ribbed sculpin

Syn.: *Triglops beani* Gilbert
- *Brachyphallus crenatus*
- *Podocotyle endophrysi*

Family Psychrolutidae

Cottunculus microps Collett — polar sculpin
- *Gonocerca* sp.

Family Agonidae

Leptagonus decagonus (Bloch and Schneider) — Atlantic poacher
- *Anisorchis opisthorchis*

Family Cyclopteridae

Careproctus reinhardti (Kryer) — sea tadpole
- *Stenakron vetustum*

Liparis pulchellus Ayres — showy snailfish
- *Lecithaster gibbosus*

Order Perciformes

Suborder Percoidei

Family Percichthyidae

Morone americana (Gmelin) — white perch
- *Apophallus venustus* metacercaria
- *Clinostomum complanatum* metacercaria
- *Diplostomum huronense* metacercaria
- *Lepocreadium setiferoides*
- *Posthodiplostomum minimum* metacercaria
- *Tylodelphys scheuringi* metacercaria

Morone chrysops (Rafinesque) — white bass
- *Allacanthochasmus artus*
- *Allacanthochasmus varius*
- *Bucephalus* sp. metacercaria
- *Clinostomum complanatum* metacercaria
- *Diplostomum spathaceum* metacercaria
- *Leuceruthrus micropteri*
- *Neochasmus umbellus*

Morone saxatilis (Walbaum) — striped bass
- *Homalometron pallidum*
- *Lepocreadium setiferoides*
- *Stephanostomum tenue*

Family Centrarchidae

Ambloplites rupestris (Rafinesque) — rock bass
- *Azygia angusticauda*
- *Azygia longa*
- *Bunodera luciopercae*
- *Centrovarium lobotes*
- *Clinostomum complanatum* metacercaria
- *Crepidostomum cooperi*
- *Crepidostomum cornutum*
- *Cryptogonimus chili*
- *Cryptogonimus chili* metacercaria
- *Cryptogonimus diaphanus*
- *Diplostomum spathaceum* metacercaria
- *Leuceruthrus micropteri*
- *Microphallus opacus*
- *Phyllodistomum superbum*
- *Posthodiplostomum minimum* metacercaria
- *Posthodiplostomum minimum centrarchi* metacercaria
- *Proterometra macrostoma*
- *Rhipidocotyle papillosa* metacercaria
- *Ribeiroia ondatrae* metacercaria
- *Tylodelphys scheuringi* metacercaria
- *Uvulifer ambloplitis* metacercaria

Lepomis gibbosus (Linnaeus) pumpkinseed

- *Apophallus venustus* metacercaria
- *Azygia angusticauda*
- *Bunodera sacculata*
- *Clinostomum complanatum* metacercaria
- *Crepidostomum cooperi*
- *Crepidostomum cornutum*
- *Diplostomum huronense* metacercaria
- *Diplostomum spathaceum* metacercaria
- "Heterophyid" = (?) *Apophallus* sp. metacercaria
- *Homalometron armatum*
- *Proterometra macrostoma*
- *Posthodiplostomum minimum* metacercaria
- *Posthodiplostomum minimum centrarchi* metacercaria
- *Tylodelphys scheuringi* metacercaria
- *Uvulifer ambloplitis* metacercaria

Lepomis macrochirus Rafinesque bluegill

- *Bunodera luciopercae*
- *Bunodera sacculata*
- *Clinostomum complanatum* metacercaria
- *Crepidostomum cornutum*
- *Posthodiplostomum minimum* metacercaria
- *Posthodiplostomum minimum centrarchi* metacercaria
- *Uvulifer ambloplitis* metacercaria

Lepomis megalotis (Rafinesque) longear sunfish

- *Posthodiplostomum minimum* metacercaria

Micropterus dolomieui Lacépède smallmouth bass

- *Apophallus brevis* metacercaria
- *Apophallus venustus* metacercaria
- *Azygia angusticauda*
- *Azygia longa*
- *Bucephalus* sp. metacercaria
- *Caecincola parvulus*
- *Centrovarium lobotes*
- *Clinostomum complanatum* metacercaria
- *Crepidostomum cooperi*
- *Crepidostomum cornutum*
- *Cryptogonimus chili*
- *Cryptogonimus chili* metacercaria
- *Diplostomum spathaceum* metacercaria
- *Leuceruthrus micropteri*
- *Maritrema obstipum* (accidental)
- *Microphallus opacus*
- *Neochasmus umbellus*
- *Phyllodistomum superbum*
- *Posthodiplostomum minimum* metacercaria
- *Posthodiplostomum minimum centrarchi* metacercaria
- *Quasimaritremopsis medius* (accidental)
- *Rhipidocotyle papillosa*
- *Rhipidocotyle papillosa* metacercaria
- *Uvulifer ambloplitis* metacercaria

Micropterus salmoides (Lacépède)	largemouth bass
Azygia angusticauda	
Bunodera luciopercae	
Caecincola parvulus	
Clinostomum complanatum metacercaria	
Crepidostomum cornutum	
Cryptogonimus chili	
Diplostomum spathaceum metacercaria	
Leuceruthrus micropteri	
Microphallus opacus	
Posthodiplostomum minimum metacercaria	
Posthodiplostomum minimum centrarchi metacercaria	
Tylodelphys scheuringi metacercaria	
Pomoxis nigromaculatus (Le Sueur)	black crappie
Clinostomum complanatum metacercaria	
Diplostomum spathaceum metacercaria	
Posthodiplostomum minimum metacercaria	
Tylodelphys scheuringi metacercaria	

Family Percidae

Ammocrypta pellucida (Putnam)	eastern sand darter
Plagioporus cooperi	
Etheostoma caeruleum Storer	rainbow darter
Diplostomum spathaceum metacercaria	
Etheostoma exile (Girard)	Iowa darter
Apophallus brevis metacercaria	
Azygia angusticauda	
Crepidostomum isostomum	
Diplostomum spathaceum metacercaria	
Posthodiplostomum minimum metacercaria	
Uvulifer ambloplitis metacercaria	
Etheostoma flabellare Rafinesque	fantail darter
Clinostomum complanatum metacercaria	
Ichthyocotylurus platycephalus metacercaria	
Etheostoma nigrum Rafinesque	johnny darter
Apophallus brevis metacercaria	
Azygia angusticauda	
Clinostomum complanatum metacercaria	
Crepidostomum isostomum	
Diplostomum spathaceum metacercaria	
Leuceruthrus micropteri	
Plagioporus cooperi	
Posthodiplostomum minimum metacercaria	
Rhipidocotyle papillosa	
Perca flavescens (Mitchill)	yellow perch
Apophallus brevis metacercaria	
Apophallus venustus metacercaria	
Apophallus sp. metacercaria	
Azygia angusticauda	

Azygia longa
Bucephalus elegans
Bunodera luciopercae
Bunodera sacculata
Caecincola sp. metacercaria
Centrovarium lobotes
Clinostomum complanatum metacercaria
Crassiphiala bulboglossa metacercaria
Crepidostomum cooperi
Crepidostomum cornutum
Crepidostomum farionis
Crepidostomum isostomum
Creptotrema funduli
Cryptogonimus chili
Diplostomum adamsi metacercaria
Diplostomum huronense metacercaria
Diplostomum spathaceum metacercaria
Diplostomum spathaceum indistinctum metacercaria
Ichthyocotylurus erraticus metacercaria
Ichthyocotylurus pileatus metacercaria
Leuceruthrus micropteri
Lissorchis kritskyi
Metorchis conjunctus metacercaria
Microphallus opacus
Neascus pyriformis metacercaria
Ornithodiplostomum ptychocheilus metacercaria
Phyllodistomum superbum
Posthodiplostomum minimum metacercaria
Posthodiplostomum minimum centrarchi metacercaria
Prosorhynchoides pusilla
Rhipidocotyle papillosa
Rhipidocotyle papillosa metacercaria
Sanguinicola occidentalis
Tylodelphys scheuringi metacercaria
Uvulifer ambloplitis metacercaria

Percina caprodes (Rafinesque) — logperch
Allopodocotyle boleosomi
Clinostomum complanatum metacercaria
Crepidostomum isostomum
Leuceruthrus micropteri
Posthodiplostomum minimum metacercaria
Tylodelphys scheuringi metacercaria

Percina copelandi (Jordan) — channel darter
Plagioporus cooperi

Stizostedion canadense (Smith) — sauger
Centrovarium lobotes
Clinostomum complanatum metacercaria
Ichthyocotylurus platycephalus metacercaria
Phyllodistomum superbum
Prosorhynchoides pusilla

Stizostedion vitreum (Mitchill) — yellow walleye

- *Apophallus venustus* metacercaria
- *Azygia angusticauda*
- *Bucephalus* sp. metacercaria
- *Bunodera sacculata*
- *Centrovarium lobotes*
- *Centrovarium lobotes* metacercaria
- *Clinostomum complanatum* metacercaria
- *Crassiphiala bulboglossa* metacercaria
- *Crepidostomum cooperi*
- *Diplostomum spathaceum* metacercaria
- *Ichthyocotylurus pileatus* metacercaria
- *Ornithodiplostomum ptychocheilus* metacercaria
- *Phyllodistomum superbum*
- *Posthodiplostomum minimum* metacercaria
- *Prosorhynchoides pusilla*
- *Rhipidocotyle papillosa*
- *Sanguinicola occidentalis*
- *Tylodelphys scheuringi* metacercaria
- *Uvulifer ambloplitis* metacercaria

Stizostedion sp.

- *Azygia longa*

Family Bramidae

Brama japonica Hilgendorf — Pacific pomfret

- *Lampritrema miescheri*

Family Sciaenidae

Aplodinotus grunniens Rafinesque — freshwater drum

- *Bunodera luciopercae*
- *Clinostomum complanatum* metacercaria
- *Cotylogaster occidentalis*
- *Diplostomum spathaceum* metacercaria
- *Homalometron armatum*
- *Microcreadium parvum*
- *Phyllodistomum superbum*
- *Sanguinicola* sp.

Family Embiotocidae

Brachyistius frenatus Gill — kelp perch

- *Lecithaster gibbosus*

Cymatogaster aggregata Gibbons — shiner perch

- *Galactosomum humbargari* metacercaria
- *Lecithaster gibbosus*
- *Telolecithus pugetensis*
- *Tubulovesicula lindbergi*
- *Neozoogonus californicus*

SUBORDER LABROIDEI

Family Labridae

Tautogolabrus adspersus (Walbaum)	cunner
Cryptocotyle lingua metacercaria	
Derogenes varicus	
Hemiurus levinseni	
Lecithaster gibbosus	
Lepidapedon elongatum	
Metadena sp.	
Podocotyle atomon	
Podocotyle reflexa	
Ptychogonimus megastoma (imm., ? accidental)	

SUBORDER ZOARCOIDEI

Family Zoarcidae

Aprodon cortezianus Gilbert	bigfin eelpout
Lecithaster gibbosus	
Parahemiurus merus	
Pseudopecoleus japonicus	
Lycodes esmarki Collett	greater eelpout
Gonocerca sp.	
Lycodes reticulatus Reinhardt	Arctic eelpout
Derogenes varicus	
Helicometra plovmornini	
Lecithaster gibbosus	
Lepidophyllum appyi	
Progonus muelleri	
Lycodes vahli Reinhardt	checker eelpout
Derogenes varicus	
Genolinea laticauda	
Helicometra plovmornini	
Lecithaster gibbosus	
Lepidapedon rachion	
Lepidophyllum appyi	
Neophasis oculatus	
Progonus muelleri	
Stenakron vetustum	
Steringotrema ovacutum	
Lycodes sp.	
Stephanostomum baccatum	
Lycodonus mirabilis Goode and Bean	chevron scutepout
Gonocerca sp.	
Macrozoarces americanus (Schneider)	ocean pout
Lepidophyllum appyi	

Family Stichaeidae

Anoplarchus purpurescens Gill	high cockscomb
Lecithaster gibbosus	
Lumpenus lumpretaeformis (Walbaum)	snakeblenny
Helicometra insolita	
Progonus muelleri	
Stenakron vetustum	
Lumpenus sagitta Wilimovsky	snake prickleback
Lecithaster gibbosus	

Family Pholidae

Pholis gunnellus (Linnaeus)	rock gunnel
Podocotyle atomon	
Pholis ornata (Girard)	saddleback gunnel
Lecithaster gibbosus	

Family Anarhichadidae

Anarhichas denticulatus Kryer	northern wolffish
Derogenes varicus	
Fellodistomum fellis	
Lepidophyllum steenstrupi	
Anarhichas lupus Linnaeus	Atlantic wolffish
Brachyenteron pycnorganum	
Caudotestis nicolli	
Derogenes varicus	
Fellodistomum agnotum	
Fellodistomum fellis	
Genolinea laticauda	
Helicometra plovmornini	
Lepidophyllum steenstrupi	
Neophasus anarrhichae	
Neophasus pusilla	
Peracreadium idoneum	
Progonus muelleri	
Steringophorus furciger	
Steringotrema ovacutum	
Anarhichas minor Olafsen	spotted wolffish
Brachyenteron pycnorganum	
Fellodistomum agnotum	
Fellodistomum fellis	
Lepidophyllum steenstrupi	
Neophasis anarrhichae	
Peracreadium idoneum	
Pseudozoogonoides microacetabulum	
Steringophorus furciger	

SUBORDER BLENNIOIDEI

Family Bathymasteridae

Ronquilus jordani (Gilbert) — northern ronquil
    Lecithaster gibbosus

Family Cryptacanthodidae

Cryptacanthodes maculatus Storer — wrymouth
    Derogenes varicus
    Stephanostomum baccatum

Lyconectes aleutensis Gilbert — dwarf wrymouth
    Lecithaster gibbosus

SUBORDER GOBIOIDEI

Family Gobiidae

Coryphopterus nicholsi (Bean) — blackeye goby
    Lecithaster gibbosus
    Pronoprymna petrowi

SUBORDER TRACHINOIDEI

Family Ammodytidae

Ammodytes americanus DeKay — American sand lance
    Derogenes varicus

Ammodytes dubius Reinhardt — northern sand lance
    Brachyphallus crenatus
    Derogenes varicus
    Lecithaster gibbosus

Ammodytes hexapterus Pallas — Pacific sand lance
    Galactosomum phalacrocoracis metacercaria
    Lecithaster gibbosus

SUBORDER SCOMBROIDEI

Family Scombridae

Scomber scombrus Linnaeus — Atlantic mackerel
    Cryptocotyle lingua metacercaria
    Halvorsenius exilis
    Podocotyle atomon

Thunnus thynnus (Linnaeus) — bluefin tuna
    Hirudinella ventricosa

Family Xiphiidac

Xiphias gladius Linnaeus — swordfish
    Hirudinella ventricosa
    Opisthadena dimidia

Order Pleuronectiformes

Family Bothidae

Citharichthys stigmaeus Jordan and Gilbert — speckled sanddab
- *Tubulovesicula lindbergi*

Scophthalmus aquosus (Mitchill) — windowpane
- *Cryptocotyle lingua* metacercaria
- *Lepidapedon rachion*
- *Stephanostomum baccatum* metacercaria

Family Pleuronectidae

Atheresthes stomias (Jordan and Gilbert) — arrowtooth flounder
- *Derogenes varicus*
- *Prosorhynchus* sp. metacercaria
- *Steganoderma formosum*
- *Stephanostomum* sp. metacercaria

Eopsetta exilis (Jordan and Gilbert) — slender sole

Syn.: *Lyopsetta exilis* (Jordan and Gilbert)
- *Opecoeloides vitellosus*
- *Parahemiurus merus*

Eopsetta jordani (Lockington) — petrale sole
- *Lecithochirium exodicum*
- *Otodistomum* sp. metacercaria
- *Parahemiurus merus*

Glyptocephalus cynoglossus (Linnaeus) — witch flounder
- *Aporocotyle simplex*
- *Cryptocotyle lingua* metacercaria
- *Derogenes varicus*
- *Genolinea laticauda*
- *Gonocerca macroformis*
- *Lecithaster gibbosus*
- *Macvicaria soleae*
- *Otodistomum* sp. metacercaria
- *Progonus muelleri*
- *Stenakron vetustum*
- *Stephanostomum baccatum* metacercaria
- *Steringophorus furciger*

Hippoglossoides platessoides (Fabricius) — Canadian plaice
- *Aporocotyle simplex*
- *Brachyphallus crenatus*
- *Cryptocotyle lingua* metacercaria
- *Derogenes varicus*
- *Genolinea laticauda*
- *Gonocerca macroformis*
- *Hemiurus levinseni*
- *Lecithaster gibbosus*
- *Neophasis burti*
- *Otodistomum* sp. metacercaria
- *Podocotyle atomon*

Progonus muelleri
Prosorhynchus squamatus
Pseudozoogonoides subaequiporus
Stenakron vetustum
Stephanostomum baccatum metacercaria
Steringophorus furciger
Steringotrema ovacutum
Zoogonoides viviparus

Hippoglossus hippoglossus (Linnaeus) — Atlantic halibut
Anomalotrema koiae
Brachyphallus crenatus
Cryptocotyle lingua metacercaria
Derogenes varicus
Genolinea laticauda
Gonocerca phycidis
Hemiurus levinseni
Lecithaster gibbosus
Lecithophyllum botryophorum
Lepidapedon rachion
Macvicaria soleae
Otodistomum sp. metacercaria
Paraccacladium jamiesoni
Podocotyle atomon
Podocotyle reflexa
Prosorhynchus squamatus
Steganoderma formosum
Stenakron vetustum
Stephanostomum baccatum
Stephanostomum baccatum metacercaria
Steringophorus furciger
Steringotrema pagelli

Microstomus pacificus (Lockington) — Dover sole
Derogenes varicus
Otodistomum sp. metacercaria
Prosorhynchus sp. metacercaria
Steringophorus brevis

Platichthys stellatus (Pallas) — starry flounder
Lecithaster gibbosus
Neolebouria tinkerbellae (exper.)
Podocotyle endophrysi
Podocotyle gibbonsia

Pleuronectes americanus Walbaum — winter flounder
Syn.: *Pseudopleuronectes americanus* (Walbaum)
Cryptocotyle lingua metacercaria
Derogenes varicus
Genolinea laticauda
Hemiurus levinseni
Lecithaster gibbosus
Macvicaria soleae
Podocotyle atomon

Steganoderma formosum
Stenakron vetustum
Stephanostomum baccatum
Stephanostomum baccatum metacercaria
Steringophorus furciger

Pleuronectes bilineatus (Ayres) — rock sole
Syn.: *Lepidopsetta bilineata* (Ayres)
Lecithaster gibbosus
Otodistomum sp. metacercaria

Pleuronectes ferrugineus (Storer) — yellowtail flounder
Syn.: *Limanda ferruginea* (Storer)
Cryptocotyle lingua metacercaria
Derogenes varicus
Gonocerca macroformis
Lecithaster gibbosus
Macvicaria soleae
Otodistomum sp. metacercaria
Podocotyle atomon
(?) *Podocotyle reflexa*
Progonus muelleri
Steganoderma formosum
Stenakron vetustum
Stephanostomum baccatum
Stephanostomum baccatum metacercaria
Steringophorus furciger

Pleuronectes putnami (Gill) — smooth flounder
Syn.: *Liopsetta putnami* (Gill)
Cryptocotyle lingua metacercaria
Derogenes varicus
Podocotyle atomon
Stephanostomum baccatum metacercaria
Steringophorus furciger

Pleuronectes vetulus (Girard) — English sole
Syn. *Parophrys vetulus* Girard
Derogenes varicus
Genitocotyle acirrus
Lecithaster gibbosus
Otodistomum sp. metacercaria
Podocotyle sinusacca
Prosorhynchus sp. metacercaria
Steringophorus brevis
Tubulovesicula lindbergi

Reinhardtius hippoglossoides (Walbaum) — Greenland halibut
Anomalotrema koiae
Brachyphallus crenatus
Derogenes varicus
Dinosoma triangulata
Dissosaccus laevis
Genolinea laticauda
Gonocerca phycidis

Hemiurus levinseni
Lecithaster gibbosus
Lecithophyllum botryophorum
Neophasis sp. metacercaria
Otodistomum sp. metacercaria
Podocotyle atomon
Podocotyle reflexa
Progonus muelleri
Prosorhynchus squamatus
Steganoderma formosum
Stenakron vetustum
Stephanostomum baccatum metacercaria
Steringophorus furciger

Order Tetraodontiformes

Family Molidae

Mola mola (Linnaeus) — ocean sunfish
Accacladium serpentulum
Accacladocoelium macrocotyle
Accacladocoelium nigroflavum
Accacoelium contortum
Dihemistephanus lydiae
Odhnerium calyptrocotyle

ACKNOWLEDGEMENTS

I am grateful to the following for reading sections of the draft manuscript: Dr. T.A. Bakke, Dr. H.L. Ching, Dr. T.H. Cribb, Prof. I. Kanev, Dr. A. Kostadinova, Dr. K. Odening, Prof. J.C. Pearson, and especially my colleague Dr. R.A. Bray. Dr. T.A. Dick, Dr. J.C. Holmes, Dr. J.R. Lichtenfels (USNM Collection), Dr. L. Margolis, and the staff of the National Museum of Natural Sciences of Canada kindly lent material. I am also indebted to Dr. L. Margolis and T. McDonald for making available to me a draft copy of their check-list, without which I should have been unaware of data from various government reports, etc. Lastly, I should like to thank my Canadian wife Patricia, through whom I claim the Canadian link qualifying me to attempt this work, for her help in checking the records.

REFERENCES

ALBERT, E., AND CURTIS, M.A., 1991. Prevalence and abundance of helminth parasites in an intensively fished population of brook trout (*Salvelinus fontinalis*) at a small subarctic lake. Can. J. Zool. 69: 691–697.

ALIFF, J.V. 1977. Digenetic trematodes from Kentucky fishes. Trans. Kentucky Acad. Sci. 38: 1–14.

AMIN, O.M. 1982a. Two larval trematodes (Strigeoidea) of fishes in southeastern Wisconsin. Proc. Helminthol. Soc. Wash. 49: 207–213.

1982b. Adult trematodes (Digenea) from lake fishes of southeastern Wisconsin, with a key to the species of the genus *Crepidostomum* Braun, 1900 in North America. Proc. Helminthol. Soc. Wash. 49: 196–206.

ANNEREAUX, R.F. 1947. Three new trematodes from marine fishes of California. Trans. Am. Microsc. Soc. 66: 249–255.

ANONYMOUS. 1978. Report on environmental studies for the McGregor diversion project. Volume 4 — Faunal transfer. Prepared by Reid, Crowther, and Partners Ltd. for British Columbia Hydro and Power Authority, Vancouver, British Columbia.

1981. Kemano completion hydroelectric development baseline studies. Volume 7. Fish Diseases and Parasites. Prepared by Envirocon Ltd. for Aluminum Co. of Canada, Vancouver, British Columbia.

1984. Environmental studies associated with the proposed Kemano completion hydroelectric development. Volume 7. Fish Diseases and Parasites Baseline Information. Prepared by Envirocon Ltd. for Aluminum Co. of Canada, Vancouver, British Columbia. 179 p. (A draft of Section C of this report was circulated in 1983 under the title "Aluminum Company of Canada, Ltd. Kemano Completion Hydroelectric Development Technical Memorandum 1881/20. Further studies of Fish Parasites and Viruses, 1980–81".)

ANTHONY, D.D. 1974. Helminth parasites of sturgeon (*Acipenser fulvescens*) from Lake Nipissing, Ontario, Canada. Proc. 3rd Int. Congr. Parasitol. 3: 1642–1643.

1976. The trematode and cestode parasites of walleye (*Stizostedion vitreum*) from Lake Nipissing, Ont. Program Abstracts, 51st Annual Meeting American Society of Parasitology. Abstr. 48:35.

1978a. Helminth parasites of lake herring (*Coregonus artedi*) from Lake Nipissing, Ontario. Program and Abstracts of the 53rd Annual Meeting of American Society of Parasitologists, 5–10 November 1978, Chicago, Illinois. p. 80–81.

1978b. Metazoan parasites of percids from Lake Nipissing, Ontario, Canada. Fourth International Congress of Parasitology, 19–26 August 1978, Warsaw, Poland. Short Communications, Section H, p. 39–40.

1982. Helminth parasites of freshwater drum (*Aplodinotus grunniens*) from Lake Nipissing, Ontario, Canada. Molecular and Biochemical Parasitology, Supplement. Parasites — their world and ours, p. 438. (Abstracts of the Fifth International Congress of Parasitology, 7–14 August 1982, Toronto, Ontario).

1983. Metazoan parasites of esocids from Lake Nipissing, Ontario. Program and Abstracts of the 58th Annual Meeting of American Society of Parasitologists, 4–8 December 1983, San Antonio, Texas. p. 41.

1984. Helminth parasites of the white bass (*Morone chrysops*) from Lake Nipissing, Ontario. Program and Abstracts of the 59th Annual Meeting of American Society of Parasitologists, 5–9 August 1984, Snowbird, Utah. p. 27.

1985. Helminth parasites of centrarchids from Lake Nipissing, Ontario. Program and Abstracts of the 60th Annual Meeting of American Society of Parasitologists, 4–8 August 1985, Athens, Georgia. p. 34.

1986. Helminth parasites of an exclusively freshwater stock of pink salmon (*Oncorhynchus gorbuscha*) from the Canadian Great Lakes. *In*: Howell, M. J. (Ed.). Parasitology — Quo Vadit? Handbook Supplement, p. 6. Program and Abstracts Sixth International Congress of Parasitology, 25–29 August 1986, Brisbane, Australia.

1987. Helminth parasites of burbot (*Lota lota*) from Lake Nipissing, Ontario. Program and Abstracts of the 62nd Annual Meeting of American Society of Parasitologists, 2–5 August 1987, Lincoln, Nebraska. p. 42–43.

APPY, R.G., AND BURT, M.D.B. 1982. Metazoan parasites of cod, *Gadus morhua* L., in Canadian Atlantic waters. Can. J. Zool. 60: 1573–1579.

APPY, R.G., AND DADSWELL, M.J. 1978. Parasites of *Acipenser brevirostrum* LeSueur and *Acipenser oxyrhynchus* Mitchill (Osteichthyes: Acipenseridae) in the Saint John River estuary, N.B., with a description of *Caballeronema pseudoargumentosus* sp. n. (Nematoda: Spirurida). Can. J. Zool. 56: 1382–1391.

ARAI, H.P. 1967a. A preliminary report on a study of the parasites of marine fishes of Burke Channel, British Columbia. Fish. Res. Board. Can. MS Rep. 925: 26 p.

1967b. Ecological specificity of parasites of some embiotocid fishes. J. Fish. Res. Board Can. 24: 2161–2168.

1969a. A new trematode of the genus *Lepidophyllum* (Digenea: Steganodermatidae) from a cottid fish, *Hemilepidotus hemilepidotus*. J. Fish. Res. Board Can. 26: 799–803.

1969b. Preliminary report on the parasites of certain marine fishes of British Columbia. J. Fish. Res. Board Can. 26: 2319–2337.

Arai, H.P., and Mudry, D.R. 1983. Protozoan and metazoan parasites of fishes from the headwaters of the Parsnip and McGregor Rivers, British Columbia: a study of possible parasite transfaunations. Can. J. Fish. Aquat. Sci. 40: 1676–1684.

Arai, H.P., Kabata, Z., and Noakes, D. 1988. Studies on seasonal changes and latitudinal differences in the metazoan fauna of the shiner perch, *Cymatogaster aggregata*, along the west coast of North America. Can. J. Zool. 66: 1514–1517.

Armstrong, H. A study of the helminths parasites of the family Macrouridae from the Gulf of Mexico and Caribbean Sea: their systematics, ecology and zoogeographical implications. PhD Thesis, Texas A&M University. 329 p.

Arnold, J.J. 1934. Some trematodes of the common bullhead *Ameiurus nebulosus* (LeSueur). Trans. Am. Microsc. Soc. 53: 267–276.

Arthur, J.R. 1983. A preliminary analysis of the discreteness of stocks of walleye pollock (*Theragra chalcogramma*) from the northeastern Pacific off Canada based on their parasites. Can. Tech. Rep. Fish. Aquat. Sci. No. 1184: 15 p.

1984. A survey of the parasites of walleye pollock (*Theragra chalcogramma*) from the northeastern Pacific Ocean off Canada and a zoogeographical analysis of the parasite fauna of this fish throughout its range. Can. J. Zool. 62: 675–684.

Arthur, J.R., and Albert, E. 1992a. Parasites of Greenland halibut (*Reinhardtius hippoglossoides*) from the Atlantic Ocean off Canada. Bull. Can. Soc. Zool. 23(2): 26. (31st Annual Meeting, Canadian Society of Zoologists, St. Francis Xavier University, 13–17 May 1992, Antigonish, Nova Scotia. Program and Abstracts).

1992b. Parasites as biological tags for Greenland halibut (*Reinhardtius hippoglossoides*) from the Atlantic Ocean off Canada. Program and Abstracts of the 67th Annual Meeting of American Society of Parasitologists, 4–8 August 1992, Philadelphia, Pennsylvania. p. 63.

1993. Use of parasites for separating stocks of Greenland halibut (*Reinhardtius hippoglossoides*) in the Canadian Northwest Atlantic Ocean. Can. J. Fish. Aquat. Sci. 50: 2175–2181.

1994. A survey of the parasites of Greenland halibut (*Reinhardtius hippoglossoides*) caught off Atlantic Canada, with notes on their zoogeography in this fish. Can. J. Zool. 72: 765–778.

Arthur, J.R., and Arai, H.P. 1980a. Studies on the parasites of Pacific herring (*Clupea harengus pallasi* Valenciennes): survey results. Can. J. Zool. 58: 64–70.

1980b. Studies on the parasites of Pacific herring (*Clupea harengus pallasi* Valenciennes): a preliminary evaluation of parasites as indicators of geographical origin for spawning herring. Can. J. Zool. 58: 521–527.

Arthur, J.R., Margolis, L., and Arai, H.P. 1976. Parasites of fishes of Aishihik and Stevens Lakes, Yukon Territory and potential consequences of their interlake transfer through a proposed water diversion for hydroelectrical purposes. J. Fish. Res. Board Can. 33: 2489–2499.

Baer, J.G. 1933. Note sur un nouveau trématode, *Clinostomum lophophallum* sp. nov. avec quelques considerations générales sur la famille des Clinostomidae. Rev. Suisse Zool. 40: 317– 342.

Baggs, E.M., and Cowan, G.I.McT. 1989. The onset of parasitism of young-of-the-year brook charr, *Salvelinus fontinalis* (Mitchill), in Big Northern Pond, Avalon Peninsula, Newfoundland, Canada. J. Fish. Biol. 35: 317–319.

Bailey, R.E., and Margolis, L. 1987. Comparison of parasite fauna of juvenile sockeye salmon (*Oncorhynchus nerka*) from southern British Columbian and Washington State lakes. Can. J. Zool. 65: 420–431.

Bailey, R.E., Margolis, L., and Groot, C. 1988. Estimating stock composition of migrating juvenile Fraser River (British Columbia) sockeye salmon, *Oncorhynchus nerka*, using parasites as natural tags. Can. J. Fish. Aquat. Sci. 45: 586–591.

Bailey, R.E., Margolis, L., and Workman, G.D., 1989. Survival of certain naturally acquired freshwater parasites of juvenile sockeye salmon, *Oncorhynchus nerka* (Walbaum), in hosts held in fresh and sea water, and implications for their use as population tags. Can. J. Zool. 67: 1757–1766.

Bakke, T.A. 1984. A redescription of adult *Phyllodistomum umblae* (Fabricius) (Digenea, Gorgoderidae) from *Salvelinus alpinus* (L.). Zool. Scr. 13: 89–99.

1985. *Phyllodistomum conostomum* (Olsson, 1876) (Digenea, Gorgoderidae): a junior subjective synonym for *P. umblae* (Fabricius, 1780). Zool. Scr. 14: 161–168.

1988. Abnormalities in adult digeneans, with special reference to *Phyllodistomum umblae* (Fabricius) (Platyhelminthes, Gorgoderidae). Zool. Scr. 17: 123–134.

Bakke, T.A., and Bailey, R.E. 1987. *Phyllodistomum umblae* (Fabricius) (Digenea, Gorgoderidae) from British Columbia salmonids: a description based on light and scanning electron microscopy. Can. J. Zool. 65: 1703–1712.

Bangham, R.V. 1941. Parasites of fish of Algonquin Park lakes. Trans. Am. Fish Soc. 70: 161–171.

1955. Studies on fish parasites of Lake Huron and Manitoulin Island. Am. Midl. Nat. 53: 184–194.

Bangham, R.V., and Adams, R.J. 1954. A survey of the parasites of freshwater fish from the mainland of

British Columbia. J. Fish. Res. Board Can. 11: 673–708.

Bangham, R.V., and Hunter, G.W. III. 1939. Studies on fish parasites of Lake Erie. Distribution studies. Zoologica (N.Y.) 24: 385–448.

Bangham, R.V., and Venard, C.E. 1946. Parasites of fish of Algonquin Park lakes. Univ. Toronto Stud. Biol. Ser. 53: 31–46. (Ont. Fish. Res. Lab. Publ. 65).

Barker, S.C., Blair, D., Cribb, T.H., and Tonion, K. 1993a. Phylogenetic position of *Heronimus mollis* (Digenea): evidence from 18S ribosomal RNA. Int. J. Parasitol. 23: 533–536.

Barker, S.C., Blair, D., Garrett, A.R., and Cribb, T.H. 1993b. Utility of the D1 domain of nuclear 28S rRNA for phylogenetic inference in the Digenea. Syst. Parasitol. 26: 181–188.

Barraclough, W.E. 1967. Data records. Number, size composition and food of larval and juvenile fish caught with a two-boat surface trawl in the Strait of Georgia, June 6–8, 1966. Fish. Res. Board Can. MS Rep. 928: 58 p.

Barraclough, W.E., and Fulton, J.D. 1967. Data records. Number, size composition and food of larval and juvenile fish caught with a two-boat surface trawl in the Strait of Georgia, July 4–8, 1966. Fish. Res. Board Can. MS Rep. 940: 82 p.

Baturo, B. 1977. *Bucephalus polymorphus* Baer, 1827 and *Rhipidocotyle illense* (Ziegler, 1883) (Trematoda, Bucephalidae): morphology and biology of developmental stages. Acta Parasitol. Polon. 24: 203–220.

Beaver, P.C. 1939. The morphology and life history of *Psilostomum ondatrae* Price, 1931 (Trematoda: Psilostomidae). J. Parasitol. 25: 383–393.

Becker, D.A., and Cloutman, D.G. 1975. Parasites of selected game fishes of Lake Fort Smith, Arkansas. Proc. Arkansas Acad. Sci. 29: 12–18.

Bennington, E., and Pratt, I. 1960. The life history of the salmon-poisoning fluke, *Nanophyetus salmincola* (Chapin). J. Parasitol. 46: 91–100.

Besprozvannykh, V.V. 1989. Biology of the trematode *Echinochasmus milvi* from the south of the Far East of the USSR. Parazitologiya 23: 237–243. (In Russian).

Beverley-Burton, M. 1987. *Ophioxenos microphagus* (Ingles, 1936) comb. n. (Digenea: Paramphistomidae) from ectotherms in western North America with comments on host-parasite relationships. Proc. Helminthol. Soc. Wash. 54: 197–199.

Beverley-Burton, M., and Margolis, L. 1982. *Ophioxenos lampetrae* sp. nov. (Digenea: Paramphistomidae) from ammocoetes of the western brook lamprey (*Lampetra richardsoni* Vladykov and Follett) in British Columbia, with comments on lamprey host-parasite relationships. Can. J. Zool. 60: 2514–2520.

Black, G.A. 1981. Metazoan parasites as indicators of movements of anadromous brook charr (*Salvelinus fontinalis*) to sea. Can. J. Zool. 59: 1892–1896.

Blair, D. 1976. Observations on the life cycle of the strigeoid trematode *Apatemon (Apatemon) gracilis* (Rudolphi, 1819) Szidat, 1928. J. Helminthol. 50: 125–131.

Bond, W.A., and Erickson, R.N. 1985. Life history studies of anadromous coregonid fishes in two freshwater lake systems on the Tuktoyaktuk Peninsula, Northwest Territories. Can. Tech. Rep. Fish. Aquat. Sci. 1336: 61 p.

Bouillon, D.R., and Dempson, J.B. 1989. Metazoan parasite infections in landlocked and anadromous Arctic charr (*Salvelinus alpinus* Linnaeus), and their use as indicators of movement to sea in young anadromous charr. Can. J. Zool. 67: 2478–2485.

Bourgeois, C.E., and Ni, I.-H. 1984. Metazoan parasites of Northwest Atlantic redfishes (*Sebastes* spp.). Can. J. Zool. 62: 1879–1885.

Bouvry, M., Brassard, P., and Rau, M.E. 1984. *Diplostomum spathaceum* dans le système nerveux des poissons. Le Naturaliste canadien (Québec) 111: 311–313.

Boyce. N.P. 1966. The parasites of central British Columbia pink salmon during their early sea life, with special notes on the trematode *Lecithaster gibbosus*. Fish. Res. Board Can. MS Rep. (Biol) 877: 8 p.

1969. Parasite fauna of pink salmon (*Oncorhynchus gorbuscha*) of the Bella Coola River, central British Columbia, during their early sea life. J. Fish. Res. Board Can. 26: 813–820.

Bradstreet, M.S.W. et al. 1986. Aspects of the biology of Arctic cod (*Boreogadus saida*) and its importance in Arctic marine food chains. Can. Tech. Rep. Fish. Aquat. Sci. No. 1491: 193 pp.

Brady, A. 1989. Studies on the taxonomy and biology of *Diplostomum* species (Digenea). PhD Thesis, University of Stirling, 277 pp.

Brassard, P., Curtis, M.A., and Rau, M.E. 1982. Seasonality of *Diplostomum spathaceum* (Trematoda: Strigeidae) transmission to brook trout (*Salvelinus fontinalis*) in northern Quebec, Canada. Can. J. Zool. 60: 2258–2263.

Bray, R.A. 1979. Digenea in marine fishes from the eastern seaboard of Canada. J. Nat. Hist. 13: 399–431.

1987a. A revision of the family Zoogonidae Odhner, 1902 (Platyhelminthes: Digenea): Introduction and subfamily Zoogoninae. Syst. Parasitol. 9: 3–28.

1987b. A revision of the family Zoogonidae Odhner, 1902 (Platyhelminthes: Digenea): Subfamily Lepidophyllinae and comments on some aspects of biology. Syst. Parasitol. 9: 83–123.

1987c. A study of the helminth parasites of *Anarhichas lupus* (Perciformes: Anarhichadidae) in the North Atlantic. J. Fish Biol. 31: 237–264.

1988. A discussion of the status of the subfamily Baccigerinae Yamaguti, 1958 (Digenea) and the constitution of the family Fellodistomidae Nicoll, 1909. Syst. Parasitol. 11: 97–112.

1990. A review of the genus *Parahemiurus* Vaz & Pereira, 1930 (Digenea: Hemiuridae). Syst. Parasitol. 15: 1–21.

BRAY, R.A., AND DES CLERS, S.A. 1992. Multivariate analyses of metrical features in the *Lepidapedon elongatum* (Lebour, 1908) species-complex (Digenea: Lepocreadiidae) in deep and shallow water gadiform fishes of the NE Atlantic. Syst. Parasitol. 21: 223–232.

BRAY. R.A., AND GIBSON, D.I. 1977. The Accacoeliidae (Digenea) of fishes from the Northeast Atlantic. Bull. Br. Mus. (Nat. Hist.) Zool. 31: 51–99.

1980. The Fellodistomidae (Digenea) of fishes from the northeast Atlantic. Bull. Br. Mus. (Nat. Hist.) Zool. 37:199–293.

1986. The Zoogonidae (Digenea) of fishes from the north-east Atlantic. Bull. Br. Mus. (Nat. Hist.) Zool. 51: 127–206.

1988. The Lepocreadiidae (Digenea) of fishes from the north-east Atlantic: review of the genus *Paralepidapedon* Shimazu & Shimura, 1984, with a description of *P. williamsi* n. sp. Syst. Parasitol. 12: 87–92.

1989. The Lepocreadiidae (Digenea) of fishes from the north-east Atlantic: review of the genus *Neolepidapedon* Manter, 1954, with a description of *N. smithi* n. sp. Syst. Parasitol. 13: 11–23.

1990. The Lepocreadiidae (Digenea) of fishes from the north-east Atlantic: review of the genera *Opechona* Looss, 1907 and *Prodistomum* Linton, 1910. Syst. Parasitol. 15: 159–202.

1991a. The Acanthocolpidae (Digenea) of fishes from the north-east Atlantic: the status of *Neophasis* Stafford, 1904 (Digenea) and a study of North Atlantic forms. Syst. Parasitol. 19: 95–117.

1991b. The Lepocreadiidae (Digenea) of fishes from the north-east Atlantic: *Profundivermis intercalarius* n. g., n. sp. from the marine fish *Coryphaenoides (Nematonurus) armatus* (Hector) (Macrouridae) from the Porcupine Abyssal Plain. Syst. Parasitol. 18: 121–125.

1995. The Lepocreadiidae (Digenea) of fishes of the north-east Atlantic: a review of the genus *Lepidapedon* Stafford, 1904. Syst. Parasitol. 31: 81–132.

BRAY, R.A., SOTO, A., AND ROLLINSON, D. 1994. The status and composition of the genus *Steringophorus* Odhner, 1905 (Digenea: Fellodistomidae), based on partial small subunit rRNA sequences. Int. J. Parasitol. 24: 433–435.

BRINKMANN, A. Jr. 1975. Trematodes from Greenland. Meddr Grnland 205: 2–88.

BROOKS, D.R. 1990. Phylogenetic systematic evaluation of a classification of the Zoogonidae Odhner, 1902 (Cercomeria: Trematoda: Digenea: Plagiorchiiformes). Syst. Parasitol. 16: 127–137.

1992. New distribution records for *Bunodera eucaliae* and *Bunodera inconstans* n. comb. (Digenea: Allocreadiidae: Bunoderinae) with discussion of their phylogenetic relationships. J. Parasitol. 78: 779–783.

BROOKS, D.R., AND MCLENNAN, D.A. 1993. Parascript. Parasites and the language of evolution. Smithsonian Institution Press, Washington and London. 429 p.

BROOKS, D.R., O'GRADY, R.T., AND GLEN, D.R. 1985. Phylogenetic analysis of the Digenea (Platyhelminthes: Cercomeria) with comments on their adaptive radiation. Can. J. Zool. 63: 411– 443.

BROOKS, D.R., BANDONI, S.M., MACDONALD, C.A., AND O'GRADY, R.T. 1989. Aspects of the phylogeny of the Trematoda Rudolphi, 1808 (Platyhelminthes: Cercomeria). Can. J. Zool. 67: 2609–2624.

BURGNER, R.L., LIGHT, J.T., MARGOLIS, L., OKAZAKI, T., TAUTZ, A.,, AND ITO, S. 1992. Distribution and origins of steelhead trout (*Oncorhynchus mykiss*) in offshore waters of the North Pacific Ocean. International North Pacific Fisheries Commission Bulletin 51: 92 p.

BYRD, E.E. 1939. Studies on the blood flukes of the family Spirorchiidae. Part II. Revision of the family and description of new species. J. Tenn. Acad. Sci. 14: 116–161.

CABLE, R.M. 1952. On the systematic position of the genus *Deropristis*, of *Dihemistephanus sturionis* Little, 1930, and of a new digenetic trematode from a sturgeon. Parasitology 42: 85–91.

CABLE, R.M., AND HUNNINEN, A.V. 1942. Studies on *Deropristis inflata* (Molin), its life history and affinities to trematodes of the family Acanthocolpidae. Biol. Bull. Mar. Lab. Woods Hole 82: 292–312.

CAIRA, J.N. 1989. A revision of the North American papillose Allocreadiidae (Digenea) with independent cladistic analyses of larval and adult forms. Bull. Univ. Neb. St. Mus. 11: 1–58.

CAMERON, T.W.M. 1936. Studies on the heterophyid trematode, *Apophallus venustus* (Ransom, 1920) in Canada. Part I. Morphology and taxonomy. Can. J. Res., Sect. D. 14: 59–69.

1937a. Studies on the heterophyid trematode *Apophallus venustus* (Ransom, 1920) in Canada. Part II. Life history and bionomics. Can. J. Res., Sect. D. 15: 38–51.

1937b. Studies on the heterophyid trematode *Apophallus venustus* (Ransom, 1920) in Canada. Part III. Further hosts. Can. J. Res. Sect D. 15: 275.

1940. *Metorchis conjunctus* (Cobbold, 1859) Looss, 1899. Rep. Proc. (3rd) Int. Congr. Microbiol., New York, 1939, p. 437–438.

1944. The morphology, taxonomy, and life history of *Metorchis conjunctus* (Cobbold, 1860). Can. J. Res., Sect. D. 22: 6–16.

1945. Fish-carried parasites in Canada. Can. J. Comp. Med. 9: 245–254, 283–286, 302–311.

CAMPBELL, R.A. 1975. *Hudsonia agassizi* gen. et sp. n. (Zoogonidae: Hudsoniinae subf. n.) from a deep-sea fish in the western North Atlantic. J. Parasitol. 61: 409–412.

1992. A new digenetic trematode, *Gibsonia borealis* sp. n. (Lepocreadiidae: Lepidapedinae), parasitic in the rattail *Macrourus berglax* from the Flemish Cap off Newfoundland. Proc. Helminthol. Soc. Wash. 59: 5–8.

CAMPBELL, R.A., AND MUNROE, T.A. 1977. New hemiurid trematodes from deep-sea benthic fishes in the western North Atlantic. J. Parasitol. 63: 285–294.

CANNON, L.R.G. 1971. The life-cycles of *Bunodera sacculata* and *B. luciopercae* (Trematoda Allocreadiidae) in Algonquin Park, Ontario. Can. J. Zool. 49: 1417–1429.

1972. Studies on the ecology of papillose allocreadiid trematodes of the yellow perch in Algonquin Park. Ont. Can. J. Zool. 50: 1231–1239.

1973. Diet and intestinal helminths in a population of perch, *Perca flavescens*. J. Fish Biol. 5: 447–457.

CARNEY, J.P., AND BROOKS, D.R. 1991. Phylogenetic analysis of *Alloglossidium* Simer, 1929 (Digenea: Plagiorchiiformes: Macroderoididae) with discussion of the origin of truncated life cycle patterns in the genus. J. Parasitol. 78: 890–900.

CAVENY, B.A., AND ETGES, F.J. 1971. Life history studies of *Microphallus opacus* (Trematoda: Microphallidae). J. Parasitol. 57: 1215–1221.

CHANDLER, A.S. 1951. Studies on metacercariae of *Perca flavescens* in Lake Itasca, Minnesota. Am. Midl. Nat. 45: 711–721.

CHEN, H.T. 1956. Studies on Chinese microphallid trematodes of the sub-family Microphallinae. Acta Zool. Sin. 8: 49–58 (In Chinese).

CHEN, H.T. *ET AL.* 1985. Fauna Sinica. Platyhelminthes Trematoda. Digenea (I). Beijing: Science Press, 697 p. (In Chinese).

CHING, H.L. 1960. Some digenetic trematodes of fishes of Friday Harbor, Washington. J. Parasitol. 46: 241–250.

1961. Redescription of the digenetic trematodes *Lepidapedon calli* and *L. pugetensis* Acena, and new host records for *L. calli* and *Eurycreadium vitellosum* Manter from fishes of Washington State. Can. J. Zool. 39: 615–621.

1979. The life cycle of *Podocotyle enophrysi* Park, 1937 (Trematoda: Opecoelidae). Can. J. Zool. 57: 1341–1344.

1984. Occurrence of the eyefluke, *Diplostomum (Diplostomum) baeri bucculentum* Dubois et Rausch, 1948 in salmonid fishes of northern British Columbia. Program and Abstracts of the 59th Annual Meeting of American Society of Parasitologists, 5–9 August 1984, Snowbird, Utah. p. 27.

1985. Occurrence of the eyefluke, *Diplostomum (Diplostomum) baeri bucculentum* Dubois et Rausch, 1948, in salmonid fishes of northern British Columbia. Can. J. Zool. 63: 396–399.

CHINNIAH, V.C., AND THRELFALL, W. 1978. Metazoan parasites of fish from the Smallwood Reservoir, Labrador, Canada. J. Fish Biol. 13: 203–213.

CHOQUETTE, L.P.E. 1947. *Phyllodistomum lachancei* sp. nov., a trematode from the ureters of *Salvelinus fontinalis* (Mitchill), with a note on its pathogenicity. Can. J. Res., Sect. D. 25: 131–134.

1948. Parasites of freshwater fish. IV. Internal helminths parasitic in speckled trout (*Salvelinus fontinalis* (Mitchill)) in rivers and lakes of the Laurentide Park, Quebec, Canada. Can. J. Res., Sect. D. 26: 204–211.

1951. Parasites of freshwater fish. V. Parasitic helminths of the muskallunge, *Esox m. masquinongy* Mitchill in the St Lawrence watershed. Can. J. Zool. 29: 290–295.

1954. A note on the intermediate hosts of the trematode *Crepidostomum cooperi* Hopkins, 1931, parasitic in speckled trout (*Salvelinus fontinalis* (Mitchill)) in some lakes and rivers of the Quebec Laurentide Park. Can. J. Zool. 32: 375–377.

CHOUDHOURY, A., AND DICK, T.A. 1993. Parasites of lake sturgeon, *Acipenser fulvescens* (Chondrostei: Acipenseridae), from central Canada. J. Fish Biol. 42: 571–584.

CHOUDHOURY, A., DICK, T.A., HOLLOWAY, H., AND OTTINGER, C. 1990. The lake sturgeon — *Acipenser fulvescens* (Chondrostei, Acipenseridae) in Canada: preliminary studies on parasitofauna and immunological parameters, p. 121–133. *In*: 1990 Interbasin Biota Transfer Study Program Proceedings. North Dakota Water Resources Research Institute, North Dakota State University, Fargo, North Dakota. 144 p.

COLLINS, J.J., AND DECHTIAR, A.O. 1974. Parasite fauna of kokanee salmon (*Oncorhynchus nerka*) introduced into Lake Huron. J. Fish. Res. Board Can. 31: 1818–1821.

CONE, D.K., AND ANDERSON, R.C. 1977. Parasites of pumpkinseed (*Lepomis gibbosus* L.) from Ryan Lake, Algonquin Park, Ontario. Can. J. Zool. 55: 1410–1423.

CONE, D.K., AND RYAN, P.M. 1984. Population sizes of metazoan parasites of brook trout (*Salvelinus fontinalis*) and Atlantic salmon (*Salmo salar*) in a small Newfoundland lake. Can. J. Zool. 62: 130–133.

CONE, D.K., MARCOGLIESE, D.J., AND WATT, W.D. 1993. Metazoan parasite communities of yellow eels (*Anguilla rostrata*) in acidic and limed rivers of Nova Scotia. Can. J. Zool. 71: 177–184.

COOPER, A.R. 1915 Trematodes from marine and freshwater fishes, including one species of ectoparasitic turbellarian. Trans. R. Soc. Can. Ser. 3, 9(4): 181–205.

1921. Trematodes and cestodes of the Canadian Arctic Expedition, 1913–18. Rep. Can. Arct. Exped. 1913–18, 9(G-H): 1–27.

COOPER, D.W. 1988. The preparation of serial sections of platyhelminth parasites, with details of the materials and facilities required. Syst. Parasitol. 12: 211–229.

COURTNEY, C.C. 1987. Ecology of helminths of catostomids from western Lake Superior. Program and Abstracts of the 62nd Annual Meeting of American Society of Parasitologists, 2–5 August, 1987, Lincoln, Nebraska. p. 29.

CRIBB, T.H. 1987a. Studies on gorgoderid digeneans from Australian and Asian freshwater fishes. J. Nat. Hist. 21: 1129–1153.

1987b. A new species of *Phyllodistomum* (Digenea: Gorgoderidae) from Australian and New Zealand freshwater fishes with notes on the taxonomy of *Phyllodistomum* Braun, 1899. J. Nat. Hist. 21: 1525–1538.

CRIBB, T.H., BRAY, R.A, AND BARKER, S.C. 1992. A review of the Family Transversotrematidae (Trematoda: Digenea) with description of a new genus, *Crusziella. Invertebr. Taxon.* 6: 909–935.

CROSSMAN, E.J. 1962. The grass pickerel *Esox americanus vermiculatus* Le Sueur in Canada. R. Ont. Mus. Life Sci. Contrib. 55: 29 p.

CURTIS, M.A. 1979. Metazoan parasites of resident Arctic char (*Salvelinus alpinus*) from a small lake on southern Baffin Island. Le Naturaliste canadien (Québec) 106: 337–338.

1981. Observations on the occurrence of *Diplostomum spathaceum* and *Schistocephalus* sp. in ninespine sticklebacks (*Pungitius pungitius*) from the Belcher Islands, Northwest Territories, Canada. J. Wildl. Dis. 17: 241–246.

1982. Host-parasite interactions in Arctic and subarctic lakes, p. 41–57. *In*: Meerovitch, E. (Ed.). Aspects of parasitology — a Festschrift dedicated to the 50th Anniversary of the Institute of Parasitology of McGill University, 1932–1982. McGill University, Montreal, Québec. 485 p.

1984. The influence of food web structure on parasite transmission to Arctic char (*Salvelinus alpinus*) in northern Canada. Parasitology 89(2): xlix. (Proceedings of the British Society for Parasitology. Spring Meeting, 4–6 April 1984, held at the University of Bristol).

1988. Determinants in the formation of parasite communities in coregonids. Finn. Fish. Res. 9: 303–312.

DAWES, B. 1947. The Trematoda of British fishes. London: The Ray Society, 364 p.

DEBLOCK, S. 1971. Contribution à l'étude des Microphallidae Travassos, 1920 (Trematoda). XXIV. Tentative de phylogenie et de taxonomie. Bull. Mus. Nat. Hist. Nat. Paris, 3 ser., 7: 359–467.

1973. Contribution à l'étude des Microphallidae Travassos, 1920 (Trematoda). XXVII. A propos d'espèces décrites au Japon par S. Yamaguti: A. Invalidation du genre *Maritreminoides* Rankin. Creation de genres satellites du genre *Maritrema*: *Quasimaritrema*, *Maritremopsis* et *Quasimaritremopsis*. Ann. Parasitol. Hum. Comp. 48: 543–557.

DECHTIAR, A.O. 1966. A new species of *Phyllodistomum* (Trematoda: Gorgoderidae) from *Coregonus clupeaformis* (Mitchill) from Lake of the Woods (Ontario). Can. J. Zool. 44: 135–140.

1972a. New parasite records from Lake Erie fish. Tech. Rep. Great Lakes Fish. Comm. 17: 20 p.

1972b. Parasites of fish from Lake of the Woods, Ontario. J. Fish. Res. Board Can. 29: 275–283.

DECHTIAR, A.O., AND BERST, A.H. 1978. Parasite fauna of splake (*Salvelinus fontinalis* × *S. namaycush*). Proc. Helminthol. Soc. Wash. 45: 249–254.

DECHTIAR, A.O., AND CHRISTIE, W.J. 1988. Survey of the parasite fauna of Lake Ontario fishes, 1961–71. In: Nepszy, S.J. (Ed.). Parasites of fishes in the Canadian waters of the Great Lakes. Tech. Rep. Great Lakes Fish. Comm. No. 51: 66– 95.

DECHTIAR, A.O., AND LAWRIE, A.H. 1988. Survey of the parasite fauna of Lake Superior fishes, 1969 to 1975. In: Nepszy, S.J. (Ed.). Parasites of fishes in the Canadian waters of the Great Lakes. Tech. Rep. Great Lakes Fish. Comm. No. 51: 1– 18.

DECHTIAR, A.O., AND NEPSZY, S.J. 1988. Survey of the parasite fauna of Lake Erie fishes, 1970–75. *In*: Nepszy, S.J. (Ed.). Parasites of fishes in the Canadian waters of the Great Lakes. Tech. Rep. Great Lakes Fish. Comm. No. 51: 49–65.

DECHTIAR, A.O., COLLINS, J.L, AND RECKAHN, J.A. 1988. Survey of the parasite fauna of Lake Huron fishes, 1961 to 1971. *In*: Nepszy, S.J. (Ed.). Parasites of fishes in the Canadian waters of the Great Lakes. Tech. Rep. Great Lakes Fish. Comm. No. 51: 19–48.

DECHTIAR, A.O., MACLEAN, J.A., AND NEPSZY, S.J. 1989. Parasites of fishes from Algonquin Park lakes. Ont. Fish. Tech. Rep. Ser. No. 29: 19 p.

DE MARTIN, J.D., AND PRATT, I. 1964. The life cycle of *Telolecithus pugetensis* Lloyd and Guberlet, 1932 (Trematoda: Monorchidae). J. Parasitol. 50: 101–105.

DICK, T.A. 1984. Parasites and Arctic charr management — an academic curiosity or practical reality?, p. 371–394. *In*: Johnson, L. and Burns, B.L. (Eds.). Biology of the Arctic charr. Proceedings of the International Symposium on Arctic charr, May 1981, Winnipeg, Manitoba. University of Manitoba Press, Winnipeg, Manitoba. 584 p.

DICK, T.A., AND BELOSEVIC, M. 1978. Parasites of Arctic char (*Salvelinus alpinus*) and their use in identifying marine (migratory) and freshwater (non-migratory) populations of char. Fourth International Congress of Parasitology, 19–26 August 1978, Warsaw, Poland. Short Communications, Section H, p. 5.

1981. Parasites of Arctic charr *Salvelinus alpinus* (Linnaeus) and their use in separating sea-run and non-migrating charr. J. Fish Biol. 18: 339–347.

DICK, T.A., AND ROSEN, R. 1981. Identification of *Diplostomum* spp. from the eyes of lake whitefish, *Coregonus clupeaformis* (Mitchill), based on experimental infection of herring gull chicks, *Larus argentatus* Pontoppidan. Can. J. Zool. 59: 1176–1179.

DICKINSON, A.B., AND THRELFALL, W. 1975. Metazoan parasites of *Fundulus heteroclitus* (Linnaeus, 1766) from insular Newfoundland. Proc. Helminthol. Soc. Wash. 42: 111–116.

1976. Some parasites in *Fundulus diaphanus* and *Pungitius pungitius* from insular Newfoundland. Proc. Helminthol. Soc. Wash. 43: 86–87.

DOBROVOLNY, C.G. 1939a. The life history of *Plagioporus lepomis*, a new trematode from fishes. J. Parasitol. 25:461–470.

1939b. Life history of *Plagioporus sinitsini* Mueller and embryology of new cotylocercous cercariae (Trematoda). Trans. Am. Microsc. Soc. 58: 121–155.

DOLLFUS, R.-P. 1960. Transformation de *Dihemistephanus lydiae* (M. Stossich, 1900) A. Looss, 1901 en *Stenocollum fragile* (Edwin Linton, 1900) J. Stafford, 1904 par macération. Ann. Parasitol. Hum. Comp. 35: 528–537.

1975. "Salmon posioning disease". Histoire de la découverte de cette prodigieuse maladie transmise par une douve de la famille Nanophyetidae R. Ph. Dollfus (1939). Actes du 95e Congrès National des Sociétés Savantes (Reims, 1970). Section des sciences. Vol. II, 267–310.

DOWSETT, J.A., AND LUBINSKY, G.A. 1966. Maturation of *Clinostomum complantum* (Trematoda) in laboratory mice. Can. J. Zool. 44: 496.

1980. *Clinostomum complanatum* or *Clinostomum marginatum*? (Trematoda: Clinostomidae). A problem of priority and synonymy. Can. J. Zool. 58: 680–682.

DUBOIS, G. 1938. Monographie des Strigeida (Trematoda). Mem. Soc. Neuchâtel. Sci. Nat. 6: 1–535.

1968. Synopsis des Strigeidae et des Diplostomatidae (Trematoda). Mem. Soc. Neuchâtel. Sci. Nat. 10: 1–258.

1970. Synopsis des Strigeidae et des Diplostomatidae (Trematoda). Deuxime partie. Mem. Soc. Neuchâtel. Sci. Nat. 10: 259–727.

DURU, C., JOHNSON, A.D., AND BLOUIN, E. 1981. *Neascus pyriformis* Chandler, 1951 (Trematoda: Diplostomatidae), redescription and incidence in fishes from Bule Creek, South Dakota. Proc. Helminthol. Soc. Wash. 48: 177–183.

EDWARDS, S.R., AND NAHHAS, F.M. 1968 Some endoparasites of fishes from the Sacramento-San Joachin Delta, California. Calif. Fish Game 54: 247–256.

ERASMUS, D.A. 1972. The biology of trematodes. London: Edward Arnold, 312 p.

ETGES, F.J. 1953. Studies on the life histories of *Maritrema obstipum* (Van Cleave & Mueller, 1932) and *Levinseniella amnicolae* n. sp. (Trematoda: Microphallidae). J. Parasitol. 39: 643–662.

1961. Contributions to the life history of the brain fluke of newts and fish, *Diplostomum scheuringi* Hughes, 1929 (Trematoda: Diplostomatidae). J. Parasitol. 47: 453–458.

EVANS, W.S. 1963. *Amphimerus pseudofelineus* (Ward, 1901) (Digenea: Opisthorchidae) and its second intermediate host in Manitoba. Can. J. Zool. 41: 649–651.

FANTHAM, H.B., AND PORTER, A. 1948. The parasite fauna of vertebrates in certain Canadian fresh waters, with some remarks on their ecology, structure and importance. Proc. Zool. Soc. London, Ser. B. 117: 609–649.

FEIZULLAEV, N.A., AND MIRZOEVA, S.S. 1983. Revision of the superfamily Clinostomoidea and an analysis of its system. Parazitologiya 17: 3–11. (In Russian).

1986. Keys to taxa of the superfamily Clinostomoidea (Trematoda). Parazitologiya 20: 195–201. (In Russian).

FISCHTHAL, J.H., CARSON, D.O., AND VAUGHT, R.S. 1982. Comparative size allometry of the digenetic trematode *Lissorchis attenuatus* (Monorchiidae) at four intensities of infection in the white sucker. J. Parasitol. 68: 314–318.

FOERSCH, W., REICHENBACH-KLINKE, H.H., SCHALLER, D., AND SCHMIDT, K. 1984. Metacercarial infection in *Umbra limi* (Teleostei) and a trematode infection in *Physa* sp. (Mollusca). Tierärztliche Praxis 12: 422–425. (In German).

FONT, W.F. 1991. Life cycle of *Amphimerus elongatus* (Trematoda: Opisthorchiidae). J. Helminthol. Soc. Wash. 58: 24–30.

FRÉCHETTE, J.L., RAU, M.E., AND WEBSTER, G.F. 1978. Les parasites des animaux sauvages du Québec. 3. Les helminthes de la barbotte brune, *Ictalurus nebulosus* dans la région du fleuve Saint-Laurent et ses principaux affluents. Le Naturaliste canadien (Québec) 105: 58.

FREDERICKSEN, D.W. 1972. Morphology and taxonomy of *Cotylogaster occidentalis* (Trematoda: Aspidogastridae). J. Parasitol. 58:1110–1116.

FRIMETH, J.P. 1987a. A survey of the parasites of nonanadromous and anadromous brook char (*Salvelinus fontinalis*) in the Tabusintac River, New Brunswick, Canada. Can. J. Zool. 65: 1354–1362.

1987b. Potential use of certain parasites of brook charr (*Salvelinus fontinalis*) as biological indicators in the Tabusinttac River, New Brunswick, Canada. Can. J. Zool. 65: 1989–1995.

FROST, N. 1940. A preliminary study of Newfoundland trout. Newfoundland Res. Bull. 9 (Fish): 1–30.

GAEVSKAYA, A.V. 1975. Two new species of trematodes, *Gonocerca macrouri* sp. n. and *Hemiurus macrouri* sp. n. from *Macrourus rupestris* L. of the north-east Atlantic. Parazitologiya 9: 457–459. (In Russian).

1989. Revision of the genus *Lecithophyllum* (Trematoda: Lecithasteridae). Zool. Zhurn. 68(9): 130–134. (In Russian).

GAEVSKAYA, A.V., AND ALESHKINA, L.D. 1983. New data on the trematodes from fishes of the Atlantic coast of Africa. Parazitologiya 17: 12–17. (In Russian).

GAEVSKAYA, A.V., AND UMNOVA, B.A. 1977. On the parasite fauna of the principal commercial fishes of the northwest Atlantic. Biol. Morya (Vladivostok) 4: 40–48. (In Russian).

GIBSON, D.I. 1976. Monogenea and Digenea from fishes. Discovery Rep. 36: 179–266.

1983. Comments on the zoogeography of some marine digeneans. *In*: Parasites and parasitic diseases of fish. First International Symposium of Ichthyoparasitology, Institute of Parasitology, Czechoslovak Academy of Sciences. p. 24–25.

1986. *Podocotyle araii* sp. nov. (Digenea: Opecoelidae) from *Sebastes* spp. off the coast of British Columbia, with a key to the species of *Podocotyle* Dujardin, 1845, occurring off the Pacific coast of North America. J. Nat. Hist. 20: 735–743.

1987. Questions in digenean systematics and evolution. Parasitology 95: 429–460.

GIBSON, D.I., AND BRAY, R.A. 1977. The Azygiidae, Hirudinellidae, Ptychogonomidae, Sclerodistomidae and Syncoeliidae of fishes from the northeast Atlantic. Bull. Br. Mus. (Nat. Hist.) Zool. 32: 167–245.

1979. The Hemiuroidea: terminology, systematics and evolution. Bull. Br. Mus. (Nat. Hist.) Zool. 36: 35–146.

1982. A study and reorganization of *Plagioporus* Stafford, 1904 (Digenea: Opecoelidae) and related genera, with special reference to forms from European Atlantic waters. J. Nat. Hist. 16: 529–559.

1984. On *Anomalotrema* Zhukov, 1957, *Pellamyzon* Montgomery, 1957, and *Opecoelina* Manter, 1934 (Digenea: Opecoelidae), with a description of *Anomalotrema koiae* sp. nov. from North Atlantic waters. J. Nat. Hist. 18: 949–964.

1986. The Hemiuridae (Digenea) of fishes from the north-east Atlantic. Bull. Br. Mus. (Nat. Hist.) Zool. 51: 1–125.

1994. Evolutionary expansion and host-parasite relationships of the Digenea. Int. J. Parasitol. 24: 1213–1226.

GIBSON, D.I., AND CHINABUT, S. 1984. *Rohdella siamensis* gen. et sp. nov. (Aspidogastridae: Rhodellinae subfam. nov.) from freshwater fishes in Thailand, with a reorganization of the classification of the subclass Aspidogastrea. Parasitology 88: 383–393.

GIBSON, D.I., MACKENZIE, K., AND COTTLE, J. 1981. *Halvorsenius exilis* gen. et sp. nov., a new didymozoid trematode from the mackerel *Scomber scombrus* L. J. Nat. Hist. 15: 917–929.

GIBSON, D.I., TASKINEN, J., AND VALTONEN, E.T. 1992. Studies on bucephalid digeneans parasitising molluscs and fishes in Finland. II. The description of *Rhipidocotyle fennica* n. sp. and its discrimination by principal components analysis. Syst. Parasitol. 24:

GLENN, C.L. 1980. Seasonal parasitic infections in mooneye, *Hiodon tergisus* (LeSeur), from the Assiniboine River. Can. J. Zool. 58: 252–257.

GORDON, D.M., AND RAU, M.E. 1982a. Possible evidence for mortality induced by the parasite *Apatemon gracilis* in a population of brook sticklebacks (*Culaea inconstans*). Parasitology 84: 41–47.

1982b. Temporal changes in the frequency distributions of *Apatemon gracilis* in brook sticklebacks (*Culaea inconstans*). Parasitology 84(1): iii. (Proceedings of the British Society for Parasitology. Joint Spring Meeting, 12–16 April 1981, held at the Leeuwenhorst Congress Centre, Noordwijkerhout, The Netherlands. Abstracts).

GORDON, D., CROLL, N.A., AND RAU, M.E. 1978. Les parasites des animaux sauvages de Québec. 1. Les parasites des poissons et des mammifères de la region de Schefferville. Le Naturaliste canadien (Québec) 105: 55.

GROOT, C., BAILEY, R.E., MARGOLIS, L., AND COOKE, K. 1989. Migratory patterns of sockeye salmon (*Oncorhynchus nerka*) smolts in the Strait of Georgia, British Columbia, as determined by analysis of parasite assemblages. Can. J. Zool. 67: 1670–1678.

HAEDRICH, R.L., AND POLLONI, P.T. 1976. A contribution to the life history of a small rattail fish, *Coryphaenoides carapinus*. Bull. South. Calif. Acad. Sci. 75:203–211.

HANEK, G., AND MOLNAR, K. 1974. Parasites of freshwater and anadromous fishes from Matamek River system, Quebec. J. Fish. Res. Board Can. 31: 1135–1139.

HANEK, G., AND THRELFALL, W. 1969a. Digenetic trematodes from Newfoundland. Canada. 1. Three species from *Gasterosteus aculeatus* Linnaeus, 1758. Can. J. Zool. 47: 793–794.

1969b. Digenetic trematodes from Newfoundland, Canada. 2. Two species from *Gasterosteus aculeatus* Linnaeus, 1758. Can. J. Zool. 47: 1086–1087.

1970a. Helminth parasites of the four-spine stickleback (*Apeltes quadracus* (Mitchill)) in Newfoundland. Can. J. Zool. 48: 404–406.

1970b. Metazoan parasites of the American eel (*Anguilla rostrata* (LeSueur)) in Newfoundland and Labrador. Can. J. Zool. 48: 597–600.

1970c. Parasites of the ninespine stickleback *Pungitius pungitius* (L.) in Newfoundland and Labrador. Can. J. Zool. 48: 600–602.

1970d. Parasites of the threespine stickleback (*Gasterosteus aculeatus*) in Newfoundland and Labrador. J. Fish. Res. Board Can. 27: 901–907.

1971. Metazoan parasites of the twospine stickleback, *Gasterosteus wheatlandi* Putnam, 1867 in Newfoundland. Am. Midl. Nat. 85: 275–276.

HARE, G.M. 1975. Atlantic salmon (*Salmo salar*) parasites as biological tags in the Miramichi River system, New Brunswick. Dissertation Abstracts International 35B: 4769.

HARE, G.M., AND BURT, M.D.B. 1975a. Identification, host sites and biology of parasites infecting juvenile Atlantic salmon (*Salmo salar*) in the Miramichi River system. Fish. Mar. Serv. Res. Tech. Rep. 581: 34 p.

1975b. Abundance and population dynamics of parasites infecting Atlantic salmon (*Salmo salar*) in Trout Brook, New Brunswick, Canada. J. Fish. Res. Board Can. 32: 2069–2074.

1976. Parasites as potential biological tags of Atlantic salmon (*Salmo salar*) smolts in the Miramichi River system, New Brunswick. J. Fish. Res. Board Can. 33: 1139–1143.

HARE, G.M., AND FRANTSI, C. 1974. Abundance and potential pathology of parasites infecting salmonids in Canadian Maritime hatcheries. J. Fish. Res. Board Can. 31: 1031–1036.

HELLER, A.F. 1949. Parasites of cod and other marine fish from the Baie de Chaleur region. Can. J. Res., Sect. D. 27: 243–264.

HICKS, F.J., AND THRELFALL, W. 1973. Metazoan parasites of salmonids and coregonids from coastal Labrador. J. Fish Biol. 5: 399–415.

HOFFMAN, G.L. 1956. The life cycle of *Crassiphiala bulboglossa* (Trematoda: Strigeida). Development of the metacercaria and cyst, and effect on the fish hosts. J. Parasitol. 42: 435– 444.

1958a. Experimental studies on the cercaria and metacercaria of the strigeoid trematode *Posthodiplostomum minimum*. Exp. Parasitol. 7: 23–50.

1958b. Studies on the life-cycle of *Ornithodiplostomum ptychocheilus* (Faust) (Trematoda: Strigeoidea) and the "self cure" of infected fish. J. Parasitol. 44: 416–421.

1960. Synopsis of Strigeoidea (Trematoda) of fishes and their life-cycles. Fishery Bull. 175 (Fish. Bull. of Fish and Wildl. Serv.) 60: 439–469.

1967. Parasites of North American freshwater fishes. University of California Press. Berkeley and Los Angeles. 486 p.

HOFFMAN, G.L., AND PUTZ, R.E. 1965. The black-spot (*Uvulifer ambloplitis*: Trematoda) of centrarchid fishes. Trans. Am. Fish. Soc. 94: 143–151.

HOFFMAN, G.L., FRIED, B., AND HARVEY, J.E. 1985. *Sanguinicola fontinalis* sp. nov. (Digenea: Sanguinicolidae): a blood parasite of brook trout, *Salvelinus fontinalis* (Mitchill), and longnose dace, *Rhinichthys cataractae* (Valenciennes). J. Fish Dis. 8: 529–538.

HOGANS, W.E. 1984. Helminths of striped bass (*Morone saxatilis*) from the Kouchibouguac River, New Brunswick. J. Wildl. Dis. 20: 61–63.

HOGANS, W.E., AND HURLBUT, T.R. 1984. Parasites of the knifenose chimaera, *Rhinochimaera atlantica*, from the Northwest Atlantic Ocean. Can. Field-Nat. 98: 365.

HOGANS, W.E., BRATTEY, J., UHAZY, L.S., AND KURLEY, P.C. 1983. Helminth parasites of swordfish (*Xiphias gladius* L.) from the northwest Atlantic Ocean. J. Parasitol. 69: 1178–1179.

HOGANS, W.E., DADSWELL, M.J., UHAZY, L.S., AND APPY, R.G. 1993. Parasites of American shad, *Alosa sapidissima* (Osteichthyes: Clupeidae), from rivers of the North American Atlantic coast and the Bay of Fundy, Canada. Can. J. Zool. 71: 941–946.

HOGUE, C.C., SUTHERLAND, D.R., AND CHRISTENSEN, B.M. 1993. Ecology of metazoan parasites infecting *Catostomus* spp. (Catostomidae) from southwestern Lake Superior. Can. J. Zool. 71: 1646–1652.

HOLL, F.J. 1929. The phyllodistomes of North America. Trans. Am. Microsc. Soc. 48: 48–53.

HOLMES, J.C. 1969 Habitat segregation in blood flukes of rockfishes from the Pacific coast. Program Abstracts of the 44th Annual Meeting American Society of Parasitology Abst. 191: 74.

1971a. Two new sanguinicolid blood flukes (Digenea) from scorpaenid rockfishes (Perciformes) of the Pacific coast of North America. J. Parasitol. 57: 209–216.

1971b. Habitat segregation in sanguinicolid blood flukes (Digenea) of scorpaenid rockfishes (Perciformes) on the Pacific coast of North America. J. Fish. Res. Board Can. 28: 903–909.

1990. Helminth communities in marine fishes. p.101–130. *In*: Esch, G., Bush, A. and Aho, J. (Eds.). Parasite communities: patterns and processes. Chapman Hall, London.

HOPKINS, S.H. 1931. Studies on *Crepidostomum*. II. The "*Crepidostomum laureatum*" of A.R. Cooper. J. Parasitol. 18: 79–91.

1934. The papillose Allocreadiidae. A study of their morphology, life histories and relationships. Ill. Biol. Monogr. 13: 1–80.

HORSFALL, M.W. 1934. Studies on the life history and morphology of the cystocercous cercariae. Trans. Am. Microsc. Soc. 53: 311–347.

HOSKINS, G.E., BELL, G.R., AND EVELYN, T.P.T. 1976. The occurrence, distribution and significance of infectious diseases and neoplasms observed in fish in the Pacific Region up to the end of 1974. Fish. Mar. Serv. Res. Dev. Tech, Rep. 707: 35 p.

HOUSTON, K.A., AND HAEDRICH, R.L. 1986. Food habits and intestinal parasites of deep demersal fishes from the upper continental slope east of Newfoundland, northwest Atlantic Ocean. Mar. Biol. 92: 563–574.

HUGHES, R.C. 1927. Studies on the trematode family Strigeidae (Holostomidae). No. VI. A new metacercaria *Neascus ambloplitis*, sp. nov. representing a new larval group. Trans. Am. Microsc. Soc. 46: 248–267.

1928a. Studies on the trematode family Strigeidae (Holostomidae). No. X. *Neascus bulboglossa* (Van Haitsma). J. Parasitol. 15: 52–57.

1928b. Studies on the trematode family Strigeidae (Holostomidae). No. IX. *Neascus van-cleavei* (Agersborg). Trans. Am. Microsc. Soc. 47: 420–341.

1929. Studies of the trematode family Strigeidae (Holostomidae). No. XIV. Two new species of diplostomula. Occas. Pap. Mus. Zool. Univ. Mich. 202: 1–29.

HUGHES, R.C., AND PISZCZEK, F.R. 1928. Studies on the trematode family Strigeidae (Holostomidae). No. XI. *Neascus ptychocheilus* (Faust). J. Parasitol. 15: 58–62.

HUNNINEN, A.V., AND CABLE, R.M. 1941. Studies on the life history of *Anisoporus manteri* Hunninen and Cable, 1940 (Trematoda: Allocreadiidae). Biol. Bull. Mar. Lab. Woods Hole 80: 415–428.

HUNTER, G.W. III, AND BANGHAM, R.V. 1932. Studies of fish parasites of Lake Erie. I. New trematodes (Allocreadiidae). Trans. Am. Microsc. Soc. 51: 137–152.

HUNTER, G.W. III, AND HAMILTON, J.M. 1941. Studies on host-parasite reactions to larval parasites. IV. The cyst of *Uvulifer ambloplitis* (Hughes). Trans. Am. Microsc. Soc. 40: 498–507.

HUNTSMAN, A.G. 1918. The Canadian plaice. Bull. Fish. Res. Can. 1: 32 p.

ICHIHARA, A. 1970. A new blood fluke *Aporocotyle theragrae* n. sp. (Digenetic trematode: Aporocotylidae) from a marine fish, *Theragra chalcogramma*. Res.Bull. Meguro Parasitol. Mus. 3: 1–4.

ILES, C. 1971. *Fistulicola plicatus* (Cestoda) and *Tristoma* spp. (Trematoda) on swordfish from the northwest Atlantic. J. Fish. Res. Board Can. 28: 31–34.

IP, H.S., AND DESSER, S.S. 1984. Transmission electron microscopy of the tegumentary sense organs of *Cotylogaster occidentalis* (Trematoda: Aspidogastrea). J. Parasitol. 70: 563–575.

JENNINGS, M.R., AND HENDRICKSON, G.L. 1982. Parasites of chinook salmon (*Oncorhynchus tshawytscha*) and coho salmon (*O. kisutch*) from the Mad River and vicinity, Humboldt County, California. Proc. Helminthol. Soc. Wash. 49: 279–284.

KABAT, A.R. 1986. Effects of trematode parasitism on reproductive output of the bivalve *Transennella tantilla*. Can. J. Zool. 64: 267–270.

KABATA, Z., AND WHITAKER, D.J. 1984. Results of three investigations of the parasite fauna of several marine fishes of British Columbia. Can. Tech. Rep. Fish. Aquat. Sci. 1303: 19 p.

KABATA, Z., MCFARLANE, G.A., AND WHITAKER, D.J. 1988. Trematoda of sablefish, *Anoplopoma fimbria* (Pallas, 1811), as possible biological tags for stock identification. Can. J. Zool. 66: 195–200.

KARLSBAKK, E. 1993. On the morphology and ecology of some digenean parasites of the four-bearded rockling (*Enchelyopus cimbrius* (L.)) (Gadidae) in Rauenfjorden, western Norway. Cand. Scient. Thesis, University of Bergen.

KENNEDY, M.J. 1979. The responses of miracidia and cercariae of *Bunodera mediovitellata* (Trematoda: Allocreadiidae) to light and gravity. Can. J. Zool. 57: 603–609.

1983. A revision of North American species of the genus *Parametorchis* Skrjabin, 1913 (Trematoda: Opisthorchiidae). Proc. Helminthol. Soc. Wash. 50: 312–317.

KENNEDY, W.A., AND SPRULES, W.M. 1967. Goldeye in Canada. Bull. Fish. Res. Board Can. 161: 45 p.

KHAN, R.A. 1987. Crude oil and parasites of fish. Parasitol. Today 3: 99–100.

KHAN, R.A., AND KICENIUK, J. 1983. Effects of crude oils on the gastrointestinal parasites of two species of marine fish. J. Wildl. Dis. 19: 253–258.

KNISKERN, V.B. 1952. Studies on the trematode family Bucephalidae Poche, 1907. Part I. A systematic review of the family Bucephalidae. Trans. Am. Microsc. Soc. 71: 253–266.

KØIE, M. 1978. On the morphology and life-history of *Stephanostomum caducum* (Looss, 1901) Manter, 1934 (Trematoda, Acanthocolpidae). Ophelia 17: 121–133.

1983. Digenetic trematodes from *Limanda limanda* (L.) (Osteichthyes, Pleuronectidae) from Danish and adjacent waters, with special reference to their life-histories. Ophelia 22: 201–228.

1984. Digenetic trematodes from *Gadus morhua* L. (Osteichthyes, Gadidae) from Danish and adjacent waters, with special reference to their life-histories. Ophelia 23: 195–222.

1985a. On the morphology and life-history of *Lepidapedon elongatum* (Lebour, 1908) Nicoll, 1910 (Trematoda, Lepocreadiidae). Ophelia 24: 135–153.

1985b. The surface topography and life-cycles of digenetic trematodes in *Limanda limanda* (L.) and *Gadus morhua* L. Summary of Doctoral Thesis, Marine Biological Laboratory, University of Copenhagen, 20 p.

1991. Aspects of the morphology and life cycle of *Lecithocladium excisum* (Digenea, Hemiuridae), a parasite of *Scomber* spp. Int. J. Parasitol. 21: 597–602.

KOROTAEVA, V.D. 1982. The fauna of the trematodes in fishes of the Order Zeiformes. Parazitologiya 16: 464–468. (In Russian).

KRYGIER, B.B., AND MACY, R.W. 1969. *Lissorchis heterorchis* sp. n. (Trematoda: Lissorchiidae) from *Catostomus macrocheilus* Girard in Oregon. Proc. Helminthol. Soc. Wash. 36: 136–139.

KRZYKAWSKI, S., AND WIERZBICKA, J. 1992. An attempt to determine systematic position of Greenland halibut, *Reinhardtius hippoglossoides* (Walbaum), from Labrador region and Barents Sea on the basis of morphometric, biologic, and parasitological studies. Acta Ichthyol. Pisc. 22: 59–79.

KUITUNEN-EKBAUM, E. 1937. Intestinal parasites of the haddock in Canadian waters. Fish. Res. Board Can. MS Rep. (Biol.) 303: 8 p.

KUNTZ, R.E. 1951. Embryonic development of the excretory system in a psilostome cercaria, a gymnocephalous (fasciolid) cercaria and in three monostome cercariae. Trans. Am. Microsc. Soc. 70: 95–118.

Kuntz, S.M., and Font, W.F. 1984. Seasonal dynamics of *Allopodocotyle boleosomi* (Pearse, 1924) n. comb. (Digenea: Opecoelidae) in Wisconsin darters (Etheostomatinae). Can. J. Zool. 62: 2666–2672.

Lachance, F. 1947. Black spot disease of bass. Part I. Distribution of the disease. Can. Fish Cult. 1: 16–21.

Landry, T., Boghen, A.D., and Hare, G.M. 1986. A comparative study of the parasites of alewife *Alosa pseudoharengus* and blueback herring *Alosa aestivalis* from the Miramichi River, N.B. Bulletin of the Canadian Society of Zoologists 17(2): 23. (25th Annual Meeting Canadian Society of Zoologists, 11–14 May 1986, University of Saskatchewan, Saskatoon, Saskatchewan. Program and Abstracts).

1992. Les parasites de l'alose d'été (*Alosa aestivalis*) et du gaspareau (*Alosa pseudoharengus*) de la rivière Miramichi, New Brunswick. Can. J. Zool. 70: 1622–1624.

Lasee, B.A., Font, W.F., and Sutherland, D.R. 1988. *Culaeatrema inconstans* gen.n., sp.n. (Digenea: Allocreadiidae) from the brook stickleback (*Culaea inconstans*) in Wisconsin and observations on parthenogenetic populations. Can. J. Zool. 66: 1328–1333.

Layman, E.M. 1930. (Parasitic worms from fishes of Peter the Great Bay.) Izv. Tikhookean. Nauchno Promysl. Stn 3: 1–120. (In Russian, with German summary).

Lee, D.L., Whitaker, D.J., and Stanley, R.D. 1990. A preliminary examination of the parasite fauna of yellowtail rockfish, *Sebastes flavidus*. Can. Tech. Rep. Fish. Aquat. Sci. 1777: 21 p.

Lehmann, D.L. 1965. Intestinal parasites of northwestern amphibians. Yb. Am. Phil. Soc. 1964: 284–285.

Leigh, W.H. 1958. The life-history of *Macroderoides spiniferus* Pearse. 1924, a trematode of the Florida spotted gar, *Lepisosteus platyrhincus*. J. Parasitol. 44: 379–387.

Leong, T.S., and Holmes, J.C. 1974. Acquistion of helminths by coho salmon, *Oncorhynchus kisutch* introduced into Cold Lake, Alberta: a comparison with helminths of native salmonid fish. Proc. 3rd Int. Congr. Parasitol. 3: 1639.

1981. Communities of metazoan parasites in open water fishes of Cold Lake, Alberta. J. Fish Biol. 18: 693–713.

Lester, R.J.G. 1975. Parasites of *Gasterosteus aculeatus* near Vancouver, British Columbia. Syesis 7: 195–200.

1977. An estimate of the mortality in a population of *Perca flavescens* owing to the trematode *Diplostomum adamsi*. Can. J. Zool. 55: 288–292.

Lester R.J.G., and Huizinga, H.W. 1977. *Diplostomum adamsi* sp.n.: description, life cycle, and pathogenesis in the retina of *Perca flavescens*. Can. J. Zool. 55: 64–73.

Lewis, F.J. 1935. The trematode genus *Phyllodistomum* Braun. Trans. Am. Microsc. Soc. 54: 103–117.

Linkletter, L.E., Lord, E.I., and Dadswell, M.J. 1977. A checklist of marine fauna and flora of the Bay of Fundy. Huntsman Mar. Lab. New Brunswick, 68 p.

Linton, E. 1901. Parasites of fishes of the Woods Hole region. Bull U.S. Fish. Com. (1899). 19: 405–492.

1940. Trematodes from fishes mainly from the Woods Hole Region, Massachusetts. Proc. U.S. Nat. Mus. 88 (No. 3078): 1–172.

Liston, J., and Hitz, R. 1961. Second survey of the occurrence of parasites and blemishes in Pacific ocean perch, *Sebastodes alutus*, May-June 1959. U.S. Fish. Wildl. Serv. Spec. Sci. Rep. Fish. 383: 6 p.

Liston, J., Peters, J., and Stern, A.J. 1960. Parasites in summer caught Pacific rockfishes. U.S. Fish. Wildl. Serv. Spec. Sci. Rep. Fish. 352: 10 p.

Lloyd, L.C. 1938. Some digenetic trematodes from Puget Sound fish. J. Parasitol. 24: 103–133.

Lloyd, L.C., and Guberlet, J.E. 1932. A new genus and species of Monorchidae. J. Parasitol. 18: 232–239.

Love, M.S., and Moser, M. 1983. A checklist of parasites of California, Oregon, and Washington marine and estuarine fishes. NOAA Tech. Rep. NMFS SSRF 777: 577 p.

Lubinsky, G.A. 1976. Ichthyoparasites of Manitoba. p. 253–289. *In*: International Garrison diversion board. Rep. Int. Joint Comm. Append. C. Biol. Rep. Attachment C.II-7.

Lubinsky, G.A., and Loch, J.S. 1979. Ichthyoparasites of Manitoba; literature review and bibliography. Fish. Mar. Serv. MS Rep. 1513: 20 p.

Lumb, S.M., Bray, R.A., and Rollinson, D. 1993. Partial small subunit (18S) rRNA gene sequences from fish parasites of the families of the Lepocreadiidae and Fellodistomidae (Digenea) and their use in phylogenetic analyses. Syst. Parasitol. 26: 141–149.

Lundahl, W.S. 1941. Life history of *Caecincola parvulus* Marshall and Gilbert (Cryptogonimidae, Trematoda) and the development of its excretory system. Trans. Am. Microsc. Soc. 60: 461–484.

Lyster, L.L. 1939. Parasites of freshwater fish. I. Internal trematodes of commercial fish in central St Lawrence watershed. Can. J. Res., Sect. D. 17: 154–168.

1940a. Parasites of freshwater fish. II. Parasitism of speckled and lake trout and the fish found associated with them in Lake Commandant, Quebec, Can. J. Res., Sect. D. 18: 66–78.

1940b. *Apophallus imperator* sp. nov., a heterophyid encysted in trout with a contribution to its life history. Can. J. Res., Sect. D. 18: 106–121.

MacCallum, W.G. 1895. On the anatomy of two distome parasites of freshwater fish. Vet. Mag. 2: 401–412.

Machida, M. 1984. Trematodes of marine fishes from depth of 200–400m off Yamagat, the Japan Sea. Mem. Natn. Sci. Mus., Tokyo 17: 101–110.

MACHIDA, M., AND ARAKI, J. 1992. Three aspidogastrean trematodes from marine fishes of Japan. Bull. Nat. Sci. Mus., Tokyo 18: 51–55.

MACLULICH, D.A. 1943. Parasites of trout in Algonquin Provincial Park, Ontario. Can. J. Res., Sect. D. 21: 405–412.

MACKENZIE, K., AND GIBSON, D.I. 1970. Ecological studies of some parasites of plaice *Pleuronectes platessa* L. and flounder *Platichthys flesus* (L.). Symp. Br. Soc. Parasitol. 8: 1–42.

MAMAEV, Y.L. 1968. Evaluation of up-to-date classification systems of the Monorchidae. Papers on helminthology presented to Academician K.I. Skrjabin on his 90th birthday. Akad. Nauk SSSR: Moscow, 239–243 (In Russian).

MANTER, H.W. 1931. Some digenetic trematodes of marine fishes of Beaufort, North Carolina. Parasitology 23: 396–411.

1934. Some digenetic trematodes from deep-water fish of Tortugas, Florida. Pap. Tortugas Lab. 28: 257–345.

1947. The digenetic trematodes of marine fishes of Tortugas, Florida. Am. Midl. Nat. 38: 257–416.

1954. Some digenetic trematodes from fishes of New Zealand. Trans. R. Soc. New Zealand 82: 475–568.

1962. Notes on the taxonomy of certain digenetic trematodes of South American freshwater fishes. Proc. Helminthol. Soc. Wash. 29: 97–102.

MANTER, H.W., AND PRITCHARD. M.H. 1960. Some hemiurid trematodes from Hawaiian fishes. Proc. Helminthol. Soc. Wash. 27: 87–102.

MANTER, H.W., AND VAN CLEAVE, H.J. 1951. Some digenetic trematodes including eight new species from marine fishes of La Jolla, Calif. Proc. U.S. Nat. Mus. 101 (No. 3279): 315–340.

MARCOGLIESE, D.J., AND CONE, D.K., 1991a. Importance of lake characteristics in structuring parasite communities of salmonids from insular Newfoundland. Can. J. Zool. 69: 2962–2967.

1991b. Do brook charr (*Salvelinus fontinalis*) from insular Newfoundland have different parasites than their mainland counterparts? Can. J. Zool. 69: 809–811.

MARGOLIS, L. 1956a. Anomalous development of vitellaria in *Hemiurus levinseni*. Can. J. Zool. 34: 207–208.

1956b. Report on parasite studies of sockeye and pink salmon collected in 1955, with special reference to the utilization of parasites as a means of distinguishing between Asiatic and American stocks of salmon on the high seas — a progress report on work being carried out as part of F.R.B.'s commitments to INPFC. Fish. Res. Board Can. MS Rep. (Biol.) 624. 36 p.

1957. A study of the parasites of sockeye and pink salmon with particular attention to their application in distinguishing between Asiatic and North American stocks of these fish on the high seas — report of results of examinations of 1956 samples. Fish. Res. Board Can. MS Rep. (Biol.) 641: 24 p.

1958a. The application of parasitological data to the problem of recognizing the continent of origin of stocks of sockeye salmon in the North Pacific Ocean and adjacent seas — a summary of three years' (1955 through 1957) research. Fish. Res. Board Can. MS Rep. (Biol.) 658: 20 p.

1958b. A new species of *Lecithophyllum* from North Pacific fishes with a consideration of the taxonomy of the genera *Lecithophyllum*, *Aponurus*, and *Brachadena* (Trematoda: Hemiuridae). Can. J. Zool. 36: 893–904.

1962. *Lampritrema nipponicum* Yamaguti (Tremadoda) from new hosts in the North Pacific Ocean, the relationship of *Distomum miescheri* Zschokke, and the status of the family Lampritrematidae. Can. J. Zool. 40: 941–950.

1963. Parasites as indicators of the geographical origin of sockeye salmon, *Oncorhynchus nerka* (Walbaum), occurring in the North Pacific Ocean and adjacent seas. Int. North Pac. Fish. Comm. Bull. 11: 101–156.

1964. *Paurorhynchus hiodontis* Dickerman, 1954 (Trematoda: Bucephalidae): a second record involving a new host and locality in Canada. Can. J. Zool. 42: 416.

1972. The trematode *Plagioporus shawi* (McIntosh, 1939) n. comb. (Opecoelidae: Plagioporinae) from the North Pacific sockeye salmon, *Oncorhynchus nerka*. An. Inst. Biol., Univ. Nac. Autón. México, Ser. Zool. Núm. Unico: 99–108.

1982. Parasitology of Pacific salmon — an overview. p. 135–226. *In*: Meerovitch, E. (Ed.). Aspects of parasitology — a festschrift dedicated to the 50th anniversary of the Institute of Parasitology of McGill University, 1932–1982. McGill University, Montreal, Québec. 485 p.

1990. Trematodes as population markers for North Pacific steelhead trout. p. 735. *In*: Bulletin de la Société Française de Parasitologie, Volume 8, Supplement 2. VII International Congress of Parasitology, 20–24 August 1990, Paris, France. Abstracts. p. 659–1310.

1992. A brief history of Canadian research from 1955 to 1990 related to Pacific salmon (*Oncorhynchus* species) on the high seas. p. 1–10. *In*: Ishida, Y., Nagasawa, K., Welch, D.W., Myers, K.W. and Shershnev, A.P. (Eds.). Proceedings of the International Workshop on Future Salmon Research in the North Pacific Ocean. Special Publication of the National Research Institute of Far Seas Fisheries 20: 79 p. Shimuzu, Japan.

MARGOLIS, L., AND ADAMS, J.R. 1956. Description of *Genolinea oncorhynchi* n. sp. (Trematoda: Hemiuridae) from *Oncorhynchus gorbuscha* in British Columbia with notes on the genus. Can. J. Zool. 34: 573–577.

MARGOLIS, L., AND ARTHUR, J.R. 1979. Synopsis of the parasites of fishes of Canada. Bull. Fish. Res. Board Can. 199: 269 p.

MARGOLIS, L., AND BOYCE, N.P. 1969. Life span, maturation, and growth of two hemiurid trematodes, *Tubulovesicula lindbergi* and *Lecithaster gibbosus* in Pacific salmon (genus *Oncorhynchus*). J. Fish. Res. Board Can. 26: 893–907.

MARGOLIS, L., AND CHING, H.L. 1965. Review of the trematode genera *Bacciger* and *Pentagramma* (Fellodistomatidae) and description of *P. petrowi* (Layman, 1930) n. comb. from marine fishes from the Pacific coast of Canada. Can. J. Zool. 43: 381–405.

MARGOLIS, L., AND MCDONALD, T.E. 1986. Parasites of white sturgeon, *Acipenser transmontanus*, from the Fraser River, British Columbia. J. Parasitol. 72: 794–796.

MARGOLIS, L., AND MORAVEC, F. 1982. *Ramellogammarus vancouverensis* Bousfield (Amphipoda) as an intermediate host for salmonid parasites in British Columbia. Can. J. Zool. 60: 1100–1104.

MARSHALL, W.S., AND GILBERT, N.C. 1905. Three new trematodes found principally in black bass. Zool. Jahrb., Syst. 22: 477–488.

MARTIN, W.E. 1938. Studies on trematodes of Woods Hole: The life cycle of *Lepocreadium setiferoides* (Miller and Northup), Allocreadiidae, and the description of *Cercaria cumingiae* n. sp. Biol. Bull. Mar. Biol. Lab. Woods Hole 75: 463–474.

1939. Studies on the trematodes of Woods Hole. II. The life cycle of *Stephanostomum tenue* (Linton). Biol. Bull. Mar. Biol. Lab. Woods Hole 77: 65–73.

MATTHEWS, R.A. 1973. The life-cycle of *Prosorhynchus crucibulum* (Rudolphi, 1819) Odhner, 1905 and a comparison of its cercaria with that of *Prosorhynchus squamatus* Odhner, 1905. Parasitology 66: 133–164.

MCALLISTER, D.J., AND MUDRY, D.R. 1983. Diseases and parasites of Alberta fishes and potential consequences of their interbasin transfer. Alberta Environmental Research Trust, Calgary, Alberta. 80 p.

MCDONALD, T.E., AND MARGOLIS, L. 1995. Synopsis of the parasites of fishes of Canada: Supplement (1978–1993). Can. Spec. Publ. Fish. Aquat. Sci. 122: 265 p.

MCFARLANE, S.H. 1934. *Stephanostomum casum* (Linton) a trematode possessing a uroproct. Trans. Am. Microsc. Soc. 53: 172–173.

1936 A study of the endoparasitic trematodes from marine fishes of Departure Bay, B.C. J. Biol. Board Can. 2: 335–337.

MCGLADDERY, S.E., AND BURT, M.D.B. 1985. Potential of parasites for use as biological indicators of migration, feeding, and spawning behaviour of northwestern Atlantic herring (*Clupea harengus*). Can. J. Fish. Aquat. Sci. 42: 1957–1968.

MCGLADDERY, S.E., MURPHY, L., HICKS, B.D., AND WAGNER, S.K. 1990. The effects of *Stephanostomum tenue* (Digenea: Acanthocolpidae) on marine aquaculture of the rainbow trout, *Salmo gairdneri*.p.305–315. *In*: Perkins, F.O., and Cheng, T.C. (Eds.). Pathology in marine science. Academic Press: San Diego, California.

MILLER, G.C. 1959. Studies on the genus *Homalometron* Stafford, 1904 (Trematoda: Lepocreadiidae) with a redescription of *H. armatum* (MacCallum, 1895). J. Parasitol. 45: 539–542.

MILLER, M.J. 1936. *Bunoderina eucaliae* gen. et sp. nov., a new papillose Allocreadiidae from the stickleback. Can. J. Res., Sect. D. 14: 11–14.

1940a. Parasites of freshwater fish. III. Further studies on the internal trematodes of fish in the central St Lawrence watershed. Can. J. Res., Sect. D. 18: 423–434.

1940b. Black spot in fishes. Can. J. Comp. Med. 4: 303–305.

1941a. A critical study of Stafford's report on "Trematodes of Canadian Fishes" based on his trematode collection. Can. J. Res., Sect. D. 19: 28–52.

1941b. The life history of *Apophallus brevis* Ransom, 1920. J. Parasitol. 27(suppl.): 12.

1942. Black spot disease of speckled trout. Rev. Can. Biol. 1: 464–471.

MOLNAR, K., HANEK, G., AND FERNANDO, C.H. 1974. Parasites of fishes from Laurel Creek, Ontario. J. Fish Biol 6: 717–728.

MORRISON, C M., MCCLELLAND, G., CORNICK, J., AND MARCOGLIESE, D. 1986. Parasites and diseases of some marine finfish off Nova Scotia. Can. Tech. Rep. Fish. Aquat. Sci. 1424: 36 p.

MORTENSEN, D.G., AND MOTHERSHEAD, P.D. 1988. Occurrence of two hemiuroid trematodes in the stomachs of juvenile Pacific salmon from the waters of southeastern Alaska and British Columbia. Trans. Am. Fish. Soc. 117: 452–455.

MUDRY, D.R., AND ANDERSON, R.S. 1976. A bibliography of the literature relating to the parasites of freshwater fishes in Alberta, 1932–1975. Can. Wildl. Serv., Calgary, Alta. 10 p.

1977. Helminth and arthropod parasites of fishes in the mountain national parks of Canada. J. Fish Biol. 11: 21–33.

MUDRY, D.R., AND MCCART, P. 1976. Metazoan parasites of Arctic char (*Salvelinus alpinus*) from the north slopes of Canada and Alaska. J. Fish. Res. Board Can. 33: 271–275.

MUELLER, J.F. 1934. Two new trematodes from Oneida Lake fishes. Trans. Am. Microsc. Soc. 53: 231–236.

MUELLER, J.F., AND VAN CLEAVE, H.J. 1932. Parasites of Oneida Lake fishes. Part II. Descriptions of new species and some taxonomic considerations, especially concerning the trematode family Heterophyidae. Bull.New York State Coll. Forestry Syracusae Univ. 5 (Roosevelt Wild Life Ann. 3): 79–137.

MYERS, B.J. 1959. Parasites from elasmobranch hosts from the Magdalen Islands region of the Gulf of St Lawrence. Can. J. Zool. 37: 245–246.

NAGATY, H.F. 1937. Trematodes of fishes from the Red Sea. Part I. Studies on the family Bucephalidae Poche, 1907. The Egyptian Univ., Fac. of Med., Publ. No. 12: 192 p.

NAHHAS, F.M., AND KRUPIN, R. 1977. Parasites of *Liparis callyodon* (Pallas) with a description of a new species, *Metadena caballeroi*. Excerta Parasitologica en Memoria del Doctor Eduardo Caballero y Caballero. Univ. Nac. Auton. Mex. Inst. Biol. Publ. Espec. 4: 261–266.

NAHHAS, F.M., AND SHORT, R.B. 1965. Digenetic trematodes of marine fishes from Apalachee Bay, Gulf of Mexico. Tulane Stud. Zool. 12: 39–50.

NASIR, P., AND DIAZ, M.T. 1971. A revision of the genus *Lecithochirium* Luehe, 1901 and a redescription of *L. monticellii* (Linton, 1898) Skrjabin & Guschanskaja, 1955. Riv. Parassit. 32: 27–36.

NEPSZY, S.J. (Ed.). 1988. Parasites of fishes in the Canadian waters of the the Great Lakes. Tech. Rep. Great Lakes Fish. Comm. 51: 106 p.

NIEWIADOMSKA, K. 1984. Present status of *Diplostomum spathaceum* (Rudolphi, 1819) and differentiation of *Diplostomum paraspathaceum* nom. nov. (Trematoda: Diplostomidae). Syst. Parasitol. 6: 81–86.

1986. Verification of the life-cycles of *Diplostomum spathaceum* (Rudolphi, 1819) and *D. pseudospathaceum* Niewiadomska, 1984 (Trematoda, Diplostomidae). Syst. Parasitol. 8: 23–31.

1988. *Diplostomum* metacercariae (Digenea) in fish of the Dgal Wielki lakes: *D. numericum* sp. n. and *D. baeri* Dubois, 1937, with comments on the synonymy of this species. Acta Parasitol. Polon. 33: 7–24.

1989. On the synonymy of *Diplostomum pseudospathaceum* Niewiadomska, 1984 (Digenea, Diplostomidae). Acta Parasitol. Polon. 34: 103–106.

NOBLE, E.R. 1973. Parasites and fishes in the deep-sea environment. Adv. Mar. Biol. 11: 121–195.

ODENING, K. 1979. A propos de la validité d'*Ichthycotylurus variegatus* (Creplin, 1825) et de la position specifique de *Tetracotyle ovata* (v. Linstow, 1877) (Trematoda: Strigeidae). Ann. Parasitol. Hum. Comp. 54: 171–183.

ODHNER, T. 1902. Mitteilungen zur Kenntnis der Distomen. 2. Zentr. Bakt. 1. Abt., Orig. 31: 152–162.

1905. Die trematoden des arktischen Gebietes. Fauna Arct. 4:289–372.

OLSON, R.E. 1970. The life cycle of *Cotylurus erraticus* (Rudolphi, 1809) Szidat, 1928 (Trematoda: Strigeidae). J. Parasitol. 56: 55–63.

OSBORN, H.L. 1903. On *Cryptogonimus* (n. g.) *chili* (n. sp.), a fluke with two ventral suckers. Zool. Anz. 26: 315–318.

1910. On the structure of *Cryptogonimus* (nov. gen.) *chyli* (n. sp.), an aberrant distome, from fishes of Michigan and New York. J. Exper. Zool. 9: 517–536.

PALSSON, J. 1986. Quantitative studies on the helminth fauna of capelin (*Mallotus villosus* (Müller) in the Northwest Atlantic for the purpose of stock discrimination. Can. Tech. Rep. Fish. Aquat. Sci. No. 1499: 21 p.

PALSSON, J., AND BEVERELY-BURTON, M. 1984. Helminth parasites of capelin, *Mallotus villosus*, (Pisces: Osmeridae) of the North Atlantic. Proc. Helminthol. Soc. Wash. 51: 248–254.

PARNELL, I.W. 1934. Fish parasites and their importance. Trans. Am. Fish. Soc. 64: 390–400.

PARUKHIN, A.M. 1976. Parasitic worms in food fish of the Southern Seas. Naukova Dumka: Kiev. 183 p. (In Russian).

PEARSE, A.S. 1924. Observations on parasitic worms from Wisconsin fishes. Trans. Wisc. Acad. Sci. Art. Lett. 21: 147–160.

PEARSON, J.C. 1973. A revision of the subfamily Haplorchinae Looss, 1899 (Trematoda: Heterophyidae). II. Genus *Galactosomum*. Phil. Trans. R. Soc. London 266: 341–447.

1986. The paranephridial system in the Digenea: occurrence and possible phylogenetic significance. In: Cremin, M. *et al.* (Eds.). Parasite lives. (Papers on parasites, their hosts and their associations to honour J.F.A. Sprent.) St. Lucia: University of Queensland Press, 56–68.

1992. On the position of the digenean family Heronimidae: an inquiry into a cladistic classification of the Digenea. Syst. Parasitol. 21: 81–166.

PEARSON, J.C., MARGOLIS, L., AND BOYCE, N.P 1978. The metacercaria of *Galactosomum phalacrocoracis* (Trematoda: Heterophyidae) from pink salmon, *Oncorhynchus gorbuscha*, from British Columbian waters. Can. J. Zool. 56: 2235–2238.

PERKINS, K.W. 1956. Studies on the morphology and biology of *Acetodextra amiuri* (Stafford) (Trematoda: Heterophyidae). Am. Midl. Nat. 55: 139–161.

PETERS, L.E. 1957. An analysis of the trematode genus *Allocreadium* Looss with description of *Allocreadium neotenicum* sp. nov. from water beetles. J. Parasitol. 43:136–142.

1961. The allocreadioid problem with reference to the excretory system in four types of cercariae. Proc. Helminthol. Soc. Wash. 28: 102–108.

PIKE, A.W., AND BURT, M.D.B. 1981. Host tissue responses of the yellow perch (*Perca flavescens*) to three species of encysted metacercariae (Platyhelminthes: Digenea). Bulletin of the Canadian Society of Zoologists 12(2): 76. (Annual Meeting Canadian Society of Zoologists, 10–13 May 1981, University of Waterloo, Waterloo, Ontario. Program and Abstracts).

1982. Tissue response of *Perca flavescens* (yellow perch) to infection with *Apophallus brevis* (Heterophyidae) (the sandgrain worm). Parasitology 84(1): xlviii. (Proceedings of the British Society for

Parasitology. Joint Spring Meeting, 12–16 April 1981, held at the Leeuwenhorst Congress Centre, Noordwijkerhoot, The Netherlands. Abstracts).

1983. The tissue response of yellow perch, *Perca flavescens* Mitchill to infections with the metacercarial cyst of *Apophallus brevis* Ransom, 1920. Parasitology 87: 393–404.

PIKE, A.W., THOMPSON, A.B., AND BURT, M.D.B. 1983. The distribution of *Apophallus brevis* metacercariae in the yellow perch *Perca flavescens*. Parasitology 87(2): xxix. (Proceedings of the British Society for Parasitology. Spring Meeting, 12–14 April 1983, held at the University of Salford. Abstracts).

PIPPY, J.H.C. 1969. Preliminary report on parasites as biological tags in Atlantic salmon (*Salmo salar*). I. Investigations 1966–1968. Fish. Res. Board Can. Tech. Rep. 134: 60 p.

1980. The value of parasites as biological tags in Atlantic salmon at West Greenland. Rapp. Proces-Verb. Réun. Cons. Int. Expl. Mer 176: 76–81.

PLATT, T.R. 1992. A phylogenetic and biogeographic analysis of the genera of Spirorchinae (Digenea: Spirorchidae) parasitic in freshwarer fishes. J. Parasitol. 78:616:629.

POOLE, B.C., AND DICK, T.A. 1983. Metacercarial distribution of *Apophallus brevis* (Heterophyidae) in yellow perch (*Perca flavescens*) from the Heming Lake study area. Can. J. Zool. 61: 2104–2109.

1985. Parasite recruitment by stocked walleye, *Stizostedion vitreum vitreum* (Mitchill), fry in a small boreal lake in Central Canada. J. Wildl. Dis. 21: 371–376.

PRATT, I., AND MCCAULEY, J.E. 1961. Trematodes of the Pacific Northwest. An annotated catalogue. Oreg. Sta. Monogr. (Zool.) 11: 188 p.

PRICE, E.W. 1931. Four new species of trematode worms from the muskrat, *Ondatra zibethica*, with a key to the trematode parasites of the muskrat. Proc. U.S. Nat. Mus. 79(4): 1–13.

PRITCHARD, M.H. 1966. A revision of the genus *Podocotyle* (Trematoda; Opecoelidae). Zool. Jahrb., Syst. 93: 158–172.

PRUDHOE, S., AND BRAY, R.A. 1982. Platyhelminth parasites of the Amphibia. Oxford University Press: Oxford. 217 p.

RANKIN, J.S. 1939. Studies on the trematode family Microphallidae Travassos, 1921. III. The genus *Maritrema* Nicoll, 1907, with description of a new species and a new genus, *Maritreminoides*. Am. Midl. Nat. 22: 438–451.

RAND, T.G., AND BURT, M.D.B. 1985. Seasonal occurrence, recruitment, and maturation of *Allocreadium lobatum* Wallin, 1909 (Digenea: Allocreadiidae) in the fallfish, *Semotilus corporalis* Mitchell, in a New Brunswick, Canada, lake system. Can. J. Zool. 63: 612–616.

RAU, M.E., AND GORDON, D.M. 1977. A technique for the demonstration of the metacercariae of *Apatemon gracilis pellucidus* (Yamaguti, 1933) in the deep tissues of the brook stickleback (*Culaea inconstans*). Can. J. Zool. 55: 1200–1201.

1978. The frequency distribution of tetracotyles of *Apatemon gracilis pellucidus* (Yamaguti, 1933) in stickleback *Culaea inconstans* (Kirtland) populations of homogeneous age and size structure. J. Fish Dis. 1: 259–263.

RAUSCH, R. 1947. Some observations on the host relationships of *Macrophallus opacus* (Ward, 1894) (Trematoda: Microphallidae). Trans. Am. Microsc. Soc. 66: 59–63.

RAWSON, D.S. 1953. The limnology of Amethyst Lake, a high alpine type near Jasper, Alberta. Can. J. Zool. 31: 193–210.

1960. Five lakes on the Churchill River near Stanley, Saskatchewan. Fish. Branch, Dep. Nat. Resources, Sask. Fish. Rep. 5: 38 p.

RAZMASHKIN, D.A. 1978. On the species position of metacercariae of the genus *Metorchis* (Trematoda: Opisthorchidae). Parazitologiya 12: 68–78 (In Russian).

REDKOZUBOVA, O.I. 1978. Parasitic fauna of flatfishes of the Atlantic coast of Canada. Abstracts of the First All-Union Congress of Parasito-coenologists. Kiev. Part 3: 133–134. (In Russian).

REIMER, L.W. 1981. Parasiten von *Reinhardtius hippoglossoides* (Walbaum), dem Schwarzen Heilbutt aus dem Nordatlantik, p. 121–123. *In*: Physiologie, Biologie und Parasitologie von Nutzfischen. (IV. Wissenschaftliche Konferenz zu Fragen der Physiologie, Biologie und Parasitologie von Nutzfischen vom 3. bis 6. September 1980 in Rostock.) Wilhelm-Pieck-Universität, Rostock, DDR. 183 p.

RICHARDSON, L.R. 1936. Observations on the parasites of the speckled trout in Lake Edward, Quebec. Trans. Am. Fish. Soc. 66: 343–356.

1937. *Raphidascaris laurentianus* sp. n. (Ascaroidea) from *Salvelinus fontinalis* (Mitchill) in Quebec. Can. J. Res., Sect. D. 15: 112–115.

1942. The parasites of fishes of Lake Wakonichi, central northern Quebec. Trans. Am. Fish. Soc. 71: 286–289.

ROBINSON, D.G. 1969. Data record. Number, size composition, weight and food of larval and juvenile fish caught with a two-boat surface trawl in the Strait of Georgia April 24–25, 1968. Fish. Res. Board Can. MS Rep. 1067: 63 p.

ROBINSON, D.G., BARRACLOUGH, W.E., AND FULTON, J.D. 1968a. Data record. Number, size composition, weight and food of larval and juvenile fish caught with a two-boat surface trawl in the Strait of Georgia May 1–4, 1967. Fish. Res. Board Can. MS Rep. 964: 105 p.

1968b. Data record. Number, size composition, weight and food of larval and juvenile fish caught with a two-boat surface trawl in the Strait of Georgia June 5–9, 1967. Fish. Res. Board Can. MS Rep.972: 109 p.

Rohde, K., and Watson, N.A. 1992. Ultrastructure of the tegument, ventral sucker and rugae of *Rugogaster hydrolagi* (Trematoda: Aspidogastrea). Int. J. Parasitol. 22: 967–974.

Rohde, K., Heap, M., Hayward, C.J., and Graham, K.J. 1992. *Calicotyle australiensis* n.sp. and *Calicotyle* sp. (Monogenea, Monopisthocotylea) from the rectum and rectal glands, and *Rugogaster hydrolagi* Schell, 1973 (Trematoda, Aspidogastrea) from the rectal glands of holocephalans off the coast of southern Australia. Syst. Parasitol. 21: 69–79.

Ronald, K. 1959. A check list of the metazoan parasites of the Heterosomata. Department of Fisheries, Quebec, Contribution 67: 152 p.

1960. The metazoan parasites of the Heterosomata of the Gulf of St Lawrence. VI. Digenea. Can. J. Zool. 38: 923–937.

Sandeman, I.M., and Pippy, J.H.C. 1967. Parasites of freshwater fishes (Salmonidae and Coregonidae) of insular Newfoundland. J. Fish. Res. Board Can. 24: 1911–1943.

Sankurathri, C.S., Kabata, Z., and Whitaker, D.J. 1983. Parasites of the Pacific hake, *Merluccius productus* (Ayres, 1855) in the Strait of Georgia, in 1974–1975. Syesis 16: 5–22.

Schell, S.C. 1973. *Rugogaster hydrolagi* gen. et sp. n. (Trematoda: Aspidobothrea: Rugogastridae fam. n.) from the ratfish, *Hydrolagus colliei* (Lay and Bennett, 1839). J. Parasitol. 59: 803–805.

Scott, J.S. 1969a. *Lampritrema nipponicum* (Trematoda) from west Atlantic argentines. Can. J. Zool. 47: 139–140.

1969b.Morphology and morphometric variation in *Lecithophyllum botryophorum* (Trematoda: Hemiuridae) in *Argentina silus*. Can. J. Zool. 47: 213–216.

1969c. Trematode populations in the Atlantic argentine, *Argentina silus*, and their use as biological indicators. J. Fish. Res. Board Can. 26: 879–891.

1973. Intestinal helminth parasites of northern sand lance (*Ammodytes dubius*). J. Fish. Res. Board Can. 30: 291–292.

1975a. Incidence of trematode parasites of American plaice (*Hippoglossoides platessoides*) of the Scotian Shelf and the Gulf of St Lawrence in relation to fish length and food. J. Fish. Res. Board Can. 32: 479–483.

1975b. Geographic variations in incidence of trematode parasites of American plaice (*Hippoglossoides platessoides*) in the northwest Atlantic. J. Fish. Res. Board Can. 32: 547–550.

1975c. Intestinal trematode parasites (Trematoda: Digenea) and food of yellowtail flounder (*Limanda ferruginea* (Storer, 1839)) from the Scotian Shelf and Gulf of St. Lawrence. Fish. Mar. Serv. Res. Dev. Tech. Rep. 584: 12 p.

1975d. Digenetic trematode parasites and food of witch flounder (*Glyptocephalus cynoglossus* L.) from the Scotian Shelf and Gulf of St Lawrence. Fish. Mar. Serv. Res. Dev. Tech. Rep. 593: 9 p.

1976. Digenetic trematode parasites and food of winter flounder (*Pseudopleuronectes americanus* (Walbaum, 1792)) from the Scotian Shelf and Gulf of St Lawrence. Fish. Mar. Serv. Res. Dev. Tech. Rep. 618: 9 p.

1981. Alimentary tract parasites of haddock (*Melanogrammus aeglefinus* L.) on the Scotian Shelf. Can. J. Zool. 59: 2244–2252.

1982. Digenean parasite communities in flatfishes of the Scotian Shelf and southern Gulf of St. Lawrence. Can. J. Zool. 60: 2804–2811.

1985a. Occurrence of alimentary tract helminth parasites of pollack (*Pollachius virens* L.) on the Scotian Shelf. Can. J. Zool. 63: 1695–1698.

1985b. Digenea (Trematoda) populations in winter flounder (*Pseudopleuronectes americanus*) from Passamaquoddy Bay, New Brunswick, Canada. Can. J. Zool. 63: 1699–1705.

1987. Helminth parasites of the alimentary tract of the hakes (*Merluccius, Urophycis, Phycis*: Teleostei) of the Scotian Shelf. Can. J. Zool. 65: 304–311.

1988. Helminth parasites of the redfish (*Sebastes fasciatus*) from the Scotian Shelf, Bay of Fundy, and eastern Gulf of Maine. Can. J. Zool. 66: 617–621.

Scott, J.S., and Bray, S.A. 1989. Helminth parasites of the alimentary tract of Atlantic halibut (*Hippoglossus hippoglossus* L.) and Greenland halibut (*Reinhardtius hippoglossoides* Walbaum) on the Scotian Shelf. Can. J. Zool. 67: 1476–1481.

Sekerak, A.D., and Arai, H.P. 1973. Helminths of *Sebastes alutus*(Pisces: Teleostei) from the northeastern Pacific. Can. J. Zool. 51: 475–477.

1974. A revision of *Helicometra* Odhner, 1902 and related genera (Trematoda: Opecoelidae), including a description of *Neohelicometra sebastis* n. sp. Can. J. Zool. 52: 707–738.

1977. Some metazoan parasites of rockfishes of the genus *Sebastes* from the northeastern Pacific Ocean. Syesis 10: 139–144.

Sekhar, S.C., and Threlfall, W. 1970a. Helminth parasites of the cunner, *Tautogolabrus adspersus* (Walbaum) in Newfoundland. J. Helminthol. 44: 169–188.

1970b. Infection of the cunner, *Tautogolabrus adspersus* (Walbaum), with metacercariae of *Cryptocotyle lingua* (Creplin, 1825). J. Helminthol. 44: 189–198.

Sey, O. 1991. Handbook of the zoology of amphistomes. CRC Press, Boca Raton. 480 p.

Sheldon, A.J. 1938. Studies on the life cycle of *Maritrema medium* (Trematoda) and a redescription of the species. J. Parasitol. 24: 259–262.

Shigin, A.A. 1968. A systematic review of metacercariae of the genus *Diplostomum* — parasites of fishes in the Volga delta and Rybinsk Reservoir. Trudy

Atrakhan. Zapov. (Sborn. Gel'mint. Rabot.) 11: 275–324. (In Russian).

1976. Metacercariae of the genus *Diplostomum* in the fauna of the USSR. Parazitologiya 10: 346–351. (In Russian).

1986. Trematodes of the fauna of the USSR. Genus *Diplostomum* metacercariae. Akad. Nauk SSSR: Moscow. 253 p.

1987. Morphology, biology and taxonomy of the maritae of *Diplostomum chromatophorum* (Brown, 1931) Shigin, 1986 (Trematoda: Diplostomidae). Trudy Gel'mint. Lab. 35: 176–182.

Shoop, W.L. 1989. Systematic analysis of the Diplostomidae and Strigeidae (Trematoda). J. Parasitol. 75: 21–32.

Shostak, A.W., Tompkins, J.C., and Dick, T.A. 1987. The identification and morphological variation of *Diplostomum baeri bucculentum* from two gull species, using metacercarial infections from least cisco from the Northwest Territories. Can. J. Zool. 65: 2287–2291.

Shvetsova, L.S. 1990. A new find of *Rugogaster hydrolagi* Schell, 1973 in *Hydrolagus ogilbyi* from the southern Pacific. Parazitologiya 24: 82–84. (In Russian).

Simer, P.H. 1929. Fish trematodes of the Lower Tallahatchie River. Am. Midl. Nat. 11: 563–588.

Sinclair, N.R. 1972a. Studies on the heterophyid trematode *Apophallus brevis*, the "sand-grain grub" of yellow perch (*Perca flavescens*). I. Re-description and resolution of synonymic conflict with *Apophallus imperator*, Lyster, 1940 and other designations. Can. J. Zool. 50: 357–364.

1972b. Studies on the heterophyid trematode *Apophallus brevis*, the "sand-grain grub" of yellow perch (*Perca flavescens*). II. The metacercaria: position, structure, and composition of the cyst; hosts; geographical distribution. Can. J. Zool. 50: 577–584.

Sindermann, C.J. 1957 Diseases of fishes of the western North Atlantic. VI. Parasites as indicators of herring movements. Maine Dep. Sea Shore Fish. Res. Bull. 27: 1–30.

1961. Parasite tags for marine fish. J. Wildl. Manage. 25: 41–47.

1965. Effects of environment on several diseases of herring from the western North Atlantic. Int. Comm. Northwest Atl. Fish. Spec. Publ. 6: 603–610.

Sinitsin, D. 1931. Studien über die Phylogenie der Trematoden. IV. The life histories of *Plagioporus siliculus* and *Plagioporus virens*, with special reference to the origin of the Digenea. Z. Wiss. Zool. 138: 409–456.

Skryabin, K.A., and Gushanskaya, L.K. 1955. Suborder Hemiurata (Markevitsch, 1951) Skryabin & Gushanskaya, 1954. Second part. *In*: Skryabin, K.I. (Ed.). Trematodes of animals and man. Principles of trematodology. Nauka: Moscow. 10: 337–643. (In Russian).

Skryabin, K.A., and Koval, V.P. 1966. Family Bunoderidae Nicoll, 1914. *In*: Skryabin, K.I. (Ed.). Trematodes of animals and man. Principles of trematodology. Nauka: Moscow. 22: 313– 456. (In Russian).

Skryabina, E.S. 1974. Helminths of the sturgeon family Acipenseridae Bonaparte, 1831. Nauka: Moscow. 167 p. (In Russian).

Smith, J.W. 1967. *Aporocotyle margolisi* n.sp. (Digenea: Aporocotylidae) from *Merluccius productus*. J. Fish. Res. Board Can. 24: 1763–1773.

Smith, R.J. 1968. Ancylid snails as first intermediate hosts of *Lissorchis mutabile* comb. n. (Trematoda: Lissorchiidae). J. Parasitol. 54: 283–285.

Stafford, J. 1900. Some undescribed trematodes. Zool. Jahrb., Syst. 13: 399–414.

1902. Notes on worms. Zool. Anz. 25: 481–483.

1904. Trematodes from Canadian fishes. Zool. Anz. 27: 481–495.

1905. Trematodes from Canadian vertebrates. Zool. Anz. 28: 681–694.

1907. Preliminary report on the trematodes of Canadian marine fishes. Contrib. Can. Biol. 1902–1905: 91–94.

Stanley, R.D., Lee, D.L., and Whitaker, D.J. 1992. Parasites of yellowtail rockfish, *Sebastes flavidus* (Ayres, 1862) (Pisces: Teleostei), from the Pacific coast of North America as potential biological tags for stock identification. Can. J. Zool. 70: 1086–1096.

Stewart, D.B., and Bernier, L.M.J. 1982. An aquatic resource survey of the islands bordering Viscount Melville Sound, District of Franklin, Northwest Territories. Lands Directorate of Environment Canada and Northern Environment Directorate of Indian and Northern Affairs, Northern Land Use Series, Background Report No. 2: 110 p.

1983. An aquatic resource survey of King William and Victoria islands, and the northeastern District of Keewatin, Northwest Territories. Lands Directorate of Environment Canada and Northern Environment Directorate of Indian and Northern Affairs, Background Report No. 3: 127 p.

1984. An aquatic resource survey of Melville Peninsula, Southampton Island, and the northeastern District of Keewatin, Northwest Territories. Lands Directorate of Environment Canada and Northern Environment Directorate of Indian and Northern Affairs, Background Report No. 4: 144 p.

Stunkard, H.W. 1930. The life history of *Cryptocotyle lingua* (Creplin), with notes on the physiology of the metacercariae. J. Morphol. Physiol. 50: 143–191.

1961. *Cercaria dipterocerca* Miller and Northrup, 1926 and *Stephanostomum dentatum* (Linton, 1900) Manter, 1931. Biol. Bull. Mar. Lab. Woods Hole 120: 221–237.

——— 1964. The morphology, life-history and systematics of the digenetic trematode, *Homalometron pallidum* Stafford, 1904. Biol. Bull. Mar. Lab. Woods Hole 126: 163–173.

——— 1972. Observations on the morphology and life-history of the digenetic trematode, *Lepocreadium setiferoides* (Miller and Northup, 1926) Martin, 1938. Biol. Bull. Mar. Lab. Woods Hole 142: 326–334.

——— 1975. The excretory system and systematics of the gasterostome trematodes. Dr. B.S. Chauhan Commemeration Volume. Zoological Society of India: Orissa. p. 1–11.

——— 1980. The morphology, life-history, and taxonomic relationships of *Lepocreadium areolatum* (Linton, 1900) Stunkard, 1969 (Trematoda: Digenea). Biol. Bull. Mar. Biol. Lab. Woods Hole 158:154–163.

——— 1981. The morphology, life history, and systematic relations of *Lasiotocus elongatus* (Manter, 1931) Thomas, 1959 (Trematoda: Digenea). Biol. Bull. Mar. Biol. Lab. Woods Hole 160: 155–160.

SUDARIKOV, V.E. 1960. Strigeidida: Diplostomatidae. *In*: Skryabin, K.I. (Ed.). Trematodes of animals and man. Principles of trematodology. Nauka: Moscow. 17: 157–530. (In Russian).

——— 1971. Order Strigeidida (La Rue, 1926) Sudarikov, 1959. Suborder Strigeata La Rue, 1926. *In*: Skryabin, K.I. (Ed.). Trematodes of animals and man. Principles of trematodology. Nauka: Moscow. 24: 69–272. (In Russian: English translation, 1982, Amerind: New Delhi).

SURBER, E.W. 1928. *Megalogonia ictaluri*, a new species of trematode from the channel catish, *Ictalurus punctatus*. J. Parasitol. 14: 269–271.

SZALAI, A.J., AND DICK, T.A. 1987. Parasites of quillback, *Carpioides cyprinus* (LeSueur), from Dauphin Lake, Manitoba, Canada. J. Parasitol. 73: 446–447.

——— 1988. Helminths of stocked rainbow trout (*Salmo gairdneri*) with special reference to *Clinostomum complanatum*. J. Wildl. Dis. 24: 456–460.

——— 1991. Role of predation and parasitism in growth and mortality of yellow perch in Dauphin Lake, Manitoba. Trans. Am. Fish. Soc. 120: 739–751.

SZUKS, H. 1975. Zum Befall von *Macrourus rupestris* (Gunnerus) aus dem Bereich von Labrador mit digenen Trematoden. Wiss. Z. Pad. Hochsch. "Liselotte Hermann" Güstrow, Math.-Naturwiss. Fak. 2: 225–231.

——— 1980. Die Verwendbarkeit von Parasiten zur Gruppentrennung beim Grenadierfischen *Macrourus rupestris*. Angew. Parasitol. 21: 211–214.

TASKINEN, J., VALTONEN, E.T., AND GIBSON, D.I. 1991. Studies on bucephalid digeneans parasitising molluscs and fishes in Finland. I. Ecological and experimental studies. Syst. Parasitol. 19: 81–94.

TAYLOR, L.H., HALL, B.K., AND CONE, D.K. 1993. Experimental infection of yellow perch (*Perca flavescens*) with *Apophallus brevis* (Digenea, Heterophyidae): parasite invasion, encystment, and ossicle development. Can. J. Zool. 71: 1886–1894.

TAYLOR, P.W. 1978. *Macroderoides trilobatus* sp. n. (Digenea: Macroderoididae) from the bowfin, *Amia calva*, and emendation of the genus. J. Parasitol. 64: 393–394.

TEDLA, S., AND FERNANDO, C.H. 1969a. Observations on the seasonal changes of the parasite fauna of yellow perch (*Perca flavescens*) from the bay of Quinte, Lake Ontario. J. Fish. Res. Board Can. 26: 833–843.

——— 1969b. Changes in the parasite fauna of the white perch, *Roccus americanus* (Gmelin), colonizing new habitats. J. Parasitol. 55: 1063–1066.

——— 1972. On the characterization of the parasite fauna of yellow perch (*Perca fluviatilis* L) in five lakes in southern Ontario, Canada. Helminthologia (Bratislava) 11: 23–33.

TEMPLEMAN, W. 1965. Rare skates of the Newfoundland and neighbouring areas. J. Fish. Res. Board Can. 22: 259–279.

——— 1973b. First records, description, distribution, and notes on the biology of *Bathyraja richardsoni* (Garrick) from the Northwest Atlantic. J. Fish. Res. Board Can. 30: 1831–1840.

THULIN, J. 1971. Parasitiska plattmaskar hos torsk. Zool. Revy. 33: 76–84.

THOMPSON, A.B., AND MARGOLIS, L. 1987. Descriptions of *Neolebouria tinkerbellae* n. sp. (Trematoda: Digenea: Opecoelidae) from experimental fish hosts, and of metacercariae of *N. tinkerbellae* and an unidentified digenean from *Pandalus jordani* (Decapoda: Penaeidae) from the Pacific coast of Canada. Can. J. Zool. 65: 188–193.

THOMPSON, P.A., AND THRELFALL, W. 1978. The metazoan parasites of two species of fish from the Port-Cartier-Sept-Iles Park, Québec. Le Naturaliste canadien (Québec) 105: 429–431.

THRELFALL, W. 1967. Some parasites recovered from the ocean sunfish *Mola mola* (L.) in Newfoundland. Can. Field-Nat. 81: 168–172.

——— 1968. A mass die-off of three spined sticklebacks (*Gasterosteus aculeatus* L.) caused by parasites. Can. J. Zool. 46: 105–106.

——— 1969. Some parasites from elasmobranchs in Newfoundland. J. Fish. Res. Board Can. 26: 805–811.

——— 1981. Metazoan parasites of two populations of rainbow smelt in Newfoundland, Canada. J. Parasitol. 67: 129–131.

THRELFALL, W., AND HANEK, G. 1970a. Metazoan parasites, excluding Monogenea from longnose and white suckers. J. Fish. Res. Board Can. 27: 1317–1319.

——— 1970b. Helminths from northern pike (*Esox lucius* L.) in Labrador. J. Parasitol. 56: 662.

——— 1970c. Metazoan parasites of salmonids and coregonids from the Avalon Peninsula, Newfoundland. J. Fish. Res. Board Can. 27: 1894–1897.

1971. Helminth parasites, exluding Monogenea from some Labrador fishes. J. Parasitol. 57: 684–685.
TIMMERS, S.F. 1979. *Alloglossidium schmidti* sp.n. (Trematoda: Macroderoididae) from hirudinid leaches in Manitoba. Proc. Helminthol. Soc. Wash. 46: 180–184.
UMNOVA, B.A. 1979. Ecological analysis of the trematode fauna of pleuronectids and gadids from the northwest Atlantic. p.109–110. *In*: Bauer, O.N. (Ed.). Tezisy Dokladov. VII All-Union Conference on Parasites and Diseases of Fish (Leningrad). Nauka: Leningrad. (In Russian).
VAN CLEAVE, H.J., AND MUELLER, J.F. 1932 Parasites of Onedia Lake fishes. Part I. Descriptions of new genera and new species. Bull. New York State Coll. Forestry Syracuse Univ. 4 (Roosevelt Wild Life Ann. 3): 9–71.
1934. Parasites of Onedia Lake fishes. Part III. A biological and ecological survey of the worm parasites. Bull. New York State Coll. Forestry Syracuse Univ. 7 (Roosevelt Wild Life Ann. 3): 161–334.
VAN DER LAND, J. 1967. A new blood fluke (Trematoda) from *Chimaera monstrosa* L. Proc. K. Ned. Akad. Wet., Ser. C, 70: 110–120.
VANDE VUSSE, F.J. 1980. Revision of *Alloglossidium* Simer, 1929 (Trematoda: Macroderoididae) and description of *A. microspinatum* sp. n. from a leach. J. Parasitol. 66: 667–670.
VOJTEK, J. 1964. The importance of life-history studies for the systematics of the genus *Apatemon* (Trematoda, Strigeidae). p.121–130. *In*: Ergens, R. and Rysavy, B. Parasitic worms and aquatic conditions. Czech. Acad. Sci.: Prague.
WALLACE, F.G. 1939. The metacercaria of *Amphimerus elongatus* Gower (Trematoda: Opisthorchiidae). J. Parasitol. 25: 491– 494.
WARD, H.B. 1938. On the genus *Deropristis* and the Acanthocolpidae (Trematoda). In: Livro Jubilar Prof. Travassos, Rio de Janeiro, 509–522.
WARD, H.B., AND MUELLER, J.F. 1926. A new pop-eye disease of trout-fry. Arch. Schiffs- u. Tropenhyg. 30: 602–609.
WATSON, R.A., AND DICK, T.A. 1979. Metazoan parasites of whitefish *Coregonus clupeaformis* (Mitchill) and cisco *C. artedii* LeSueur from Southern Indian Lake, Manitoba. J. Fish Biol. 15: 579–587.
1980. Metazoan parasites of pike, *Esox lucius* Linnaeus, from Southern Indian Lake, Manitoba, Canada. J. Fish Biol. 17: 255–261.
WATSON, T.G. 1981. Growth of *Metorchis conjunctus* (Cobbold, 1860) Looss, 1899 (Trematoda: Opisthorchiidae) in the bile ducts of various definitive hosts. Can. J. Zool. 59: 2014–2019.
WHITAKER, D.J. 1985. A parasite survey of juvenile chum salmon (*Oncorhynchus keta*) from the Nanaimo River. Can. J. Zool. 63: 2875–2877.
WIERZBICKA, J. 1988. Parasite fauna of the blue halibut (*Reinhardtius hippoglossoides* Walbaum, 1792) from selected districts of its distribution area. Akademia Rolnicza, Szczecin, Poland. 87 p. (English translation: Canadian Translation of Fisheries and Aquatic Sciences 5520: 116 p.).
1991a. Parasite fauna of the Greenland halibut, *Reinhardtius hippoglossoides* (Walbaum, 1792) from Labrador. Acta Ichthyol. Pisc. 21: 21–29.
1991b. An analysis of parasite fauna of Greenland halibut, *Reinhardtius hippoglossoides* (Walbaum, 1792) in different age groups. Acta Ichthyol. Pisc. 21: 31–41.
WILES, M. 1975. Parasites of *Fundulus diaphanus* (LeSueur) (Pisces: Cyprinodontidae) in certain Nova Scotian freshwaters. Can. J. Zool. 53: 1578–1580.
WILSON, K.A., AND RONALD, K. 1967. Parasite fauna of the sea lamprey (*Petromyzon marinus* von Linné) in the Great Lakes region. Can. J. Zool. 45: 1083–1092.
WOLFGANG, R.W. 1954a. Studies of the trematode *Stephanostomum baccatum* (Nicoll, 1907). I. The distribution of the metacercaria in eastern Canadian flounders. J. Fish. Res. Board Can. 11: 954–692.
1954b. Studies of the trematode *Stephanostomum baccatum* (Nicoll, 1907). II. Biology with special reference to the stages affecting the winter flounder. J. Fish. Res. Board Can. 11: 963–987.
1955a. Studies of the trematode *Stephanostomum baccatum* (Nicoll, 1907). III. Its life cycle. Can. J. Zool. 33: 113–128.
1955b. Studies of the trematode *Stephanostomum baccatum* (Nicoll, 1907). IV. The variation of the adult morphology and the taxonomy of the genus. Can. J. Zool. 33: 129–142.
WOLFGANG, R.W., AND MYERS, B.J. 1954. *Gonocerca macroformis* sp. nov. (Derogenetinae: Hemiuridae) from the ovary of cod. Can. Zool. 32: 25–29.
WOODHEAD, A.E. 1929. Life history studies on the trematode family Bucephalidae. Trans. Am. Microsc. Soc. 48: 256–275.
1930. Life history studies on the trematode family Bucephalidae. No. II. Trans. Am. Microsc. Soc. 49: 1–17.
1931. The germ cell cycle in the trematode family Bucephalidae. Trans. Am. Microsc. Soc. 50: 169–188.
WORLEY, D.E., AND BANGHAM, R.V. 1952. Some parasites of fishes of the upper Gatineau River valley. Ohio. J. Sci. 52: 210–212.
WRIGHT, B.J., ALBERT, E., AND CURTIS, M.A. 1989. The effect of population depletion on the growth, feeding behavior and parasite community of brook trout (*Salvelinus fontinalis*) in a subarctic lake. Bulletin of the Canadian Society of Zoologists 20(2): 63. (28th Annual Meeting Canadian Society of Zoologists, 16–20 May 1989, York University, Toronto, Ontario. Program and Abstracts).
WRIGHT, R.R. 1879 Contributions to American helminthology. Proc. Can. Inst., New Ser. 1: 54–75.

YAMAGUTI, S. 1934. Studies on the helminth fauna of Japan. Part 21. Trematodes of fishes, I. Jpn. J. Zool. 3: 249–541.

1938. Studies on the helminth fauna of Japan. Part 21. Trematodes of fishes, IV. Published by the author: Kyoto. 139 p.

1940. Studies on the helminth fauna of Japan. Part 31. Trematodes of fishes, VII. Jpn. J. Zool. 9: 35–108.

1953. Systema helminthum. Part 1. Digenetic trematodes of fishes. Published by the author: Tokyo. 405 p.

1958. Systema helminthum. Vol. I. Digenetic trematodes. Pt. I. Interscience Publishers Inc.: New York. 900 p.

1971. Synopsis of digenetic trematodes of vertebrates. Vol. I. Keigaku Publishing Co.: Tokyo. 1074 p.

ZHUKOV, E.V. 1960. Endoparasitic worms of the fishes in the Sea of Japan and the South Kuril shallow waters. Trudy Zool. Inst. Leningr. 28: 3–146. (In Russian).

ZUBCHENKO, A.V. 1975. On the parasite fauna of macrourids in the North Atlantic. Trudy Polyar. Nauchno-issled. Proekt. Inst. Morsk. Khoz. Okeanogr. 35: 234–238. (In Russian).

1976. On the question of the existence of separate populations of the blunt-nosed macrurus (*Macrurus rupestris* Gunner) from the north-west Atlantic in the light of its parasitological data. In: Gaevskaya, A.V. (Ed.). Summaries of Reports from the Second All-Union Symposium on Parasites and Diseases of Marine Animals. AtlantNIRO, Kaliningrad, p. 29–30. (In Russian).

1978. New species of trematodes from fishes of North Atlantic. Parazitologiya 12: 116–120. (In Russian).

1980. Parasitic fauna of Anarhichadidae and Pleuronectidae families of fish in the Northwest Atlantic. Int. Comm. N.W. Atlant. Fish. 6: 41–46.

1981a. Atlantic fauna of some Macrouridae in the Northwest Atlantic. J. Northw. Atl. Fish. Sci. 2: 67–72.

1981b. Using parasitological data in studying rock grenadier (*Coryphaenoides rupestris* Gunner) local groupings. *In*: Bauer, O.N. et al. (Eds.). Symposium on parasitology and pathology of marine organisms, Leningrad. (Abstract), 25–32. (In Russian: English version - NOAA Tech. Rep. 25: 19–23).

1984. Ecological peculiarities of the parasite fauna of some Alepocephalidae. *In*: Ecologo-parasitological studies of the northern seas. Murmanskii Morskoi Biologicheskii Institut, Apatity. pp. 77–81. (In Russian).

1985a. Parasitic fauna of American plaice (*Hippoglossoides platessoides*) from the Northwest Atlantic. J. Northw. Atl. Fish. Sci. 6: 165–171.

1985b Use of parasitological data in studies of the local groupings of rock grenadier, *Coryphaenoides rupestris* Gunner. p.19–23. *In*: Hargis, W.J. Jr (Ed.). Parasitology and pathology of marine animals of the World Ocean. U.S. Department of Commerce, NOAA Technical Report NMFS 25: iv + 135 p.

INDEX TO PARASITES

INDEX OF HOSTS